Techniques in Protein Chemistry II

TECHNIQUES IN PROTEIN CHEMISTRY II

Edited by

Joseph J. Villafranca
Department of Chemistry
Pennsylvania State University
University Park, Pennsylvania

ACADEMIC PRESS, INC.
Harcourt Brace Jovanovich, Publishers
San Diego New York Boston
London Sydney Tokyo Toronto

Academic Press Rapid Manuscript Reproduction

This book is printed on acid-free paper. ∞

Academic Press, Inc.
San Diego, California 92101

United Kingdom Edition published by
Academic Press Limited
24–28 Oval Road, London NW1 7DX

International Standard Book Number 0-12-721958-7 (Hardcover)
International Standard Book Number 0-12-721957-9 (Paperback)

PRINTED IN THE UNITED STATES OF AMERICA
91 92 93 94 9 8 7 6 5 4 3 2 1

Contents

Section II
Peptide Sequence Analysis

Section IV
Protein Structure/Function Analysis

Section V
NMR/X-Ray Workshop

Section VI
Mass Spectrometry Workshop

Foreword

This is the third volume of what is rapidly becoming an annual event. The basic aim has been to extract a set of papers focusing on new methods and innovative applications of established ones from the wealth of new information presented at the annual Protein Society symposium and to impart this practical information to protein scientists as quickly as possible. Based on the reception of the first two volumes our goal is being achieved, and a large number of protein scientists are pleased that these books are becoming a regular feature of The Protein Society's activities.

The success of the volume is based on the integration of many different components and efforts: the good science presented at the symposium, the editor's selection and organization of the individual papers, and the commitment of the editor, authors, and publisher to meet deadlines so that the book can be published in timely fashion. We are grateful to those involved in successfully completing this third volume in the series, which is based on the fourth annual symposium of The Protein Society held in San Diego, August 11–15, 1990. A very special expression of thanks and congratulations is extended to the volume editor, Joseph Villafranca and his associates, who started working on the project the day all of the more than 470 symposium abstracts arrived last spring and who have now seen it through to a successful conclusion.

Finn Wold
President
The Protein Society

Preface

Once again the excitement in the field of "protein science" was evident by the large attendance at the fourth annual symposium of The Protein Society. For this year's "techniques" book, I read all the abstracts (over 470) submitted for poster presentation to begin the process of selecting the six to seven dozen to give to the associate editors for the final invitation of the approximately fifty manuscripts to be included. After the invitations were sent, over 90% of the invitees responded positively. Two other sources of invited papers were used: the Mass Spectrometry and NMR/X-Ray Workshops and the posters selected at the meeting by the associate editors. To my delight all manuscripts were submitted to me for review by early September and were returned to me by the authors with revisions (when requested) by the end of September. This collection of superb manuscripts includes worldwide representation from academe, research institutes, and industry.

Special thanks are due to the people who made the book "happen": the associate editors Ken Williams, Juliette Lecomte, Brad Gibson, and John Smith; the president of the society Finn Wold; the secretary/treasurer of the society Ken Walsh; the symposium manager Teresa Spellman; the staff of Academic Press; my secretary Gwen Corman; my wife Beth who produced the front cover photograph; and all the cooperative authors of the manuscripts. I am confident that the quality of the book will continue under the next editor.

Joseph J. Villafranca

Acknowledgments

The Protein Society acknowledges with thanks the following organizations who through their support of the Society's program goals contributed in a meaningful way to the fourth annual symposium and thus to this volume.

AAA Laboratory
Advanced Chem Tech
Applied Biosystems
AVIV Associates
Beckman Instruments, Inc.
Biosym Technologies, Inc.
Bristol-Meyers Squibb Pharmaceutical Research Institute
E.I. DuPont de Nemours and Co.
Eli Lilly and Company
Finnigan Corporation
Immuno-Dynamics, Inc.
JASCO Inc.
Merck, Sharp & Dohme Research Laboratories
The Merck Foundation
Milligen/Biosearch
Molecular Dynamics, Inc.
PE/SCIEX
Pharmacia LKB Biotechnology
Porton Instruments, Inc.
Savant Instruments, Inc.
Tripos Associates, Inc.
Triton Biosciences, Inc.
Vydac, The Separations Group, Inc.

SECTION I

Peptide/Protein Separation Methods and Analysis

1

BENEFITS OF AUTOMATION IN THE SEPARATION OF BIOMOLECULES BY HIGH PERFORMANCE CAPILLARY ELECTROPHORESIS

G. McLaughlin, R. Palmieri, and K. Anderson
Beckman Instruments, Inc., PO Box 10200, Palo Alto, CA 94304 USA

I. INTRODUCTION

The advent of automated high performance capillary electrophoresis (HPCE) systems presents new alternatives to the labor-intensive and irreproducible aspects of classical electrophoresis.Many factors should be investigated to optimize an electrophoretic separation. An investigator may not have the time or desire to perform all the experiments required to optimize a method. In addition, post run data analysis has continued to be a problem for most scientists. This is especially the case with data intensive automated experiments. Some of the major areas that need to be investigated to resolve these issues are:

A. Separation parameters (ionic strength, voltage, temperature, use of organic modifiers, Ph, and capillary dimensions)

B. Sampling strategies (repetitive injections, reproducibility, mobility markers, microvolume sampling, and rapid peak identification)

C. Post run data analysis (time/mobility corrections, molecular weight and base number estimates, theoretical plate number, etc.)

The purpose of this study was to show the benefits of automating methods in the separation of biomolecules by HPCE. In some cases the experimental parameters investigated were based on previously published reports but extended using automated techniques [1-16]. A portion of this work was devoted to simplifying post run data analysis, the generation of reports and publication-quality outputs, and visualization of key effects.

Techniques in Protein Chemistry II

II. MATERIALS AND METHODS

The capillary electrophoresis system used was the P/ACE™System 2000 (Beckman Instruments, Palo Alto, CA) including an IBM PS/2™ model 50Z computer. System operation was controlled through P/ACE Control Software (Beckman Instruments) operating in a Windows™ (Microsoft Corp., Redmond, WA) environment. Data acquisition and analysis were carried out using System Gold™Chromatography Software (Beckman Instruments, San Ramon,CA). The control software automates all system functions, including sampling, rinses, power supply, etc. It joins numerous steps (up to 20) to form a method. Additionally, it can link methods from 1-99 together in sequences for extended periods of operation.

The P/ACE System 2000 has a 34-position, random access autosampler consisting of two concentric rings that position 4.5 ml vials under the inlet and outlet ends of the capillary. A vial setup for a pH study is shown in Fig. 1. In this experiment vials 6-9 and 27-30 (run buffers), 31-34 (rinse buffers), and 11-20 (peptide samples) were used. The peptide mixture was injected in one run and then spiked with the individual components in successive runs for identification. The buffers were then replaced with a different buffer at another pH and the experiment was repeated. All runs in this study were performed in a completely automated manner.

The instrument uses a recirculating liquid with Peltier-temperature control of the capillary. Fused silica capillaries of either 50 um and 75 um inner diameter (id) x 375 um outer diameter (od) x 56.9 cm total length (tl) x 50 cm length to the detector (td) were provided in temperature controlled cartridges (Beckman Instruments).

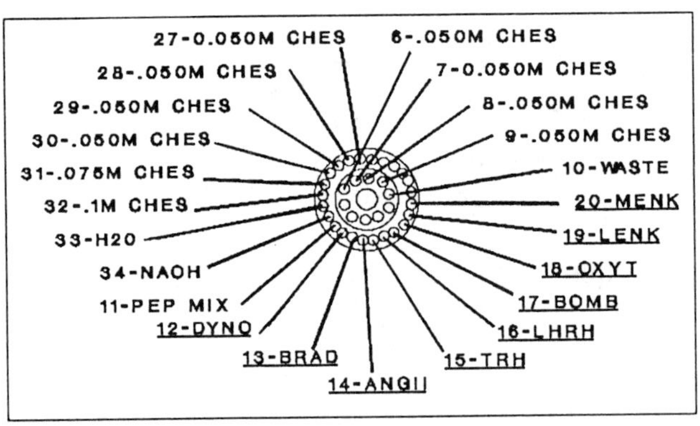

Fig. 1. Autosampler setup for one Ph experiment.Instrument control allows single or multiple injections and random buffer access.

Some post run data analysis, customized report generation, and graphics production were performed by direct data translation or by use of macros. Software programs used were Lotus 123™ (Lotus Development, Cambridge, MA), Excel™ (Microsoft Corp., Redmond, WA), Harvard Graphics™ (Software Publishing, Mountain View, CA), and Graph Plus™ (Micrografx, Richardson, TX). The computer used for data analysis was an IBM model PS/2 55SX (386, 40 mB hard drive, and 2 mB RAM).

General capillary electrophoresis conditions were: constant voltage of 20-30 kV; set temperature of 19-49°C; current limit at 250 uA; 5 sec. pressure injection (approx. 25 nL out of 20-40 uL in microvial); 3-30 min. run times; and detection by UV absorbance at 200 nm. Specific conditions used for each set of experiments are included on the figures.

The fused-silica capillaries were washed initially with 0.1M NaOH (15 min.). Prior to each run the capillary was washed with 0.1M NaOH (3 min.), with equilibration buffers at the same pH but at a concentration of 0.05M and 0.025M higher than used for the separation (3 min. each), and finally with separation buffer (8 min.). Run buffer vials were carefully filled to 4.6 ml, used without caps, and changed every fourth run to minimize ion depletion.

The buffers used were electrophoresis, HPLC or analytical grade and included monosodium dihydrogen phosphate from J.T.Baker (Phillipsburg, NJ) and sodium borate, sodium hydroxide, and urea from Beckman Instruments. Good's buffers were acquired from Boehringer-Mannheim, Indianapolis, IN) and used in pH studies (data not shown). The buffers were diluted to the desired molarity using MilliQ™ (Millipore Corp., Bedford, MA) deionized water. The buffer was prepared by mixing equal-molar concentrations of the acid and base components to give the desired pH, and filtered (0.45um).

Bioactive peptides listed in Table 1 were purchased from Sigma Chemical (Saint Louis, MO). Stock peptide solutions (5 mg/ml) were dissolved in 0.1% trifluoroacetic acid (TFA) and stored at 5°C. Peptides analogs listed in Table 2 were synthesized by Genentech as described previously [17]. Peptide mixtures were prepared at concentrations of 10-500 ug/ml in 0.1% TFA.

A poly d(A)40-60 oligonucleotide standard was purchased from Pharmacia, LKB (Piscataway, NJ). Oligonucleotides standards were prepared at 185 ug/ml in water and stored at 5°C. Gel filled capillaries (polyacrylamide) were prepared at Beckman Instruments (Palo Alto, CA).

#	PEPTIDE NAME AND SEQUENCE	MW	ABV.
1	**BRADYKININ:** ARG-PRO-PRO-GLY-PHE-SER-PRO-PHE-ARG-OH	1060	BRAD
2	**ANGIOTENSIN II :** ASP-ARG-VAL-TYR-ILE-HIS-PRO-PHE-OH	1046	ANGII
3	**TRH:** p-GLU-HIS-PRO-NH2	362	TRH
4	**LHRH:** p-GLU-HIS-TRP-SER-TYR-GLY-LEU-ARG-PRO-GLY-NH2	1182	LHRH
5	**BOMBESIN :** p-GLU-GLN-ARG-LEU-GLY-ASN-GLN-TRP-ALA-VAL-GLY-HIS-LEU-MET-NH2	1620	BOMB
6	**LEUCINE ENKEPHALIN:** TYR-GLY-GLY-PHE-LEU-OH	556	LENK
7	**METHIONINE ENKEPHALIN:** TYR-GLY-GLY-PHE-MET-OH	574	MENK
8	**OXYTOCIN:** CYS-TYR-ILE-GLN-ASN-CYS-PRO-LEU-GLY-NH2	1007	OXYT
9	**DYNORPHIN:** H_2N-TYR-GLY-GLY-PHE-LEU-ARG-ARG-ILE-ARG-PRO-LYS-LEU-LYS-TRP-ASP-ASN-GLN-OH	2147	DYNO

TABLE 1. Bioactive peptides used as samples

#	PEPTIDE NAME AND SEQUENCE	MW	ABV.
10	**GEN-1:** ARG-HIS-ASP-ALA-ALA-ALA-TYR-LEU-LEU-OH	1029	GEN1
11	**GEN-2:** LEU-LEU-TYR-ALA-ALA-ALA-ASP-HIS-ASP-OH	1029	GEN2
12	**GEN-3:** HIS-ARG-ALA-LEU-ASP-TYR-ALA-ALA-LEU-OH	1029	GEN3
13	**GEN-4:** HIS-ALA-LEU-ASP-TYR-ALA-ARG-LEU-ALA-OH	1029	GEN4
14	**GEN-5:** ASP-HIS-ALA-TYR-LEU-LEU-ALA-ALA-ARG-OH	1029	GEN5

TABLE 2. Synthetic peptide analogs used as samples

III. RESULTS AND DISCUSSION

A. Optimization of Separation Conditions

1. Ionic strength

As ionic strength increased lower mobilities were observed. This effect was clearly seen in the 50 um capillary (Fig. 2A) where mobilities decreased linearly over the buffer range of 0.025M to 0.125M before leveling off. By contrast migration times in the 75 um capillary increased up to 0.075M (Fig. 2B). Above 0.075M migration times actually decreased. As ionic strength was increased from 0.025M, the zeta potential and

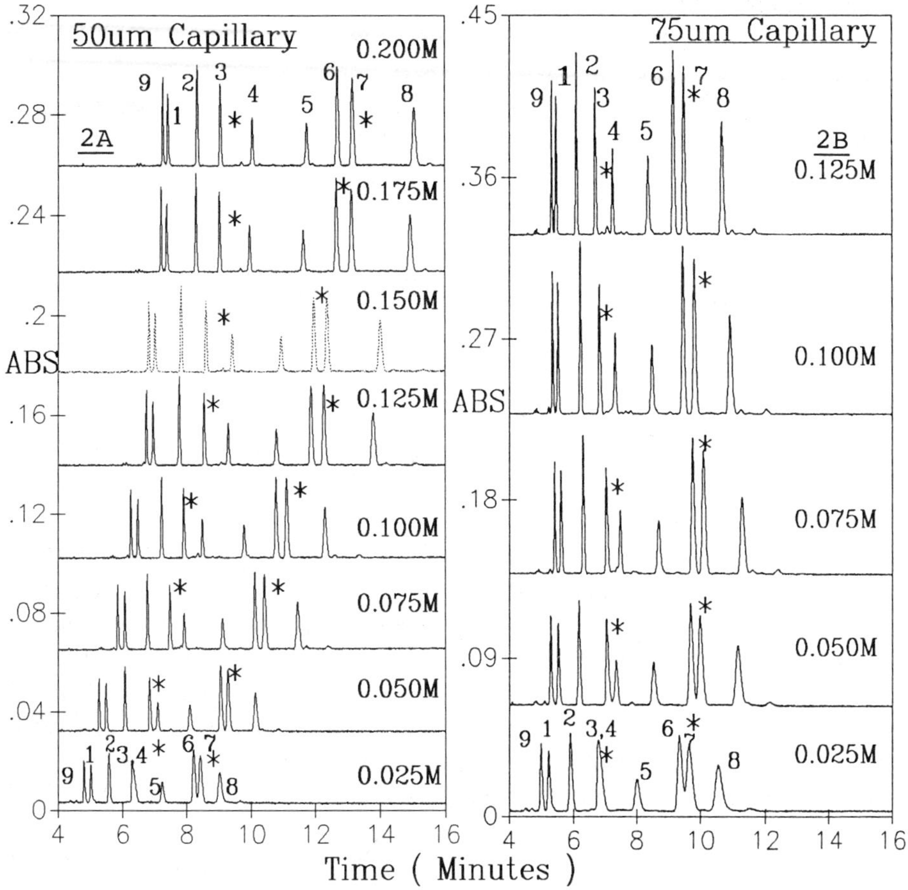

Figs. 2A-B. Effect of ionic strength on peptide separations. Conditions: 30 kV, 20°C and 0.025-0.200M NaH_2PO_4, pH 2.44.

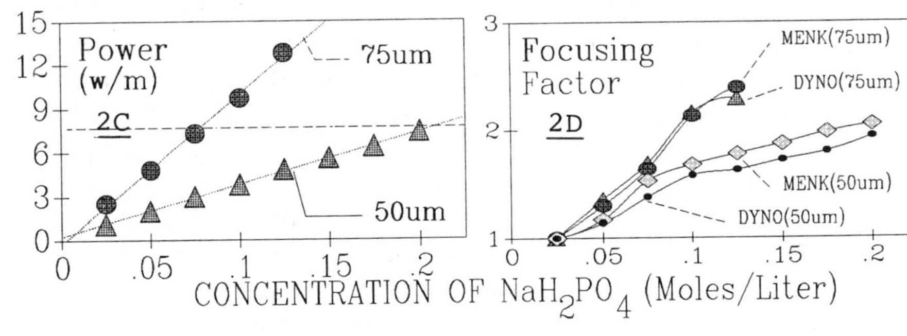

Figs. 2C-D. Effect of ionic strength on power and focusing electroosmotic flow decreased and lower mobilities were observed [3]. At higher ionic strengths (>0.075M) the results in the 75 um capillary can be explained by decreases in buffer viscosity due to increased heating. The dashed line in Fig. 2 represents the power limit under these conditions.

Equations 1-4 show the quantitative relationships for mobility, ionic strength, current, and other factors in HPCE [10,18]:

$$E = I R \qquad (1)$$

$$W = I^2 R \qquad (2)$$

$$I = [\, \pi r^2 \epsilon E c_i \zeta_i \,] / [\, \eta \,] \qquad (3)$$

$$v_{eo} = [\, \epsilon \zeta E \,] / [\, 4\pi\eta \,] = \mu_{eo} E \qquad (4)$$

where E = applied voltage, I = current, R = resistance, W = power, r = capillary internal radius, ϵ = dielectric constant, c_i = molar concentration of species i, ζ_i = zeta potential of i, η = viscosity, v_{eo} = electroosmotic flow velocity, and μ_{eo} = the apparent mobility of the medium under electroosmotic flow. The Ohm's law (Eq. 1) and power (Eq. 2) plots (Fig. 2C) were observed to be linear at all buffer concentrations for the 50 um and at up to 0.075M for the 75 um capillary. Thus smaller diameter capillaries allowed use of higher ionic strengths (Eq. 3) because of better heat dissipation. Other critical observations are shown in Figs. 2A and 2B.

a. As ionic strength differences between the sample (0.1% TFA) and run buffer increased, the peak height increased. This focusing effect was the result of localized differences in potential gradient [8,19]. Focusing appeared greater in 50 um than 75 um capillaries. The focusing factor (peak height$_{iM}$/peak height$_{0.025M}$) vs. buffer concentration plot (Fig. 2D) showed this effect. The focusing effect can increase sensitivity and load.
b. Higher ionic strength gave better resolution, reduced wall interactions, and increased mass recovery (Fig. 2A-B, see *). The increase in resolution over the range of 0.025-0.075M was 1.3x greater for the 50 um capillary relative to the 75 um capillary.
c. Low ionic strength buffers (< 0.035 M) resulted in peak broadening and coelution, although shorter migration times were observed.
d. Automation of the Ohm's law and power analyses provided a convenient way to determine ionic strength limits of the system.
e. Effective temperature control was important at high ionic strengths because of the high power densities generated (up to 12.8 W/m).

2. Voltage

The borate buffer showed a nearly linear current-voltage relationship over the range of 0 to 30 kV (Fig. 3 inset). Maximum efficiency is obtained just prior to the point where the curve deviates from linearity [20]. At low voltages, peaks broadened due to diffusion (Fig. 3). Higher voltages yielded more plates, faster times and sharper peaks, as predicted by Eq. 5 [10,21]:

$$N = [\mu_{eo}V] / [2D] \qquad (5)$$

where N = theoretical plate number, V = voltage, and D = diffusion coefficient. Excessive voltage causes non-controlled Joule heating and a lower number of theoretical plates. The conditions for maximum resolution were obtained when current and Joule heating are not limiting factors. Temperature control of the system is very important in attempting to use higher voltages and is required to gain the higher-field strength benefits.

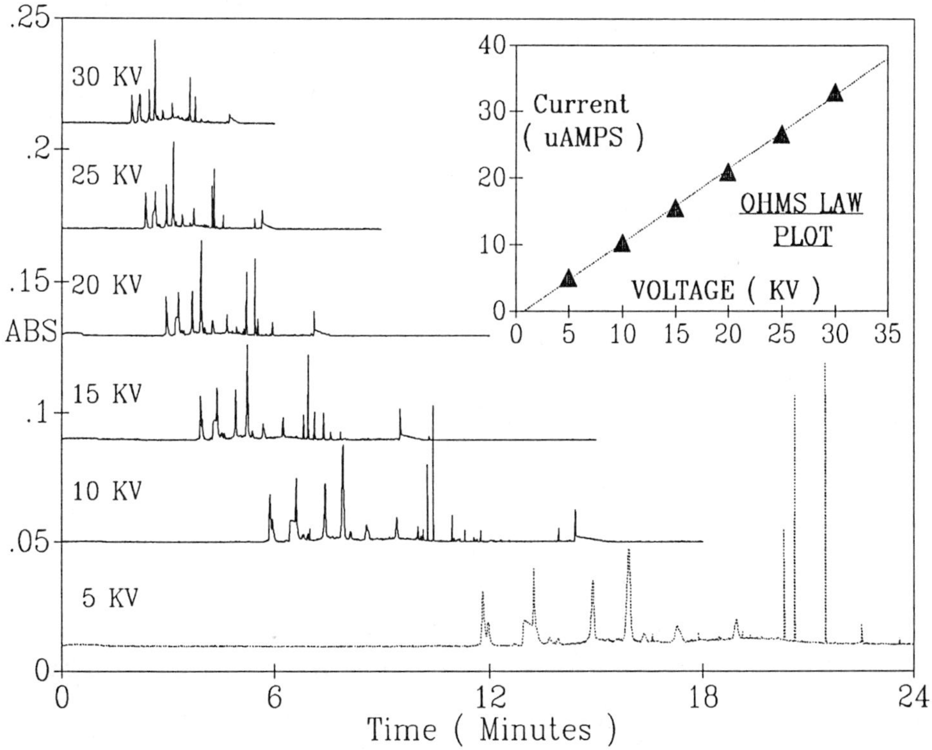

Fig. 3. Effect of voltage on migration time and current (inset). Conditions: 5-30 kV, 25°C, 75 um (id), 0.1M Na Borate, pH 8.3.

3. Temperature control

As the temperature was increased from 19 to 49°C, the peptides moved to shorter times (Fig. 4A). While the current increased linearly with temperature (Fig. 4B), there was a parallel decrease in the viscosity (Fig. 4C; refs. [22,23]). Given that migration time was inversely proportional to the viscosity and thermal equilibrium existed (i.e. viscosity was constant), data sets were corrected for temperature dependence. Migration times were corrected using bradykinin (reference marker) and a reference state at 24°C, using the following correction (Eq. 6):

$$MT_{COR} = [\, MT_{REF(t)}/MT_{MM(t)} \,] \times [\, MT_{MM(24°C)} \,] \qquad (6)$$

where MT= migration time, MT_{COR} = corrected MT, $MT_{REF(t)}$ = MT at temperature (t), $MT_{MM(t)}$ = MT of a mobility marker (MM) at temperature (t), and $MT_{MM(24°C)}$ = MT of the mobility marker at 24°C. Small changes in resolution and efficiency were observed that can be advantageous.

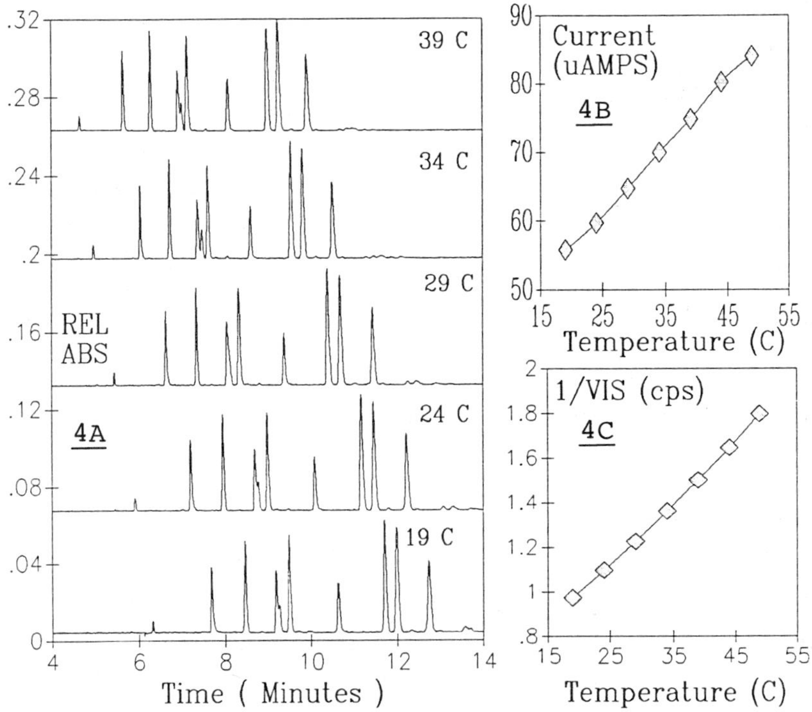

Fig. 4A-C. Effect of temperature on time, current, and viscosity. Conditions: 20 kV, 19-49°C, 75 um (id), and 0.05M NaH_2PO_4, pH 2.52.

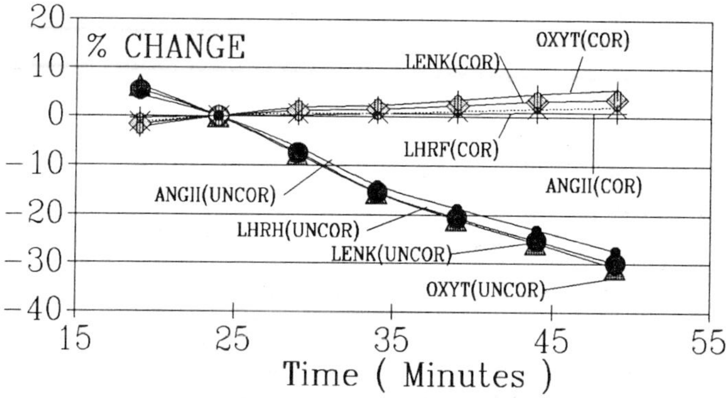

Fig. 4D. Temperature correction effect on apparent migration time.

The corrected and uncorrected results are shown in Fig. 4D. The correction substantially reduced the temperature dependence. However, as the migration time increased, the error in the corrected value increased. Thus a single reference peak apparently was not sufficient to correct the whole data set. This was not surprising since viscosity changes were significant over the temperature range studied. The importance of temperature control has been described by Nelson et al [24].

4. Organic modifiers

Addition of acetonitrile resulted in slight increases followed by steady decreases in migration times (Fig. 5A). Longer migration times were seen over the same concentration range with methanol (Fig. 5B). Migration time behavior paralleled changes in viscosity (Fig. 5C) [25,26]. Viscosity increased slightly up to 20% acetonitrile and decreased at higher percentages. The viscosity of methanol solutions increased up to 50%.

Eq. 4 shows that mobility depends not only on viscosity but also on the dielectric constant, zeta potential, and on other factors, such as permittivity [27,28]. Plots of mobility vs. percent modifier (data not shown) indicated that mobility was directly proportional to viscosity. This observation is consistent with the results of Vanorman et al [27].

Changes in zeta potential, dielectric constant, permittivity, and net charge may also influence results and must be considered. Organic additives offer a convenient way to modify these parameters. Their effect on activity, recovery, precipitation of biomoloecules and buffers are some potential problems that need to be considered.

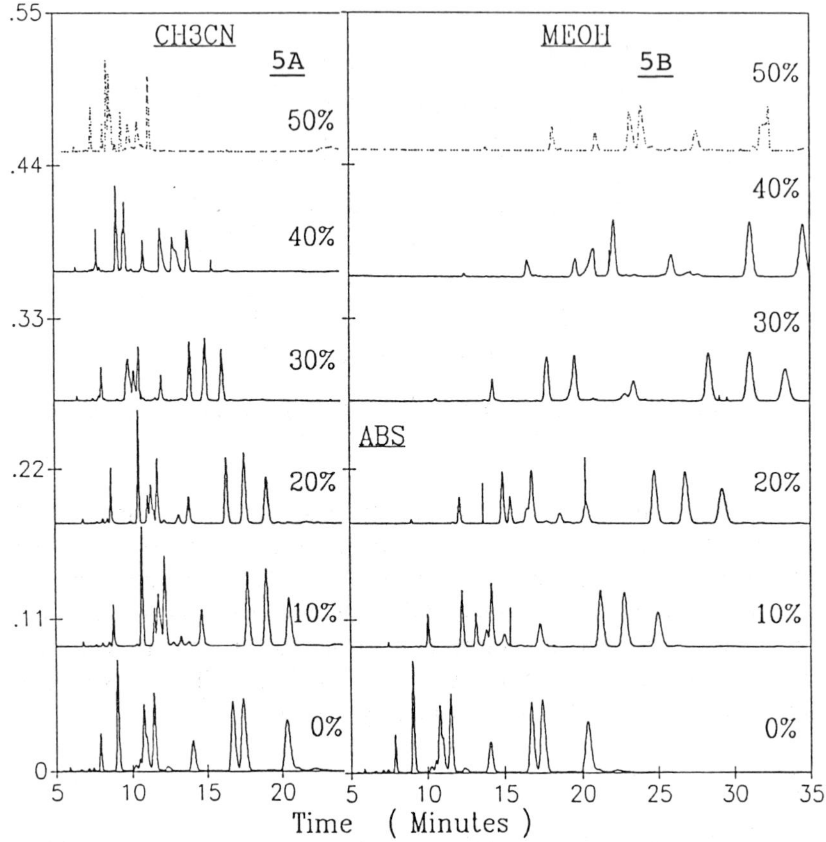

Fig. 5A-B. Effect of organic modifiers on peptide mobility. Conditions were: 25 kV, 30°C, 75 um (id), and 0.05M NaH_2PO_4, pH 2.44 (before addition of organic solvent).

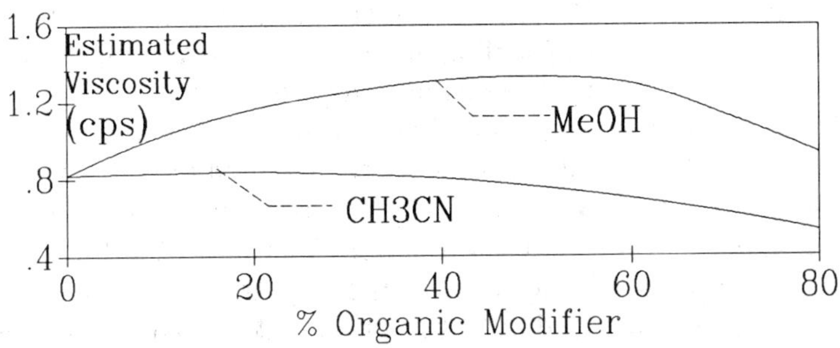

Fig. 5C. Plot of estimated viscosity vs. percent organic modifier for acetonitrile and methanol in 0.05M NaH_2PO_4 at 30°C.

5. pH

A series of five peptides were synthesized with the same amino acid composition but different primary sequence [17]. As the pH was increased from 2.5 to 3.1 (Fig. 6), resolution was obtained for this set of closely related analogs. Separation can be attributed to differences in the intrinsic pKa of the C- terminal carboxyl group. Although this titration covered the region of the pKa, no peak splitting was observed. This was likely due to a rapid equilibrium between the various species. In certain cases peptides moved more slowly than might have been predicted, e.g. peptide 14. This suggested that the side chain pKa of the N-terminal aspartic group was lower than the average pKa which is approximately 4.5 [29]. These results suggest that small differences in pKa can be the basis of separating closely related peptides by capillary electrophoresis. Thus, it is equally important to investigate the high and low pH regions. The pH range of 2.4-8.5 was investigated for this set of peptides (data not shown).

An effective approach for the resolution of unknown peptide mixtures should involve titration using automated step changes across the pKa's of the potential side chain groups.

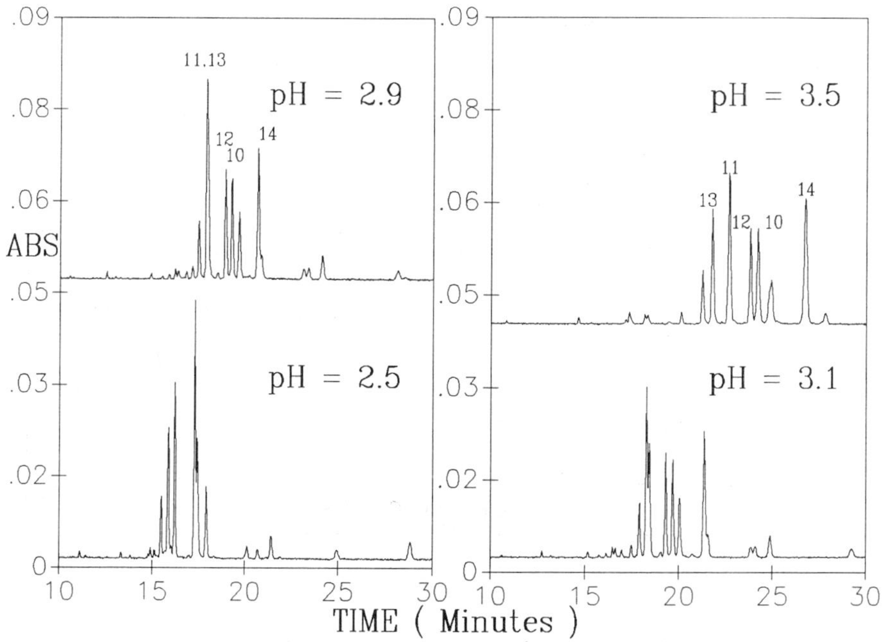

Fig. 6. Determination of optimal pH. Conditions were: 15 kV, 25°C, 75 um (id), 0.10M NaH_2PO_4, pH 2.5 to 3.1.

B. Sampling

1. Rapid peak identification

Several instrument features were utilized for rapid peak identification, including the ability to load samples from either end of the capillary, to reverse the polarity, and to program multiple injections. Fig. 7 depicts a series of automated runs using reversed polarity and injection onto the short end of the capillary, thus giving an effective length of 6.9 cm. and rapid analysis times. Using multiple injections and varying injection times allowed discrimination of closely eluting components. Although short capillaries (short effective length to the detector) offered less theoretical efficiency and resolution, they were adequate for peak identification and

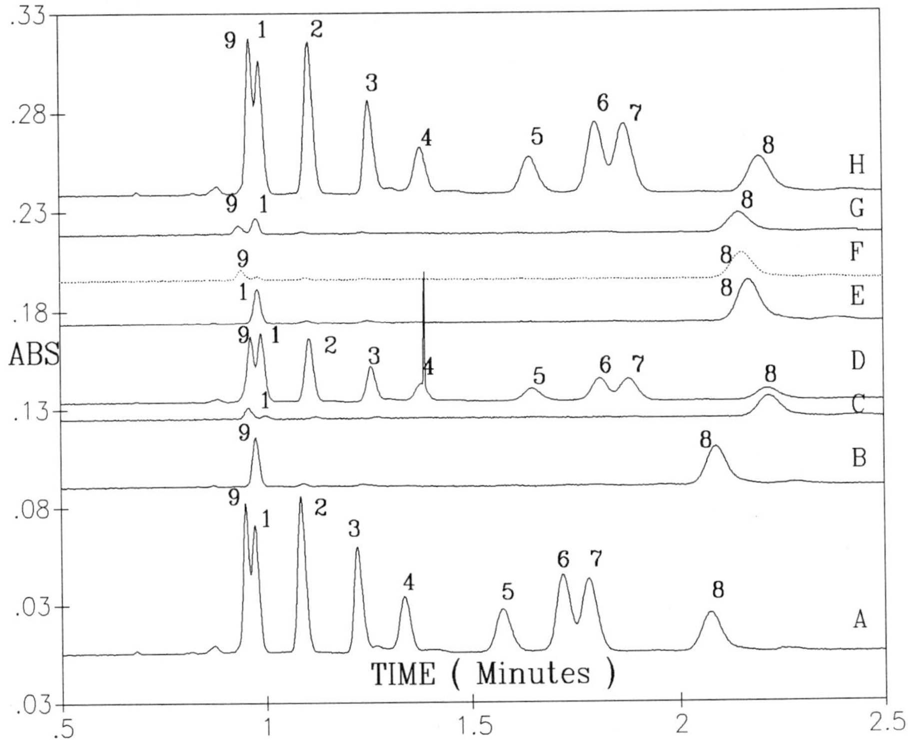

Fig. 7. Rapid peak identification and methods development. Conditions: 25 kV (reversed polarity), 20°C, 75 um (id) x 6.9 cm (td), 0.05M NaH_2PO_4, pH 2.44. and electrokinetic injection for 1 to 5 sec.

shortened run times. The total time for these 8 runs was less than 60 minutes, including setup, programming, rinses, and separations. Use of short length capillaries can be a very effective means of performing rapid methods development or complete analytical analyses of few components where large numbers of theoretical plates are not needed.

2. Reproducibility and mobility correction

An overlay of every 4th run of 15 consecutive runs is displayed in Fig. 8A. Peak area precision (data not shown) was 1.5 to 2.3%. The plot of percent relative standard deviation (RSD) vs. migration time (Fig. 8B) indicated that reproducibility of less than 1% could be obtained. However, the percent RSD increased with migration time, indicating that mobility was not constant. Mobility correction using the method of Moring et al. was applied [30-31]. The results (Fig. 8C) indicated that correction for mobility with a single marker (Dynorphin) improved precision but did not eliminate the variation with time. Multiple mobility markers may be required.

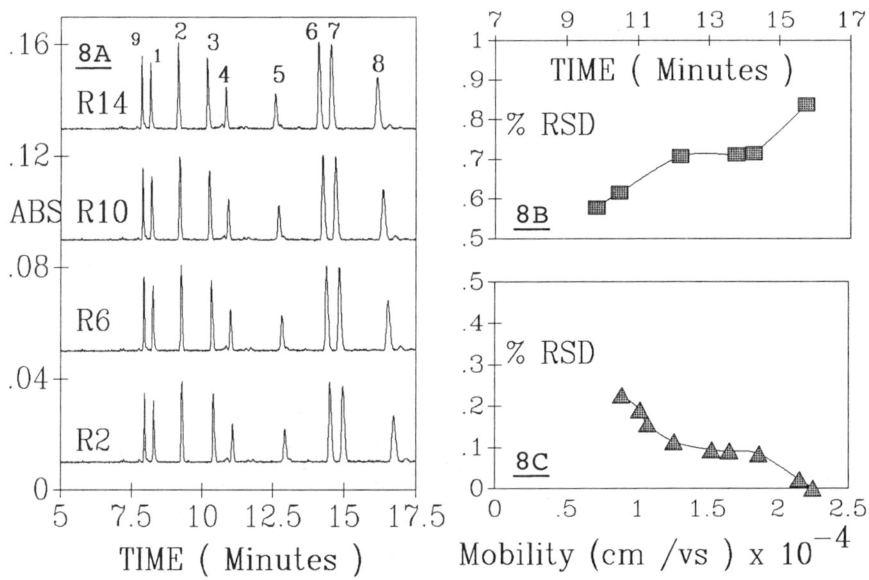

Fig. 8 A - C. Comparison of migration time and mobility reproducibility. Conditions were: 23.7 kV, 25°C, 50 um (id) 0.075M NaH_2PO_4, pH 2.44, and 5 sec. injected from a capped vial (4.6 ml). Results are averages from 15 consecutive runs.

C. Automated Post Run Data Analysis

A combination of commercially available software that allowed intra-computer data transfer using standard translation (ASCII) was used to produce the desired outputs, reports and publication-quality printouts. Lotus 123 was used for statistical analysis, combining, and offsetting the axes of large (4,000-30,000 pts.) data sets (e.g. Fig. 5A and B). Excel was used for statistical analysis of small data sets (<4,000 pts.) and to produce special multitasking analysis macros (e.g. Fig. 9). However, most figures were produced with Graph Plus because of its ability to create scalable, high-quality images with custom text from large data sets (e.g. Fig. 7).

1. Theoretical plate calculation

In the example shown in Fig. 9, an Excel macro was used to calculate theoretical plate number by the method of Foley et al. which corrects for peak asymmetry [32]. The peak parameters were obtained from data files using P/ACE control software operating in a multitasking mode. The operator could view both programs simultaneously. The spreadsheet prompted the operator to input a certain parameter. The parameter was automatically transferred to the spreadsheet. Typical theoretical plate number values were 90,000 plates/meter for 50 um capillaries. The macro can also generate statistical summaries for individual or multiple data sets.

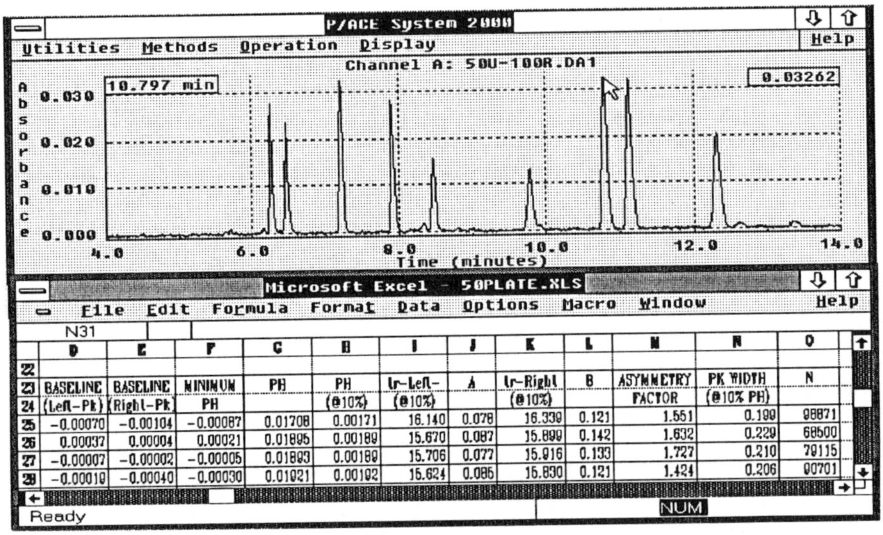

Fig. 9. Post run data analysis using Windows multitasking capability.

2. Determination of apparent molecular weight/base number

The use of capillary gel electrophoresis (CGE) in the analysis of oligonucleotides [2,33], proteins [10, 34], and DNA sequencing [35-36] has been described recently. Excel macros were used to analyze the data from a polyadenylic acid separation, which has been described previously by Rampal, et al [37]. In this example the macro plotted absorbance vs. time (Fig. 9A), produced calibration plots (Fig. 9b), calculated apparent molecular weight and base number, and generated a custom report (Fig. 9C). The use of macros and spreadsheet programs allows the convenient performance of calculations not available in standard data analysis programs. Additionally, macros have been applied to isoelectric focusing (data not shown) and has been very useful to determine the apparent pI of analytes.

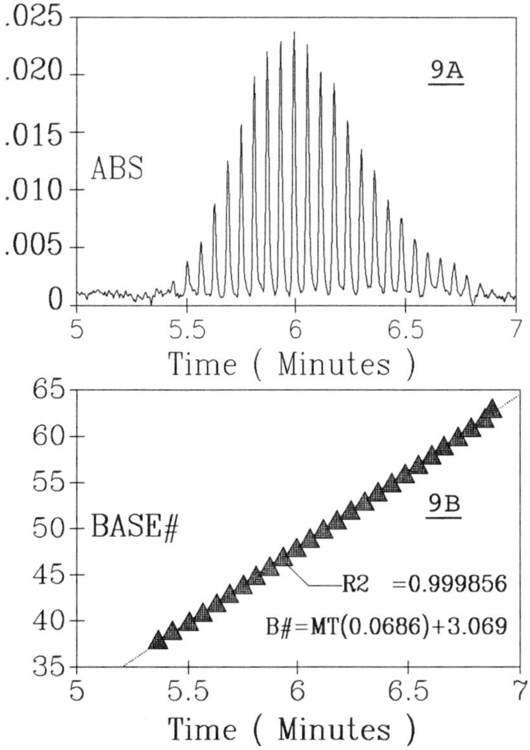

	Time	Base	Conc.	Peak	Area	Peak
#	(min.)	Name	ug/ml	Area	%	Ht.
1	5.367	pdA38	.08	.00221	.398	.00009
2	5.43	pdA39	.06	.00155	.279	.00011
3	5.505	pdA40	.29	.00795	1.433	.00032
4	5.567	pdA41	.35	.00974	1.756	.00049
5	5.63	pdA42	.51	.01408	2.539	.00081
6	5.69	pdA43	.79	.02191	3.951	.00118
7	5.753	pdA44	.97	.02697	4.862	.0015
8	5.814	pdA45	1.25	.0348	6.274	.00189
9	5.875	pdA46	1.38	.03829	6.902	.00214
10	5.936	pdA47	1.48	.04104	7.398	.00223
11	5.997	pdA48	1.46	.04063	7.324	.00229
12	6.058	pdA49	1.55	.04289	7.732	.0022
13	6.119	pdA50	1.4	.03879	6.994	.00198
14	6.18	pdA51	1.38	.03833	6.909	.00187
15	6.241	pdA52	1.16	.03218	5.802	.00157
16	6.303	pdA53	1.05	.02903	5.234	.00131
17	6.364	pdA54	.99	.02753	4.962	.00113
18	6.425	pdA55	.79	.02181	3.932	.0009
19	6.485	pdA56	.69	.01908	3.44	.00075
20	6.546	pdA57	.51	.01411	2.544	.00057
21	6.606	pdA58	.54	.01496	2.696	.00046
22	6.664	pdA59	.47	.01312	2.364	.00041
23	6.727	pdA60	.34	.00944	1.702	.00036
24	6.783	pdA61	.26	.00728	1.312	.00026
25	6.843	pdA62	.19	.00536	.967	.00016
26	6.877	pdA63	.06	.00163	.294	.00002

Fig. 9A-C. Capillary gel electrophoresis of pd(A) 40-60: electropherogram, calibration plot, and report. Conditions: 30 kV, 75 um (id), 7.5 cm (td), 20°C, TRIS-borate-urea (polyacrylamide gel).

VI. CONCLUSIONS

Automated methods development for the separation of biomolecules by capillary electrophoresis offers the following advantages:

- rapid HPCE parameter optimization
- repetitive operations and better reproducibility
- time savings and increased sample throughput
- minimization of data transfer and calculation errors
- ability to apply correction algorithms
- simplification of report and publication-quality printout generation

ACKNOWLEDGMENTS

The authors wish to thank Dr. A. Guttman (Beckman Instruments) for providing capillary gel electrophoresis data.

REFERENCES

1. Gordon, M., Huang, X., Pentoney, S., and Zare, R. (1988). Science 242, 224-228.
2. Guttman, A., Cohen, A., Heiger, D. and, Karger, B. (1990). Anal.Chem. 62, 137-141.
3. Jorgenson, J. and Lukacs, K. (1981). J. Chromotogr. 218, 209-216.
4. Burolla, V., Pentoney, S., and Zare, R. (1989). ABL Dec., 56-64.
5. Ewimg, A., Wallingford, R., and Olefirowicz, T. (1989). Anal. Chem. 61, 292a.
6. Lauer, H. and McManigill, D. (1986). Anal. Chem. 58, 166-170.
7. Terabe, S., Utsumi,H., Otsuka, K. and Ando, T. (1986). HRC & CC. 9, 666-670
8. Mikkers, F., Everaerts, F., and Verheggen, T. (1979). J. Chromatogr. 169, 1-10.
9. McCormick, R. (1988). Anal. Chem. 60, 2322-2328.
10. Cohen, A., Paulus, A. and Karger, B. (1987). Chromatographia. 24, 15-24.
11. Kuhr, W. and Yeung, E. (1988). Anal. Chem. 60, 1832-1834.
12. Hjerten, S., Elenbring, K., Kilar, F. and Liao, J. (1987). J. Chromatogr. 403, 47-61.
13. Gassmann, E., Kuo, J. and Zare, R. (1985). Science 239, 813.
14. Tsuda, T., Nakagawa, G., Sata, M. and Yagi, K. (1983). J. App. Biochem. 5, 330-336.
15. Jorgenson, J., Rose, D., and Kennedy, R. (1988). Amer. Lab. Apr., 32-41.
16. Pospichal, J. Gebauer,P. and Bocek, P. (1989). Chem. Rev. 89, 419-430.

17. Field, M., Palmieri, R, Keck, R., Ohms, J., and O'Connor, J. (1990). HPCE'90, Int. Symp. High Perform. Capillary Electrophor., 2nd, San Francisco, Abstr. # P214.
18. Wallingford, R. and Ewing, A. (1990)."Advances in Chromatography", Vol. 29, pp. 3-19. Marcel Dekker, Inc., New York.
19. Beckers, J. and Everaerts, F. (1990). J. Chromatogr. 508, 19-26.
20. Sepaniak, M., and Cole, R. (1987) Anal. Chem. 59, 472-476.
21. Jorgenson, J., and Lukacs, K. (1983). Science 222, 266-272.
22. Hardy, R. and Cottingham, R. (1949). J. Res. NBS 42, 573.
23. Kell, G., (1972). "CRC Handbook of Chemistry and Physics", R.Weast, ed., p. F-5. The Chemical Rubber Co., Cleveland.
24. Nelson, R., Paulus, A., Cohen, A., Guttman, A., and Karger, B. (1989). J. Chromatogr. 480, 11-127.
25. Kirkland, J. and Snyder, L. (1979). "Introduction to Modern Liquid Chromatography", 2nd edition, p. 838. Wiley-Interscience, New York.
26. National Bureau of Standards, (1972). "CRC Handbook of Chemistry and Physics", R. Weast, ed., p. F-36. The Chemical Rubber Co., Cleveland
27. Vanorman, B., Olefirowicz, T., Liverside, G., Ewing, A. and McIntire, G. (1990). HPCE'90, Int. Symp. High Perform. Capillary Electrophor., 2nd, San Francisco, Abstr. # P116.
28. Hunter, J., (1981), "Zeta Potential in Colloid Science, Principles and Applications.", Academic Press, New York.
29. Palmieri, R., Akiyama, T., Holladay, L., and Ulfelder, K. (1990) Pitts. Conf. on Anal. Chem. and App. Spectro., 41st, New York, Abstr. # 1121.
30. Moring, S., Colburn, J., Grossman, P. and Lauer, H. (1989). LC-GC 8, 34-46.
31. Grossman, P., Colburn, J., and Lauer, H. (1989). Anal. Biochem. 179, 29-33.
32. Foley, J. and Dorsey, J. (1983). Anal.Chem. 55, 730-737.
33. Paulus, A. and Ohms, J, (1990). J.Chromatogr. 507, 113-123.
34. Karger, B., Cohen, A., Rush, R., Guttman, A., and Heiger, D. (1990). HPCE'90, Int. Symp. High Perform. Capillary Electrophor., 2nd, San Francisco, Abstr. # ML103.
35. Cohen, A., Najarian, D., Gilbert, T., and Karger, B. (1990). HPCE'90, Int. Symp. High Perform. Capillary Electrophor., 2nd, San Francisco, Abstr. # TL105.
36. Drossman, H., Luckey, J., Kostichka, A., and Smith, L. (1990). HPCE'90, Int. Symp. High Perform. Capillary Electrophor., 2nd, San Francisco, Abstr. # TL107.
37. Rampal, S., Guttman, A. and O'Brien, R. (1990). HPCE'90, Int. Symp. High Perform. Capillary Electrophor., 2nd, San Francisco, Abstr. # P120.

2

Two-Dimensional Electrophoresis of Proteins from Cultured Human Colonic Carcinoma Cells: Development of a Database of Amino Acid Sequences

Larry D. Ward, Ji Hong and Richard J. Simpson

Joint Protein Structure Laboratory
Ludwig Institute for Cancer Research (Melbourne Branch)
and The Walter and Eliza Hall Institute of Medical Research
P.O. Royal Melbourne Hospital, Victoria 3050, Australia

I INTRODUCTION

High-resolution two dimensional electrophoresis (2-DE) is an extremely powerful technique for resolving and mapping the very complex mixture of proteins found in cells and body fluids (1). The 2-DE gel pattern or map which describes overall gene expression can be analysed by computer-based scanning systems to provide quantification and loci (e.g., pI and M_r) information for fractionated proteins. This approach allows the ready comparison of cellular protein expression in response to growth factors (2), changes in cell cycle (3), transformation (4) and development (5). Thus, 2-DE provides a potentially important approach towards obtaining a description of cellular and subcellular functions in molecular terms.

The availability of 2-DE gel-databases of identified proteins greatly enhances this approach since unknown proteins can be readily identified on the basis of their 2-DE gel behaviour (6). Such databases have expanded rapidly in recent times due to developments in protein sequencing technology (for recent reviews see refs. 7 and 8 and refs. therein) that permit direct sequence analysis of 2-DE resolved proteins.

One of the major interests in our laboratory is the control of cell proliferation and differentiation, with particular emphasis on the cells of the gastrointestinal tract. Several cell lines from human colonic carcinomas which retain growth factor responsiveness and the ability to produce differentiated cells *in vitro,* have been isolated (9). To gain some insight into the biochemical mechanisms underlying the growth factor-induced morphological changes in one of these cell lines (LIM 1863), we are using 2-DE to investigate associated changes in cellular protein patterns.

Techniques in Protein Chemistry II

Here we report an update on our protein database for the human colon carcinoma cell line LIM 1863 and present an improved experimental approach for obtaining both N-terminal and internal amino acid sequence information for gel-resolved proteins.

II EXPERIMENTAL

A. Two-dimensional gel electrophoresis

Sample preparation (in 9.0M urea, 2% (v/v) nonidet P-40, 2% total ampholytes (LKB) and 50mM DTT) and two-dimensional electrophoresis were performed as previously described (10).

B. N-terminal Sequence Analysis of Gel-resolved Proteins

Two methods were examined for isolating proteins from acrylamide gels for sequence analysis:

1. *Electroblotting onto polyvinylidene difluoride (PVDF) membranes* as previously described (10,11,12).

2. *Passive elution from acrylamide gels.* Proteins were visualized in acrylamide gels - (1.5mm thick) without staining - by high-resolution dynamic imaging (HRDI) (13), a phase-contrast method that relies on protein-associated differences in refractive index. HRDI was performed using a Molecular Dynamics (Sunnyvale, CA) Autophoresis™ instrument. Polyacrylamide gels were poured using acrylamide and piperazine diacrylamide (BioRad) as crosslinker. Using the above reagents and polymerization times of 20 min, gels of uniform refractive index, as monitored by HRDI, were observed. Proteins were passively eluted from gel slices by soaking with 20mM Tris-HCl buffer, pH 7.4 containing 0.02% Tween 20 (3 x 1ml) at 25°C for ~20h. For sequence analysis, proteins were concentrated and desalted either by microbore column RP-HPLC (7,14) or inverse-gradient RP-HPLC (7,15).

C. Internal Sequence Analysis of Gel-resolved Proteins

Peptides were generated from gel-resolved proteins by *in situ* proteolytic digestion in gel slices followed by acid extraction and separation of peptides by microbore column RP-HPLC. Two different approaches were employed, depending upon whether proteins were visualized in the gel by staining with Coomassie brilliant blue R-250 (CBR250) or by HRDI.

1. *Proteins visualized by CBR250.* After electrophoresis acrylamide gels were stained with 0.1% CBR250 in 50% methanol/10% acetic acid for up to 12 h and then destained in 7% acetic acid/12% methanol. Protein spots were excised, placed in an Eppendorf tube and then washed (2 x 1ml) in 50% aqueous n-propanol containing 3% SDS (for ~3 h at 25°C or until gel was clear) in order to remove CBR250. The gel slices were then washed extensively (24h) with water - to remove SDS and other acrylamide-related contaminants - dried by centrifugal lyophilization and then rehydrated in 100µl of 100mM NH_4HCO_3, 0.5mM $CaCl_2$, 0.02% Tween 20 containing 2µg trypsin. Following proteolytic digestion (37°C, 12h) peptide fragments were extracted from the gel slice with 100µl of 1% TFA (4h) followed by 70% TFA (2 x 4h) and 50% TFA, 50% acetonitrile (2 x 4h). Tween 20 (0.02%) was included in all solvents. The combined eluents were evaporated to near dryness and then diluted to 1ml with 0.1% TFA. Peptides were purified by microbore column RP-HPLC as described elsewhere (14)

2. *Proteins visualized by HRDI.* Following electrophoresis, proteins were visualized in the gel by HRDI and then excised. After fixing (50% methanol, 10% acetic acid), gel slices were washed extensively with water (24h) to remove gel-related artifacts. Generation of peptides for sequence analysis by *in situ* proteolytic digestion was performed as outlined above.

D. Amino Acid Sequence Analysis

Sequence analysis of proteins and peptides was performed using Applied Biosystems sequencers (models 470A and 477A) equipped with on-line phenylthiohydantoin (PTH) amino acid analyzers (models 120A) and an improved sample transfer device that allows the injection of the total PTH-amino acid derivative (16).

III RESULTS AND DISCUSSION

A typical 2-DE protein pattern of whole cell lysate (150μg total protein) from the human colon cancer cell line LIM 1863 - using pH 3-10 carrier ampholytes in the first dimension and CBR250 staining - is presented in Fig. 1. Proteins #1-13 were identified by sequence analysis following electroblotting onto PVDF membrane. Sequence data, along with the coordinates pI and M_r of identified proteins are summarised in Table 1. Sequencing yields from 4 identical PVDF membrane electroblots were in the range 3-20 pmol. It should be noted that these yields were obtained using commercial protein sequencers that had been fitted with an improved transfer device designed in this laboratory (16), that permits analysis of >95% of PTH-amino acid derivative (compared to 40-45% for unmodified versions of the same commercial sequencers).

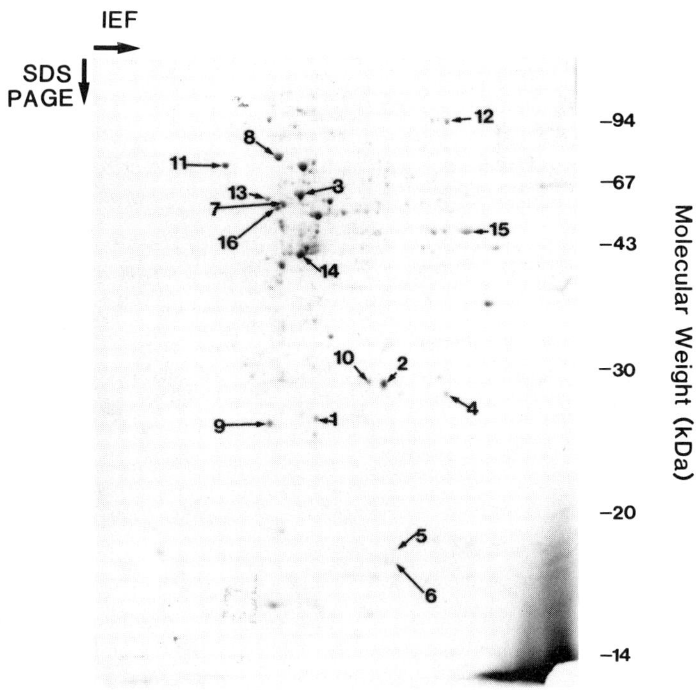

Fig. 1 2-DE gel protein pattern of cell lysate (150μg total protein) of human colonic cell line (LIM 1863). Protein was detected by Coomassie Blue R250 staining. The indicated proteins were identified by protein sequence analysis (Table 1).

Table 1. Identification of 2-DE resolved human colonic carcinoma cell line (LIM 1863) proteins by sequence analysis.

Protein[a]		Yield[b] (pmol)	M_r of protein (kDa)	M_r of[c] peptides (kDa)	pI	Identification by[d,e] protein database search
# 1:	P P Y T V V Y F P V R G R X A A X X M X L A D	3	26		6.3	Glutathione-S-transferase π
# 2:	G K G D P K K P R G K M X X F	4	30		6.8	HMGI
# 3:	A K D V K F G A D A R A L M L Q G V D L L A D A V A V T M	9	66		6.2	Ribosomal protein A - related
# 4:	A P S X K F F V G G N X K M	5	28		7.7	TPI
# 5:	A R T K Q T A R K S T G G K A P R K Q L A T K A	10	17		7.1	Histone H3
# 6:	P E P A K S A P X X K K G X K K	4	16		7.0	Histone H2B
# 7:	M X E X I S I H V G Q A G V Q I G N A X X E L Y X L	9	59		5.9	α-Tubulin
# 8:	X E E D K K E D V G X V V G I D L G	9	75		5.8	78 kDa Hsp
# 9:	M I I Y R K L I S H D E M F S D I Y	20	25		5.7	Hypothetical 21kDa protein
#10:	G K G D P K K P R G K M X X Y A F F V Q T	11	30		6.6	HMGI
#11:	E P A V Y F K E Q F L D	8	62		4.9	Ro/SS Antigen
#12:	V N F T V D Q I R A I M D X K X N I	4	94		7.7	Elongation factor-2
#13:	D A P E E E D X V	6	62		5.6	β-PHase on PDI
#14:	No Sequence	-	48		6.2	Actin
	M A P A A S S X X L E K	4	-	17		"
#15:	No Sequence	-	54		7.8	α Enolase
	X M I L P V X A X N F X E A M X I X A E	1		22		"
	X X I L P V G A	1	-	18		"
#16:	Not determined	-	58		5.7	β Tubulin
	M Y E I V N L A A Y D I G N Q I G A K F R E V I	11	-	25	-	"

a. Proteins isolated by 2-DE (Fig. 1). Four protein spots from identical gels (150μg total protein/gel) were taken for sequence analysis (#1-13) and 10 protein spots for #14-16. X, indicates positions in the sequence where residues could not be identified unambiguously. **b.** Quantitation of PTH-amino acid derivatives in each degradation cycle was determined by measurement of peak area and comparison with a 12.5 pmol PTH-amino acid standard mixture. Initial yield is the extrapolated value (pmol) of the PTH-amino acid in cycle 1 determined from a least-squares plot of the data. **c.** Internal sequence was obtained using in situ V8 digestion, electrophoretic separation of fragments and electroblotting onto PVDF (10). **d.** HMGI, High mobility group; β-PHase, β subunit of prolyl-4-hydroxylase; PDI, protein disulfide isomerase; TPI, triose phosphate isomerase; Hsp, heat shock protein. **e.** Proteins were identified by comparison with proteins in the following databases: NBRF Protein (version 22.0), NBRF Provisional Proteins (version 40.0), EMBL Protein (version 21.0), GenBank Protein Translation (version 61.0), Protein Research Foundation (PRF) of Japan Protein (version 89/4). This table was adapted from Ref. 10.

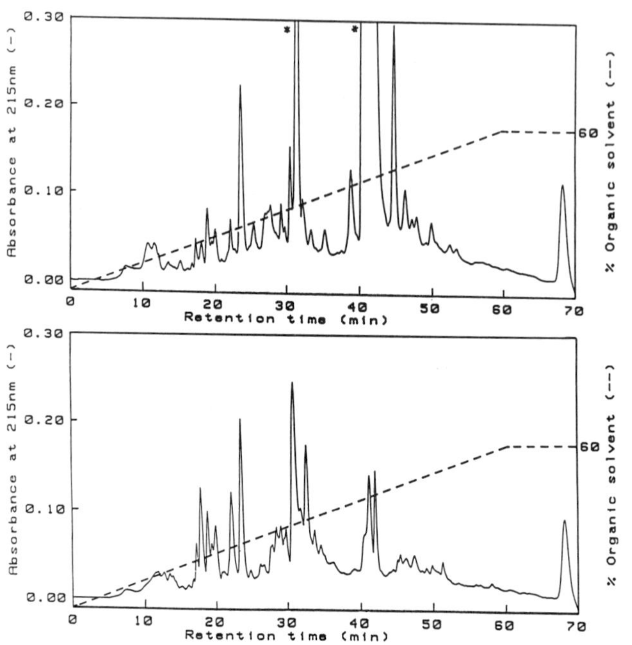

Fig. 2 Microbore RP-HPLC peptide mapping of a tryptic digest of human growth hormone. Digestion was performed *in situ* in the gel slice (see Methods) after CBR250 staining, *Panel A:* without, and *Panel B:* with a 3% SDS-50% propanol wash step to remove CBR250. Asterisks (*) indicate peaks due to CBR250.

Electroblotted proteins often contain appreciable amounts (50-100 pmol) of contaminating free amino acids that persist during the first 5-6 cycles of the Edman degradation. Since such background levels seriously interfere with sequence interpretation, we routinely wash CBR250-stained PVDF membranes with dionized water (10-12h) to reduce contaminating amino acid levels to <0.5 pmole.

Of the 22 protein spots taken for N-terminal sequence analysis, 9 were refractory to Edman degradation. In this regard, internal amino acid sequence information is particularly useful, for both identifying N^{α}-blocked proteins as well as gene-cloning strategies.

We have evaluated two general strategies for obtaining internal amino acid sequence data from gel-resolved proteins namely:- Method 1 detergent elution of electroblotted proteins from PVDF membranes, separation of proteins from the detergent mixture by inverse-gradient-RP-HPLC, proteolytic digestion of protein, and RP-HPLC separation of fragments (11) - Method 2 *in situ* proteolytic (or chemical)

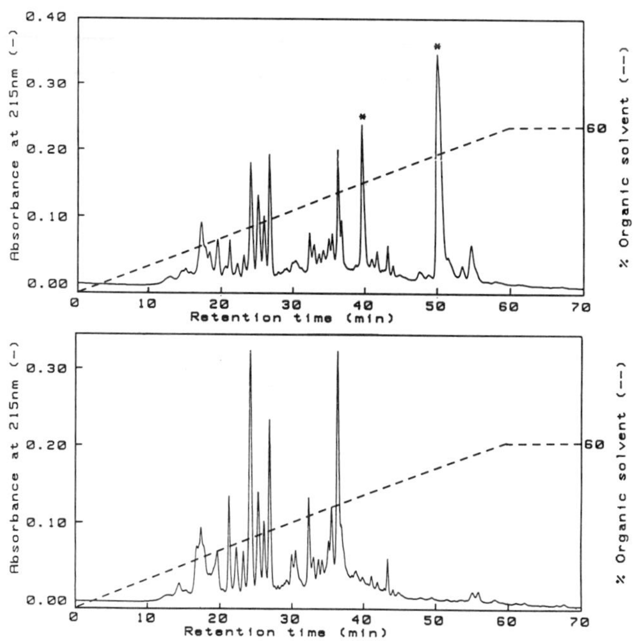

Fig. 3 Microbore RP-HPLC peptide mapping of a tryptic digest of β-lactoglobulin. Digestion was performed *in situ* in the gel slice after visualization by CBR250 staining (A) or by high resolution dynamic imaging (B). Asterisks (*) indicate peaks due to CBR250.

digestion of the protein in the gel matrix. Peptides were either - (a) extracted from the matrix and separated by RP-HPLC (*Method* 2a) or - (b) electrophoretically separated and isolated by electroblotting onto PVDF (*Method* 2b).

In our experience, higher yields are obtained by *Method* 2a. Typical *in situ* digests for β-lactoglobulin (βLG) and human growth hormone (hGH), using this approach, are shown in Fig. 2. For βLG, initial yields of 100-400 pmol for isolated peptides were obtained when 20μg (1100 pmol) was loaded onto the gel (12). The lower yields obtained with *Methods* 1 and 2b appear to be due to large losses encountered during electrotransfer and, in the case of *Method* 2b, the incomplete nature of the digest (partial digestion is inherent in this procedure in order to obtain fragments of a size suitable for separation by SDS-PAGE).

One disadvantage of *Method* 2a, however, is the high absorbance of the CBR250 stain which coextracts with peptides and can interfere with the RP-HPLC peptide mapping procedure. Peaks marked with asterisks in Figs. 2a and 3a are related to

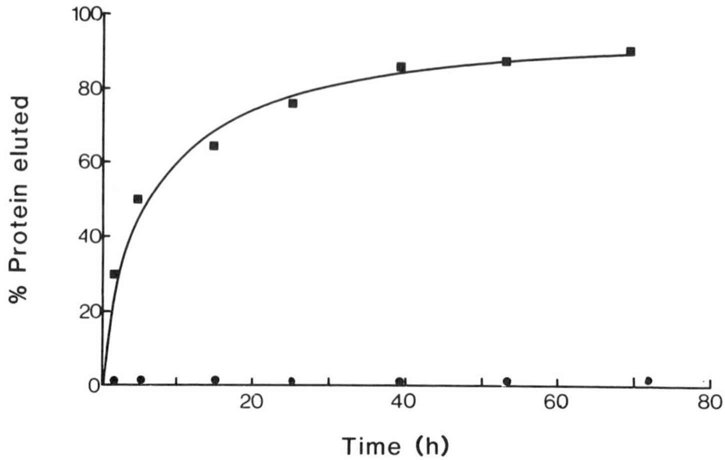

Fig. 4 Elution of protein from SDS-polyacrylamide gel. ^{125}I-BSA (20μg), separated by SDS-PAGE on 10% acrylamide gel, was detected by Coomassie Blue R250 staining (●) or high-resolution dynamic imaging (■), excised from the gel and placed in elution buffer (20mM Tris-HCl buffer, pH 7.4, 0.02% Tween-20) as described in Methods section. Eluted protein was detected by radiochemical detection.

CBR250 staining (12h). The intensity of such peaks can be markedly reduced if the time of CBR250 staining is shortened (~10-15 mins) to minimise the CBR250/protein ratio. This is, however, not always possible. When protein is being purified by multiple electrophoretic separations the low quantity of purified protein in each electrophoretic run may require high CBR250 protein ratios in order to detect the protein of interest.

A procedure designed to remove CBR250 artefacts peaks is presented in Fig. 2b. Gel pieces containing CBR250-stained hGH were washed with 3% SDS, 50% n-propanol prior to extensive washing with water (24h). This approach has previously been used to remove CBR250 from CBR250-stained proteins in gels as a method of protein quantification (17). It is thought that SDS competes for the hydrophobic binding sites of CBR250 on the protein surface while n-propanol fixes the protein in the gel during the extraction procedure (17). It can be seen that the CBR250 artefact peaks in Fig. 2a are removed by this washing step (Fig. 2b). Interestingly, the yield of peptides is also increased suggesting that CBR250 binding may sterically hinder protease digestion. A similar finding was obtained when proteins were visualized by HRDI, a recently-developed technology (13) that allows the direct visualisation of protein without the need for fixing and staining (see Figs. 3a,b for βLG).

Another advantage of visualizing gel-resolved proteins by HRDI (for the purpose of purifying proteins for sequence analysis) is that protein can be readily recovered from gel slices, in high yield, by passive elution. This is illustrated in Fig. 4 for the elution of bovine serum albumin (BSA) from a 10% polyacrylamide gel following visualization by either HRDI or CBR250. Following visualization by HRDI, 90% of BSA was recovered from the gel by passive elution, whereas <1% was eluted when BSA was detected by staining with CBR250. Proteins that have been passively eluted from gel slices (after visualization by HRDI) can be readily concentrated and salt detergent-exchanged (by RP-HPLC or inverse-gradient RP-HPLC (7)) for the purpose of structural analysis. The advantage of this approach, relative to *in situ* proteolytic digestion, is that recovered protein is readily amenable to further manipulation (e.g., reduction/alkylation). Such as manipulation is often a necessary prerequisite for peptide mapping and subsequent internal amino acid sequence analysis.

In summary, 2-DE when combined with the protein sequencing methodologies presented in this study allows the rapid separation and identification, by amino acid sequence analysis, of cellular proteins. Although only relatively abundant proteins can be directly sequenced when working with whole cell lysates, the advantage of this approach is that trace-abundant proteins can be readily enriched (18) prior to the final 2-DE purification procedure.

IV ACKNOWLEDGEMENTS

The authors are indebted to Gavin E. Reid and Robert L. Moritz for preparing the figures and to Colleen Backus for preparing the manuscript.

V REFERENCES

1. O'Farrel, P.H. (1975) J. Biol. Chem., **250**, 4007-4021.
2. Bravo, R., In: Levine, A.J., Van de Woude, G.F., Topp, W.C. and Watson, J.D. (Eds.), *Cancer Cells/1, The Transformed Phenotype,* Cold Spring Harbor Laboratory, (1984), pp. 147-151.
3. Mathews, M.B., Bernstein, R.M., Franza, B.R. and Garrels, J.I. (1984) Nature, **309**, 374-376.
4. Garrels, J.I. and Franza, B.R. (1989) J. Biol. Chem., **264**, 5299-5312.
5. Sung, Z.R., In: Celis, J.E. and Bravo, R. (Eds.), *Two Dimensional Electrophoresis of Proteins,* Academic Press Inc., Orlando, Florida, (1984), pp. 397-413.
6. Klose, J. (1989) Electrophoresis, **10**, 140-152.

7. Simpson, R.J., Moritz, R.L., Begg, G.S., Rubira, M.R. and Nice, E.C. (1989) Anal. Biochem., **177**, 221-236.
8. Shively, J.E., Paxton, R.J. and Lee, T.D. (1989) Trends Biochem. Sci., **14**, 246-252.
9. Whitehead, R.H., Jones, J.K., Gabriel, A. and Lukies, R.E. (1987) Cancer Res., **47**, 2683-2689.
10. Ward, L.D., Hong, J., Whitehead, R.H. and Simpson, R.J. Electrophoresis, In Press.
11. Simpson, R.J., Ward, L.D., Reid, G.E., Batterham, M.P. and Moritz, R.L. (1989) J. Chromatogr., **476**, 345-361.
12. Ward, L.D., Reid, G.E., Moritz, R.L. and Simpson, R.J. (1990) In: Villafranca, J.J. (Ed.), *Current Research in Protein Chemistry: Techniques, Structure and Function,* Academic Press, pp 179-190.
13. Johnson, R.F., Manian, B.S., Picket, S.C. and Barker, D.L. (1990) In: Villafranca, J.J. (Ed.), *Current Research in Protein Chemistry: Techniques, Structure and Function,* Academic Press, pp 263-269.
14. Simpson, R.J., Moritz, R.L., Rubira, M.R. and Van Snick, J. (1988) Eur. J. Biochem., **176**, 187-197.
15. Simpson, R.J., Moritz, R.L., Nice, E.C. and Grego, B. (1987) Eur. J. Biochem., **165**, 21-29.
16. Begg, G.S. and Simpson, R.J. (1989) In: Hugli, T.E. (Ed.), *Techniques in Protein Chemistry,* Academic Press, pp 79-88.
17. Ball, E.H. (1986) Anal. Biochem., **155**, 23-27.
18. Bauw, G., Van Damme, J., Puype, M., Vandekerckhove, J., Gesser, B., Ratz, G.P., Lauridsen, J.B. and Celis, J.E. (1989) Proc. Natl. Acad. Sci. USA, **86**, 7701-7705.

3

Techniques for Improved Micropreparative Fraction Collection

Yngve Hareland
Pharmacia LKB Biotechnology AB, S-751 82 Uppsala, Sweden

I. Introduction

Many biologically active substances such as transcription factors, neuroactive substances, receptors and growth factors are expressed only in trace quantities and pose major problems for their isolation and purification.

A new micropurification system from Pharmacia LKB Biotechnology, SMART™ System, based on total integration and miniaturization, enables preparative work in the µg and ng range. By using narrow bore columns together with a low dead volume system, smaller peak volumes are generated which thus give a much higher mass sensitivity and peak concentration (1-3).

Conventional HPLC columns and systems for the separation of biological macromolecules generate peak volumes in the range 0.5 - 2 ml. Collection of these peaks is relatively simple and is typically performed by collecting drops, either by time or volume. However, using small bore columns (1.6 - 3.2 mm I.D.) together with the compact micropurification system, peak volumes in the range 10 - 200 µl are obtained (Fig. 1). Collecting these peaks, or fractions of the peaks, cannot be achieved with high reproducibility and accuracy using traditional fraction collectors. This is because the volume of one drop can exceed the peak volume generated by this new micropurification system. The volume of an eluent drop depends on a number of factors (surface tension, needle geometry, etc,) but is typically around 40 µl.

Techniques in Protein Chemistry II

In this study, a new fraction collecting technique is used to solve the problem of collecting small peak volumes. Instead of collecting complete drops, the fraction collector needle is lowered into the vial, touches the bottom and wipes off the fraction (Fig. 2). Reproducibility and accuracy with this technique, here called "Wipe Mode", are measured down to 5 µl fractions and a comparison is made with standard Drop Mode collection. Since the needle remains in contact with the fraction during fractionation, it may transfer liquid from one fraction to the next. This potential risk of volumetric carry over is evaluated.

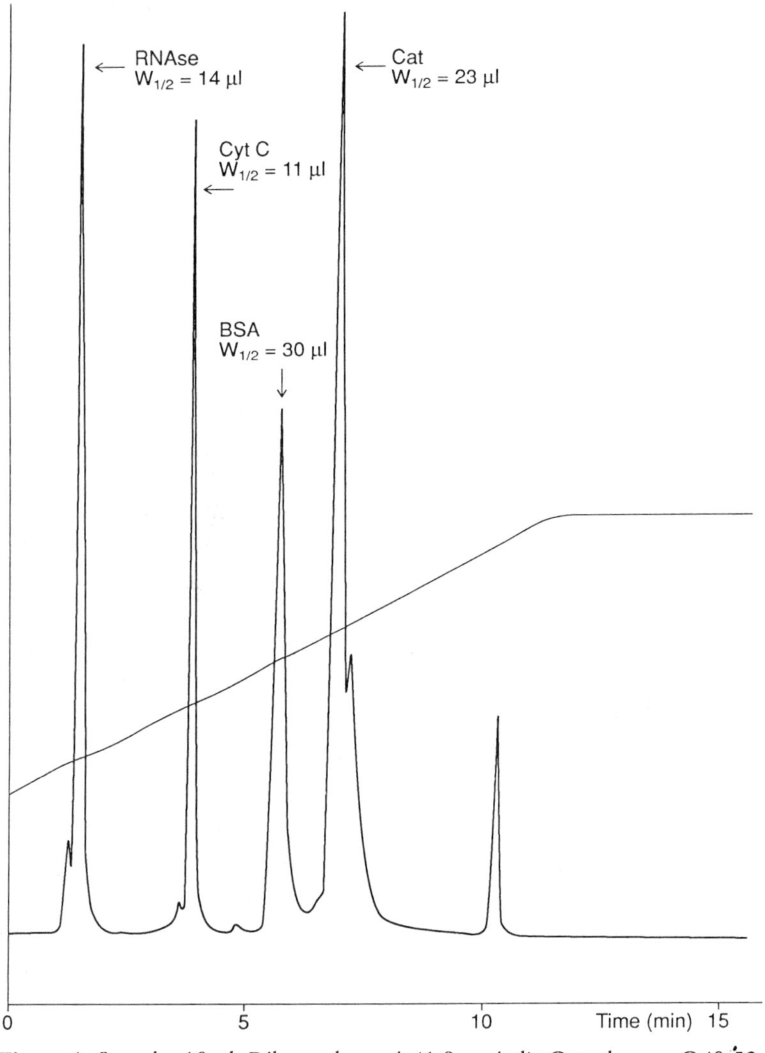

Figure 1. Sample: 10 µl, Ribonuclease A (1.8 mg/ml), Cytochrome C (0.53 mg/ml), Bovine Serum Albumin (2.1 mg/ml) and Catalase (1.9 mg/ml); column: prototype µRPC, 3 µm, C2/C18, 2.0 x 100 mm; flow rate: 117 µl/min; gradient: 25 - 55 % B in 13 minutes; eluent A: H_2O, 0.1 % TFA; eluent B: Acetonitrile, 0.1 % TFA; detector: 280 nm, 0.15 AUFS.

Wipe mode

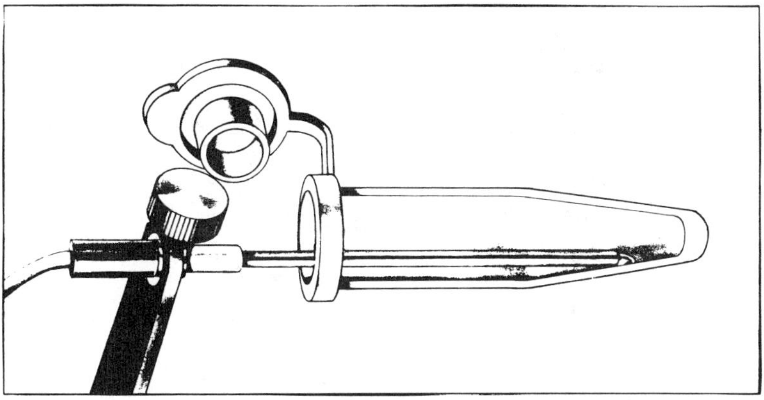

Figure 2. The fraction collector needle goes down into the vial and wipes off the fraction.

II. Material and Methods

The system used was SMART System equipped with the μFraction Collector. The fraction collector employs a high speed arm with a titanium fraction needle (I.D. 0.2 mm, O.D. 0.5 mm) that moves in and out of the vials. The time taken to move between vials in Wipe Mode is 0.5 seconds. The vials used for fraction collection were Microtube 0.5 ml polypropylene with a push cap (Sarstedt) . A prototype μRPC column (3 μm, C2/C18, 2.0 x 100 mm) was mounted in the system. The eluents in the chromatographic run were A: H_2O (Milli Q, Millipore), 0.1 % TFA (Merck), B: Acetonitrile (Lab Scan, Super Gradient), 0.1 % TFA.

To study reproducibility and accuracy in Wipe Mode, fractions of 250 μl, 100 μl, 25 μl and 5 μl were collected. Water was used as eluent and the flow rates were 250 μl/min for 250 μl fractions, 100 μl/min for 100 μl fractions, 50 μl/min for 25 μl fractions and 10 μl/min for 5 μl fractions. The vials were weighed on a Mettler AE 166 analytical balance before and after the fractionation. For comparison 25 μl fractions were collected in both Wipe Mode and the traditional Drop Mode.

By determing how much of one fraction was transferred by the fractionation needle to the next fraction, volumetric carry over was measured. Vials were prefilled with 0.1 % TFA in water and in every third vial a marker was added. The marker was uracil at a concentration of 20 mg/l. Fractionation time was 1 minute and flow rate of eluent A was 5 μl/min. The prefilled volumes and the collected fractions gave total volumes of 250 μl, 100 μl and 25 μl. After fractionation the amount of marker in each vial was analysed by reversed phase chromatography. The amount of uracil transferred into a vial immediately following a vial prefilled with the marker was used as a measurement of volumetric carry over.

To study the dependence of fraction collector needle surface area in relation to volumetric carry over the same experiment as above was performed with 200 µl and 20 µl fractions, using a needle with an O.D. of 1.6 mm, made of PEEK (polyetheretherketone).

III. Results and Discussion

The reproducibility and accuracy obtained in studies of Wipe Mode and Drop Mode as previously described is shown in Figure 3 and Table 1. In Figure 4 and Table 2 the data from volumetric carry over studies in Wipe Mode are presented.

The results show that the new fraction collecting technique, Wipe Mode, can collect fractions down to 5 µl with good accuracy and reproducibility. Accurate fractionation is no longer limited by the volume of a drop.

The potential risk of volumetric carry over is caused by the needle being in contact with the fraction during fractionation. It can be kept low if the dimensions of the needle are optimized. A needle with a large surface area gives a relatively larger carry over. With an outer diameter of 0.5 mm the surface area is kept to a minimum and even with fractions as large as 250 µl the carry over is only 0.38 µl ± 0.09 µl (n=5). A PEEK needle with O.D. 1.6 mm gave a carry over of 5.2 µl ± 1.6 µl (n=5) for 200 µl fractions . Using a titanium needle gives the physical stability of a metal without sacrificing corrosion or solvent resistance.

The fraction collector is controlled by the software of this new micropurification system. With automatic peak collection a level or a slope can be specified for accurate peak cutting. The software compensates for the delay volume, 10 µl, between the UV detector and the fraction collector. It is important that this delay volume is small in order to minimize band broadening effects. However, there is one advantage in having this volume not equal to zero. If the operator wants to collect peaks by controlling the fraction collector manually through feed commands, the delay volume allows time for noting whether a "true" peak is eluting.

In Drop Mode it is impossible to control peak cutting by level sensing and to match the drop with the peak position. With the Wipe Mode technique the start and end of the fractions is fully controlled by the movement of the needle.

The Wipe Mode technique offers interesting possibilities. One development may be the collection of fractions directly onto solid phase matrices, such as membranes.

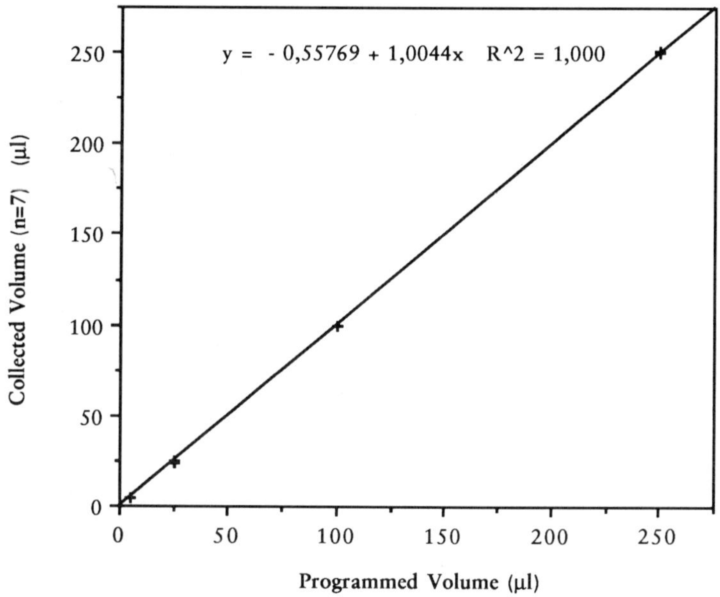

Figure 3. Reproducibility and accuracy in Wipe Mode.

Table 1. Reproducibility and accuracy in Wipe Mode and Drop Mode.

Technique	Programmed volume (µl)	Collected volume (n=7) (µl)	Collected/ Programmed volume (%)
Wipe Mode	5	4.7 ± 0.1	94.0
-"-	25	24.4 ± 0.1	97.6
-"-	100	99.8 ± 0.2	99.8
-"-	250	250.6 ± 0.3	100.2
Drop Mode	25	23.3 ± 4.9	93.2

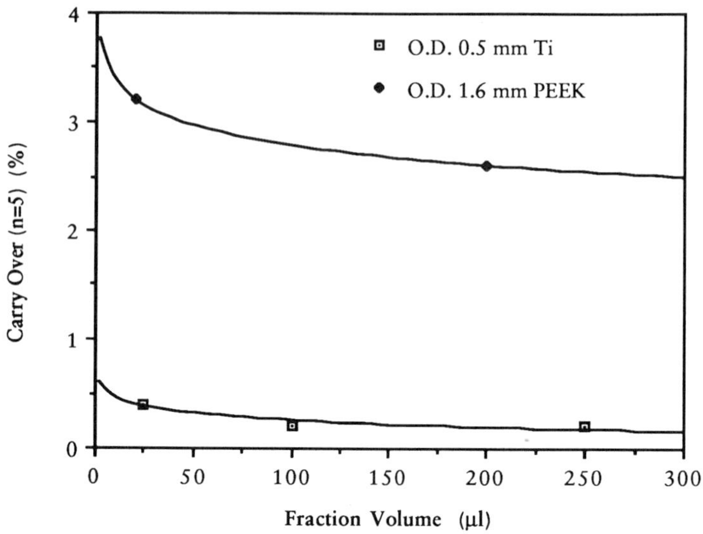

Figure 4. Volumetric carry over in Wipe Mode. Standard needle compared to larger O.D. PEEK needle.

Table 2. Volumetric carry over in Wipe Mode.
Standard needle compared to larger O.D. PEEK needle.

Fraction Collector Needle	Fraction volume (μl]	Carry over(n=5) (μl)	(%)
Ti, O.D. 0.5 mm	25	0.11 ± 0.01	0.4
-"-	100	0.23 ± 0.04	0.2
-"-	250	0.38 ± 0.09	0.2
PEEK, O.D. 1.6 mm	20	0.6 ± 0.04	3.2
-"-	200	5.2 ± 1.6	2.6

References

1. Scott, R. P. W. (1985). J. Chromatogr. Sci. **23**, 233.
2. Sagliano, Jr., N., Hsu Shih-Hsien, Floyd, T. R., Raglione, T. V., and Hartwick, R. A. (1985). J. Chromatogr. Sci. **23**, 238.
3. Schlabach, T. D., and Wilson, K. J. (1987). J. Chromatogr. **385**, 65.

4

INCREASED SENSITIVITY TO ANALYZE BRAIN TISSUE CONSTITUENTS: USE OF CAPILLARY ELECTROPHORESIS COUPLED TO FLUORESCENCE MICROSCOPY[1]

Norberto A. Guzman[2,3], Maria A. Trebilcock[3], and Juan P. Advis[4]

[3]Princeton Biochemicals, Inc., Princeton, New Jersey 08543, and [4]Department of Animal Sciences, Rutgers University, New Brunswick, New Jersey 08903.

I. INTRODUCTION

Over the past decade, protein and peptide characterization has benefited from a number of substantial technological advances that have dramatically increased both the level of resolution and sensitivity that can be achieved. An emerging technique now being tested for its high resolution power and its high detection sensitivity is capillary electrophoresis (CE) (1-5). Although CE has been characterized as having a high mass sensitivity, it is also a low concentration sensitivity technique. For this reason, in order to make CE a sensitive technique to be used in biomedical applications, improvements in the detection system are underway in several laboratories using UV and fluorescence detection paradigms (6-9).

---

[1]This work was supported in part by grants from the Wetterberg Foundation, New Jersey Experimental Station-Hatch 06108, and USDA 89-37240-4587 to JPA.

[2]Present address: Pharmaceutical Research and Development, Hoffmann-La Roche Inc., 340 Kingsland Street, Nutley, NJ 07110-1199; (201)-235-4816

Techniques in Protein Chemistry II

We have previously used capillary electrophoresis coupled to UV detection to separate and characterize various proteins and peptides from human and animal tissues (1,2;10-12). However, due to the inherent sensitivity of UV as a detection source, only macromolecules present in large concentrations were able to be analyzed. This is of particular importance when refering to the analysis of brain constituents produced in limited quantities. In an attempt to overcome the detection and sensitivity problems, we have used CE coupled to fluorescence analysis using an epillumination fluorescence microscope equipped with a mercury lamp and a computer driven single photon detection system. A significant improvement in the determination of fluorescence proteins and peptides derived from brain tissues was achieved, when compared with UV detection. Specifically, this technique was used to analyse the content of brain constituents (probably neuropeptides) from the medial preoptic area (mPOA) and median eminence (ME), discrete hypothalamic areas rich in peptidergic-containing neuronal cell bodies and terminals, respectively. In this paper we report an improvement in the detection of Luteinizing Hormone-Releasing Hormone (LHRH) in samples from these areas. Our results show that LHRH can be analyzed at the picogram level when tagged to fluorescein isothiocyanate (FITC) and detected by CE coupled to fluorescence microscopy. This improvement was at least 2-3 orders of magnitude greater than our previous analysis using a CE system coupled to UV detection (11,12). In addition, the fluorescence detection system described here was able to define new compounds not previously observed with UV detection.

II. MATERIALS AND METHODS

Reagents and Samples: High purity grade, potassium hydroxide, sodium tetraborate, hydrochloric acid and FITC-isomer I were obtained from Sigma Chemical Company (St. Louis, MO). Homogeneous synthetic peptides were purchased from Peninsula Laboratories (Belmont, CA). Acetonitrile HPLC grade was obtained from EM Science (Gibbstown, NJ). Millex disposable filter units (0.22 μm) and Sep-Pak cartridges were purchased from Millipore Corporation (Bedford, MA). Fused-silica capillary columns were obtained from Scientific Glass Engineering (Austin, TX) and Polymicro Technologies (Phoenix, AZ). Triply distilled and deionized water was used for the preparation of buffer solutions. Capillary electrophoresis buffers were routinely degassed and sonicated under vacuum after filtration.

Samples for brain tissue constituents were obtained from discrete microdissections of the mPOA and ME as previously reported (11,12). For comparison purposes, some samples were pooled and concentrated as previously described (11,12). Localized organ perfusion sampling techniques are presently a convenient system for **in vivo** determination of brain constituents (11-16). Figure 1 shows a push-pull cannula (PPC) inserted in the brain cortex of a freely-moving animal. In some of our animals, the PPC was aimed at the mPOA and in others at the ME, areas rich in LHRH neuronal cell bodies or terminals, respectively.

Instrumentation for CE: The system used was modified from that previously described (2) and shown in Figure 2. The principles of adaptation of fluorescence microscopy to capillary electrophoresis have been explained elsewhere (8) and they are summarized in Figure 3. The modified system we currently use consists of a laboratory-made capillary electrophoresis instrument with on-column fluorescence detection. The latter is composed of: **a)** a Nikon Model Optiphot-2 epillumination fluorescence microscope (Nikon Inc. Instrument Group, Garden City, NY), and **b)** a computerized single photon detector (Photon Technologies International, Inc., South Brunswick, NJ). In addition, the excitation lamp was a 50-W, high-pressure mercury lamp (Model HB-1010 1AF), the

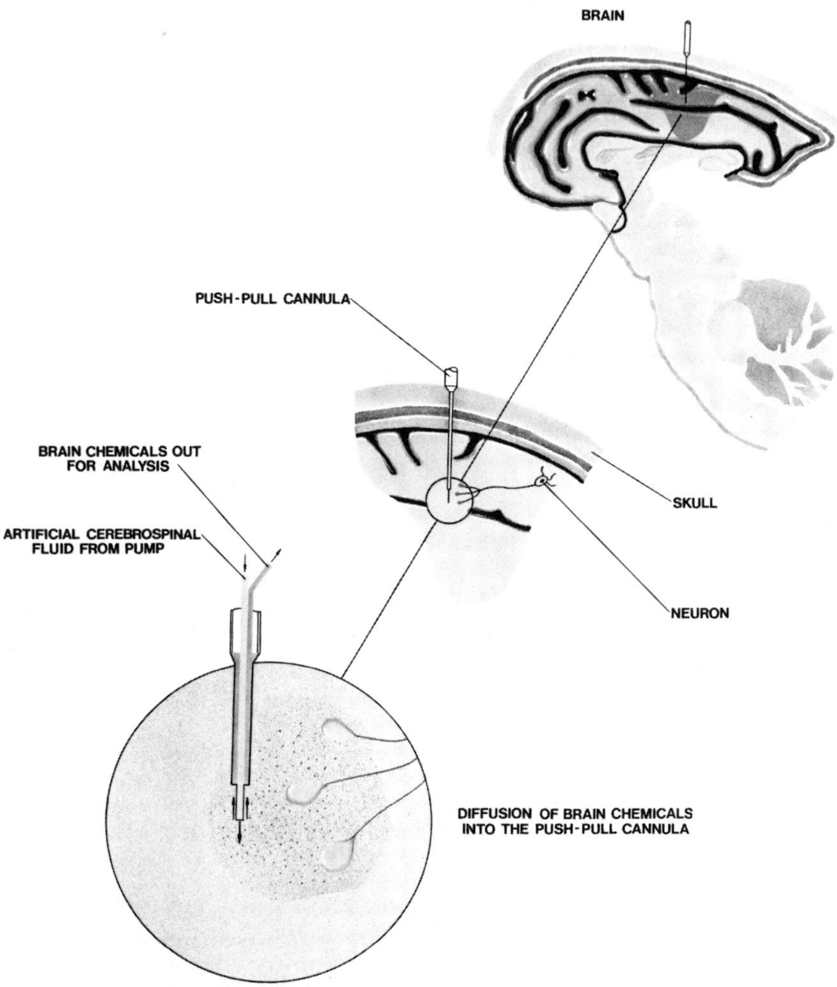

Figure 1. A Schematic Diagram of a Portion of the Brain Depicting an Inserted Push-Pull Cannula in a Cortical Area Rich in Peptidergic-Containing Neuronal Terminals.

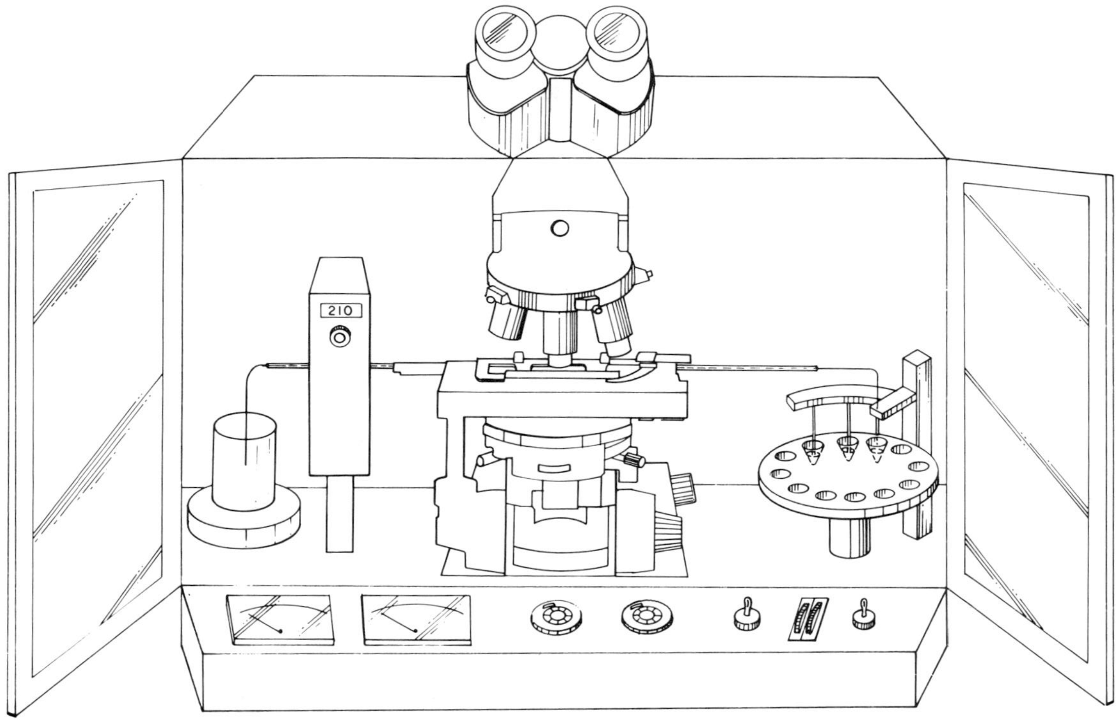

Figure 2. A Schematic Diagram of an Automated Integrated Capillary Electrophoresis (CE) Apparatus with On-Column Fluorescence Microscope Detection System (Reproduced with permission from reference 2. Copyright 1990 American Chemical Society). For the modifications made to the CE system, refer to section on Material and Methods.

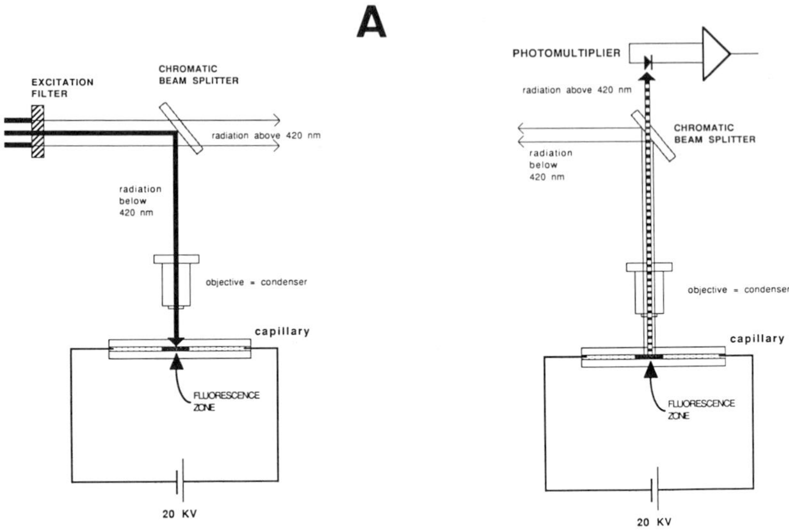

Figure 3. The Principles of Adaptation of Fluorescence Microscopy. **A.** Excitation of the sample with an epillumination fluorescence microscope. The UV beam is shown as the thick black line reflected at right-angles by the chromatic beam splitter. The fluorescence zone is indicated by the dark dotted area in the center of the capillary. **B.** Emission and measurement of fluorescence. The emitted light is indicated by the thick dashed line that crosses the chromatic beam splitter and is detected by the photodiode (Reproduced with permission from reference 8. Copyright 1990 Elsevier Science Publishers). For the modifications we made to our CE system, see section on material and methods.

band-pass excitation filter was a 450-490 nm filter, and the emission filter was a 515 nm interference filter.

Procedure. Samples were derivatized prior to injection into the capillary column as follows: **a)** A 5.5 x 10^{-4} M of FITC solution was prepared by dissolving 1 mg of FITC-isomer I in 5 ml of acetone containing a trace amount of pyridine; **b)** increasing concentrations of FITC-isomer I (4, 40, and 400 μg) were mixed in equal volumes with 20 μg of LHRH, and vortexed for 30 sec. The mixture was then allowed to react at various times (in the dark) at room temperature; **c)** injections were performed electrokinetically for 20 sec at 15 kV and the sample components were then separated at 20 kV. The capillary column (75 μm x 100 cm) was primed with either 0.025, 0.050, or 0.075 M sodium tetraborate buffer, pH 8.3, containing 0.10 M LiCl. In order to assure optimal performance of the capillary column, the column was cleansed by a cleaning cycle of potassium hydroxide-water-buffer using a vacuum pump to aspirate the solutions. Briefly, two minutes of 0.1 N potassium hydroxide were followed by two minutes of water, and finally equilibrated for 3 minutes with sodium tetraborate buffer. Criteria to qualitatively identify brain constituents was based on comparing the migration time of the electropherogram peaks from samples with: a) the migration time of the LHRH standard processed under identical conditions to those of the samples, and b) the migration time of electropherogram peaks from samples spiked with known amount of LHRH standard (just prior to injection).

III. RESULTS

Figure 4 shows the optimization conditions of the conjugation of FITC to brain sample constituents. Approximately 4 ng of FITC, and an incubation time of 3 hr were needed to obtain maximum fluorescence intensity.

In addition, the effect of electrolyte concentration on the electrophoretic behaviour of FITC-tagged LHRH was studied. For this purpose, 0.025, 0.050, and 0.075 M sodium tetraborate buffer solutions, pH 8.3, were used. The results shown in Figure 4, demonstrated a decrease in the overall mobility of LHRH with increasing concentration in the running electrolyte solution. This is due to a reduction in the electroosmotic flow (17), as a consequence of an increase in the viscosity of the electrolyte solution, and possibly a decrease in both the thickness of the double layer and the zeta potential of the capillary wall (18).

Figure 5 depicts an electropherogram of LHRH standard after conjugation with FITC. In this figure, LHRH corresponds to peak number 1, FITC-isomer I corresponds to peak number 2, and peak number 3 probably represent a minor quantity of the second isomer of FITC or some other impurities present in the FITC sample. Similarly, this conjugation technique was applied to mPOA (Figure 6) and ME (Figure 7) samples. As shown in these figures, LHRH was detected using CE coupled to fluorescence microscopy detection (CE-CFMD) at the picogram level. Furthermore, CE-CFMD was able to define additional compounds.

IV. DISCUSSION

Only in the last decade has capillary electrophoresis advanced to the point where it has attracted considerable attention among scientists. Capillary electrophoresis has become a useful analytical tool applicable to the analysis of many small molecules as well as complex macromolecules (1-5). However, the sensitivity aspect of CE is still a weak point. The sensitivity of a separation method depends on both the loading capacity of the system and the nature of the available detectors. Since capillary electrophoresis is a technique which utilizes (usually) nanoliter quantities, there is a significant limitation in the

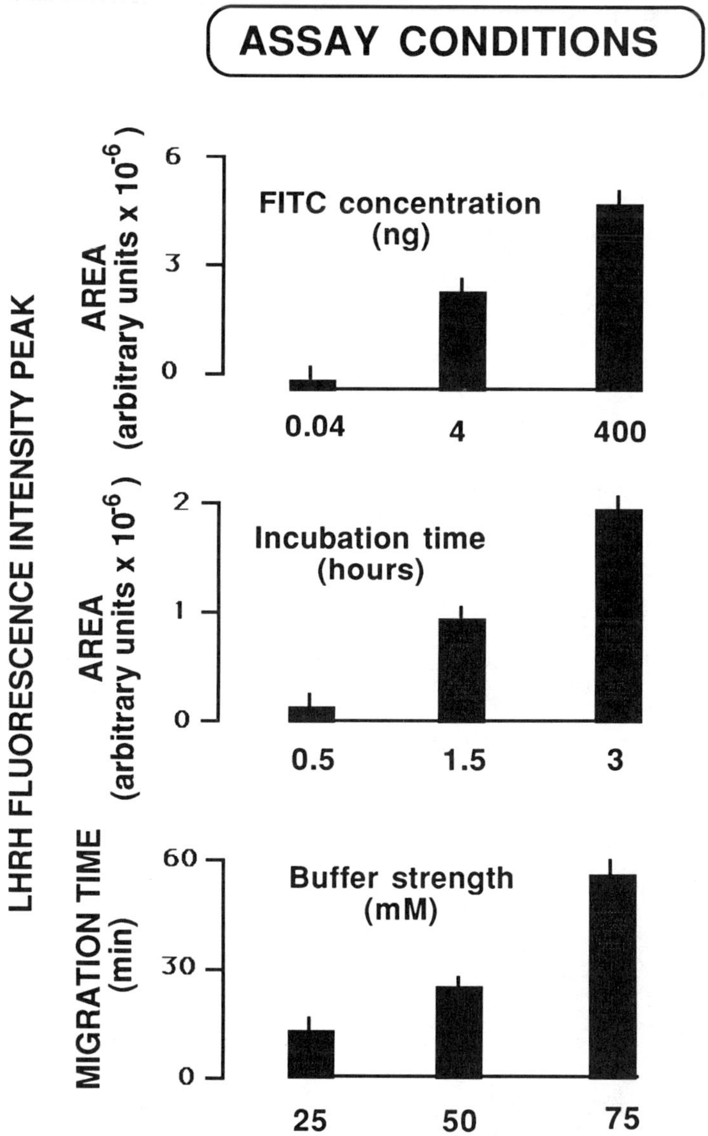

Figure 4. Optimization of Assay Conditions for the Conjugation of FITC to Luteinizing-Hormone Releasing Hormone (LHRH). Each bar represents the mean ± SEM of 6 determinations.

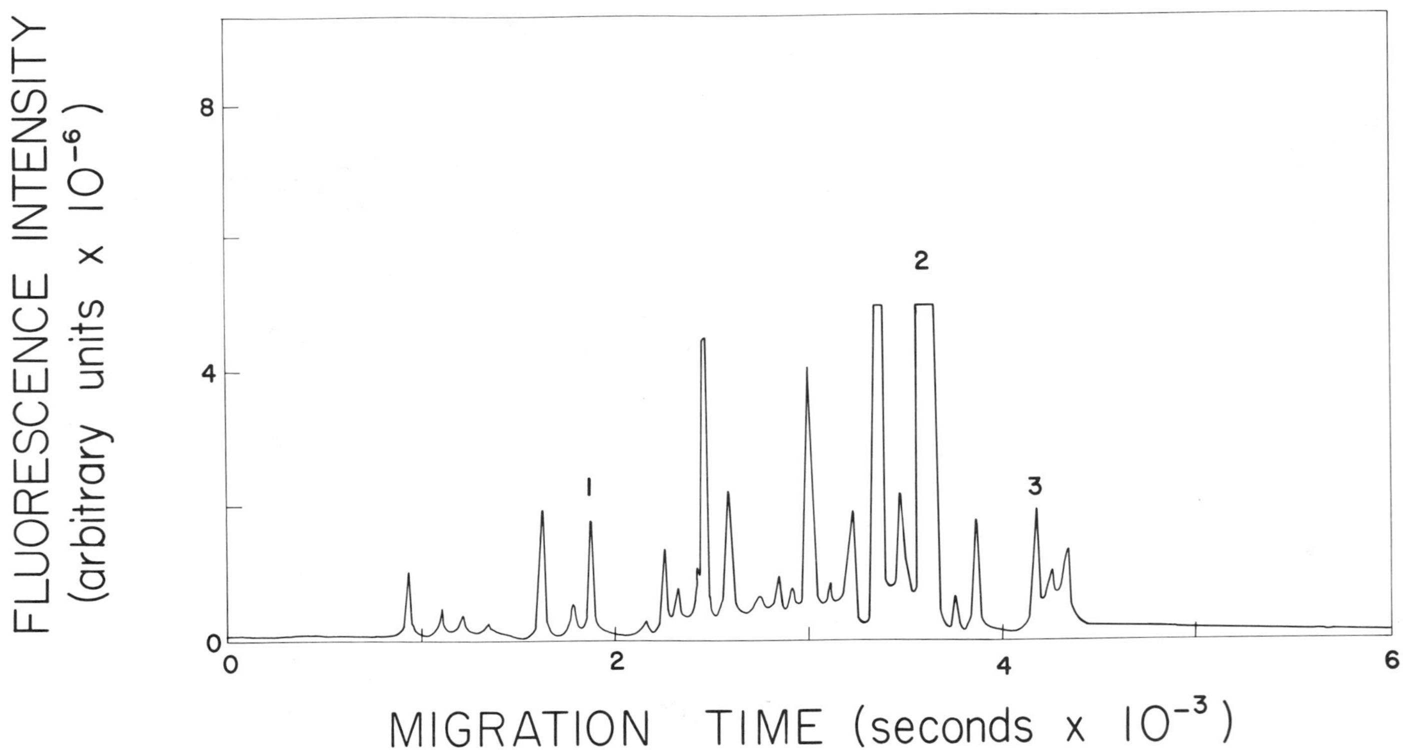

Figure 5. Electropherogram of an LHRH Standard Conjugated to FITC. Peak 1 represents LHRH. Peak 2 represents FITC-isomer 1. Peak 3 represents either a second isomer of FITC or a contaminant.

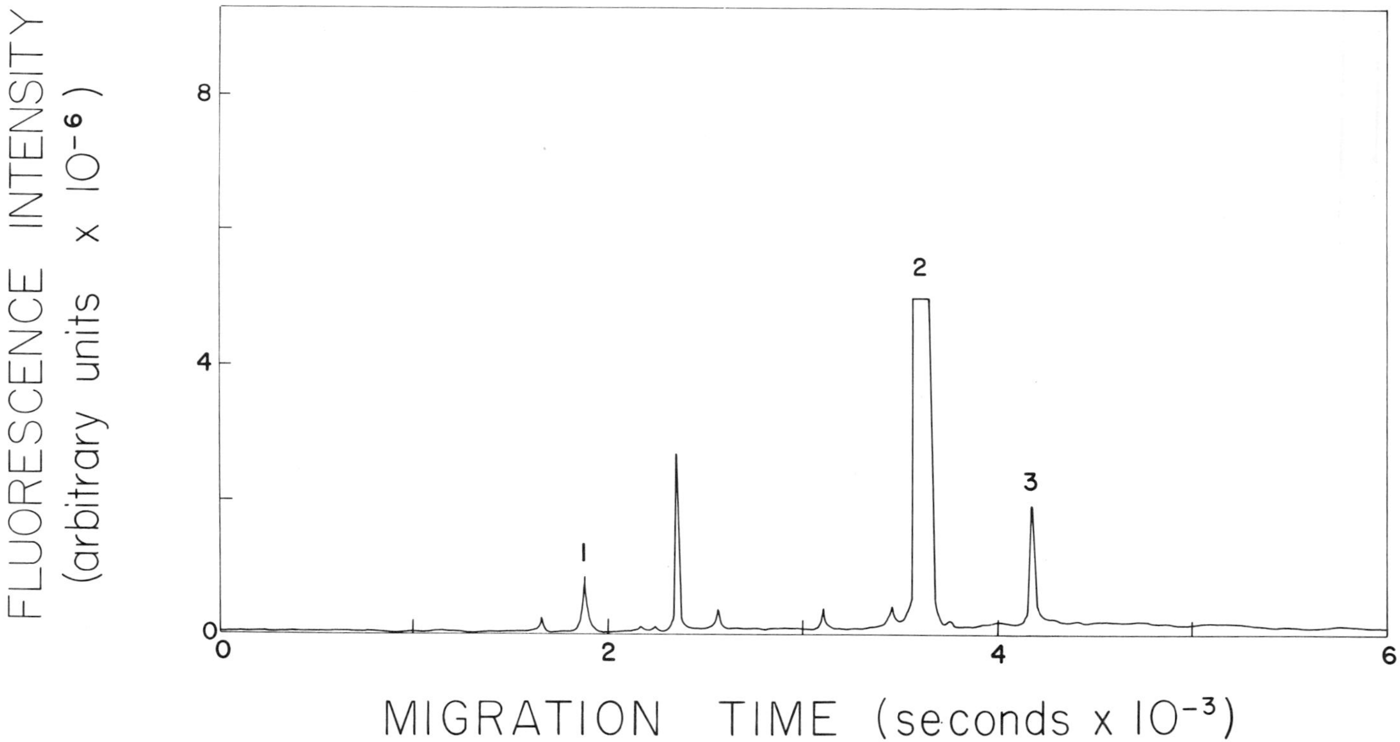

Figure 6. Electropherogram of a Medial Preoptic Tissue Sample Conjugated to FITC. The identification of peaks 1, 2 and 3 as described in Figure 5. The other peaks are unknown.

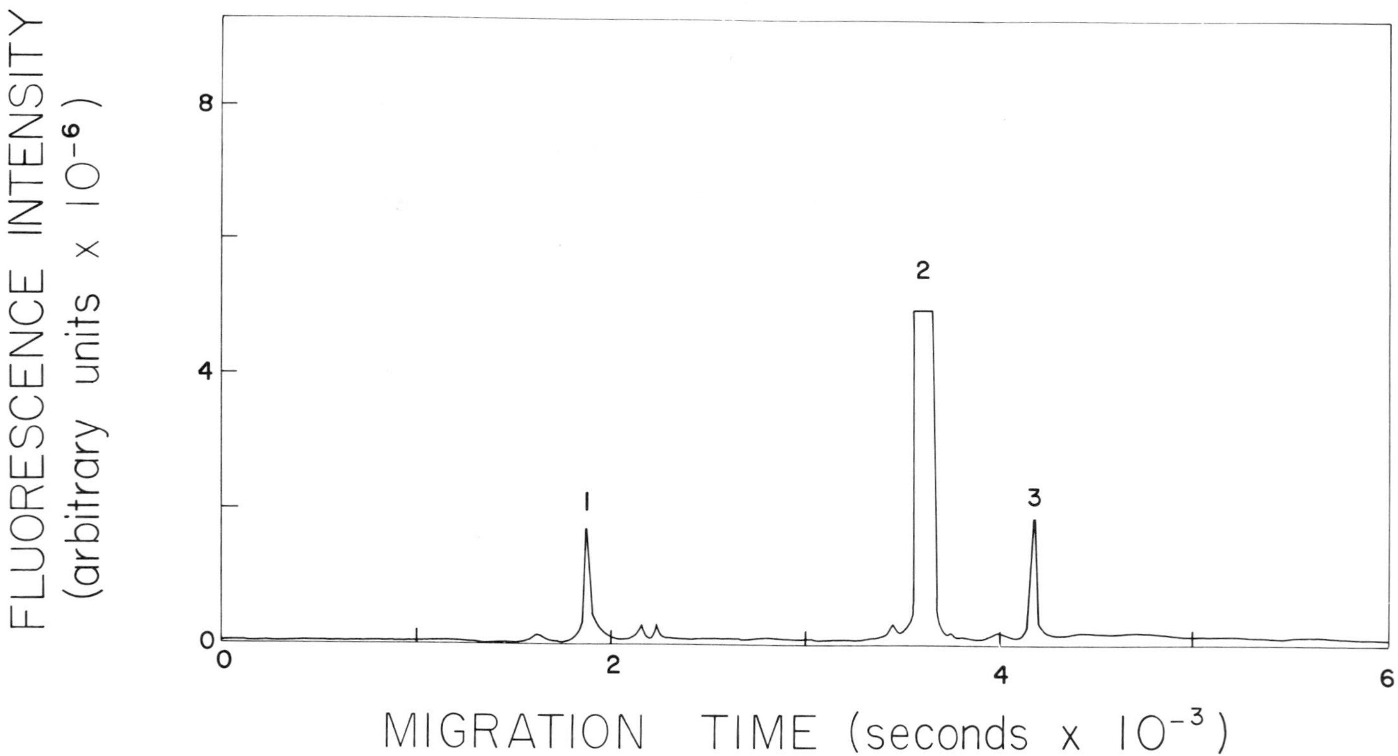

Figure 7. Electropherogram of a Median Eminence Tissue Sample Conjugated to FITC. The identification of peaks 1, 2 and 3 as described in Figure 5. The other peaks are unknown.

loading of material into the capillary column. If the amount of material injected is increased (to achieve an increased sensitivity), the broadening of the peak is also increased in a proportional fashion. Therefore, it is necessary to keep a balance between the amount of material injected, the maximum sensitivity achievable for that particular amount of sample, and the shape of the curve. This balance provides a maximum optimization gain of the system as reflected in a high number of theoretical plates and high sensitivity.

On the other hand, the availability of adequate detectors is also a limiting factor in the sensitivity for CE. Laboratory-made on-column fluorescence detectors, equipped with noncoherent light sources, have been specially designed for CE (19-21). In these detectors, either a mercury or mercury-xenon arc lamp provides the source of excitation. Lenses located at a distance of 1 cm from the capillary are used to focus both the excitation radiation on the capillary and the emitted light on the photodetector. Similar designs have been incorporated into commercially available fluorescence detectors for CE (22). In all these instruments, the fluorescence is collected at right-angles to the main axis of the capillary and the excitation beam. Conversely, the fluorescence microscope has an epillumination configuration in which the excitation and the detection of fluorescence occur at the same side of the capillary in a colinear arrangement. In addition, the focusing and aligment of the capillary can be guided visually to obtain optimum signal response. Furthermore, because of the large numerical aperture objectives of the fluorescence microscope and the high-quality optical components, it is possible to improve the efficiency of the measuring device. This improved efficiency is at least 2-3 orders of magnitude when compared with UV detection. In addition, the objective is kept at 0.33 mm from the capillary, compared with 1 cm or more for other detectors.

Previously, we have concluded that CE coupled to UV detection at 210 nm can separate and characterize a specific neuropeptide (LHRH) from discrete hypothalamic tissue samples (11,12). However, in order to obtain (using CE) a mass sensitivity comparable to the upper range of radioimmunoassays (RIA) for this neuropeptide (1 pg/μl), it was necessary to follow the time consuming process of pooling and concentrating samples prior to injection into the capillary column. Although this preinjection process was a major disadvantage of the CE technique (unlike RIA) it was able to resolve and quantitate several components at the same time.

Our results show that (when CE-CFMD was used) it is possible to reduce the number of pooling and concentrating steps of the samples during the experimental procedure. These experimental manipulations were necessary to obtain the comparable RIA sensitivity levels needed to determine the concentration of LHRH present in tissues and biological fluids. Thus, CE-CFMD provides a significant step forward with the departing from the use of radioactive material needed to detect trace amounts of analytes. Further optimization of the detection system for CE-CFMD is still underway in our laboratory. It is possible now to reach 10^{-6}-10^{-9} M concentrations with noncoherent fluorescence detection (1-8). Additionally, preinjection steps may be necessary to reach the level of sensitivity of RIA (10^{-9}-10^{-11} M), a technique that is commonly used without pretreatment steps of the samples. We are currently assessing if the use of capillary electrophoresis-coupled to laser-induced fluorescence microscopy detection (CE-LIFMD) is a necessity for the determination of brain sample constituents present in minute amounts. The possibility exists that the use of computerized single photon detection technologies might dispense the use of hazardous laser technological approaches. More recently, experiments have proven the usefulness of CE-CFMD in the analysis of brain constituents present in push-pull samples (23,24).

Although fluorescence microscopy has been known since shortly after the turn of the century, it was not until about 40 years ago that scientists began in earnest to explore the many possibilities of utilizing fluorescence microscopes in biological and industrial applications. The introduction of fluorescence microscopy to the analytical chemistry field will be a useful tool needed to improve the detection sensitivity level of many analytes, specifically those found in trace amounts in biological systems. More recently, the use of CE-LIFMD has been explored (9). The merging of these two techniques may well be the ultimate step needed to replace RIA or enzyme-linked immunosorbent assays (ELISA). In addition, when using CE-LIFMD, it has been possible to achieve sensitivity levels in the subattomole concentration range (9). In fact, it has been possible to determine less than 300 molecules of an analyte (9).

Finally, an attractive characteristic of the merging of CE with fluorescence microscopy is the visualization of the separated zones of the different analytes, of which photographs or video tape records may be made. As seen in Figure 8, various points of fluorescence intensity of the LHRH electropherographic peak are depicted by the corresponding photograph.

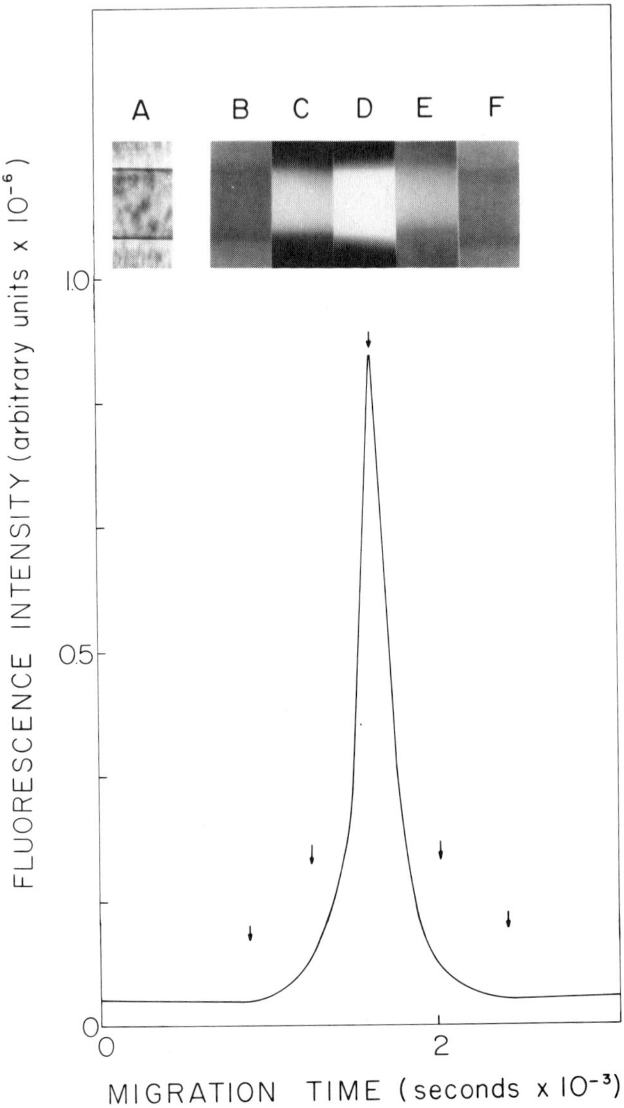

Figure 8. Visualization of the Separated Capillary Electrophoresis Zones. Sequential photomicrographs were obtained (with 50% of the emission beam split) during an FITC-derivatized LHRH peak (**B** through **F**) at the times indicated by the arrows. The photomicrograph on the far left (**A**) was obtained under bright light conditions. It clearly shows the capillary wall delimiting the inner diameter of the column.

V. REFERENCES

1. Guzman, N.A., L. Hernandez and B.G. Hoebel. 1989. Biopharm. Manufact. 2: 22-37.
2. Guzman, N.A., L. Hernandez and S. Terabe. 1990. In: **Analytical Biotechnology. Capillary Electrophoresis and Chromatography** (Horvath, C. and J.G. Nickelly, Eds.), ACS Symposium Series 434, Chapter 1, pp. 1-35, American Chemical Society, Washington, D.C.
3. Karger, B.L., A.S. Cohen and A. Guttman. 1989. J. Chromatogr. **492:** 585-614.
4. Wallingford, R.A. and A.G. Ewing. 1990. In: **Advances in Chromatography. Biotechnological Applications and Methods** (Giddings, J., E. Grushka and P.R. Brown, Eds.), Volume 29, pp. 1-76, Marcel Dekker, Inc., New York and Basel.
5. Kuhr, W.G. 1990. Anal. Chem. **62:** 403R-414R.
6. Cheng, Y.-F. and N.J. Dovichi. 1988. Science (Washington, D.C.) **242:** 562-564.
7. Gordon, M.J., X. Huang, S.L. Pentoney, Jr., and R.N. Zare. 1988. Science (Washington, D.C.) **242:** 224-242.
8. Hernandez, L., R. Marquina, J. Escalona and N.A. Guzman. 1990. J. Chromatogr. **502:** 247-255.
9. Hernandez, L., N. Joshi, J. Escalona and N.A. Guzman. Second International Symposium on High Performance Capillary Electrophoresis, Abstract ML-110, San Francisco, CA, January 29-31.
10. Guzman, N.A. and L. Hernandez. 1989. In: **Techniques in Protein Chemistry** (Hugli, T.E., Ed.), Chapter 44, pp. 456-467, Academic Press, Inc., San Diego.
11. Advis, J.P., L. Hernandez and N.A. Guzman. 1989. Peptide Res. **2:** 389-394.
12. Guzman, N.A., L. Hernandez and J.P. Advis. 1990. In: **Current Research in Protein Chemistry: Techniques, Structure, and Function** (Villafranca, J.J., Ed.), Chapter 19, pp. 203-216, Academic Press, Inc., San Diego.

13. Phillipu, A. 1984. In: **Measurement of Neurotransmitter Release In Vivo** (Marsden, C.A., Ed.), pp. 3-39. Willey and Sons, New York.
14. Tossman, U. and U. Ungerstedt. 1986. Acta Physiol. Scand. **128:** 9-14.
15. Benveniste, H. 1989. J. Neurochem. **52:** 1667-1679.
16. Freiherr, G. 1990. J. NIH Res. **2:** 93-95.
17. Bushey, M.M. and J.W. Jorgenson. 1989. J. Chromatogr. **480:** 301-309.
18. van Oss, C.J. 1979. Sep. Purif. Methods **8:** 119-124.
19. Jorgenson, J.W. and K.D. Lukacz. 1981. Anal. Chem. **53:** 1298-1301.
20. Guthrie, E.J. and J.W. Jorgenson. 1984. Anal Chem. **56:** 483-487.
21. Green, J.S. and J.W. Jorgenson. 1986. J. Chromatogr. **352:** 337-341.
22. Brownlee, R.G. and S.W. Compton. 1988. Am. Biotechnol. Lab. **6:** 10-15.
23. Advis, J.P. and N.A. Guzman. 1990. Fourteenth International Symposium on Column Liquid Chromatography, Abstract L-14.03, Boston, Massachusetts, May 20-25.
24. Advis, J.P., M.A. Trebilcock and N.A. Guzman. 1991. Peptide Res., submitted.

5

WHY NOT ETHANOL-BASED SOLVENTS FOR RP-HPLC OF PEPTIDES AND PROTEINS?

RICK FELDHOFF

Department of Biochemistry
University of Louisville - School of Medicine
Louisville, KY 40292

I. INTRODUCTION

A review of the literature or a survey of presentations at national meetings suggest that a consensus has developed among protein chemists who routinely employ reverse-phase high performance liquid chromatography (RP-HPLC) in their research endeavors (1,2). Although it is unlikely that a "universal" set of conditions will ever be defined, the current "standard" conditions for analytical RP-HPLC appear to be C_{18} alkylsilane bonded to macroporous (300 Å pores) silica beads of 5-7 μ diameter which have been "end-capped" with a trimethysilane.

In reality, C_8 derivatization may provide the best general purpose matrix for both peptides and proteins, but most researchers purchase their columns, and there is a natural reluctance to undertake costly comparisons when procedures are already working reasonably well. Pre-packed columns of 4.6 x 250 mm are still the most popular choice, however, column lengths and diameters and particle size are slowly decreasing with little effect on resolution and a substantial savings in time and eluant usage (2,3).

A variety of water-miscible solvents (acetonitrile, alcohols, tetrahydrofuran) have been tested and shown to contribute little to selectivity in the RP-HPLC of peptides and proteins (1,4). Acetonitrile (ACN) containing 0.1% trifluoroacetic acid (TFA) is, by default, the clear choice for organic solvent because of the low UV absorbance of this reagent below 230 nm where peptide bonds are monitored and because of the low viscosity of ACN/water mixtures. Lower viscosity translates into faster diffusion (which can influence resolution) and lower system operating pressures. As will be shown below, neither the low UV absorbance of ACN nor its low viscosity make it an irreplaceable organic solvent.

Techniques in Protein Chemistry II

TFA at 0.1% (w/v) promotes partial denaturation and solubilization of proteins which is related to adsorption and desorption phenomena. Additionally, 0.1% TFA maintains the pH ≈2 and serves as an ion-pair agent to help reduce free silanol:protein interactions. Neither higher nor lower concentrations of TFA provide better resolution.

Typically, gradient elution to 40-80% ACN will permit the satisfactory resolution of most peptides and many proteins. Methanol (MeOH) is occasionally employed, but I can find no good evidence to support the use of this weak, polar and toxic solvent. In contrast, proteinaceous solutes which are difficult or impossible to elute with ACN have been resolved with n-propanol (PrOH), isopropanol and ACN/propanol mixtures (1,5,8,9). PrOH is a potent hydrophobic eluant which has been reported to be about twice as effective as ACN with all tested peptides and proteins (6,8,9).

Intuitively, one might expect the C_3 chain of PrOH to enhance the solubilization and elution of hydrophobic peptides/proteins and it is somewhat surprising that PrOH and ethanol (EtOH) have not been utilized more. Furthermore, EtOH and PrOH are essentially non-toxic in normal laboratory usage while methyl cyanide (ACN) is metabolized to cyanide which is extremely toxic with low concentrations causing fatigue, weakness, headache, vertigo, nausea and/or vomiting. Even the trace amounts which volatilize during solvent preparation, reservoir changes, sample handling, loading, and the manipulation of collected fractions can cause eye and mucous membrane irritation. Additionally, solvent disposal in many research laboratories is becoming a question of economic and environmental responsibility.

The goal of the present study was to develop a "non-toxic" ethanol-based solvent system which would provide resolution similar to or better than ACN and with equivalent retention times such that a switch-over to the new system would not undermine years of experience and data collection or require "re-inventing the wheel".

II. MATERIALS AND METHODS

Apparatus: Chromatography was accomplished on a Waters HPLC system consisting of a 680 controller, 6000A and M-45 pumps, U6K manual injector with a 2 mL loading loop, 480 UV detector and a Linear chart recorder. The void volume of the system for gradients (mixing head to detector) is 5.3 mL and actual solvent concentrations for solute elution have been determined based on this displacement. In all experiments, the flow rate was constant at 1.0 mL/min. All solvents contained 0.1% TFA (w/v) and were prepared by adding 5 mL of a 20% TFA stock solution per liter of eluant. The 20% TFA solution (100 g/500 mL) has proven to be more convenient and cost-effective than 1 g ampoules of TFA.

Chemicals: EtOH, PrOH, TFA, hexane, pyridine and Alberta Peptide Institute (API) standards (7) were obtained from Pierce (Rockford, IL). MeOH was purchased from Fisher (Cincinnati, OH). ACN was from EM Science (Gibbstown, NJ). Toluene, the alkylphenone series (propio-, butyro-, valero- and hexanophenone) and the alkylsilanes were obtained from Aldrich (Milwaukee, WI). Dichloromethane was from Burdick & Jackson (Muskegon, MI).

Column and Rederivatization: A column (4.6 x 250 mm) containing 6.5 micron, macroporous (300 Å) spherical silica bonded with a C_{18} ligand was purchased from SynChrom (Lafayette, IN) and used for 3 months prior to the start of these experiments. It was necessary to rederivatize the column because of incomplete resolution of the peptide standards and an aberrant plate count run which suggested ligand solubilization from the top of the column. The procedure for on-line derivatization described by Hodges and co-workers (7) was followed except for making the silanization solutions to 10% pyridine and 10% alkylsilane in dichloromethane.

III. RESULTS AND DISCUSSION

Similar to most protein chemistry laboratories, we have used ACN/0.1% TFA for the majority of our separations with occasional use of MeOH and isopropanol. However, ACN is metabolized to cyanide and MeOH to formaldehyde with impairment of the central nervous system. ACN and MeOH can also be absorbed through the skin. The long-term effects of chronic low-level exposure are not known and it would probably be best if protein chemists did not provide the study population. In our laboratory, eye irritation caused by ACN occurs before odor threshold levels (40 ppm) are reached. Unfortunately, this also encompasses the range for permissible occupational exposure level (PEL). However, since I have no doubt that no one understands what PEL means in terms of cell damage and/or death, it appears that 0 ppm of toxic solvents is a reasonable goal.

A review of the literature provided less than satisfying results and suggested the need for a careful comparison of ACN with the C_1, C_2 and C_3 alcohol series. The API peptide standards, an alkylphenone series and human serum albumin (HSA) were chosen as representative solutes. The peptide standards are probably the single best choice for evaluating column and solvent performance. The mixture consists of 5 synthetic decapeptides (S1-S5) varying at the 3rd and 4th residues as -A-G-, -G-G-, -A-G-, -V-G-, and -V-V-. All contain an acetylated N-terminus and a C-terminal amide except S1 which has a free α-amino group to help assess silanol exposure and is otherwise identical to S3.

As shown in Fig. 1 the used column did not provide satisfactory resolution of the 5 peptide standards and was rederivatized before proceeding with solvent evaluation. Following rederivatization, the API peptides were satisfactorily resolved, retained on the column nearly twice as long and had higher and sharper peaks. The alkylphenone standards were retained 50% longer with narrow band-width peaks more than twice as high as those prior to rederivatization. These data confirm that on-line derivatization is a convenient method to reclaim ligand-degraded columns.

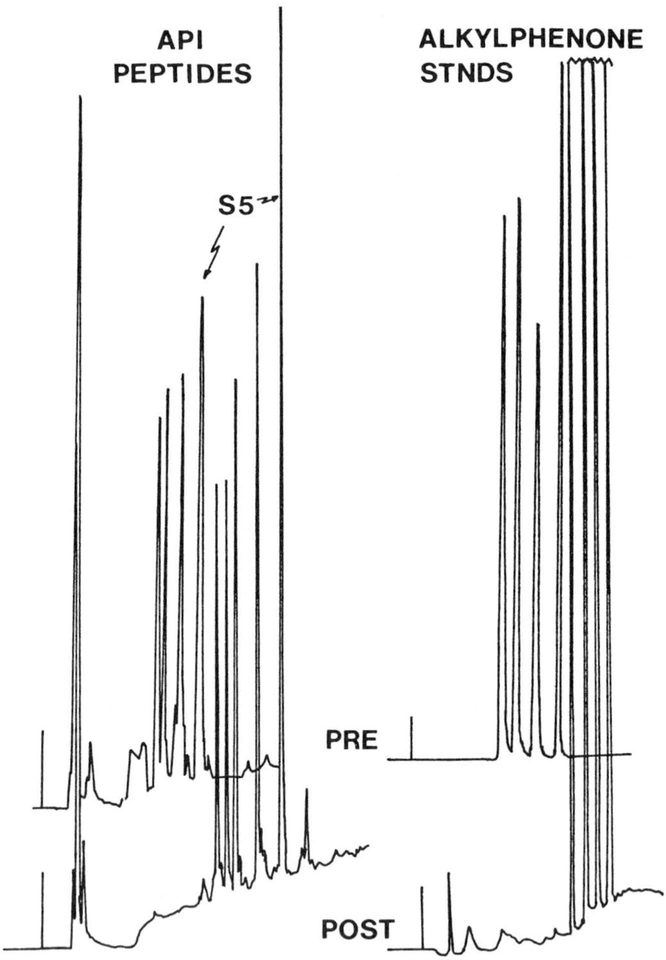

Figure 1. Effect of on-line derivatization on the resolution of peptide and alkylphenone standards. The left panel depicts identical runs of the API standard prior to and following rederivatization. An ACN gradient of 1%/min was employed with detection at 220 nm. The S5 peaks had absorption maxima of approximately 0.04 and 0.07. The right panel shows the separation of C_3, C_4, C_5 and C_6 alkylphenones pre- and post-derivatization using identical 2%/min ACN gradients at 1 AUFS at 225 nm.

Each of the four solvents was subsequently evaluated for UV absorbance at 225 nm and for effects on system operating pressure at a full range of concentrations. The data summarized in Fig. 2 were collected following 15 min reequilibration at each new setting. For EtOH and PrOH the system operating pressure increased more than 2-fold over ACN and represents one of the minor disadvantages of using these solvents. However, plastic fittings and tubing have simplified making leakproof seals and the higher system operating pressures are compatible with ordinary HPLC systems.

The absorbances for each solvent are also summarized in Fig. 2 and show the alcohols to be about three times higher than ACN. This has not proven to be a problem for our needs, but requirements for more sensitive detection may require computer processing of data. It was somewhat surprising that all of the alcohols had similar background absorbances. Since the density of each alcohol is nearly identical, the molarity of a given volume of MeOH is about twice that of PrOH, and it was assumed that the UV absorbance would be related to the -OH content. Since this does not appear to be true, perhaps better manufacturing can decrease the low UV absorbance to facilitate peptide detection.

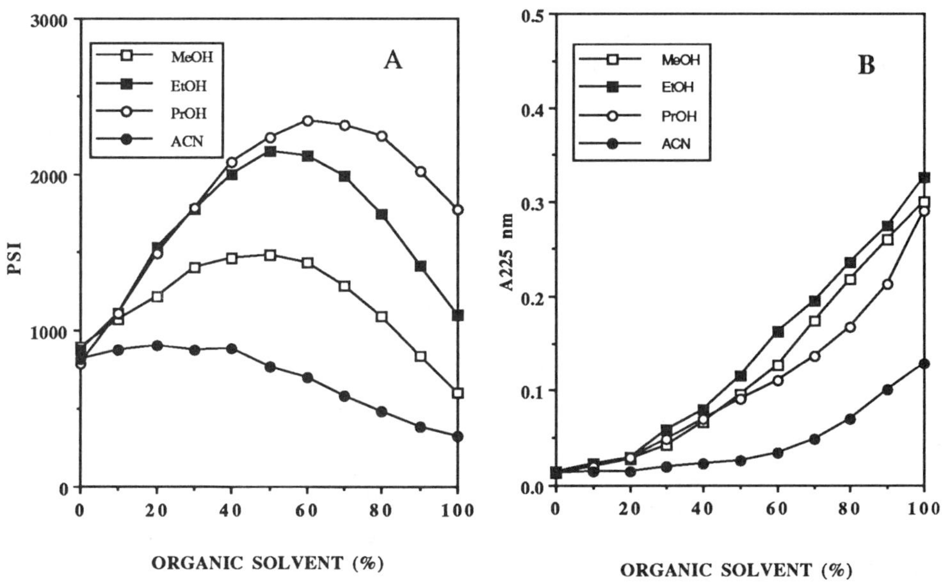

Figure 2. Effect of solvent concentration on (A) system operating pressure and (B) absorbance at 225 nm. Readings were taken after 15 min reequilibration at each setting at which time a stable baseline had been established.

We have found that the single most important factor in RP-HPLC gradient elution is the % increase in organic solvent/min. Lower % increases generally provide better separations, but the cost in time must always be considered. For solvent evaluation, each solute was run with three different gradients providing solvent increases of 0.5, 1 and 1.5%/min for the API peptides and 1, 2, and 3% for the alkylphenones and HSA. The retention time relationships for multi-component solutes within runs and between runs were very consistent. Therefore, only the solvent concentrations required to elute S5, hexanophenone and HSA were calculated for each gradient. Fig. 3 depicts the data for the preceding solutes and each solvent using 1%/min gradients. The inter-relationships held true for all other conditions and demonstrated that PrOH eluted proteins and peptides at 30-50% less solvent than ACN, while EtOH and MeOH required 10-30% and 60-80% more, respectively.

In an early study, Mahoney and Hermodson (8) prepared CNBr fragments from human hemoglobin and compared the elutropic efficacy of MeOH, EtOH, PrOH, 2-propanol, ACN, tetrahydrofuran and dioxane. The relative retention time relationships for the various solvents were similar to those reported above. However, some CNBr peptides did not elute with MeOH, EtOH or ACN, while all peptides were eluted with the propanols or propanol mixtures with ACN and tetrahydrofuran. It was concluded that ACN or EtOH should be used for early eluting peptides and a "more effective solvent" for those eluting late in broad, poorly-resolved peaks.

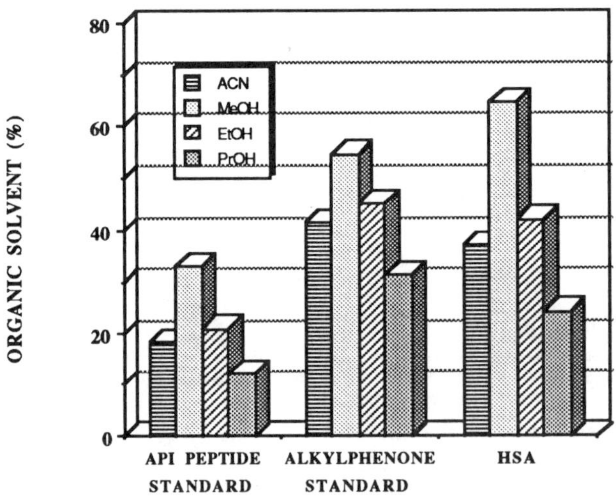

Figure 3. Comparison of solvent elutropic efficiency for various solutes. These data are typical, but represent only the experiments performed at 1% organic solvent/min. The same relationships were consistently observed in all other experiments and were used to establish the ranges for relative solvent efficiency.

Many aspects of solvent usage in RP-HPLC were subsequently reviewed in the CRC Handbooks of HPLC (9) and the interested reader is referred to the section on "Protein Separations" in Vol. II. An excellent, more recent HPLC peptide/protein review which includes a solvent comparison has been published by Simpson and Nice (10). Their data for individual solvents also support the conclusions reached in the present study. However, I can find no evidence that anyone has tried to develop or evaluate a non-toxic solvent system with high resolving power based on PrOH and EtOH.

As noted above, resolution with EtOH and PrOH was similar to that for ACN and suggested that a mixture of these alcohols, with EtOH predominating, might provide retention times similar to ACN. Trial runs of the API standard with 2:1, 3:1 and 4:1 mixtures of EtOH:PrOH led to a final 5:2 formulation and the results depicted in Fig. 4. The API peptides were resolved with ACN and 5:2, EtOH:PrOH, at 0.5, 0.75 and 1%/min increases. This direct comparison was made over a two-day period to help validate the results.

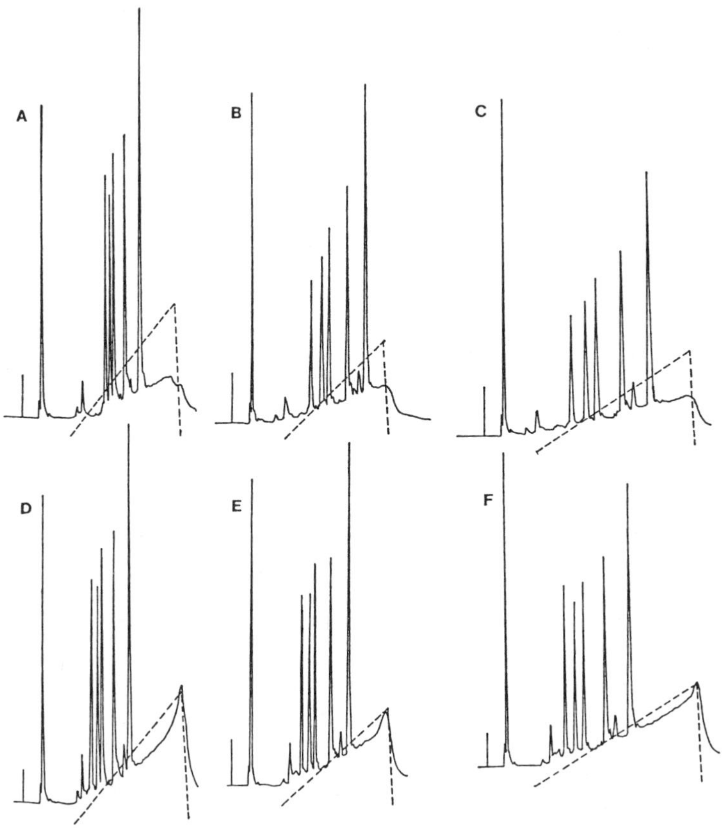

Figure 4. Comparison of ACN with a 5:2 mixture of EtOH:PrOH. The top panels (A,B,C) show ACN gradients of 1, 0.75 and 0.5%/min. The bottom panels (D,E,F) depict corresponding gradients run with 5:2 EtOH:PrOH. The only difference between the duplicates (A,D; B,E; C,F) is the solvent. The column void volume peak adjacent to each panel letter corresponds to about 0.05 AU at 220 nm.

The data (Fig. 4) show, unequivocally, that the 5:2 mixture provides better resolution under all conditions and that direct substitution of this mixture for ACN will improve peptide separations and decrease retention times by about 20% (Fig. 4A & 4D) which will provide more available instrument time. Alternatively, the relative retention times for ACN and the 5:2 mixture can be kept fairly uniform by substituting a 0.75%/min increase of 5:2 for each 1%/min increase of ACN normally used and apparent resolution will be even further improved (Fig. 4A & 4E). A comparison of peak band-widths and heights is most easily made for the 0.5%/min runs of the two solvents (Fig. 4C & 4F). The data in Fig. 4 were collected at 0.1 AUFS/220nm and show a tolerable background absorbance for the 5:2 EtOH:PrOH solvent.

It took considerably more effort than expected to eliminate ACN from our laboratory/office research environment, but that was in large part due to a desire to allow other protein chemists to make an informed decision. A 5:2 mixture of EtOH:PrOH is clearly a superior solvent to ACN in resolving power, toxicity and cost. Our system is also running "cleaner" in that starting pressures have remained constant for 3 months despite not using an in-line filter and, although the column is slowly degrading, in a fairly typical manner, the resolution has remained quite good (e.g., the Fig. 4 data was collected 3 months after rederivatization). Both of these observations may be related to the potent solubilizing properties of n-propanol in RP-HPLC (5,6,8). We are also assuming that it will be easier to recycle or dispose of alcohol waste, but this has not been fully investigated. The minor disadvantages of higher system operating pressures and background absorbance seem to be more than offset by the apparent advantages.

IV. SUMMARY

In the present study MeOH, EtOH, PrOH and ACN were compared on a rederivatized SynChrom RP-P 4.6 X 250mm column using peptide standards, albumin, and an alkylphenone series. Each sample was separated with ≥9 gradients (e.g., 0.5, 1, 2, 3% organic phase/min) for each solvent (all at 0.1% TFA) using a constant flow rate of 1 mL/min. In general, PrOH eluted peptides/protein at 30-50% less solvent than ACN, whereas EtOH and MeOH required 10-30% and 60-80% more, respectively. Resolution with EtOH and PrOH was similar to that for ACN, but with different retention times. Various EtOH:PrOH mixtures were subsequently evaluated with a 5:2 ratio at 1% organic/min providing better resolution and slightly lower retention times than ACN at 1%/min. Retention times similar to ACN at 1%/min are obtained for the 5:2 mixture at 0.75%/min with superior resolution. The data suggest that a 60 or 90 min gradient, to a final concentration of 45%, of the 5:2 mixture of EtOH:PrOH is a good initial choice for enzyme digests of proteins and 30 or 45 min to 45% for large CNBr fragments and protein mixtures. For previously used columns it is imperative to flush solvent lines and run 3 rapid gradients (e.g., 0-100% B in 20 min) with the column outlet disconnected from the detector to reequilibrate the system and prevent adsorption to the flow cell of hydrophobic material which may be released from frits and the matrix by the PrOH. The relatively large increases in system operating pressures and UV baseline for alcohols require more care in gradient design. However, these disadvantages appear to be more than offset by the excellent resolution, non-toxic nature, and lower cost of EtOH/PrOH mixtures.

V. ACKNOWLEDGEMENTS

Conversations, many years ago, with Dr. Bill Morgan (Univ. of Missouri, Kansas City) regarding the potential usefulness of n-propanol in peptide separations provided the initial stimulus for this study. The author greatly appreciates the assistance of Dr. Scot Kimball (Hershey Med. Center of Penn. State Univ.) in reformatting the final text and two figures just prior to submission.

VI. REFERENCES

1. Chicz, RM & Regnier, FE (1990). Meth. Enzymol. 182: 392-421.
2. Dong, MW, Gant, DJ & Larsen, BR (1989). BioChromat. 4: 19-34.
3. Majors, RE (1988). LC-GC 6: 298-302.
4. O'Hare, MJ & Nice, EC (1979). J. Chromatogr. 171:209-226.
5. Power, SD, Lochrie, MA & Poyton, RO (1983). J. Chromatogr. 266: 585-598.
6. Rubinstein, M (1979). Anal. Biochem. 98: 1-7.
7. Mant, CT, Parker, JMR, & Hodges, RS (1986). LC-GC 4: 1004-1009.
8. Mahoney, WC & Hermodson, MA (1980). J. Biol. Chem. 255: 11199-11203
9. Hancock, WS, ed., (1984). Handbook of HPLC for the Separation of Amino Acids, Peptides, and Proteins, Vols. I & II, CRC Press, Boca Raton, FL.
10. Simpson, RJ & Nice, EC (1989). Receptor Biochem. Methodol. 14: 201-244.

6

A NON-LABEL TECHNOLOGY FOR REAL-TIME BIOSPECIFIC INTERACTION ANALYSIS

Lars G. Fägerstam
Pharmacia Biosensor AB, S-751 82 Uppsala, Sweden

I. INTRODUCTION

For the understanding of molecular recognition and biological function, the characterization of biospecific interactions is essential. This has led us to develop an automated biosensor based system that enables monitoring of interactions in real time. The non-invasive detection principle used, surface plasmon resonance (SPR), has previously been applied to concentration analysis of biomolecules (Liedberg, et al., 1983) and for monitoring immunocomplex formation (Flanagan and Pantell, 1984; Cullen et al., 1984; Mayo and Hallock, 1989). We have combined SPR detection with miniaturized liquid handling and a controlled sensor surface chemistry which enables: 1) label free detection of multiple sequential interactions; 2) determination of association and dissociation rate constants; 3) affinity ranking; 4) concentration analysis in crude samples and 5) identification and classification. This presentation briefly describes, the detection principle, the instrument and some application examples.

II. SPR DETECTION PRINCIPLE

SPR detection (Fig. 1) senses changes in the optical properties of the medium close to *one side* of a thin (50 nm) metal film through an evanescent field, created by the interaction of collective electron oscillation or *plasmons* in the film with light incident on the *other side*. No light passes through the medium. The optical device is at one side of the metal film and the medium at the other. SPR is observed as a decrease in reflectance of the metal film at an angle of incidence at which total reflection of the light normally occurs. When all else is kept constant, a change in the refractive index of the medium will cause a proportional change of the SPR angle. This change could be caused by e.g. a protein. By monitoring changes in this angle, events that affect the refractive index of the medium, can be followed. To exploit the penetration depth of the evanescent field, we have extended the detection volume by modification of the metal surface with a carboxymethylated dextran. This allows gentle immobilization of molecules and, by dispersing of the molecules, it provides an environment suitable for interaction studies. Depending on the molecular species immobilized, the selectivity of the surface and also the recovery from crude samples of molecules can be controlled. Since the monitoring of events in the SPR detection volume is continuous, interactions between molecules can be studied as they occur.

Techniques in Protein Chemistry II

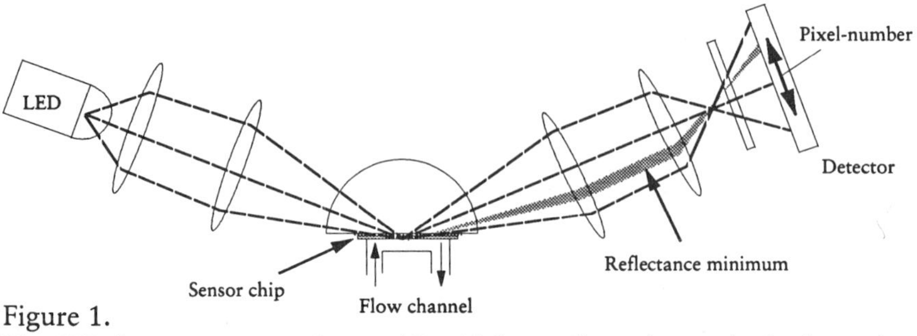

Figure 1.
The SPR detector, sensor chip and liquid flow cell are shown docked together. Light from the LED is coupled to the sensor chip via a prism. The position of the reflectance minimum is monitored continuously by a photodiode array. When a molecule introduced via the flow cell is captured at the sensor surface, the position of the reflectance minimum will change. In the instrument, a wedge-shaped light beam and a two-dimensional detector array, enables monitoring of four flow channels.

III. INSTRUMENTATION AND OPERATION

BIAcore™, the system for biospecific interaction analysis consists of a processing unit, reagents, sensor chips and a personal computer for control and evaluation. The processing unit contains the SPR monitor, an Integrated μ-Fluidic Cartridge (IFC) (Fig. 2) that, together with an autosampler, controls delivery of sample plugs into a transport buffer that continuously passes over the sensor chip surface. The exchangeable Sensor Chip CM5 accommodates four 60 nanoliter flow cells that are created by docking the sensor chip against the IFC.

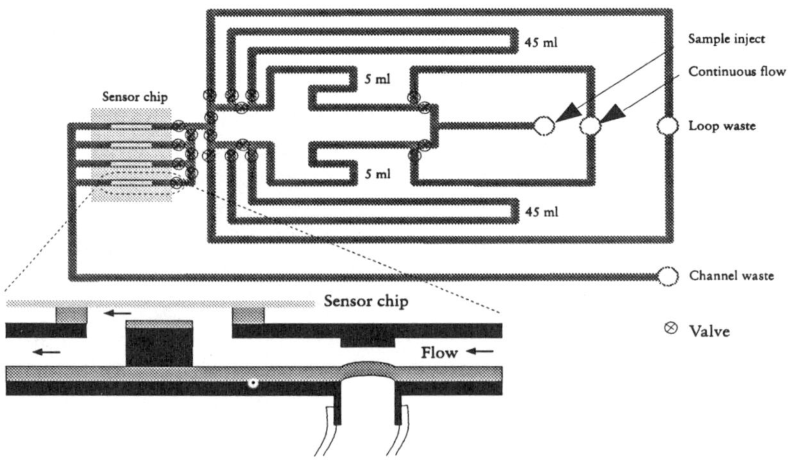

Figure 2.
The Integrated μ-Fluidic Cartridge (IFC) contains pneumatically activated microvalves that control the sample loops and choice of flow cell.

The optical changes which are monitored continuously over time are registered as a *sensorgram.* The abscissa of the sensorgram is denoted the *resonance signal* and is indicated in *resonance units (RU).* 1000 RU corresponds to a 0.1° shift in the surface plasmon resonance angle and for the average protein this corresponds to a surface concentration change of approximately 1 ng/mm². The SPR detector in BIAcore covers a range of 3° or 30 000 RU. The resonance signal at a certain point in time will be the sum of the contributions from the sensor surface, the molecules and the buffer. The amount of bound molecules can be monitored continuously when the bulk refractive index (RI) is kept constant. Otherwise the amount may be quantified by readings taken between sample injections, where the transport buffer of constant RI passes over the sensor surface.

The first experimental step is to immobilize a ligand to the sensor surface. The ligand may be one of the reactants in the interaction of interest, or e.g. an antibody directed towards one of the reactants. The immobilization is performed by the instrument once the autosampler is loaded with the necessary reagents. As shown in Figure 3, each step in the immobilization can be monitored.

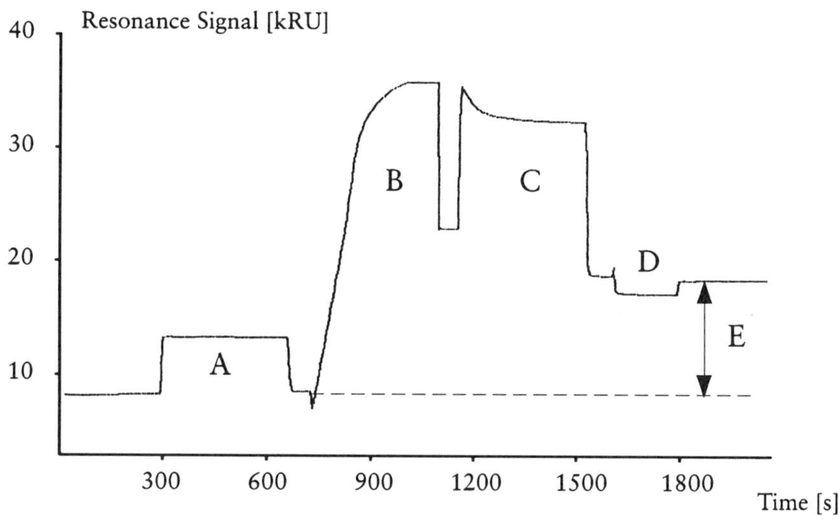

Figure 3.
This sensorgram shows all steps during the immobilization of rabbit anti-mouse Fc (RAMFc) antibodies to the sensor chip. A buffer flow of 5 µl/min continuously passes over the sensor surface. A 30 µl pulse of activation chemicals (A) is injected followed by 30 µl of a 30 µg/ml solution of RAMFc (B). A deactivation reagent (C) is then injected. The RAMFc-surface is then conditioned with 15 µl of 100 mM HCl (D). The amount of RAMFc immobilized, 10.5 kRU, is indicated at (E).

IV. APPLICATIONS

A biospecific interaction is a dynamic course of events that can be described in terms of association and dissociation rates, the participating molecules' affinity for each other, and also by the stoichiometry of the molecular complex formed. In some cases, the biospecific interaction is used as a means to classify or quantitate molecules. Figure 4 shows a sensorgram of the interaction between a mouse monoclonal antibody (MAb) in unprocessed hybridoma culture supernatant with a sensor surface to which rabbit anti-mouse Fc (RAMFc) antibodies have been immobilized. The signal obtained for this MAb and others can, under conditions of constant flow rate and injected volume, be used to determine its concentration by comparison with a standard curve. After the MAb binds to the RAMFc-surface, a series of subclass specific rabbit anti-mouse antibodies are injected. In this case the anti-IgG1 gives a response. After injection of the last subclass specific reagent, the sensor surface is regenerated down to the RAMFc level, and is then ready for the next cycle. This illustrates several features of the technique: 1) the selectivity of the sensor surface can be controlled through the immobilized molecule, thus allowing crude samples to be used; 2) concentration determinations in crude samples can be made if a standard curve is constructed; 3) different unlabeled molecules can be injected in a series and each interaction quantified; 4) the sensor surface can be regenerated as many times as the immobilized ligand allows.

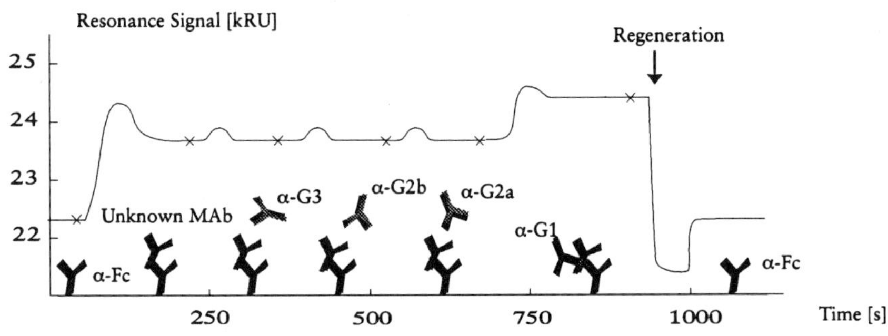

Figure 4.
A MAb of unknown subclass is injected over a sensor surface to which RAMFc previously has been immobilized. A series of subclass-specific antibodies are then injected. Here, the anti-IgG1 reagent gives a response identifying the MAb as belonging to this subclass. The surface is then regenerated and is ready for the next analysis.

Monoclonal antibodies are extensively used to elucidate the molecular topography of proteins by epitope mapping. The term 'epitope mapping' is most frequently used to denote the binding pattern of a panel of MAbs relative to each other by testing the ability of binary combinations of MAbs to bind the antigen simultaneously. Since SPR detection does not require labeling, not only can the binary binding pattern be obtained, but also multi-site binding patterns, simply by injecting a series of MAbs. Figure 5 shows an example of the sequential binding to the antigen of five MAbs, previously shown not to interfere mutually in two-site binding experiments.

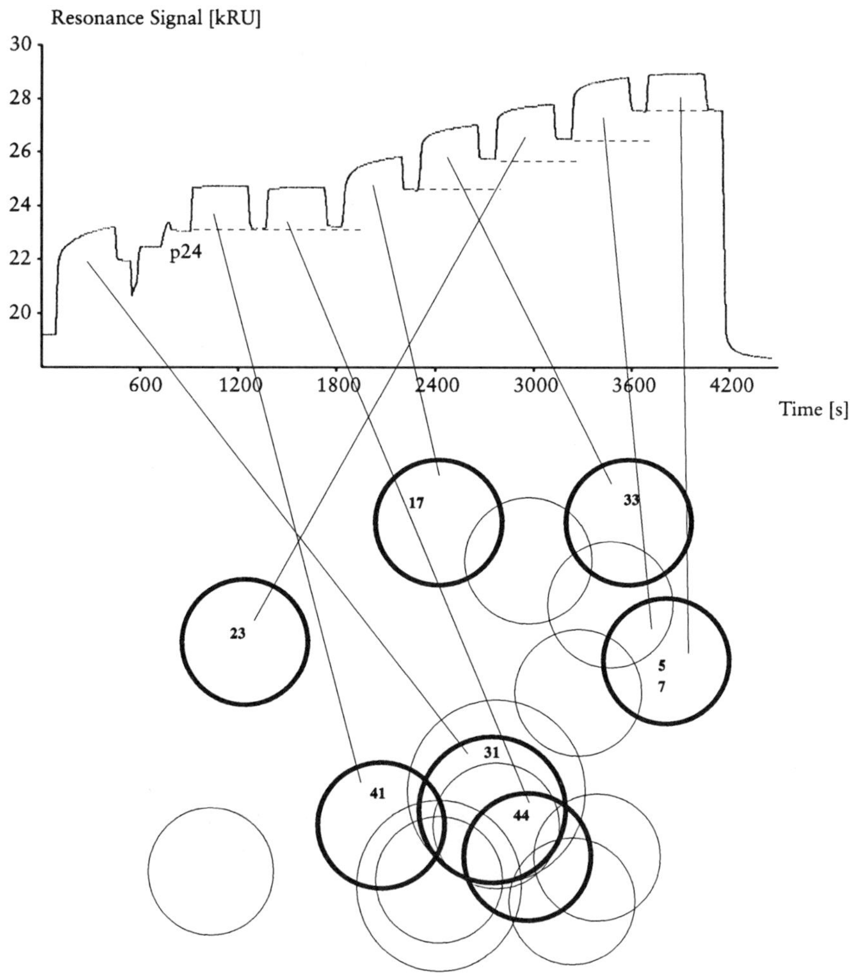

Figure 5.
This sensorgram shows an example of the sequential injection of 7 MAbs including the capturing MAb representing 5 epitopes as indicated in the map below. Overlapping circles denote MAb groups that cannot bind concurrently.

According to Langmuir theory, the interaction between an immobilized molecule and its partner in free solution can be described by:

$$d\Gamma/dt = k_{ass}\, C_o\, (\Gamma_{max} - \Gamma) - k_{diss}\, \Gamma$$

where: Γ is the surface concentration of captured molecules; C_o is the concentration of molecules in solution, Γ_{max} is the total binding capacity of the surface and is determined in a separate saturation experiment. k_{ass} and k_{diss} are the association and dissociation rate constants, respectively. Figure 6a shows a sensorgram where a MAb in free solution reacts with surface immobilized HIV-1 core protein p24. By plotting $d\Gamma/dt$ versus Γ (Fig. 6b), k_{ass} can be calculated from the intercept with the $d\Gamma/dt$ axis. $d\Gamma/dt$ is obtained directly as the slope of the response signal. The Langmuir equation is valid only if the rate of mass transport of the solute to the surface is not limiting. In the experiment shown, the initial slope of the sensorgram is constant, i.e. the decreasing number of free binding sites on the surface does not affect the binding rate. This reflects a mass transport limitation in the binding rate and as can be seen in figure 6b it persists until the surface is about 70 % saturated.

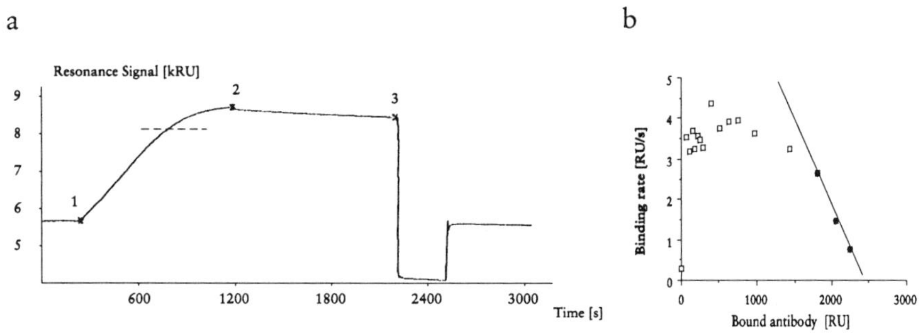

Figure 6.

a) The sensorgram shows the interaction of a MAb in free solution with surface immobilized HIV-1 core protein p24. Remark points 1 and 2 indicate the start and end of the MAb injection, respectively. At remark point 3 the surface is regenerated. The dotted line indicates the saturation level above which the binding rate is limited by the antibody-antigen reaction.

b) In this plot of $d\Gamma/dt$ versus Γ, the surface concentration (Γ) of bound antibody is expressed in resonance units (RU). As can be seen the binding rate rapidly decreases when the surface concentration of antibody has reached about 1800 RU corresponding to 70 % of a Γ_{max} of 2400 RU.

From the fact that the SPR response correlates to the protein mass present within the detected volume, the quality of e.g. MAb preparations can be investigated by saturation experiments. Figure 7 shows saturation curves obtained by binding MAbs to a RAMFc-surface and then injecting the antigen, here recombinant HIV core protein p24 at different concentrations. If two p24 molecules of MW = 24,000 can be bound by each MAb of MW = 156,000, the mass ratio between the proteins in the complex should be 2x24,000/156,000 = 0.31. In Figure 7, most of the MAb preparations show a ratio close to this value. One of the preparations however seem to contain a large proportion of inactive antibodies.

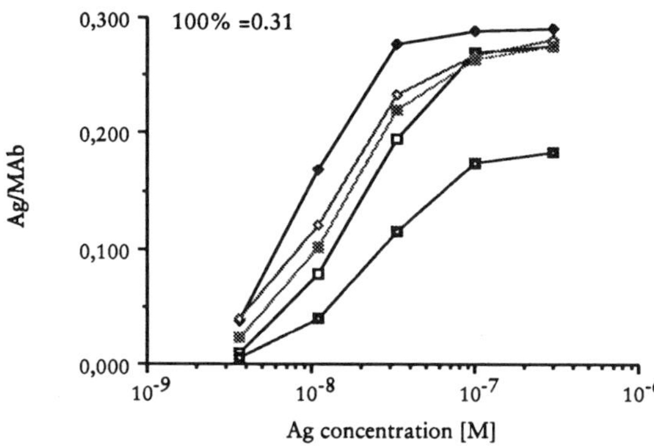

Figure 7.
This plot shows the mass ratio between MAbs bound by a RAMFc-surface and the antigen p24 at different concentrations. Notice how one of the preparations show a lowered saturation mass ratio indicating loss of antigen binding ability for a large fraction (40 %) of the preparation.

V. ACKNOWLEDGEMENTS

This presentation is based on the work of many researchers both within and outside Pharmacia AB.

VI. REFERENCES

Cullen, D.C., Brown, R.G.V. and Lowe, C.R. (1984) Biosensors 3, 211-225.

Flanagan, M.T. and Pantell, R.H. (1984) Electron. Letters 20, 968-970.

Liedberg, B., Nylander, C. and Lundström, I. (1983) Sensors and Actuators 4, 299 - 304.

Mayo, C.S. and Hallock, R.B. (1989) J. Immunol. Meth. 120, 105-114.

SECTION II

Peptide Sequence Analysis

7

Chemistry and Sequencing of Covalently Immobilized Proteins and Peptides

David W. Andrews, Mark Girardi, and Jonathan Mark

MilliGen/Biosearch, 186 Middlesex Turnpike, Burlington, MA 01803

I. Introduction

Protein chemists devote substantial amounts of time and energy to the preparation of proteins and their fragments for sequencing. As levels of starting materials have declined from grams to micrograms, methods have emerged to facilitate manipulation of these small amounts. We wish to report our exploitation of covalent immobilization techniques to provide a simple and efficient methodology for performing chemistry on protein and peptide samples in the 1-100 pmol range. These methods allow for rapid and straightforward sequencing of peptides.

We have investigated performing chemistry on proteins and peptides covalently bound to Sequelon-DITC™ and Sequelon-AA™ membranes (Coull *et al.*, 1989), as well as proteins blotted onto Immobilon-P™ membranes, and subsequently immobilized using the Sequenet™ process.(Pappin *et al.*, 1990). We have discovered that enzymatic and chemical modifications can be performed on these immobilized proteins and peptides.

II. Materials & Methods

Proteins and peptides. Antibodies 45-020 and 2397 were kind gifts of Dr. Michael N. Margolies (Massachusetts General Hospital, Boston,MA). Peptides were purchased from Sigma Chemical (St. Louis,MO).

Protein Sequencing. Sequencing of proteins and peptides was accomplished on a MilliGen/Biosearch 6600 ProSequencer using standard protocols and chemicals. Attachments to Sequelon-Aryl Amine™ membranes, and SequeNet™ immobilizations of blotted samples were performed with MilliGen/Biosearch kits using the prescribed protocols.

Pyroglutamate aminopeptidase reaction. Immobilon blots to be digested with enzymes were "blocked" with polyvinylpyrrolidone (PVP-40, Sigma) as described (Aebersold *et al.*, 1987). Briefly, after staining and destaining, blots were neutralized with a 1-2 min wash in 200 μM NaOH, followed by reaction with a 0.5% solution of PVP-40 in 0.1M acetic acid at 37°C for 30 minutes. The blots were then washed extensively (10 X 1 ml) with H_2O. Peptides were attached to Sequelon-Aryl Amine™ membranes by the established protocol.

Blots or membrane disks suspected to contain proteins with pyroglutamic acid at their amino terminus were treated with pyroglutamate aminopeptidase (Boehringer Mannheim Biochemicals), as described (Podell and Abraham, 1978). Blots were submerged in a 1.5 ml Eppendorf tube in buffer (50 mM phosphate/10 mM EDTA/ 5 mM DTT/ 5% glycerol/ pH 8.0), and 0.5 mg of enzyme was added. After flushing with argon and vortexing, the tube was placed at 4°C for 12 hours. Then, an additional 0.5 mg of enzyme was added,

Techniques in Protein Chemistry II

and the tube was incubated at 37°C for 12 hours. The blot(s) were then washed extensively (10 X 1 ml) with H_2O, lyophilized to dryness, and stored at -20°C until placed in the sequencer. Blots treated this way were then treated with the SequeNet™ process, and subjected to sequencing on the MilliGen/Biosearch 6600 ProSequencer.

De-acetylation. The peptide to be de-acetylated was first attached to a Sequelon-Aryl Amine™ membrane, and the acid hydrolysis procedure of Ozols was followed. The membrane was placed in a 1.5 ml Eppendorf tube, and wet with 5 µl of 20% trifluoroacetic acid (TFA). An additional 25 µl of 20% TFA was placed in the bottom of the tube. The tube was placed inside a Waters Pico-Tag bottle, and taken through 3 cycles of evacuation and flushing with nitrogen. After the final vacuum and sealing, the bottle was placed in an oven at 110°C for 20 minutes. After withdrawing the bottle from the oven, it was left until cool, and the membrane was removed and washed once with H_2O, then placed in the sequencer.

Asp/Pro Cleavage. The hydrolysis method used is the one described by Tarr (1986). The protein was blotted onto Immobilon™ using standard techniques. After staining and de-staining, the protein was covalently immobilized using the SequeNet™ process, and then washed with MeOH, H_2O, and MeOH. The blot was then placed in a 1.5 ml Eppendorf tube, and wet with 5 µl of 0.1% trifluoroacetic acid (TFA). Another 25 µl of the same TFA solution was placed in the bottom of the Eppendorf tube, and the Eppendorf tube was placed in a Waters Pico-Tag™ bottle. The bottle was taken through 3 cycles of evacuation and flushing with nitrogen. After the final vacuum and sealing, the bottle was placed in an oven at 110°C for 30 minutes. After withdrawing the bottle from the oven, it was left until cool, and the membrane was removed and washed once with H_2O, then placed in the sequencer.

TNBS blocking. At the first sequencing cycle, a special protocol was used which first delivers a trinitrobenzenesulfonic (TNBS) acid solution (20 mg/ml in CH_3CN), along with an equal amount of coupling buffer, in order to react with all amino-terminii except proline (J.D. Dixon & R. Laursen, personal communication). Subsequent cycles reflect the usual protocol for the MilliGen/Biosearch 6600 ProSequencer.

III. Results and Discussion

Enzymatic de-blocking. Blockage of the amino-terminus can be occasioned by numerous unhappy circumstances, but the two most amenable to enzymatic remedy are pyrrolidone carboxylic acid (or pyroglutamic acid, pGlu) and acetylated amino terminii.

In each case, a commercially available enzyme preparation purportedly removes the offending amino-terminal amino acid, leaving the second available for sequencing. With respect to pyrrolidone carboxylic acid, the enzyme (pyroglutamate amino peptidase) and procedure (Podell and Abraham, 1978) have been cited several times in the literature (Margolies *et al.*, 1981, Andrews and Capra, 1981), but its application has been infrequent and unsatisfying. Dissolution of reduced and alkylated proteins is sometimes a problem, and preparation of the resulting product for sequencing can be difficult. Use of a solid-phase approach eliminates the most pressing problems and simplifies the procedure.

We have found that simple immersion of a blocked protein on a suitably treated Immobilon™ blot in a buffered solution of pyroglutamate aminopeptidase for several hours is sufficient to cleave pGlu from a mouse antibody kappa light chain. Cleavage was demonstrable by sequencing of the blot after treatment with the SequeNet™ procedure (Table I). Similarly, we

Table I. Sequence of 45-020, blocked mouse antibody light chain after pyroglutamate aminopeptidase cleavage (50 pmol of protein applied to gel). Initial yield=14% from gel, after blotting and cleavage. Repetitive yield=92.3%.

Cycle	Amino Acid	Yield (pmol)
1	Ala	7
2	Val	5
5	Gln	5
8	Ala	4

removed pGlu from the peptide neurotensin, which had been coupled to a Sequelon-AA™ membrane. The peptide was then sequenced to its carboxyl-terminus (Table II).

Chemical de-blocking. Acetylated amino-terminii block an unknown (but large) number of proteins from eukaryotic sources. Attempts at removal of this blockage have been, in general, fruitless. Recently, a chemical method has been used with some success (Ozols, 1989; de-blocking of acetylated Ala), as well as an enzymatic procedure (Jones *et al*., 1986), which apparently only works on short peptides. We have used a modification of the chemical procedure to deblock an acetylated peptide, which had been attached to a Sequelon-AA™ membrane. Success was confirmed by sequencing the resulting peptide to its carboxyl-terminus (Table III). As a control, the same peptide was attached to a Sequelon-AA™ membrane and sequenced without acid treatment. No sequence was observed (data not shown).

While it is more straightforward to de-block an acetylated peptide before attachment, the described procedure more nearly approximates what a researcher might do to sequence an attached peptide with which he had no initial success. In fact, hydrolysis before attachment is more efficient (data not shown), since one does not risk hydrolyzing the attachment bonds, as well as the acetyl amide. It remains for us to demonstrate the generality of this method. That is, will the procedure work on peptides and proteins with something other than an acetylated serine at the amino-terminus?

Asp/Pro cleavage. Exposure of denatured proteins to dilute acid occasions a remarkably specific cleavage at Asp-Pro bonds (Piszkiewicz *et al*., 1970) This method has been used on a number of proteins (Poulsen *et al*., 1972; Jauregui-Adell & Marti, 1975; Inglis, 1983), and since this linkage is relatively rare, large fragments are produced. The revealed amino terminus is proline, so that blocking procedures can be implemented to insure that

Table II. Sequence of neurotensin after pyroglutamate aminopeptidase cleavage (103 pmol applied to disk). Initial yield=29% after attachment and cleavage. Repetitive yield=98%.

Cycle	Amino Acid	Yield (pmol)
1	Leu	30
2	Tyr	27
11	Ile	20
12	Leu	25

Table III. Sequence of alpha-melanocyte stimulating hormone after chemical cleavage of amino-terminal acetylated serine by the method of Ozols (94 pmol applied to disk). Initial yield=15% after attachment and acid hydrolysis. Repetitive yield=91.4%.

Cycle	Amino Acid	Yield (pmol)
1	Ser	-
2	Tyr	14
12	Pro	4
13	Val	5

sequencing begins only at the desired site. We have successfully used this technique on covalently immobilized rabbit antibody kappa chain, followed by blocking and subsequent sequencing (Table IV).

These examples demonstrate the utility of performing chemistry on covalently immobilized proteins and peptides. The results presented describe both enzymatic (pyroglutamate aminopeptidase) and chemical transformations (de-acetylation and Asp/Pro cleavage). We feel that, in addition to these methods, there are obvious extensions to these approaches that will help protein sequencers obtain more data from their immobilized samples.

Table IV. Mild acid Asp/Pro cleavage on 2397 rabbit antibody light chain (50 pmol applied to gel). Initial yield=30% from gel, after blotting, acid cleavage, and TNBS reaction. Repetitive yield=98%.

Cycle	Amino Acid	Yield (pmol)
1	Pro	15
2	Val	14
7	Leu	14

IV. References

Andrews, D.W. and J.D. Capra 1981. *Biochemistry* **20**, 5822—5830.

Coull, J.M., J.D. Dixon, R.A. Laursen, H. Köster, and D. Pappin. 1989, in *Methods in Protein Sequence Analysis, Proceedings of the 7th international Conference*, Berlin, July 3-8, 1988. ed. Brigitte Wittmann-Liebold. Springer-Verlag, Berlin, 1989.

A.S. Inglis.1983. *Methods Enzymology* **91**, 325-332.

Jauregui-Adell, J. & J. Marti.1975. *Anal. Biochem.* **69**, 468-473.

Jones, W., L. Manning, and J. Manning. 1986. *Biochem. Biophys. Res. Comm.***139**, 244-250.

Margolies, M.N., A. Marshak-Rothstein, and M.L. Gefter. 1981. *Molecular Immunology* **18**, 1065-1077 .

Ozols, J. 1989. *Biochim. Biophys. Acta* **997**, 121-130.

Pappin, D.J.C., J.M. Coull. and H. Köster. 1990. New approaches to covalent sequence analysis, in *Current Research in Protein Chemistry*, (ed. J. Villafranca) 191-202 (Academic Press, San Diego, 1990).

Piszkiewicz, D., M. Landon, and E.L. Smith. 1970. *Biochem. Biophys. Res. Comm.* **40**, 1173-1178 .

Podell, D.N. and G.N. Abraham. 1978 *Biochem. Biophys. Res. Comm.* **81**, 176-185

Poulsen, K., K.J. Fraser, and E. Haber.1972. *Proc. Natl. Acad. Sci. USA* **69**, 2495-2500.

Tarr, G. 1986., in *Methods of Protein Microcharacterization*, ed. J.E. Shively. Humana Press, Clifton, New Jersey.

8

REAL TIME MONITORING OF SYNTHETIC PEPTIDE PURIFICATIONS USING ULTRAFAST PEPTIDE ANALYSIS

Kerry D. Nugent[1], Kerry L. Spear[2] and Pia M. Johnson[2]

[1] Michrom BioResources, Inc., Livermore, CA 94550

[2] Triton Biosciences Inc., Alameda, CA 94501

I. INTRODUCTION

As a result of the advances in technology for automated peptide synthesis, the number of new, biologically significant peptides being produced has increased significantly over the past few years. Even though modern instrumentation and chemistries are capable of producing cycle yields of greater than 99%, the final product in a multi-residue peptide is often a crude mixture with 10-50% of the desired product. Since many of these synthetic peptides are being developed for therapeutic or diagnostic purposes, a purification scheme which allows a high yield of relatively pure (95-99%) material is required.

Over the past few years, reversed-phase high performance liquid chromatography (RP-HPLC) has become the method of choice for the preparative purification of synthetic peptides (1-3), since it offers a single step, high resolution purification in a volatile buffer system. Using semipreparative columns in a gradient elution mode (1,2), or analytical columns in a sample displacement mode (3), gram level quantities of highly purified product can readily be isolated.

Techniques in Protein Chemistry II

RP-HPLC is also the method of choice for analyzing the fractions from preparative runs to determine pooling which optimizes both yield and purity. Often, the analytical conditions employ the same packing and mobile phase gradient as the preparative separation (1,3), and can therefore create a misleading indication of purity since impurities which coelute with the product during the preparative run will most likely coelute in the analytical run (4). One way of overcoming this problem is to use a different mobile phase system and/or a column with a different selectivity in the RP-HPLC analytical assays of the preparative fractions (2,5). A second more definitive but more complex approach would be to use two orthogonal modes of chromatography (i.e. ion exchange and reversed-phase) in the preparative purification and the analytical assays to allow a more confident assessment of final product purity (4).

When HPLC is used for the preparative purification of a crude synthetic peptide mixture, the most time consuming step is the analytical assay of fractions to determine where the product of interest elutes and how to pool for maximum purity and yield. Often ten or more analytical assays are required for each preparative run, especially when samples are very complex or the preparative system is overloaded for maximum throughput. One option for minimizing total analytical time is to develop an isocratic separation for screening the peptide of interest in the preparative RP-HPLC fractions. This approach is often difficult to use, since retention times and selectivity for large molecules can change dramatically with small changes in solvent composition (2,6).

Another option for reducing the total analytical time is to employ ultrafast HPLC for real time monitoring of synthetic peptide purifications. This technique works with new types of HPLC column packings (both nonporous and macroporous) which have been developed to allow high resolution separations of proteins and peptides in less than 10 minutes (7,8). When these new packings are used in a microbore HPLC based analyzer which has been optimized for high speed separations, a total gradient analysis time of under five minutes from injection to injection can readily be achieved (9). Ultrafast HPLC has been shown to be very useful for real time monitoring of proteins during the purification of recombinant DNA derived products (10,11), and this paper will show that it is also quite useful for real time monitoring of synthetic peptide purifications.

Transforming growth factor-alpha (TGF-α) is an ideal candidate to test this new approach, since it is a large peptide (50 amino acid residues) whose synthesis results in a fairly complex crude mixture with a final product yield of less than 50%. Since TGF-α, as shown in **Figure 1**, contains six cysteine residues, it was first purified in the reduced form, and then folded to create the three correct disulfide links necessary for final purification and biological activity. Timely analysis of the acyclic TGF (1-50) preparative RP-HPLC fractions is necessary to prevent oxidation of the cysteine residues and to prevent potential acid-induced product clips.

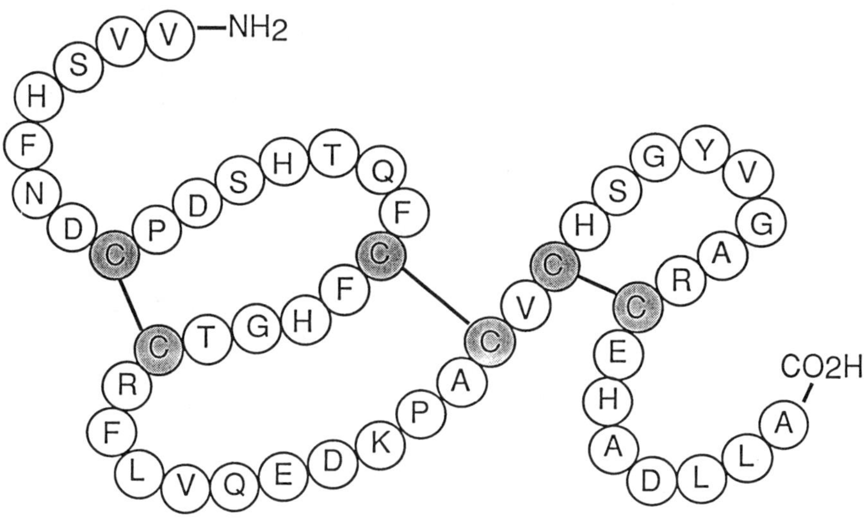

Figure 1. Structure of TGF-α

This paper will compare conventional analytical RP-HPLC with ultrafast RP-HPLC for the analysis of purification fractions of crude, reduced TGF (1-50). An ultrafast assay of the correctly folded synthetic TGF-α will also be shown, as well as an example of varying pH in RP-HPLC to aid in final product purity assessment. Finally, the potential for using ultrafast HPLC to monitor proper peptide and protein folding is explored.

II. MATERIALS AND METHODS

A. Synthesis of TGF (1-50)

Side chain protected TGF (1-50) was assembled in a step wise fashion from residue 50 (alanine) via fluorenylmethoxycarbonyl (FMOC) methodology (12), using a Milligen/Biosearch Model 9500 peptide synthesizer. The synthesis was performed on a 0.4 mmol scale and the initial residue (alanine) was derivatized through a hydroxymethylphenoxyacetic acid linker to the methyl benzhydrylamine polystyrene resin (FMOC-alanine-PAC support; 0.33 mmol/gram loading). Coupling time per residue varied from one to four hours. After completion of the synthesis, the peptide was cleaved from the resin support by the action of trifluoroacetic acid (TFA) containing thioanisole, 1,2-ethanedithiol and anisole as cation exchange scavengers. Under these conditions, all side chain protected groups were removed.

B. Chromatographic Instrumentation

Crude, reduced TGF (1-50) was purified on a 5 x 25 cm, 15-20μ, 300Å Vydac C18 column, using a Gilson 303 - 100SC binary gradient HPLC system with 0.05% Trifluoroacetic acid (TFA)/water as solvent A and 0.05% TFA/Acetonitrile (ACN) as solvent B.

Initial screening of the purified fractions was done on a 0.46 x 15 cm, 5μ, 300Å Vydac C18 column, using a Hewlett-Packard 1090 DR-5 binary gradient HPLC system with 0.05% TFA/water as solvent A and 0.05% TFA/ACN as solvent B.

All of the ultrafast separations were run on a 0.1 x 5 cm, 8μ, 4000Å Polymer Labs PLRP-S column, using a Michrom Ultrafast Microprotein Analyzer (UMA) with 0.1% TFA/water as solvent A and 0.1% TFA/ACN as solvent B.

C. Chemicals and Reagents

HPLC grade acetonitrile (ACN) was obtained from Burdick and Jackson. Water was purified prior to use with a Millipore Milli-Q water purification system. Sequanal grade trifluoroacetic acid (TFA) was obtained from Pierce Chemical Co. Zwittergen C-8 surfactant was obtained from Cal-Biochem.

III. RESULTS AND DISCUSSION

A. Comparison of Conventional HPLC and Ultrafast HPLC

A 50 mg sample of crude, reduced TGF (1-50) was separated on a semipreparative (5 x 25 cm) Vydac C18 column as shown in **Figure 2.** The eluant was collected in 50 ml fractions for subsequent analysis and pooling of the appropriate fractions. From the initial analytical work, the reduced peptide was expected to elute between 15 and 20 minutes, so fractions 16 through 19 were used for further analytical assays.

Conventional RP-HPLC conditions similar to the semipreparative conditions were used to analyze the fractions from the crude, reduced

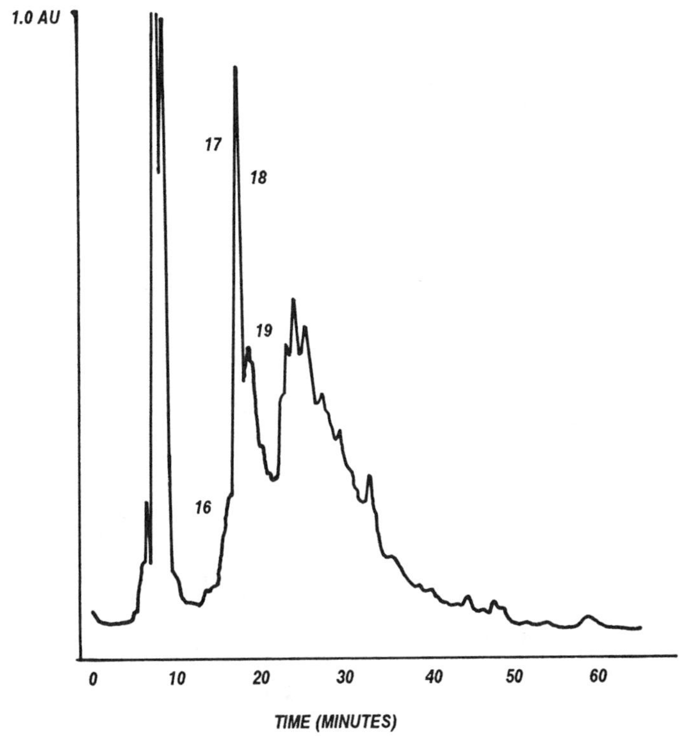

Figure 2. Purification of Crude, Reduced TGF (1-50)
The conditions were a 5 x 25 cm, 15-20μ Vydac C18 300Å column using a 1%/min linear gradient from 20 - 55% ACN in 0.05% TFA at 75 ml/min and room temperature with detection at 214 nm.

peptide purification. Using an analytical (0.46 x 15 cm) Vydac C18 column on a HP 1090 LC, the majority of the acyclic TGF (1-50) eluted in fractions 17 and 18, with a retention time of 22.4 minutes, as shown in **Figure 3**. The total analysis time for each of these fractions was 35 minutes.

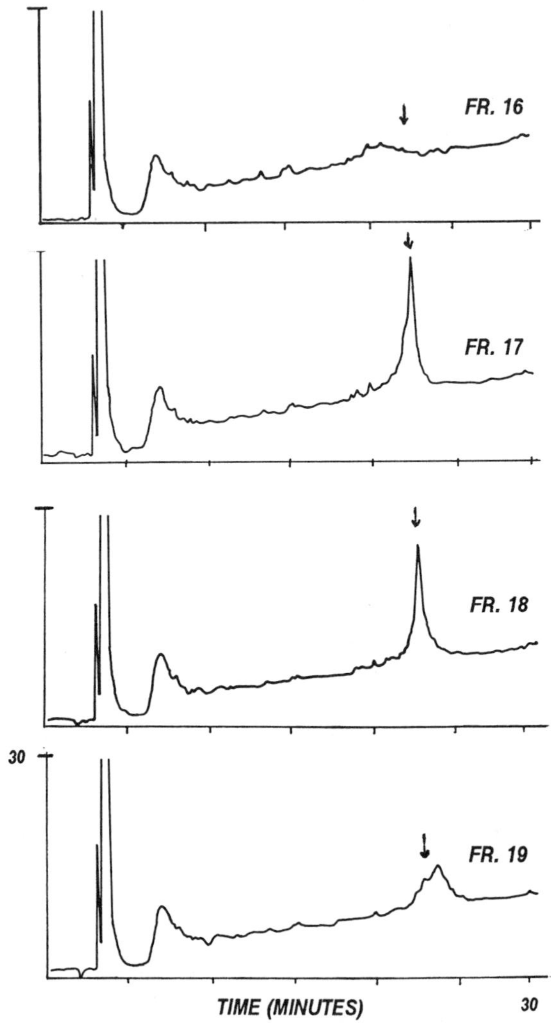

Figure 3. Conventional RP-HPLC of Preparative Fractions
Analysis of fractions 16-19 from the run shown in Figure 2 using an HP1090 LC. The conditions were a 0.46 x 15 cm, 5μ Vydac C18 300Å column using a 1%/min gradient from 20-55% ACN in 0.05% TFA at 1.0 ml/min and room temperature with detection at 214 nm.

The same fractions were then analyzed on the UMA using a microbore (0.1 x 5 cm) PLRP-S 4000Å column. **Figure 4** shows that the **UMA** was able to provide results comparable to those obtained from the conventional analytical assays shown in Figure 3, with a total analysis time of two minutes from injection to injection for each of the fractions. The use of the microbore system also gave much better

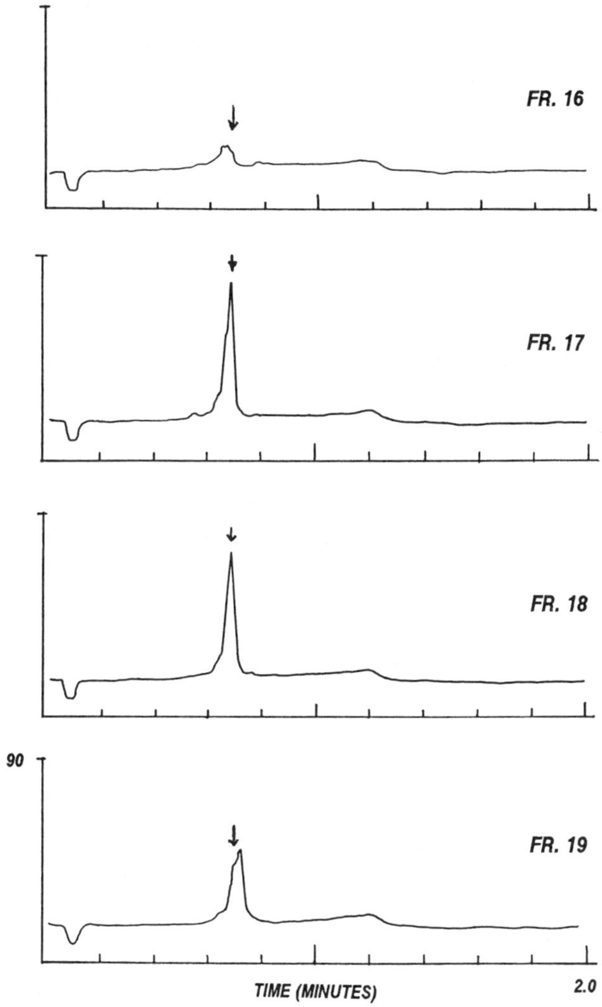

Figure 4. Ultrafast RP-HPLC of Preparative Fractions
Analysis of fractions 16-19 from the run shown in Figure 2 using the UMA. The conditions were a 0.1 x 5 cm, 8μ PLRP-S 4000Å column using a 20%/min linear gradient from 15-35% ACN in 0.1% TFA at 0.5 ml/min and 40°C with detection at 214 nm.

sensitivity, allowing direct injection of these very dilute fractions, while minimizing the baseline noise and drift problems seen in Figure 3.

B. Comparison of Reduced and Folded TGF-α

It should be noted that the peaks in both Figures 3 and 4 are much broader than expected for a peptide of this size under these types of RP-HPLC gradient conditions. Since the TGF (1-50) is an acyclic peptide, this apparent broadening may be due to slow, conformational changes occurring during the separation (6). When the purified material (fractions 17 and 18) was properly folded to form the correct structure of TGF-α (Figure 1), the three disulfide bonds greatly increase the conformational restrictions in the peptide. Using

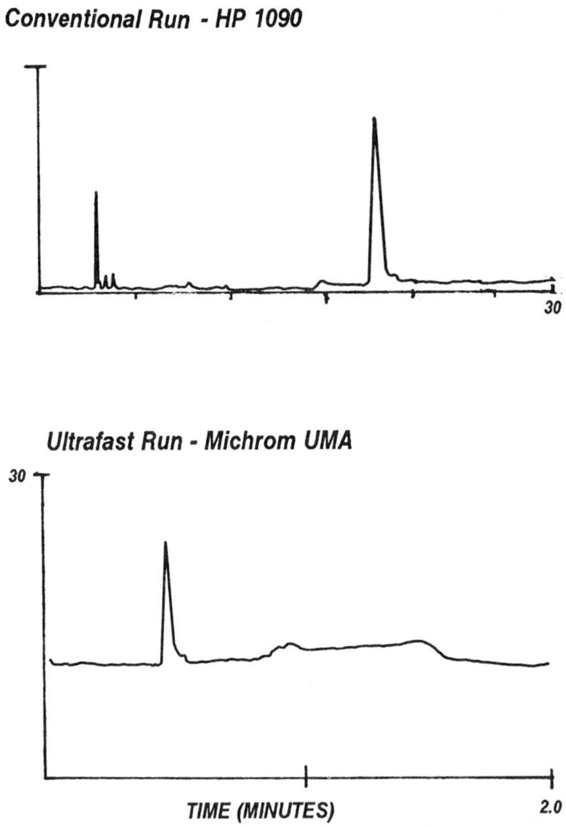

Figure 5. Analysis of Purified, Folded TGF-α
Upper. Separation of TGF-α on an HP 1090 LC using the same conditions as in Figure 3. Lower. Separation of TGF-α on the UMA using the same conditions as in Figure 4.

the same conditions as in Figure 3 or Figure 4, folded TGF-α elutes as a single sharp peak, as shown in **Figure 5**. As expected, the folded peptide elutes much earlier than the reduced peptide, which is more hydrophobic in its unfolded state (13).

C. Ultrafast HPLC Separation of Acyclic TGF Conformers

Since the peak broadening in Figures 3 and 4 was thought to be due to slow conformational changes of the acyclic TGF (1-50) occurring on the column during RP-HPLC separation, we decided to

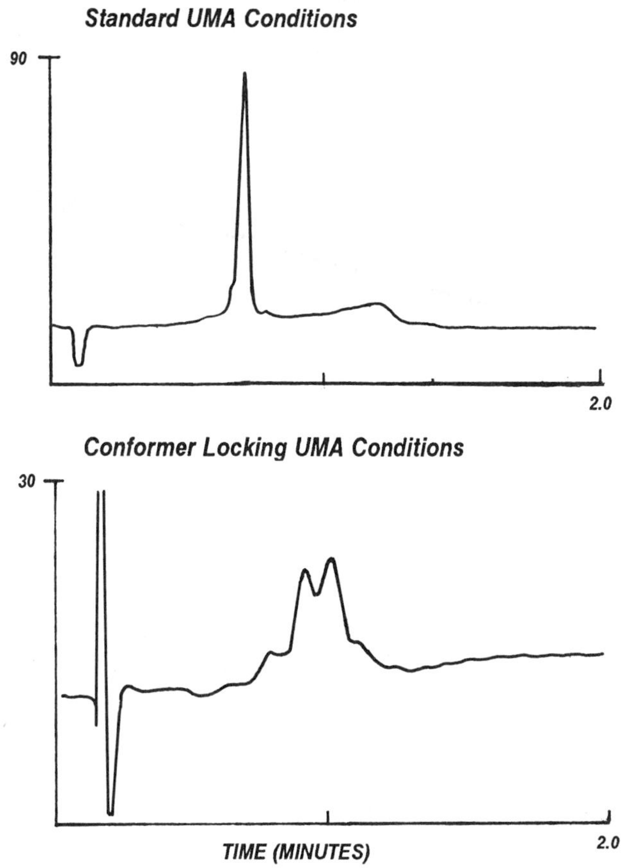

Figure 6. Separation of Acyclic TGF (1-50) Conformers
Upper. Fraction 18 run using conditions identical to Figure 4. Lower. Fraction 18 run using a 0.1 x 5 cm, 8µ PLRP-S 4000Å column using a 5%/min linear gradient from 20-30% ACN in 0.1% TFA/0.05% Zwittergen C8 at 0.5 ml/min and 60°C with detection at 214 nm.

test this hypothesis by using a reversed-phase separation system which has previously been found to minimize on-column conformational changes of certain proteins (13). Although fraction 18 from the semipreparative separation shown in Figure 2 gives a broad peak when run under the conditions described in Figure 4, (see **Figure 6A**), this fraction yields a final folded TGF-α product that is greater than 99% pure. Using a microbore (0.1 x 5 cm) PLRP-S 4000Å column heated to 60°C, with 0.05% Zwittergen C8 added to the mobile phase, resolution of conformers could be achieved as shown in **Figure 6B**. These data suggest that the UMA may be quite useful in monitoring peptide and protein folding studies, especially where peak broadening due to on column conformational conversions could obscure differences in actual folded conformers.

D. Varying pH in Ultrafast HPLC for Purity Assessment

Under acidic conditions, TGF-α can be clipped between residues 7 and 8. This can result in the presence of an N-terminal deleted peptide impurity, TGF (8-50) in the final product. Although minute amounts of this fragment have been seen in protein sequence analysis of pure products produced either synthetically or by recombinant techniques, it cannot be resolved from the parent material using a conventional RP-HPLC separation (See **Figures 7Aand 7B**).

Since Horvath has shown that optimum resolution, but different selectivity, is achieved at the extremes of pH in the RP-HPLC of proteins (14), a high pH RP-HPLC separation was investigated. Taking advantage of the wide pH range of both the UMA and the polymeric column, an assay was quickly developed to resolve the tricyclic TGF (8-50) fragment from the parent TGF-α at pH 11, as shown in **Figure 7C and 7D**. With this rapid assay, the degree of formation of this fragment can be determined to a level of 0.1%. In the synthetic material purified in this study (which was left in the pH 2.1 fractions for several weeks), the final purified TGF-α contained about 1% of the tricyclic TGF (8-50).

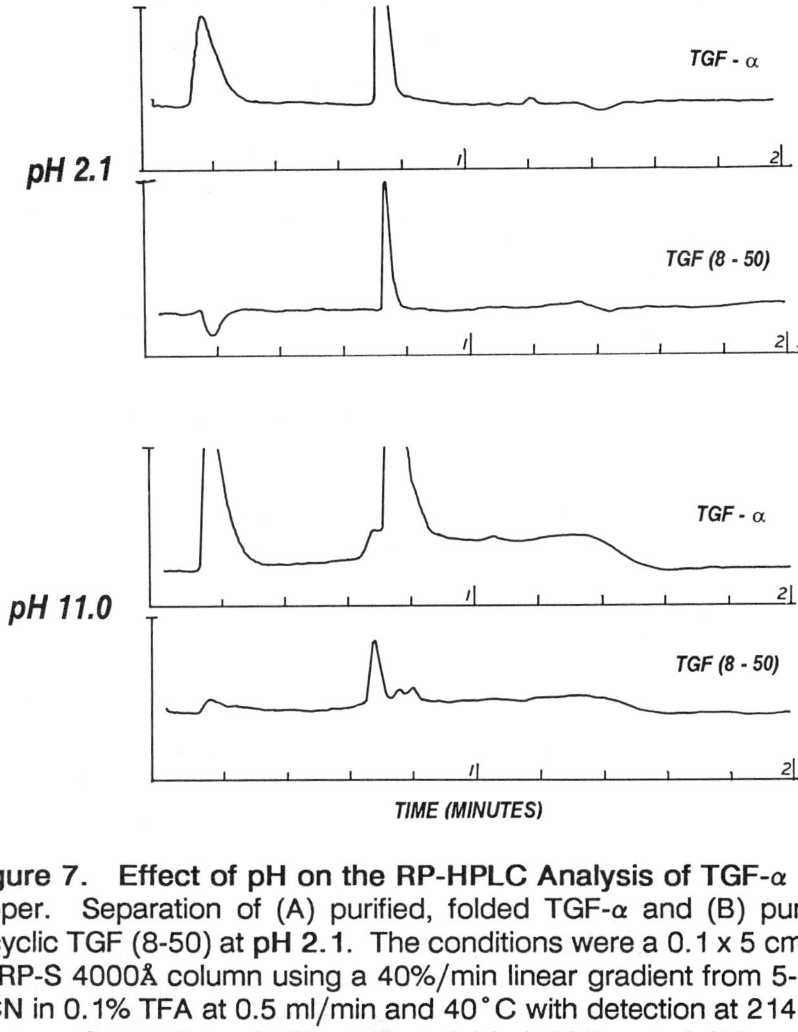

Figure 7. Effect of pH on the RP-HPLC Analysis of TGF-α
Upper. Separation of (A) purified, folded TGF-α and (B) purified tricyclic TGF (8-50) at **pH 2.1**. The conditions were a 0.1 x 5 cm, 5μ PLRP-S 4000Å column using a 40%/min linear gradient from 5-45% ACN in 0.1% TFA at 0.5 ml/min and 40°C with detection at 214 nm. Lower. Separation of (C) purified, folded TGF-α and (D) purified tricyclic TGF (8-50) at **pH 11**. The conditions were identical to those above except that the mobile phase was buffered at pH 11 using 50 mM Sodium Borate in place of the 0.1% TFA.

IV. CONCLUSIONS

Ultrafast HPLC is a powerful technique for rapid screening of fractions from a synthetic peptide purification. Total analysis times can be reduced 10-20 fold over conventional analytical HPLC, with minimal loss in resolution. Since the analysis time for assaying fractions is no longer the limiting step, the resolution in the preparative RP-HPLC purification step can be optimized (i.e. lowering the slope of the gradient during the elution of the peptide of interest) and a larger number of smaller sized fractions can be collected. Using the UMA to analyze these smaller fractions in real time will allow more precise and timely decisions on pooling of fractions, thereby improving total yield and purity.

Since the polymeric columns used in this study have a slightly different selectivity than the Vydac C18 material used in the purification step, this technique may offer a more precise assessment of peptide purity than the conventional analytical system using a separation system with exactly the same selectivity as the preparative run. The versatility of the UMA can also be utilized to run complimentary mobile phase conditions, such as the high pH runs described earlier, to further enhance confidence in final product purity. The UMA can also be used to give a true orthogonal check of purity, by either running the purification on an ion exchange HPLC column and monitoring the fractions by ultrafast RP-HPLC (4), or by using ultrafast ion exchange HPLC to monitor the conventional RP-HPLC purification (10).

Finally, the UMA may also prove to be a useful tool for peptide and protein folding studies. By using the built-in automated sample injector, certain slow folding reactions may be automatically sampled every two to five minutes, thus allowing on line analysis in real time. Future work with the UMA will include real time monitoring of TGF-α folding, as well as the use of the UMA as an on line monitor for synthetic peptide purifications. These ultrafast applications, together with the UMA's utility as a high sensitivity microbore HPLC system (15), make this a very powerful tool for the synthetic peptide chemist.

REFERENCES

1. Rivier, J.,McClintock, R., Galyean, R. and Anderson, H., J. Chromatogr., **288**, 303 (1984).

2. Hoeger, C., Galyean, R., Boublik, J., McClintock, R. and Rivier, J., BioChrom., **2**, 134 (1987).

3. Hodges, R.S., Burke, T.W. and Mant, C.T., J. Chromatogr., **444**, 349 (1988).

4. Andrews, P.C., Peptide Res., **1**, 93 (1988).

5. Gou, D., Mant, C.T. and Hodges, R.S., J. Chromatogr., **386**, 205 (1987).

6. Snyder, L.R. and Stadalius, M.A., in High Performance Liquid Chromatography - Advances and Perspectives, Horvath, Cs., ed., Academic Press, New York, **3**, 195 (1984).

7. Dong, M.W., Gant, J.R. and Larsen, B.R., BioChrom., **4**, 19 (1989).

8. Rozing, G.P. and Goetz, G., J. Chromatogr., **476**, 3 (1989).

9. Nugent, K.D., Am. Biotech. Lab., **8(7)**, 24 (1990).

10. Nugent, K.D., in Current Research in Protein Chemistry, Villafranca, J.J., ed., Academic Press, New York, 233 (1990).

11. Nugent, K. and Olson, K., BioChrom., **5**, 101 (1990).

12. Fields, G.B. and Noble. R.L., Int. J. Pept. Prot. Res., **35**, 161 (1990).

13. Nugent, K.D., Burton, W.G., Slattery, T.K., Johnson, B.F. and Snyder, L.R., J. Chromatogr., **443**, 381 (1988).

14. Maa, Y. and Horvath, Cs., J. Chromatogr., **445**, 71 (1988).

15. Nugent, K.D. and Nugent, P.W., Biochrom, **5**, 1 (1990).

9

Sequencing a Protein with an Irregularly Repeating DNP Pattern

S. Hua, M. Vestling, C. Murphy, C. Fenselau
Department of Chemistry and Biochemistry, University of Maryland Baltimore County, Baltimore, MD 21228
B. Spengler, Y. Pan, R. Wang, R. J. Cotter
Department of Pharmacology and Molecular Sciences, Johns Hopkins University, Baltimore, MD 21205
J. Theibert, J. H. Collins
Department of Biological Chemistry, School of Medicine and Medical Biotechnology Center, Maryland Biotechnology Institute, University of Maryland, Baltimore, MD 21201

I. INTRODUCTION

Staphylococcus aureus V8 protease, originally characterized by Drapeau (1,2), exhibits high specificity for cleavage of peptide bonds to which glutamic acid or aspartic acid contribute the carbonyl group. The protease is commercially available and is widely used for mapping or sequencing proteins. Carmona and Gray (3) have published an inclusive amino acid sequence derived from cDNA sequencing while Drapeau, somewhat earlier and working with the protein itself, has reported a sequence for the first 215 amino acids and an amino acid analysis of the C-terminal region. No cysteines or modified residues were detected, and the C-terminal region was found to contain an irregularly repeating pattern of DNP and NNP residues.

In the 215 residues preceeding the repeating pattern the sequences determined from the protein and from the gene disagreed in 12 places. We have already reported the use of fast atom bombardment and tandem mass spectrometry to provide a third opinion on the sequence of this first three-quarters of the protein (4). Our sequence was in agreement with all residues on which the earlier studies agreed. Among the 12 disputed residues, we identified amino acids consistent with the protein sequence at four positions, and with the gene sequence at eight positions. These conclusions are summarized in Table I.

Techniques in Protein Chemistry II

Table I

Summary of the reported differences in the first 215 amino acid residues of *S. aureus* V8 protease.

Residue	Protein	cDNA	MS
41	-	V	V
57	-	H	H
77	D	N	N
125	T	V	V
161	N	D	D
191	Q	E	Q
193	D	N	D
199	N	-	-
201	V	N	N
202	N	V	V
213	N	I	N
214	N	H	N

In order to complete our characterization of the protein and sequence the repeating domain, we employed a new strategy of mass balance, in which the molecular weight is first determined for each protein or peptide under study. This provides an endpoint for the subsequent investigation, and facilitates recognition of peptides missing from protease maps and unaccounted portions of the sequence.

The C-terminal region which is rich in prolines is expected to have a distinctive three dimensional structure (2, 3, 5) with several possible functions having been proposed (2, 3). In order to determine the irregular repeating sequence of *S. aureus* V8 protease's C-terminal region, we have augmented Edman degradation with four mass spectrometry techniques, matrix assisted laster desorption (LDMS), plasma desorption (PDMS), fast atom bombardment (FAB) and high performance tandem mass spectrometry.

II. EXPERIMENTAL

S. aureus V8 protease was purchased from Worthington Chemical Co. Endoproteinase Asp-N (sequence grade) was obtained from Boehringer Mannheim, α-chymotrypsin, cyanogen bromide, trifluoroacetic acid (TFA), and formic acid from Sigma Chemical Co., and $H_2{}^{18}O$ from MCB, Inc.

High performance liquid chromatography (HPLC) was carried out with Aquapore C18, C8, and C4 RP-300 columns, using gradients made from 0.1% TFA/water and 0.1% TFA/acetonitrile.

Matrix assisted UV laser desorption mass spectra were obtained on a modified CVC Products model 2000 time-of-flight mass spectrometer (6). A Quantel International model YG 660 frequency quadrupled Nd:YAG laser was used, with an output energy of 15 mJ/pulse at 266 nm and a pulse width of 5 ns. Filters, a quartz lens, a post-accelerating detector, a 100 MHz oscilloscope/transient recorder and a computer were added. Peptide solutions were prepared as 10^{-5} M in ultrapure water, mixed 1μL:1 μL with 50 mM nicotinic acid in water, and deposited on the probe tip.

The plasma desorption mass spectra were obtained with a Bio-Ion Nordic BIN 10K mass spectrometer with a 10 μCi ^{252}Cf source. The samples (~15 μg) were dissolved in 0.1% TFA and deposited on a nitrocellulose foil.

Fast atom bombardment and tandem mass spectra were obtained with a JEOL HX110/HX110 tandem mass spectrometer equipped with a DA5000 data system and either a FAB source operated at 6 kV or a Cs ion gun operated at 23 kV. The accelerating voltage was 10 kV. The collision gas was He, and the collision cell was floated to 4 kV. The peptides (3-6 μg)were dissolved in 0.1% TFA and measured in either thioglycerol or *m*-nitrobenzyl alcohol. Unit resolution was used for molecular weight determinations, and for tandem analysis resolution in both MS1 and MS2 was 1000.

Edman sequencing was carried out on either a Milligen model 6600 sequencer or on an ABI model 470A sequencer. We thank Dr. George Klain of Millipore/Milligen for analyses with the Milligen model 6600. Amino acid compositions were determined by the method of Heinrikson and Meredith (7).

III. RESULTS AND DISCUSSION

Our laser desorption analysis provided a molecular weight of 29,768, using an instrumental system (6) with an estimated uncertainty in this mass range of ± 1 % or ±300 Daltons. Although this does not define the exact number of amino acids in the polymer, it immediately indicates that the protein we examined is larger than that predicted from the cDNA derived sequence which gives a calculated molecular weight of 29,024. Cyanogen bromide cleavage provided five polypeptides, one of which resulted from incomplete cleavage. As summarized in Table II, the two lighter peptides were weighed with unit resolution by FAB, and the three heavier products were again weighed by laser desorption. The laser desorption mass spectrum shown in Figure 1 contains molecular ions for CNBr I [1-117] and incompletely cut peptide [1-143]. Table II shows that the polypeptide containing the carboxyl terminus (CNBr IV) exceeds that predicted by the gene sequence by

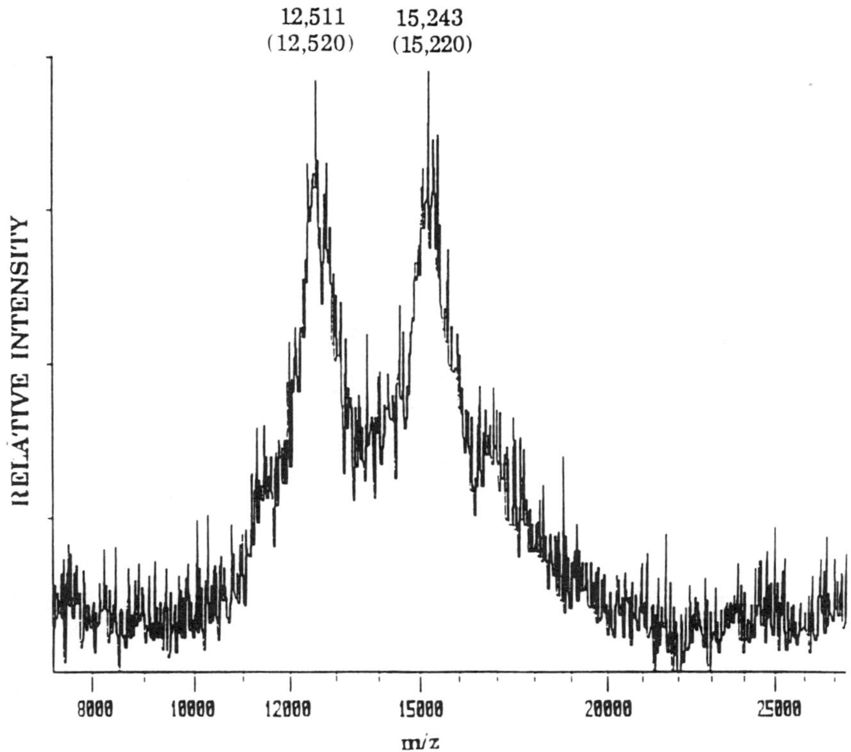

Figure 1. The matrix assisted UV laser desorption mass spectrum of CNBr I and CNBr (I and II).

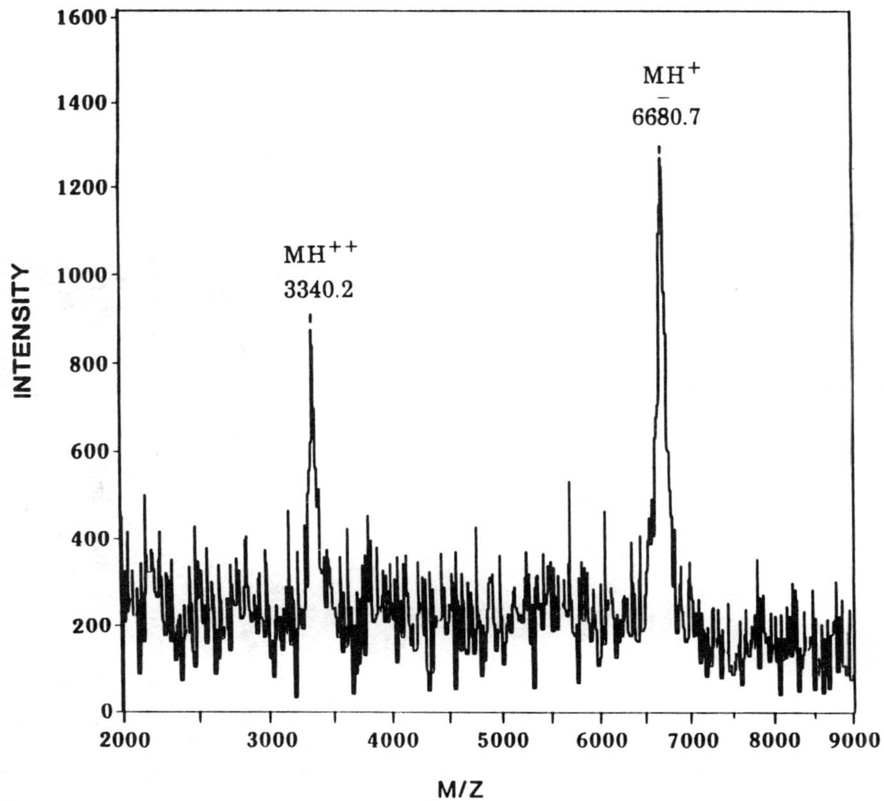

Figure 2. The plasma desorption mass spectrum of the large chymotryptic peptide from CNBr IV.

Table II

Products of Cyanogen Bromide Cleavage of *S. aureus* V8 Protease:

	calc.** MH^+	observed m/z	technique
[1-117]*	12,520 (MW)	12,511	LDMS
[1-143]*	15,220 (MW)	15,243	LDMS
[118-143]*	2,701.32 (IW)	2,701.4	FABMS
[144-158]*	1,692.91 (IW)	1,693.4	FABMS
[159-end]	12,020 (MW)	13,154	LDMS

* Homoserine lactone
**based on the cDNA derived sequence
IW = monoisotopic mass, MW = average mass (ref. 10)

about 1000 daltons. The molecular weight of the protein reconstructed from the molecular weights of the four cyanogen bromide products, corrected for the elimination of water and the conversion of homoserine lactone to methionine, is 30,150, consistent with the molecular weight observed for the total protein.

A chymotryptic digest of polypeptide CNBr IV provided a peptide of mass 6680, determined by plasma desorption mass spectrometry. A plasma desorption spectrum is shown in Figure 2. Based on measurements of several preparations made through three months, the uncertainty is estimated to be ±1 mass unit. This peptide was assigned as [216-end], and was isolated by reverse phase high performance liquid chromatography in quantity sufficient for further analysis by both Edman degradation and tandem mass spectrometry. Cleavage of CNBr IV with trypsin provided a large peptide [208-end] that was found by plasma desorption to weigh 7655. This is in agreement with the chymotrypsin experiment and confirms that the mass of [216-end] is 6680 ± 1. The mass of CNBr IV obtained by adding the mass of [216-end] to that of the known sequence [159-215] is 13,005 which agrees to within 0.15 % with the mass observed (Table II) for the intact CNBr IV.

The carboxy terminal peptide [216-end] was subjected to a number of different kinds of protease and chemical digestions. Amino acid analyses determined Asx (38.0), Pro (19.3), Ala (1.03), Glx (0.67), His (0.97), Gly (0.46), and Ser (0.30). Digestion with immobilized carboxypeptidase Y with analysis by PDMS failed to identify any shortened peptide products. Digestion with 70% formic acid, with the objective of cleaving DP bonds, did not produce new peptides. Digestion was achieved with endoproteinase Asp-N. Following fractionation by HPLC, the digest was mapped by fast atom bombardment. The protonated molecular ions detected are summarized and assigned in Table

III. The specificity of this protease for aspartic acid appears to be high; all peaks were assigned to peptides with N-terminal aspartic acid. One of these peptides, MH^+ = 670, appeared by

Table III

FAB Map of [216-end] Endoproteinase Asp-N Digestion Products

MH^+	peptide	location*
670	DNPNNP	several places
693	DHPNNP	[219-224]
808	DDHPNNP	[218-224]
996	DNPDNPNNP	[225-233]
1020	DHPNNPDNP	[219-227]
1124	DQPNNPNNPD	[267-276]
1134	DDHPNNPDNP	[218-227]
1320	ANDDHPNNPDNP	[216-227]
1321	$(DNP)(NNP)_3$	[252-263]
1322	$(DNP)_2(NNP)_2$	[228-239]
1648	$(DNP)_3(NNP)_2$	[228-239]
		[234-248]
1972	$(XNP)_6$	
1973	$(DNP)_3(NNP)_3$	[246-263]
2300	$(DNP)_4(NNP)_3$	[225-245]
		[228-248]
2625	$(XNP)_8$	
2951	$(XNP)_9$	

*See Figure 3.

HPLC to be produced in significantly higher amounts than the others, reflecting the repeating pattern DNPNNP. Peptides were detected comprising $(XNP)_n$, where n = 2 through 9, with fast atom bombardment. Both N- and C-termini were detected in peptides incorporating non-repeating sequences with distinctive masses. Because the overlaps could not be recognized across the repeating section, it was difficult to reconstruct the extent of the repeating sequence from these small peptides.

Edman degradation carried out using a Milligen/Bioresearch 6600 Prosequenator, in which the protein under study is covalently attached at its carboxyl end, provided a sequence of 48 residues, 216-263, which is shown in Figure 3, and indicated an irregular repeating pattern of NNP and DNP, commencing with position 222, with evidence for both D and N at positions 246, 249, 255, and 258.

This information allowed the assignment of D-terminal peptides in the endoproteinase Asp-N map, as shown in Table III.

```
         10        20        30        40        50
         |         |         |         |         |
VILPNNDRHQITDTTNGHYAPVTYIQVEAPTGTFIASGVVVGKDTLLTNK

         60        70        80        90       100
         |         |         |         |         |
HVVDATHGDPHALKAFPSAINQDNYPNGGFTAEQITKYSGEGDLAIVKFS

        110       120       130       140       150
         |         |         |         |         |
PNEQNKHIGEVVKPATMSNNAETQVNQNITVTGYPGDKPVATMWESKGKI

        160       170       180       190       200
         |         |         |         |         |
TYLKGEAMQYDLSTTGGNSGSPVFNEKNEVIGIHWGGVPNQFDGAVFINE

        210       220       230       240       250
         |         |         |         |         |
NVRNFLKQNIEDNNFANDDHPNNPDNPDNPNNPDNPNNPDNPNNPDNPDN

        260                  270
         |                    |
PDNPNNPNNPNNPD[229]DQPNNPNNPD
```

Figure 3. The current working sequence for *S. aureus* V8 protease.

Peptides with MH^+ = 670, 808, 996, 1020, 1124, and 1134 were also characterized by tandem mass spectrometry with collisional activation. This allowed confirmation of the chemically determined sequence and of the assignment of peaks in Table III. A mass spectrum of DNPDNPNNP is shown in Figure 4, obtained by collisional activation on our four sector tandem mass spectrometer. The sequence shown across the top is based on b series ions in the spectrum. Some other peaks are identified as well, following the nomenclature of Biemann (8).

The method of Rose (9) was used to locate a carboxyl terminal peptide in the product mixture produced by incubation with endoproteinase Asp-N. When the protease reaction was carried out in an approximately 1:1 mixture of $H_2{}^{18}O$ and $H_2{}^{16}O$,carboxyl termini produced by proteolysis were formed with the incorporation of mixtures of ^{18}O and ^{16}O. Their molecular ions appear as doublets two mass units apart, apparent in the partial map shown in Figure 5. Peptides containing the original carboxyl terminus contain only ^{16}O, and their protonated molecular ions are recorded unchanged. One such protonated molecular ion is observed in Figure 5, that of mass 1124.5.

The peptide with MH^+ = 1124.5 was sequenced from this mixture by tandem mass spectrometry and was purified by HPLC for sequencing by Edman degradation. Its sequence is shown in Figure 3, residues [267-276].

The sum of the masses of peptide [216-263] and the carboxy terminal peptide [267-276] equals 6336 daltons, short by 344 mass units of the weight (6680 daltons) of the chymotryptic peptide from which they are derived. The first residue in the 344 piece should be D based on the specificity of endoproteinase Asp-N, leaving 229 mass units to be identified. This section of the C-terminal region is still under study.

Extended repeating motifs such as that studied here are usually characterized by reading the cDNA sequence. In this report, a sequence of 14 repeating tripeptides is defined from the protein itself using a strategy of mass balance that combines mass spectrometry, selective proteolysis, HPLC, and Edman degradation.

Acknowledgements

This work was supported by grants from the National Science Foundation (DIR 8913656, DIR 8914549) and the National Institutes of Health (GM 21248).

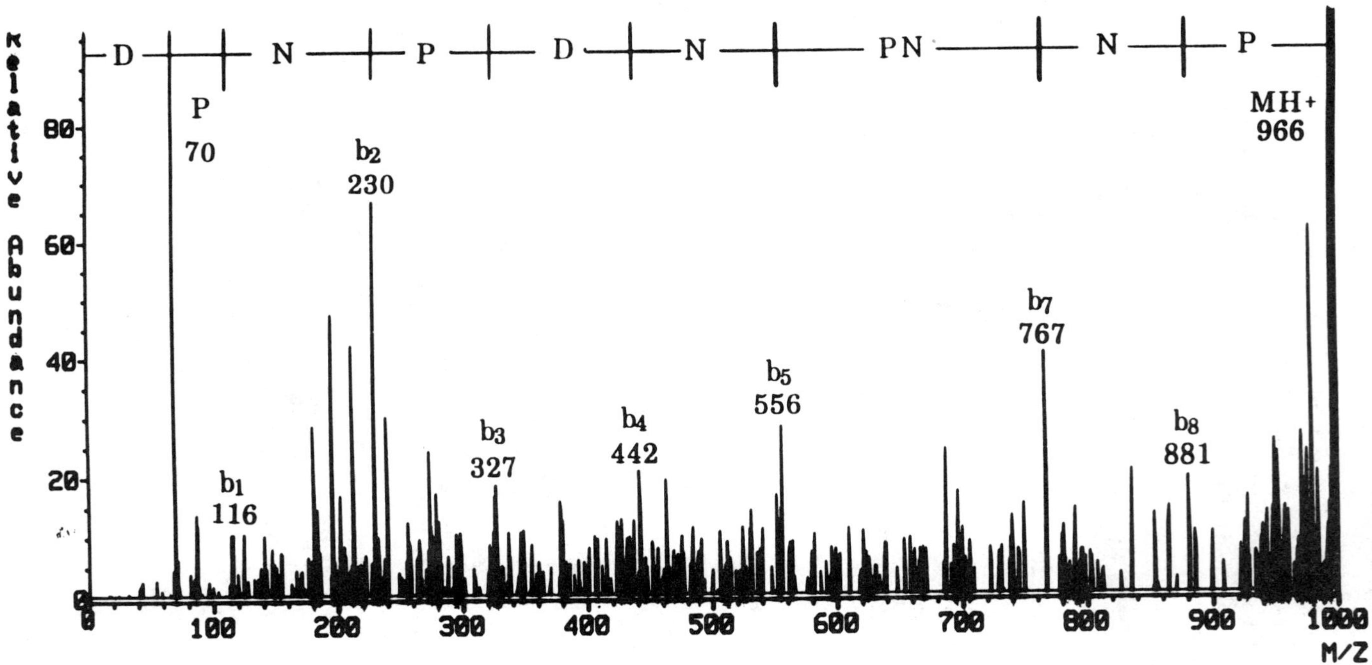

Figure 4. The tandem mass spectrum of DNPDNPNNP.

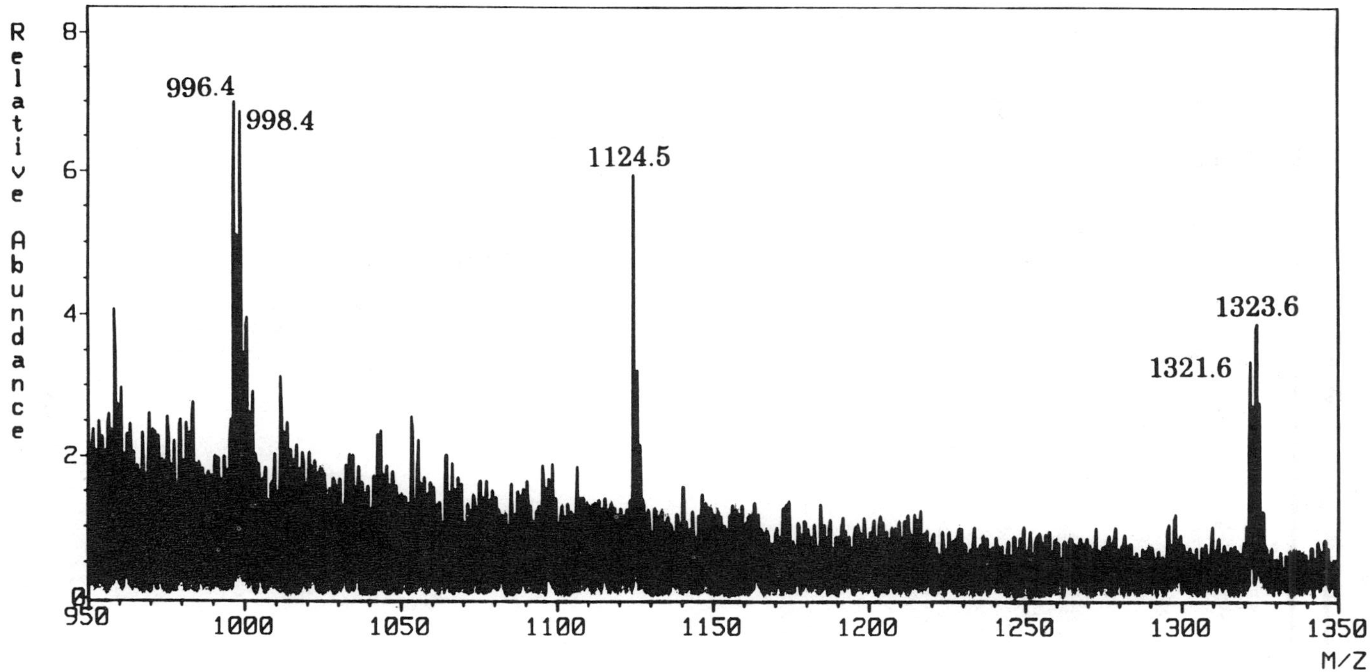

Figure 5. Part of the fast atom bombardment mass spectrum of the peptides from endoproteinase Asp-N digestion in $H_2{}^{18}O/H_2{}^{16}O$.

REFERENCES

1. Drapeau, G. R. (1978) *Can. J. Biochem.* **56**, 534-544.
2. Drapeau, G. R. (1978) *J. Biol. Chem.* **253**, 5899-5901.
3. Carmona, C. and Gray, G. L. (1987) *Nucleic Acids Res.* **15,** 6757.
4. Vestling, M., Hua, S., Murphy, C., Smith, P., and Fenselau, C. (1990) *J. Prot. Chem.* **9**, 320-321.
5. Matsushima, N., Creutz, C. E., and Kretsinger, R. H. (1990), *Proteins: Structure, Function and Genetics* **7**, 125-155.
6. Spengler, B. and Cotter, R. J. (1990) *Anal. Chem.* **62**, 793-796.
7. Heinrikson, R. L. and Meredith, S. C. (1984) *Anal. Biochem.* **136**, 65-74.
8. Biemann, K. (1988) *Biomed. Environ. Mass Spectrom.* **16**, 99-101.
9. Rose, K., Savoy, L. A., Simona, M. G., Offord, R. E., and Wingfield, P. (1988) *Biochem. J.* **250,** 253-259.
10. Yergey, J., Heller, D., Hansen, G., Cotter, R. J., and Fenselau, C. (1983) *Anal. Chem.* **55,** 353-356.

10

Studies On Carboxyl Terminal Degradations

David H. Hawke
Victoria L. Boyd

Research and Development, Protein Products
Applied Biosystems Inc.
850 Lincoln Centre Drive
Foster City, CA 94404

I. INTRODUCTION

Interest in carboxy terminal sequencing has been increasing in the last few years. A variety of chemical approaches to this problem have been studied since the Schlack and Kumpf report in 1926 (1), but the isothiocyanate degradation, detailed in Figure 1, has continued to interest generations of protein chemists.

Recent work in this area continues to focus on the isothiocyanate chemistry and has resulted in a number of interesting reports. Coupling is currently being performed with HNCS/Ac_2O (2) or TMSITC/Ac_2O (3), although a new reagent, a phosphoryl mercapto-benzothiazole derivative (4), has been described. In this report, we briefly describe some of our preliminary findings directed toward carboxyl terminal sequencing: the synthesis of the thiohydantoins, and the development of a solid phase system for studying chemistry (SPDS, solid phase development system). This work has facilitated new chemical studies, including the discovery of another new coupling reagent: benzoyl isothiocyanate (BITC). Its potential utility is supported by the C-terminal sequence determination of a short peptide, leucine enkephalin.

Techniques in Protein Chemistry II

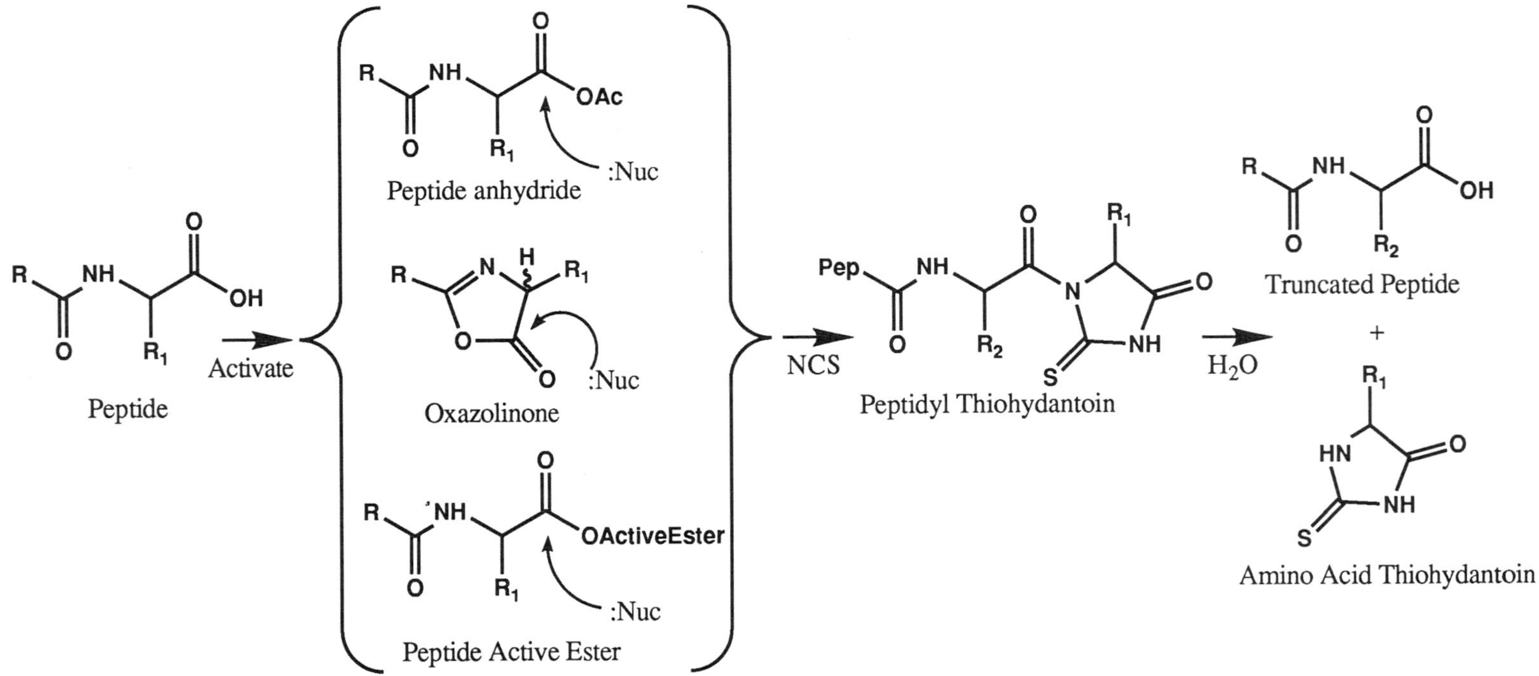

Figure 1. "Classical" mechanism for the isothiocyanate chemistry. Activation is classically performed with anhydrides, such as acetic anhydride. NCS reagents used to complete the coupling reaction include thiocyanate salts, HNCS, and TMSNCS. The intermediate peptidyl isothiocyanate is generally assumed. Other coupling reagents (see text) may react similarly. Cleavage requires the net addition of water across the peptidyl-thiohydantoin bond, releasing the TH for analysis, and revealing the next-in carboxyl group for the next round of degradation.

II. METHODS

General. Solvents were either ABI Sequencer or HPLC grade, other chemicals were reagent grade from Aldrich. Protected amino acid derivatives were from ABI or Bachem Biosciences. Dipeptides were from Bachem Biosciences or Sigma, larger peptides were from Sigma.

HPLC. Chromatography was performed on various columns, 2.1mm by 22 or 25 cm, on a Model 120A (ABI) or equivalent. Injections were made manually, by online injection, or a by a Model 231 Autoinjector (Gilson). Solvents were A=0.1% TFA in H2O, B=0.1% TFA in 70/30 MeCN/H2O. Typical gradients were: 5 min 0% B, linear 0-50%B in 30 min, 50-90% B in 10 min, 20 min at 90%, at RT and 200 µl/min. Columns were equilibrated for 11-15 min.

TH Synthesis, general procedure (5). The Fmoc-amino acids were dissolved in acetonitrile and an equivalent of solid WRK and DIEA were then added. After 2-4 hours at room temperature the solution became homogeneous and the amino acid-WRK activated ester intermediate was the sole product as determined by HPLC. The semi-stable intermediate was concentrated under reduced pressure, dissolved in dichloromethane (DCM), dried (MgSO4), and filtered. A slight excess of trimethylsilyl isothiocyanate (TMSITC) was added, and the reaction mixture stirred overnight at ambient temperature. A 50-80% yield of the purified Fmoc-protected thiohydantoin was obtained after column chromatography (silica gel, 9:1, DCM:MeOH). Deprotection of the Fmoc-group with 20% piperidine in acetonitrile, followed by column chromatography, and a final crystallization typically resulted in 15% overall yields of the amino acid thiohydantoins. Other protecting groups if present (tBu on Ser, Thr, Asp, Glu; e-Boc on Lys; Mtr on Arg) were removed last with trifluoroacetic acid (TFA, *extremely corrosive*)

SPDS. Leucine enkephalin was covalently attached to isothiocyanato-polyamide resin in an aqueous solution of acetonitrile containing 5-10 % pyridine. The reaction was carried out at 45 C for 4-18 hours in just enough solvent to wet and suspend the resin. Resin loadings of 200-400 nmol/mg were typical.

The peptide GGFL was rapidly released by treatment of the peptidyl resin (0.1-2 mg) with TFA at 60 C for 10 minutes. Products were extracted immediately with aqueous acetonitrile. The HPLC trace of this experiment is shown in figure 2A.

Sequencing. Automated carboxy terminal sequencing experiments were performed using a prototype 477A (ABI) Protein Sequencer. Standard interfacing to other system components (A/D converter and online LC) was employed.

PDMS. Molecular ions were determined using a BioIon™20 Biopolymer Mass Analyser (ABI). Samples were applied to nitrocellulose-coated targets (as supplied, ABI) and dried under a stream of nitrogen at ambient temperature. Data was collected for 1,000,000 counts (ca. 10-15 min) with 14.5 KV acceleration potential in the positive ion mode. Data was processed in the usual way on a PDP-11 computer using the program DAS. Peptide molecular weights were calculated using the program MacProMass (T. Lee, Duarte CA).

III. RESULTS AND DISCUSSION

A. *Thiohydantoin Synthesis*

The synthesis of the thiohydantoins is described in more detail elsewhere (5). Our basic strategy was based on two concepts:

1. The carboxyl group is first activated as shown in figure 1, and subsequently reacts with the thiocyanate reagent to give the thiohydantoin; and,

2. Reactive groups ($\alpha{-}NH_2$, side-chains) should be protected during the reaction, and then deprotected when desired. This approach has enabled us to prepare all of the amino acid thiohydantoins.

B. *Solid Phase Development System (SPDS)*

The advantage of attaching peptides to solid supports, as exemplified by the popularity of solid-phase peptide synthesis, is that solution phase reactants, impurities, and

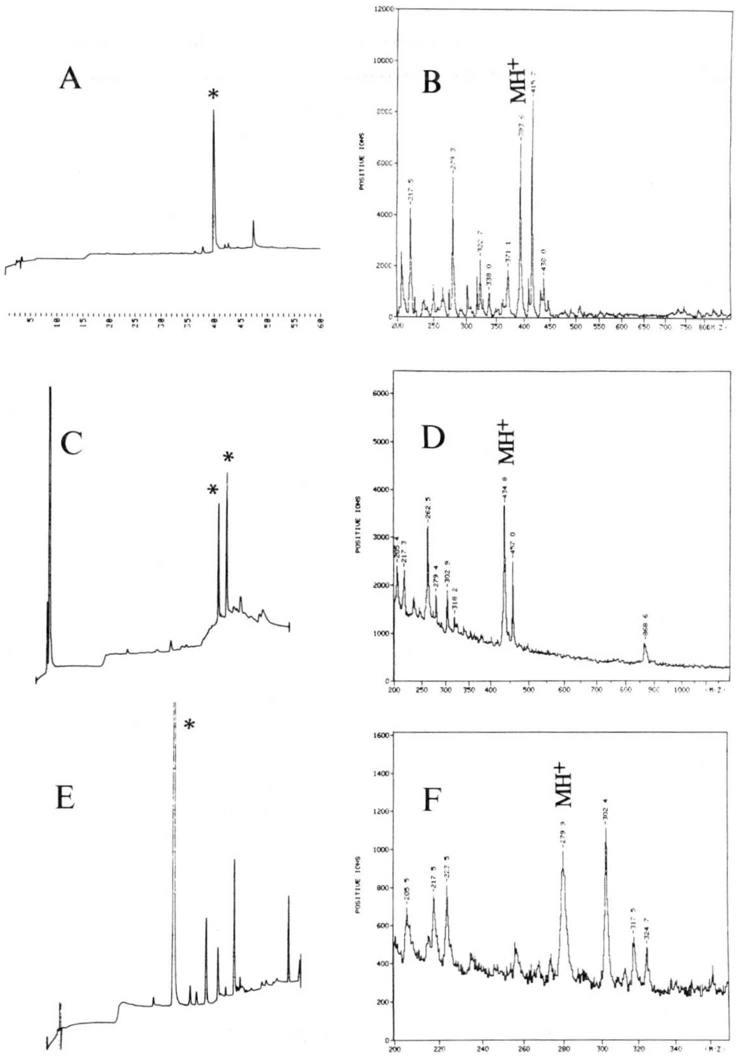

Figure 2. Cleavable system results for Leucine Enkephalin attached to isothiocyanato-resin. HPLC (with monitoring at 214 nm) and the corresponding PDMS data are shown after attachment (see text) and:

A. Simple detachment with TFA. The PDMS in **B** shows the expected MH^+ (393.6 amu, calculated 393.5 for GGFL) and $+Na^+$ ions.

C. One round of coupling with BITC, then detached with TFA. The PDMS, **D**, is consistent with the peptidyl thiohydantoin GGFL-TH (MH^+ at 434.8 amu, calculated 434.3).

E. One round of coupling, then TH cleavage (TEA/TFE/water), followed by detachment. The PDMS, **F**, shows the expected MH^+ for the peptide GGF (279.9 amu, calculated 280.3).

byproducts may be conveniently separated from the immobilized species by washing. If the peptide could then be released from the resin under mild conditions, the usual peptide analytical tools should detect any modifications. We therefore decided to develop such a tool to aid in our C-terminal chemical studies. Since the coupling conditions we have recently discovered are basic, and cleavage is also rapid in base (2), an acid cleavable link would be ideal.

Isothiocyanate resin was investigated, and as expected, leucine enkephalin was readily attached, and subsequently released (*sans* the amino terminal residue) upon acid treatment. The released peptide was identified as the expected GGFL by PDMS (figure 2a and b), amino terminal sequencing and AAA (not shown). The expected intermediate peptidyl thiohydantoin could also be isolated. Peptidyl resin was reacted with BITC, washed, and acid cleaved. The HPLC of the cleavage product, shown in figure 2c, shows very little residual starting material (<5%). The peptidyl thiohydantoin appears as two peaks, as recently reported by others (6). Finally, resin bearing the peptidyl thiohydantoin was cleaved with aqueous triethylamine, washed, and peptides released with acid for further analysis. The major product is the des-Y des-L peptide, indicated by the LC, figure 2e, and MS, figure 2f.

C. Automated C-Terminal Sequencing

Leucine enkephalin was attached to polyamide resin using a non-cleavable urea linkage. A prototype 477A was modified to accomodate the reagents for C-terminal sequencing experiments. We have studied a number of cleavage conditions, as will be summarized in a later report, but found that piperidine (2) in TFE/water appears to be a fairly good cleavage reagent. Figure 3 shows the C-terminal sequence analysis of Leucine enkephalin for four cycles. Good efficiency is indicated by the low stepwise lag exhibited by both the leucine and phenylalanine signals in cycles one and two.

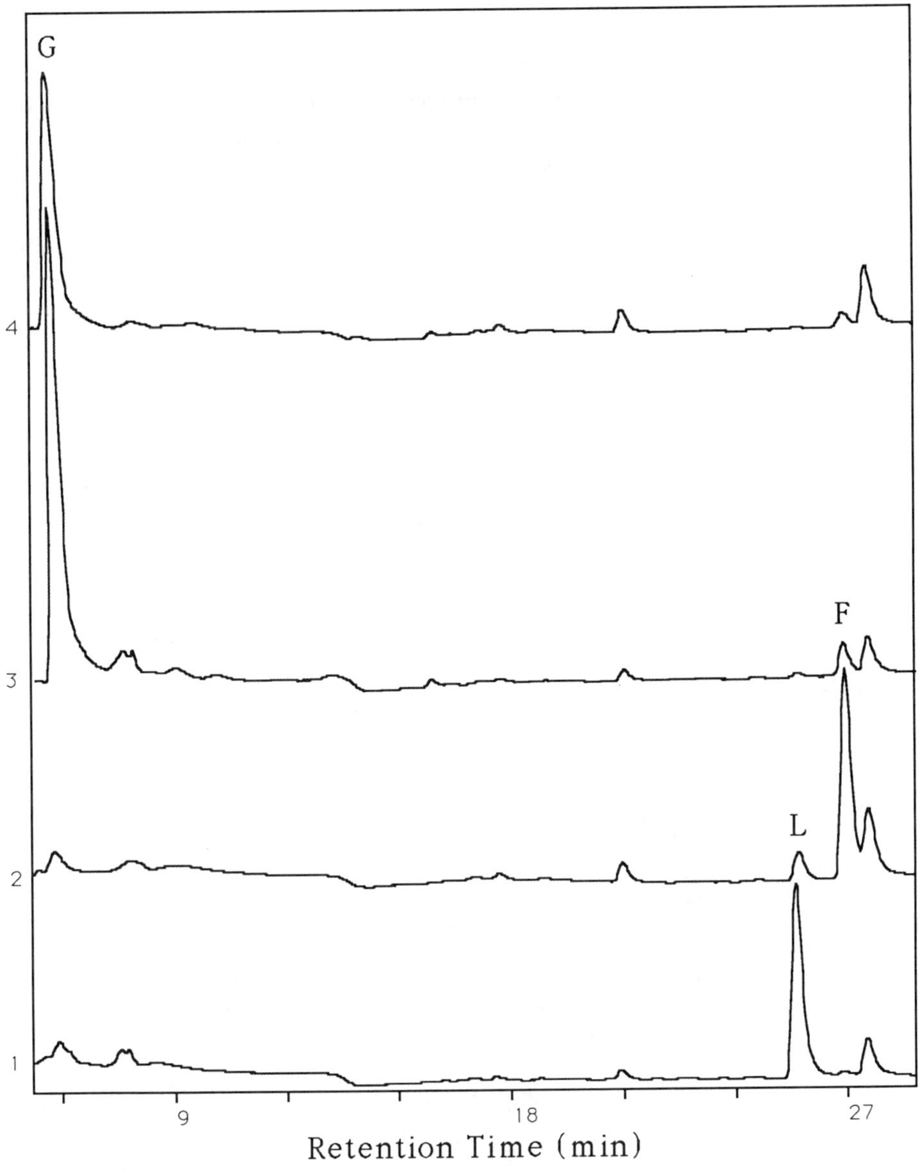

Figure 3. C-terminal sequence analysis of Leucine Enkephalin using BITC. Chromatograms of the first four degradation cycles are shown, cycle one is lowest, cycle two just above, etc, as indicated to the right of the trace. The elution positions of the TH's of L, F, and G are indicated. Thiohydantoins were monitored at 269 nm. The sequence is L-F-G-G.

IV. CONCLUSIONS

We have developed a new synthetic route for the thiohydantoins, established a solid phase development system (SPDS) to aid in chemical studies, discovered a new reagent for performing isothiocyanate chemistry, and demonstrated its potential utility by sequencing a short peptide. The stepwise efficiency appears to be quite reasonable, which is encouraging. Further work is required to establish the ultimate utility of BITC for C-terminal sequencing.

ACKNOWLEDGEMENT

The authors would like to thank the ABI R&D staff and particularly T. Geiser, R. Mattaliano, and P-M. Yuan for their assistance and support.

REFERENCES

1. Schlack, P. und Kumpf, W. , Hoppe-Seyler's Z. Physiol. Chemie **154** (1926) 126-170.
2. Inglis , A.S., Wilshire, J.F.K., Casagranda, F. and Laslett, R.L., Proceedings of the Seventh International Conference on Methods in Protein Sequence Analysis (book), ed. B. Wittmann-Liebold, Springer-Verlag (1989), 164-167.
3. Hawke, D.H., Lahm, H.-W., Shively, J.E. and Todd, C.W., Anal. Biochem., **166** (1987) 298-307.
4. Miller, C.G. and Shively, J.E., Abstract T188, The Third Symposium of the Protein Society, 29 Jul - 2 Aug 1989, Seattle, WA.
5. Boyd, V.L., Hawke, D.H. and Geiser, T.G., Tetrahedron Letters, **27** (1990) 3849-52.
6. Bailey, J.M. and Shively, J.E., Biochemistry **29** (1990) 3145-56.

11

Carboxy-Terminal Sequencing: Automation and Application to the Solid-Phase

Jerome M. Bailey and John E. Shively
Division of Immunology, Beckman Research Institute
of the City of Hope, Duarte, CA 91010

I. Introduction

The development of a reliable chemical method for the sequential degradation of a protein or peptide from the carboxy-terminus is a goal of our laboratory. Although several methods have been proposed (1,2), conversion of the C-terminal amino acid into a thiohydantoin has been the most promising. This reaction, originally observed by Johnson and Nicolet (3), was first applied to the sequential degradation of proteins from the carboxy-terminus by Schlack and Kumpf (4). These authors reacted ammonium thiocyanate, dissolved in acetic acid and acetic anhydride, with N-benzoylated peptides to form carboxy-terminal 1-acyl-2-thiohydantoins. The derivatized amino acid was then cleaved with aqueous sodium hydroxide. The main disadvantages of the procedure have been the severity of the conditions required for both the complete derivatization of the C-terminal amino acid, and for the subsequent cleavage of the peptidylthiohydantoin derivative into a shortened peptide and an amino acid thiohydantoin derivative. Cleavage by aqueous hydroxide is not desirable due to the rapid lability of the released thiohydantoin under these conditions (5). As demonstrated by Johnson and Nicolet (3) the thiohydantoin amino acids are generally stable to strong acid. Application of 12N HCL for the cleavage of peptidylthiohydantoins (6), allowed only 2 to 3 cycles of degradation to be performed, presumably due to the instability of the peptide sample under the strongly acidic conditions. Since the work of Schlack and Kumpf (4), numerous groups have reduced the severity of the conditions required in order to apply

Techniques in Protein Chemistry II

this chemistry to the sequential degradation of proteins from the carboxy-terminus. The most notable of these improvements include the introduction of acetohydroxamate for the rapid and specific cleavage of peptidylthiohydantoins at room temperature and neutral pH (7), and the introduction of trimethylsilyliso-thiocyanate (TMS-ITC) for the derivatization of the C-terminal amino acid to a thiohydantoin (8). The use of TMS-ITC significantly improved the yields of peptidylthiohydantoin formation and reduced the number of complicating side products caused by polymerization of the free isothiocyanate ions. However, even with the use of these new reagents for the sequential degradation of a peptide from the C-terminus, no more than a few cycles could

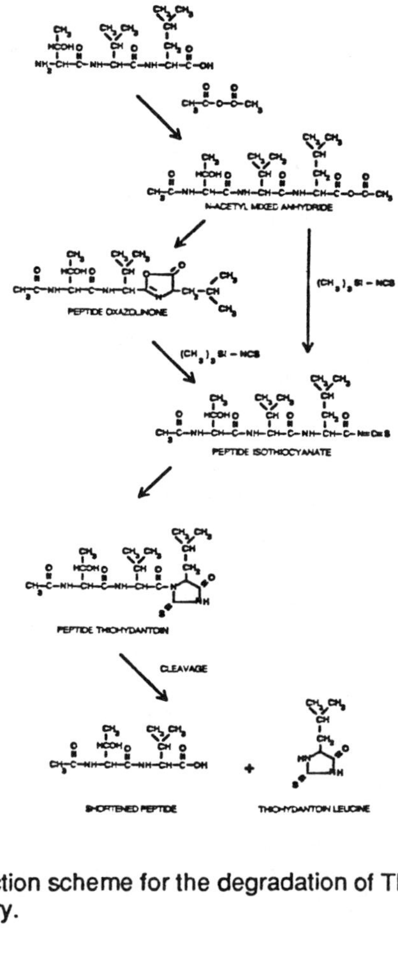

Fig. 1. Reaction scheme for the degradation of Thr-Val-Leu using the thiocyanate chemistry.

be obtained (9). Recent work from our laboratory examined the hydrolysis of peptidylthiohydantoins with acetohydroxamate and found that although the hydrolysis was rapid and specific, most of the shortened peptide was blocked to further sequencing by the formation of a stable acetohydroxamate ester at the C-terminus (10). Hydrolysis with an aqueous solution of triethylamine, as originally proposed by Kubo et al. (11), was found to be more suitable for a repetitive degradation from the C-terminus.

The need to apply the chemistry for C-terminal sequencing to the solid phase was recognized by a number of groups (12-15). Solution phase methods required laborious gel filtrations for purification of the peptidylthiohydantoin and evaporations to remove reagents (7). By covalently coupling a polypeptide to a solid support and thereby providing the ability to simply rinse away excess solvents without loss of the sample, the time required for a single cycle of degradation could be reduced to hours rather than the several days required by solution methods.

This report describes the application of the thiocyanate chemistry to the automated C-terminal degradation of peptides covalently coupled to PVDF (polyvinylidene difluoride). An alternative method for the selective hydrolysis of peptidylthiohydantoins covalently coupled to PVDF is also presented.

II. Experimental

The synthesis and characterization of the amino acid thiohydantoins was performed as described (10). Reverse phase HPLC separation of the eighteen thiohydantoin amino acid derivatives (180 pmoL) is shown in Fig. 2. This separation was performed on a Phenomenex Ultracarb 5 ODS(30) column (2.0 mm x 25 mm) on a Beckman 126 Pump Module with a Shimadzu (SPD-6A) detector. The column was eluted for 2 min with solvent A (0.1% trifluoroacetic acid in water) and then followed by a discontinuous gradient to solvent B (80% acetonitrile, 10% methanol in water) at a flow rate of 0.15 mL/min at 22°C. The gradient used was as follows: 0% B for 2 min., 0-6% B over 3 min., 6-35% B over 35 min., and 35-50% B over 10 min.

The instrument used for automated C-terminal sequencing was constructed by the Biomedical Instrumentation Services of the City of Hope. This instrument is an N-terminal sequencer based on the design described by Hawke et al. (16) and includes the addition of a continuous flow reactor (CFR) and an on-line HPLC for detection of the released thiohydantoin amino acids (17).

The peptides YGGFL and KVILF were covalently coupled to carboxylic acid modified PVDF (PVDF-COOH). This support is commercially available from the Pall Corporation (Long Island, NY). A piece of PVDF-COOH (2 mm x 10 mm) was placed in a continuous flow reactor (17). The membrane carboxylic groups were activated by application of 2 µL of a 2 mmoL/mL solution of dicyclohexylcarbodiimide in dimethylformamide. After a two hour reaction, the activated membrane was rinsed with 2 mL of dimethylformamide and then dried under a stream of argon. Peptide (YGGFL or KVILF) (2 µL of a 14 nmoL/µL solution in dimethylformamide) was carefully layered on the dry membrane and allowed to react overnight. The membrane was washed with methanol (5 mL) and dried under a stream of argon. The yield of the coupled peptide was determined by amino acid compositional analysis of the derivatized membrane.

III. Results and Discussion

The reaction sequence used for the degradation of a peptide by the thiocyanate chemistry is outlined in figure 1. The current state of the chemistry outlined above allows the stepwise degradation of peptides containing most of the naturally occurring amino acids (with the exception of proline) by manual manipulation in the solution phase (10). However, these methods are not appropriate for extended C-terminal degradation since the shortened peptide must be isolated at each step. In addition to being time consuming, the sample losses inherent in each separation step will limit the number of cycles obtainable for a given amount of starting material. The initial activation of the C-terminal carboxylate with acetic anhydride causes both racemization of the C-terminal amino acid and partial acetylation of internal amino acid side chains (particularly Thr, Ser, and Tyr) resulting in the formation of multiple peptidylthiohydantions further complicating manual solution phase C-terminal analysis. Elimination of the purification step of the peptidylthiohydantoin is not practical as there are UV absorbing side products formed during the derivatization of the C-terminal amino acid that will subsequently interfere with identification of the C-terminal thiohydantoin amino acid after cleavage of the peptidylthiohydantoin. Application of this chemistry to the solid phase has the potential to solve these problems. The advantages of covalently immobilizing a peptide or protein to a solid support include: elimination of sample washout thereby resulting in high initial and repetitive yields, the ability to use reagents and solvents optimal for derivatization and washing, the ability to efficiently wash the sample to remove reaction by-products resulting from thiohydantoin formation, thereby creating a potential for a low chemical background, and the elimination of the time consuming intermediate purification of the peptidylthiohydantoin potentially resulting in a much shorter time for each degradation cycle.

Previous laboratories have utilized chemically activated glass beads for covalent coupling of peptides and proteins for C-terminal sequence analysis (12,13,14). We have chosen PVDF membranes for our initial experiments. PVDF membranes have been shown to be the preferred support for electroblotting proteins and peptides from SDS gels for subsequent sequence analysis (18,19) and, unlike glass supports, the PVDF membranes are inert to most solvents commonly used in protein sequence analysis, particularly the basic conditions utilized in this laboratory for hydrolysis of the C-terminal peptidylthiohydantoins.

A successful method for C-terminal protein sequencing will involve identification of the released thiohydantoin amino acids. As with identification of the released phenylthiohydantoin amino acids from the sequential N-terminal degradation based on the Edman procedure, HPLC is our method of choice. In general thiohydantoin amino acids are more polar than their corresponding phenylthiohydantoin derivatives. This makes a reverse phase separation problematic for the more polar thiohydantoin amino acid derivatives, such as aspartate, glycine and asparagine. These analogues were weakly retained on a Beckman ODS column (10). The retention of these analogues was somewhat improved on a high carbon load Phenomenex Ultracarb 30 column (figure 2).

An automated instrument capable of C-terminal sequencing has been constructed in our laboratory. This instrument is an N-terminal sequencer based on the design described by Hawke et al. (16) and includes the addition of a continuous flow reactor (CFR) and an on-line HPLC for detection of the released thiohydantoin amino acids (17). The program used for automated sequencing consisted of two 5 min. activation steps with acetic anhydride, three 10 min. derivatization steps with 30% TMS-ITC in acetonitrile (v/v) each followed by a 1 min. wash with acetonitrile, and one 10 min. base hydrolysis step with 2% aqueous triethylamine all performed at 50°C. Initial results with leucine enkephalin (5.6 nmoL) covalently coupled to PVDF-COOH revealed problems in the hydrolysis of the peptidylthiohydantoin by dilute aqueous triethylamine. Although dilute aqueous triethylamine can catalyze cleavage of peptidylthiohydantoins with

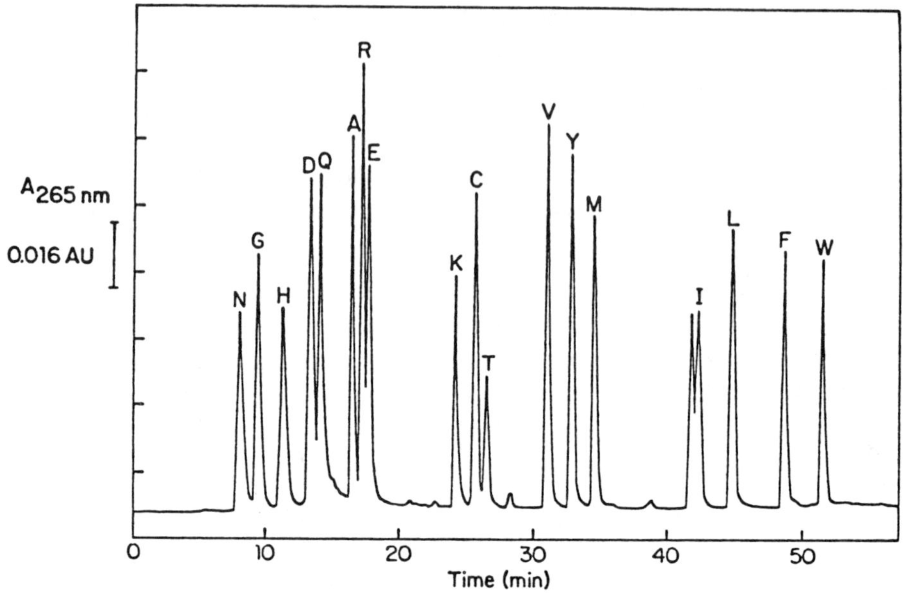

Fig. 2. Separation of the thiohydantoin amino acids (180pmoL).

a $t_{1/2}$ of less than 1 min. at 50°C in the solution phase, in the solid phase no hydrolysis of the peptidylthiohydantoins occurs. This solution was not able to effectively "wet" the PVDF strip. It was found that for cleavage to occur on covalently attached peptidylthiohydantoin to PVDF, a minimum of 30% (v/v) of an organic solvent had to be included in the aqueous triethylamine solution. Of the organic solvents tested in the solution phase (acetonitrile, dimethylformamide, acetone, dimethylsulfoxide, ethanol, methanol, trifluoroethanol) all significantly slowed the rate of cleavage, with methanol having the least effect on the rate of hydrolysis. Table I shows the effect on the rate constants for cleavage of peptidylthiohydantoin by 2% aqueous triethylamine when varying concentrations of methanol and acetonitrile are included in the hydrolysis reaction. Peptidylthiohydantoin synthesized from the model peptide Thr-Val-Leu was used for these studies.

Application of a 2% triethylamine solution in 30% aqueous methanol for hydrolysis of the peptidylthiohydantoin during the automated degradation of leucine enkephalin (650 pmoL) covalently coupled to PVDF-COOH is shown in fig. 3. The program used consisted of two 5 min. activation steps with acetic anhydride, three 10 min. derivatization steps with 30% TMS-ITC in acetonitrile (v/v) each followed by a 1 min. wash with acetonitrile, and six 8 min. base hydrolysis steps with 5% triethylamine in 30% aqueous methanol at 60°C. The cleavage solution after each base hydrolysis step was washed over to a separate flask held at 50°C and dried with a stream of argon. The thiohydantoin amino acid was then taken up in solvent A (0.8% trifluoroacetic acid) and injected directly into the HPLC column. The separation of the amino acid thiohydantoins was performed on a Beckman ODS Ultrasphere column (2.0 mm x 25 mm) as described previously (10).

Table I Hydrolysis of Peptidylthiohydantoins by Dilute aqueous triethylamine in the presence of methanol and acetonitrile.

2% Triethylamine in Water		
Temperature	Rate Constant	$T_{1/2}$
37°C	0.49 min^{-1}	1 min
22°C	0.15 min^{-1}	5 min
2% Triethylamine in Aqueous Methanol at 50°C		
% Methanol	Rate Constant	$T_{1/2}$
30	0.084 min^{-1}	8 min
50	0.028 min^{-1}	25 min
80	0.0069 min^{-1}	100 min
2% Triethylamine in Aqueous Acetonitrile at 50°C		
% Acetonitrile	Rate Constant	$T_{1/2}$
30	0.028 min^{-1}	25 min
50	0.013 min^{-1}	53 min
80	0.0055 min^{-1}	126 min

The first cycle allowed degradation of 90% of the C-terminal leucine coupled to the membrane (as determined by amino acid compositional analysis) with 65% of the expected thiohydantoin Phe in cycle 2. The UV absorbing background prevented identification of thiohydantoin glycine expected in the third degradation cycle.

Although extension of the time for base hydrolysis of the peptidylthiohydantoin for longer than the 50 min. employed here results in increased yields of hydrolysis, it also results in increased instability of the released thiohydantoin, negating any benefit. The alternative was to find another reagent capable of rapid and specific hydrolysis of C-terminal peptidylthiohydantoins in an organic solvent. Stark (7) studied a number of different cleavage reagents and found that oxygen containing nucleophiles were the best choice of reagents to effect this reaction. Although acetohydroxamate, originally proposed as a result of the study by Stark (7), is an excellent cleavage reagent for the first amino acid,

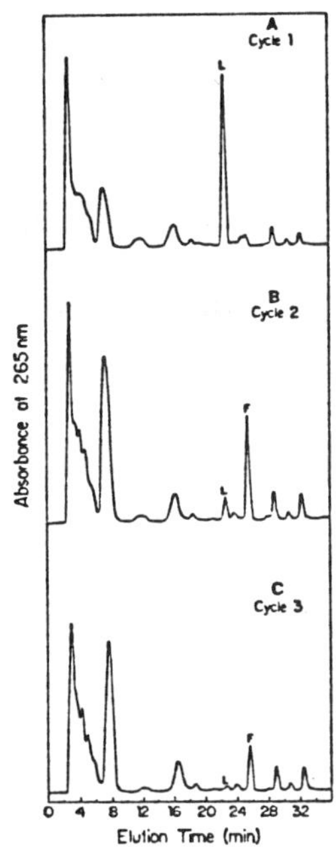

Fig. 3. C-Terminal sequencing of YGGFL covalently coupled to PVDF.

it forms a stable peptidyl hydroxamate ester which is difficult to remove, and which effectively blocks the shortened peptide from further sequencing (10). It seems that, in general, any carbon based reagent that is a good nucleophile, and thus a good cleavage reagent, will also be a poor leaving group thereby blocking much of the shortened peptide from further sequencing. Ideally a cleavage reagent should possess the following characteristics: 1) it should be able to cleave peptidyl-thiohydantoins in a volatile, water miscible organic solvent, thus eliminating the problems of incompatibility of PVDF membranes with water, 2) the reaction should be rapid and specific, 3) the shortened peptide should be capable of continued degradation, 4) the released thiohydantoin amino acid should not be destroyed by this reagent, 5) this reagent should not absorb light in the same range as is used for detection of the released thiohydantoin amino acid derivatives. Sodium trimethylsilanolate (figure 4), commercially available from Petrarch (Huls), in alcoholic solvents seems to possess all of these characteristics.

Cleavage of peptidylthiohydantoins in the solution phase with a 0.05M solution of sodium trimethylsilanolate in 100% methanol or trimethylsilylethanol is complete in less than 5 min. Application of this new cleavage reagent (0.05 M sodium

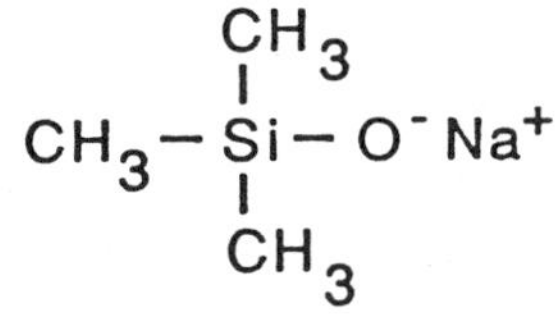

Fig. 4. Sodium Trimethylsilanolate.

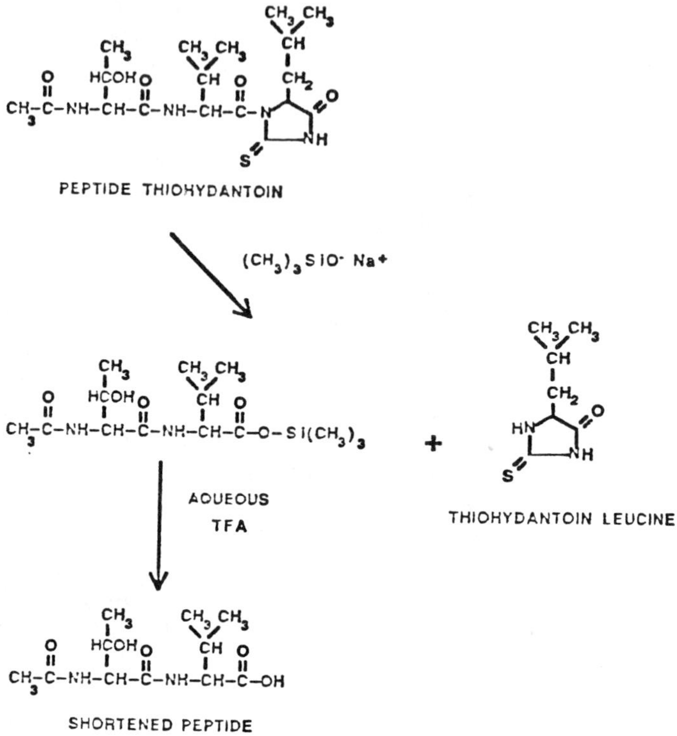

Fig. 5. Proposed mechanism for cleavage by sodium trimethylsilanolate.

trimethylsilanolate in trimethysilylethanol) to the automated sequencing of 1.2 nmoL of leucine enkephalin (YGGFL) covalently coupled to PVDF-COOH (Pall Corp.) is shown in figure 6. Four cycles of degradation were possible with a cycle time of 95 min. The fifth residue, tyrosine (Y), was not cleaved. This is the residue covalently attached to the PVDF. The details of the program used are given in Tables II and III. Application of this chemistry to another peptide (KVILF, 1.2 nmoL) covalently coupled to PVDF-COOH is shown in figure 7. Like YGGFL, four cycles of degradation were possible and the last residue, in this case lysine (K), which is covalently attached to the PVDF was not degraded.

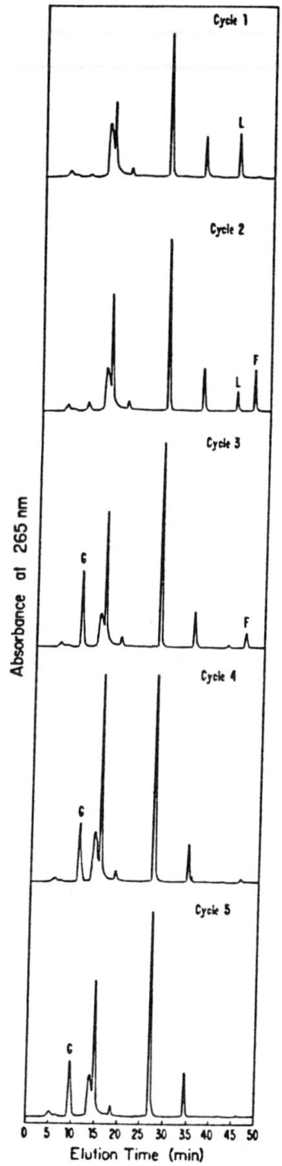

Fig. 6. C-Terminal sequencing of YGGFL (1.2 nmol) covalently attached to PVDF-COOH.

Table II Composition of Reagents and Solvents

R1	Acetic anhydride
R2	30% TMS-ITC in acetonitrile
R3	0.05 M Sodium trimethylsilanolate in 2-trimethylsilylethanol
R4	Methanol
S1	Acetonitrile
S2	0.8% Trifluoroacetic acid in water
S3	Methanol
S4	------

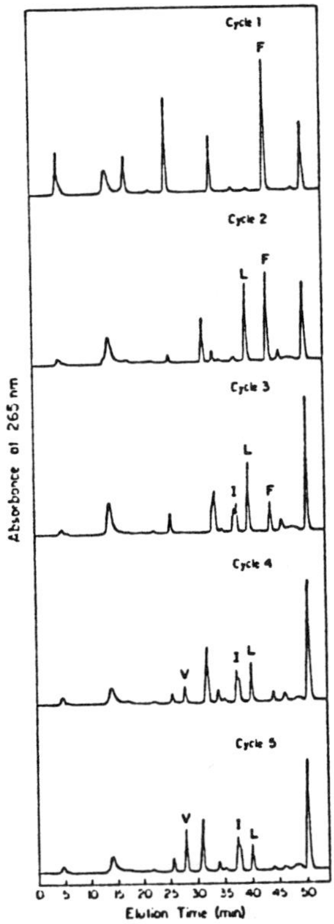

Fig. 7. C-Terminal sequencing of KVILF (1.2 nmol) covalently attached to PVDF-COOH.

Table III C-Terminal Sequencer Program Summary

Continuous Flow Reactor (CFR) (70°C)	Conversion Flask (CF) (50°C)	Duration (min)	Total Time (min)
Dry	Dry	3.0	3.0
R1 reaction		5.0	8.0
R1 reaction		5.0	13.0
S1 rinse		0.5	13.5
R2 reaction		10.0	23.5
S1 rinse		1.0	24.5
R2 reaction		10.0	34.5
S1 rinse		1.0	35.5
R2 reaction		10.0	45.5
S3 rinse		1.0	46.5
S1 rinse		1.0	47.5
S1 rinse		1.0	48.5
S1 rinse		1.0	49.5
R3 reaction		10.0	59.5
R3 to CF		0.5	60.0
	Dry	10.0	70.0
R4 delivery		0.2	70.2
	Dry	8.3	78.5
R4 delivery		0.2	78.7
	Dry	8.3	87.0
S2 delivery		0.1	87.1
S2 reaction		0.5	87.6
S2 to CF		0.5	88.1
S2 delivery		0.1	88.2
S2 reaction		0.5	88.7
S2 to CF		0.5	89.2
	inject	1.0	90.2
S3 rinse		1.0	91.2
S1 rinse		1.0	92.2
S1 rinse		1.0	93.2
S1 rinse		1.0	94.2
	R4 rinse	1.0	95.2

A limitation of the thiocyanate method is the lack of reactivity of certain amino acids, in particular aspartate and proline, to form thiohydantoins when they are at the carboxy-terminus of polypeptides. Peptides with C-terminal aspartate and proline were found not to be derivatized by TMS-ITC (10). All other amino acids examined could be quantitatively derivatized. Hydrolysis of all peptidylthiohydantoins by aqueous triethylamine was rapid and specific with the exception of asparagine. Under the basic conditions of cleavage the side chain amide group partially cyclized to form a C-terminal cyclic amide preventing further degradation.

Current work is directed toward optimization of hydrolysis by sodium trimethylsilanolate and toward the sequencing of longer peptides and proteins. Work is also in progress toward the design of a reagent capable of derivatizing the problematic amino acids, aspartate and proline.

Acknowledgments

We thank the Tosoh Corporation for funding this research, Mike Ronk for assistance with the instrumentation used for automation of the C-terminal chemistry, Dr. Harvey Brandwein of the Pall Corporation for providing us with the PVDF membranes used in this study, Dr. Gabor Dibo for covalently coupling the peptides used in this study to PVDF-COOH, and Jim Sligar for performing the amino acid compositions.

References

1. Ward, C.W. (1986) Practical Protein Chemistry -- A Handbook (Darbre, A., Ed.) pp.491- 525, John Wiley and Sons, Ltd.

2. Rangarajan, M. (1988) Protein/Peptide sequence Analysis: Current Methodologies (Brown, A.S., Ed.) pp. 135-144, CRC Press, Bora Raton, FA.

3. Johnson, T.B., and Nicolet, B.H. (1911) J. Am. Chem. Soc. **33**, 1973-1978.

4. Schlack, P., and Kumpf, W. (1926) Z. Physiol. Chem. **154,** 125-170.

5. Scoffone, E., and Turco, A. (1956) Ric. Sci. **26**, 865-871.

6. Cromwell, L.D., and Stark, G.R. (1969) Biochem. **8**, 4735-4740.

7. Stark, G.R. (1968) Biochemistry **7,** 1796-1807.

8. Hawke, D.H., Lahm, H.-W., Shively, J.E., and Todd, C.W. (1987) Anal. Biochem. **166,** 298-307.

9. Miller, C.G., Kong, C.-T., and Shively, J.E. (1989) Techniques in Protein Chemistry (Hugli, T. E., Ed.) pp. 67-78, Academic Press, Inc.

10. Bailey, J.M., and Shively, J.E. (1990) Biochemistry. **29,** 3145-3156.

11. Kubo, H., Nakajima, T., and Tamura, Z. (1971) Chem. Pharm. Bull. **19**, 210-211.

12. Williams, M. J., and Kassell, B. (1975) FEBS Lett. **54**, 353-357.

13. Meuth, J.L., Harris, D.E., Dwulet, F.E., Crowl-Powers, M.L., and Gurd, F.R.N. (1982) Biochem. **21**, 3750-3757.

14. Rangarajan, M., and Dardre, A. (1976) Biochem. J. **157**, 307-316.

15. Inglis, A.S., Wilshire, J.F.K., Casagranda, F., and Laslett, R.L. (1989) Methods in Protein Sequence Analysis (Wittmann-Liebold, B., Ed.) pp.137-144, Springer-Verlag.

16. Hawke, D.H., and Harris, D.C. and Shively, J.E. (1985) Anal. Biochem. **147**, 315-330.

17. Shively, J.E., Miller, P., and Ronk, M. (1987) Anal. Biochem. **163**, 517-529.

18. Matsudaira, P. (1987) J. Biol. Chem. **262**, 10035-10038.

19. Tous, G.I., Fausnaugh, J.L., Akinyosoye, O., Lackland, H., Winter-Cash, P., Vitorica, F.J., and Stein, S. (1989) Anal. Biochem. **179**, 50-55.

12

EVALUATION OF COMPACT PROTEIN SEQUENCER[1]

Jimmy Calaycay[2], Miro Rusnak[3], and John E. Shively[4].
Beckman Research Institute of the City of Hope, Duarte, CA 91010

I. INTRODUCTION

Protein structural analysis is commonly performed by automated Edman chemistry using the gas phase approach developed by Hewick et. al. in 1981 (1). In spite of the obvious simplicity of the chemical method, the commercial instruments are large, complicated, and expensive. These limitations have prevented many protein chemistry laboratories from obtaining their own protein sequencers, relegating such instruments to large core facilities. In view of this problem, we decided to build a compact, economical protein sequencer with reduced reagent and solvent consumption, and performance equivalent to existing instruments.

II. MATERIALS AND METHODS

A schematic of the instrument is shown in **Figure 1**. The fluid flow path is minimized by the use of a hexagonal, zero-dead volume valve connected directly to a continuous flow reactor (CFR) containing the sample immobilized on a polyvinylidinedifluoride (PVDF) strip (1 x 10 mm). The hexagonal valve allows six fluid inputs to a common channel which can be emptied by delivery of an inert gas. Output from the CFR flows into a transfer valve which leads to the conversion flask

---

[1] This research was partially supported by a grant from Tosoh corp.

[2] Current address, Merck, Sharp, and Dohme, Rahway , NJ.

[3] Department of Biomedical Instrumentation.

[4] Division of Immunology.

Techniques in Protein Chemistry II

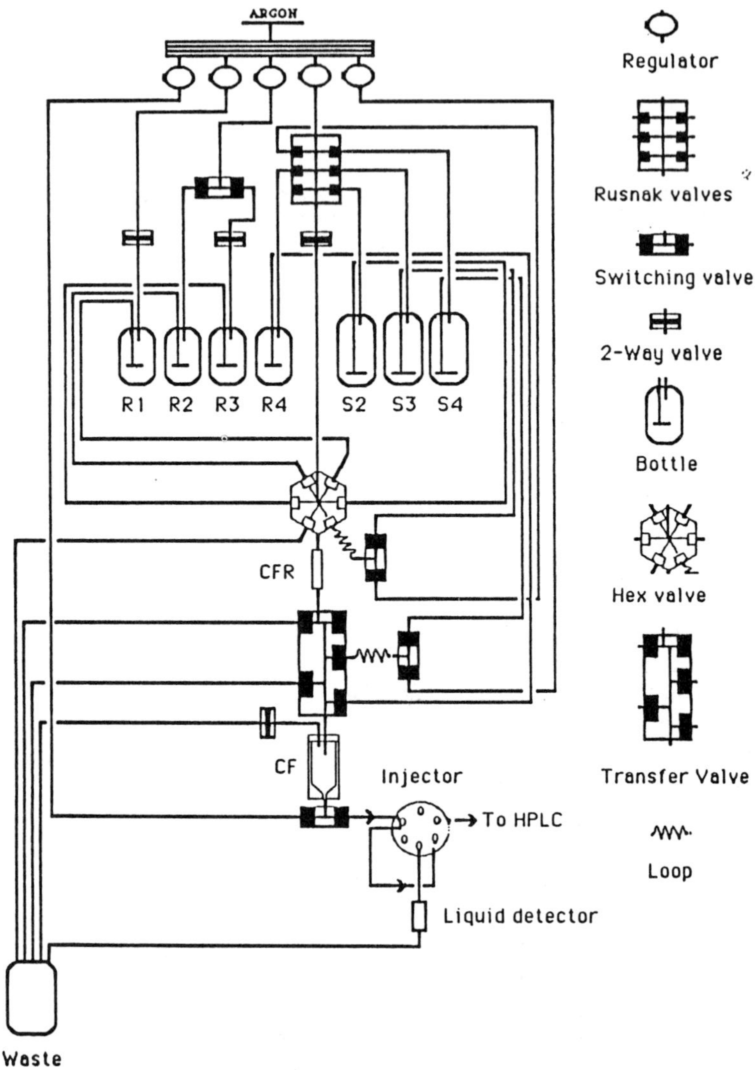

Fig. 1. Sequencer schematic. Key to components is shown on right.

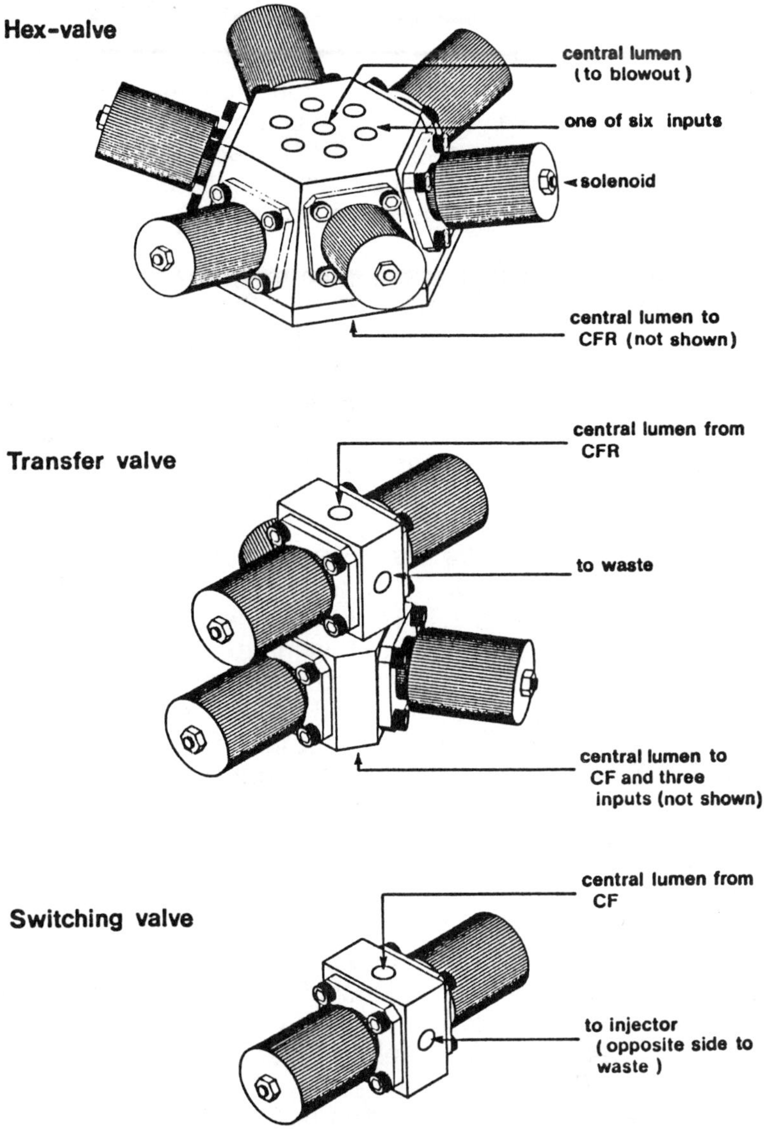

Fig. 2. Specialized valves. The hexagonal valve allows six inputs to a common lumen. The CFR is connected between this valve and the one shown below, the transfer valve. The transfer valve allows sample to flow to waste or to the CF immediately below the valve. It also allows the delivery of up to three fluids to the CF. The CF drains to the switching valve which allows transfer of sample to waste or to the autoinjector.

(CF) and allows the input of three fluids to the CF. As a result of this arrangement, the CF has only one input line, one vent line, and a drain line from the bottom. The CF is constructed from Kel-F and has an internal volume of 30 uL. Output from the CF is directed via a switching valve to either waste or a Rheodyne autoinjector connected to an on-line HPLC (Beckman System Gold). Fluid delivery from the sequencer to the autoinjector is detected by a sonic device to allow injection of >90% of the sample.

The overall dimensions of the instrument are 25 cm x 38 cm x 44 cm (h x w x d). The fluid flow path from the hexagonal valve to the switching valve (20 cm) is vertical, facilitating the draining of fluids from the lines. The specialized valves used are variations on the Rusnak valve described previously (2). Features include all wetted parts made from Kel-F and Teflon, multiple input/output capability, zero-dead volume construction, and solenoid actuated (24 VDC). A detail of the three new valves described is shown in **Figure 2**.

Details of the CFR have been previous described (3,4). The CFR is constructed from Teflon, PFA, and/or Tefzel tubing. The input lines are 1/16 in. OD and fit tightly into the reactor which has a slightly undersized 1/16 in. ID. For this instrument we prefer input lines of Tefzel for stiffness, and the reactor is made from translucent PFA in order to observe fluid deliveries. The overall length of the reactor is 2 cm, allowing room for a PVDF strip up to 1 cm in length. Delivery of phenylisothiocyanate (PITC) is limited to 2 uL, just wetting the PVDF strip. The reduction of reagent delivery and tubing volumes to a minimum dramatically reduces background noise.

Other critical deliveries are ensured by including fixed-volume loops for S3 and S4. S3 transfers the anilinothiazolinone (ATZ) derivative from the CFR and includes an internal standard for calibrating the sample volume and for reference on the on-line HPLC analysis. S4 transfers the phenylthiohydantoin (PTH) derivative from the CF to the on-line HPLC. Details of the the sample loop/valving system are shown in **Figure 1**. The composition of reagents and solvent is shown in **Table 1**.

The HPLC system employs a 2.1 x 250 mm C-18 reversed phase column and UV detection at 269 nm. No special effort is used to subtract backgrounds or otherwise computer enhance the data.

TABLE I. Composition and usage of reagents and solvents.

Reagent/solvent		Usage per cycle
R1	5% PITC in heptane	2 µL
R2	2% TEA in water	gas
R3	TFA	gas
R4	TFA/water/MeOH (1.5/28.5/70)	60 µL
S2	MeCN/Ethyl acetate (15/85)	4 x 45 µL
S3	MeCN/Butyl chloride (15/85)	4 x 60 µL
S4	Water/MeCN (10/90)	140 µL

R3, S2, S3, and S4 contain 4 µg DTE per 100 mL.

RESULTS AND DISCUSSION

The performance of the instrument was tested by the sequence analysis of 20 picomoles of horse apomyoglobin. Chromatograms for the first 16 cycles are shown in **Figure 3**. The background peaks are extremely low, with diphenylthiourea (DPTU) about 5 picomoles on cycle 1, and <5 picomoles on subsequent cycles. Yields of PTH amino acids (corrected for % analyzed) are shown in **Figure 4**. The initial and repetitive yields are 60% and 92%, respectively. Similar analyses (not shown) were performed for β-lactoglobulin A at the 6, 10, and 20 picomole level with initial yields of 40-50%, and repetitive yields of 97-98%.

In terms of sequencer sensitivity, it can be concluded that this instrument performs equivalent to the larger, more expensive commercial instruments. With further improvements in the Edman chemistry/HPLC system (eg. fluorescent derivatives, ref. 5) which take advantage of the reduced background noise and solvent/reagent consumption of this system, it may be possible to achieve further gains in sequencer sensitivity. Furthermore, this compact instrument takes up less lab bench space, and due to its more economical design may increase the accessibility of these instruments to the everyday protein lab.

ACKNOWLEDGEMENT

The authors are grateful to Dr. Ray Paxton for a critical reading of the manuscript.

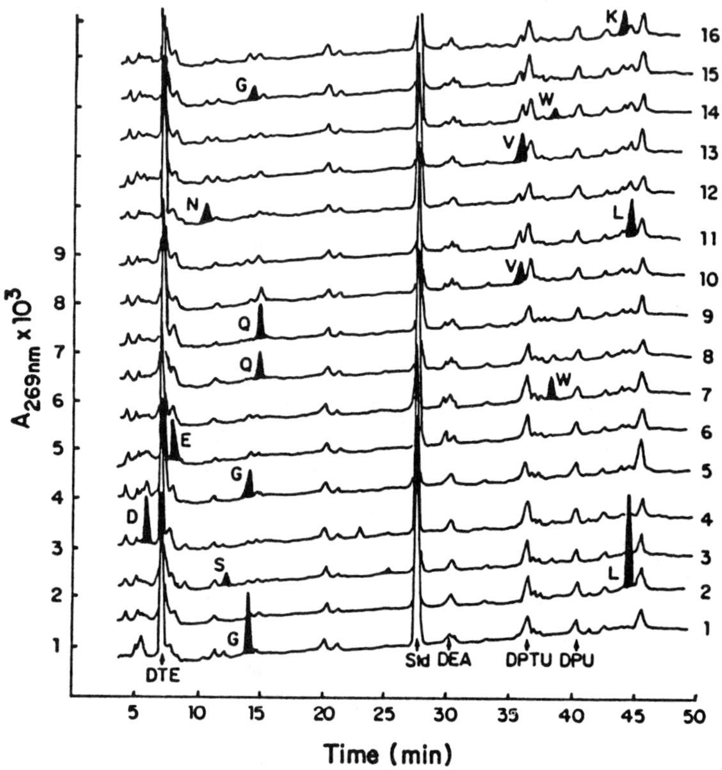

Fig. 3. Sequence analysis of horse apomyoglobin. Twenty picomoles of sample were analyzed (16 cycles shown).

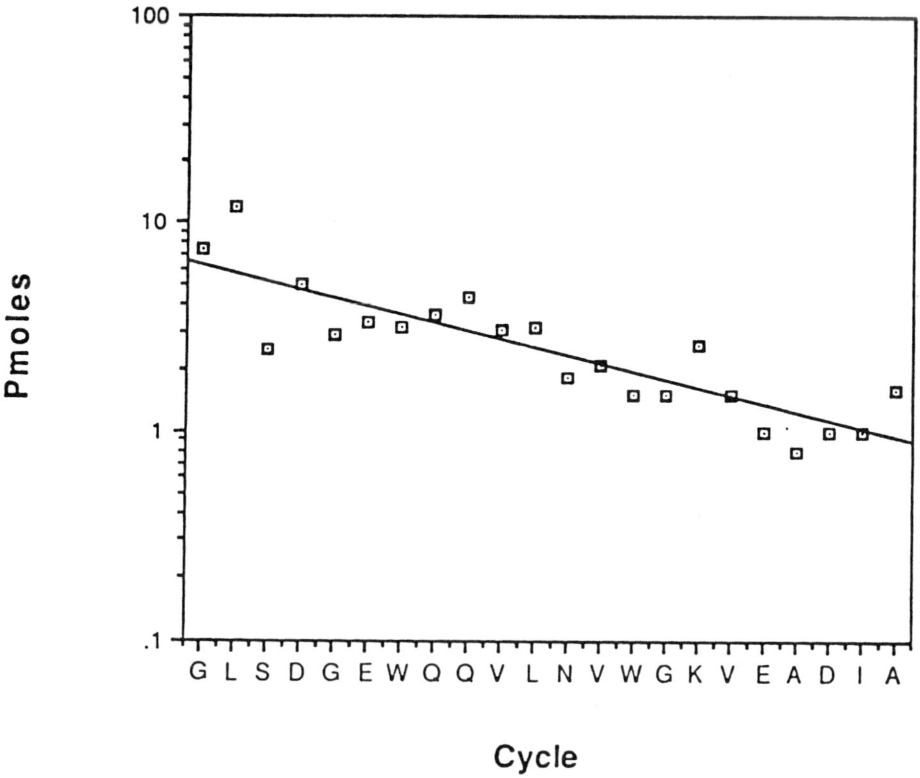

Fig. 4. Sequence analysis for horse apomyoglobin. Yields are shown for 20 picomoles analyzed.

REFERENCES

1. Hewick, R.M., Hunkapiller, M.W., Hood, L.E., and Dryer, W.J. (1981) **J. Biol. Chem. 256,** 7990-7997.
2. Hawke, D.H., Harris, D.C., and Shively, J.E. (1985) **Anal. Biochem. 147,** 315-330.
3. Shively, J.E., Miller, P., and Ronk, M. (1987) **Anal. Biochem. 163,** 517-529.
4. Xu, Q-Y., and Shively, J.E. (1989) **Anal. Biochem. 170,** 19-30.
5. Tsugita, A., Kamo, M., Jone, C.S., and Shikama, N. (1989) **J. Biochem. 106,** 60-65.

13

AMINO ACID ANALYSIS 1990: THE THIRD COLLABORATIVE STUDY FROM THE ASSOCIATION OF BIOMOLECULAR RESOURCE FACILITIES (ABRF)

George E. Tarr
Dana-Farber Cancer Institute
Boston MA 02115

Raymond J. Paxton
Beckman Research Institute/City of Hope
Duarte CA 91010

Yu-Ching E. Pan
Hoffmann-La Roche, Inc.
Nutley NJ 07110

Lowell H. Ericsson
University of Washington
Seattle WA 98195

John W. Crabb
W. Alton Jones Cell Science Center
Lake Placid NY 12946

I. INTRODUCTION

This is the third consecutive year in which the ABRF has sent peptide or protein samples to its member facilities for amino acid analysis (1-3). We have determined the accuracy and precision of analyses by core facilities of routine and moderately challenging samples, correlating results as much as possible with the various methods employed. This establishes a baseline of performance that individuals can compare with their own and, hopefully, suggests avenues for improvement. As we have noted previously (2), amino acid analysis is deceptively simple and straightforward in concept, but perhaps the most difficult common analytical method in all of protein chemistry.

Three samples were sent to each core facility: a control sample of unspecified protein, the same mixed with Tris salt, or adsorbed to PVDF. We expected the first to pose few difficulties, the second to test the ability to handle buffer contamination, and the third to offer a challenge to cope with a new and still experimental (for most) but increasingly common medium.

Techniques in Protein Chemistry II

II. MATERIALS AND METHODS

A. Choice of sample

To assess analytical accuracy for samples as commonly submitted to core facilities, a protein of high purity with regard to both contaminating proteins and non-peptide components was required. After examining several proteins (Sigma), we chose horse apomyoglobin (Cat No A-8673, Lot 11F5965): it was salt-free and pure by SDS-PAGE and sequencing.

B. Preparation and distribution

The ABRF-90AAA1 and 90AAA2 samples were aliquoted into small plastic tubes (yellow and blue, respectively). Each got 4 µg of protein plus 500 pmol norleucine as an internal standard. 90AAA2 samples also received 600 µg of Tris chloride. Samples were dried in a Savant Speed Vac. PVDF (Millipore) for 90AAA3 was cut into 5.5 mm discs with a paper punch. After soaking overnite in 100% methanol they were batch washed 3x with 30% methanol and 1x with water. Each disc was transferred damp to 267 µL horse apomyoglobin (1 µg) in a 1.5 mL sterile cryotube (Sarstedt), which was capped and shaken gently overnight at 4°C. Each disc was removed with forceps, washed 3x with water and vacuum dried in an orange plastic tube. Analysis of protein remaining in the tubes showed the adsorbtion to be about 90% efficient. One tube of 90AAA1, 2 of 90AAA2 and 3 discs of 90AAA3 were sent by regular mail to participating facilities.

Each site was asked to fill out a data sheet reporting pmol of each amino acid found in 90AAA1 and one tube of 90AAA2, both split into 3 replicates, and 90AAA3 for a total of 9 analyses, and answer questions about instrumentation, general methods, and procedures specific to each type of sample.

C. Data reduction

The University of Washington Biostatistics Service Group adapted the Statistical Package for Social Sciences (SPSS) software to perform initial calculations, with further processing by the subcommittee using other software. The amount of protein predicted by each reported residue pmol was found by dividing by the known number per molecule for that residue (2). Individual amounts were compared with the average value for a given analysis and any that deviated more than 15% were discarded; a mean amount was calculated from the remaining "relevant residues". Residues/mol were then determined by dividing pmol AA by this mean. Recoveries were calculated from the mean, MW and dilution factors for the analyses.

Precision means the agreement among the various determinations, while accuracy measures the agreement of the result found with the true result. We express both here as a percent, that is, agreement per value, in order to minimize magnitude effects. However, this exaggerates the impression of error for residues of very low abundance, here Arg, Tyr and Met. Standard deviation was that appropriate for small data sets:

% Std Deviation = 100 x Std Dev / Mean Residue Value
Average % Std Deviation = Sum (% SD Each Residue) / 16
% Error = 100 x (Experimental - True Residue Value) / True
Average % Error = Sum (Absolute % Error) / 16

III. RESULTS AND DISCUSSION

A. Instruments and methods

By Aug 1990 47 responses were received, but 3 were duplications and 3 more of 90AAA1 were excluded (filtered out) because the data gave < 4 relevant residues or average errors > 55%. By the same criteria, the data set for 90AAA2 was reduced to 35 sites and 90AAA3 to 30. Original site numbering is retained in the figures to facilitate comparisons. About half the sites used precolumn methods, half postcolumn methods, and most used either PITC or ninhydrin based chemistry (Table I).

B. Precision, accuracy, recovery

Fig. 1 shows average % standard deviations and % errors (ordered low to high) by site for each sample. Postcolumn and

Table I. Summary of instrumentation and methods AAA 1990

Instrument		Chemistry	
9	Applied Biosystems 420	1	Dabsyl Cl
12	Beckman 6300 / 7300	1	Fluram
2	Beckman HPLC	13	Ninhydrin
1	Hewlett Packard 1090	5	OPA
1	Homemade	21	PITC
1	LKB 4400		
1	Perkin Elmer	Procedure	
1	St. Johns Assoc.		
1	Varian HPLC	23	Precolumn deriv
12	Waters HPLC / Pico-Tag	18	Postcolumn
41	Total sites (90AAA1 filtered)		

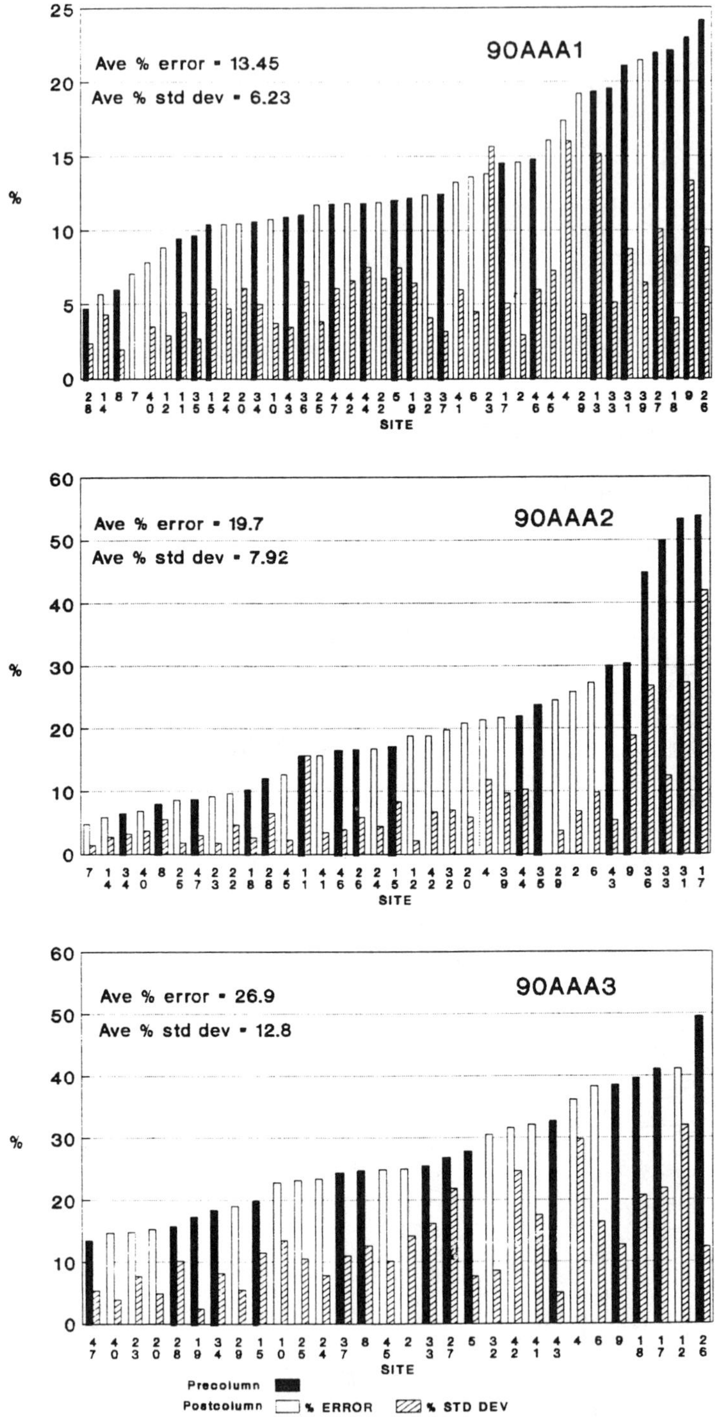

90AAA1
Ave % error = 13.45
Ave % std dev = 6.23
%
SITE
90AAA2
Ave % error = 19.7
Ave % std dev = 7.92
%
SITE
90AAA3
Ave % error = 26.9
Ave % std dev = 12.8
%
SITE
Precolumn
Postcolumn
% ERROR
% STD DEV

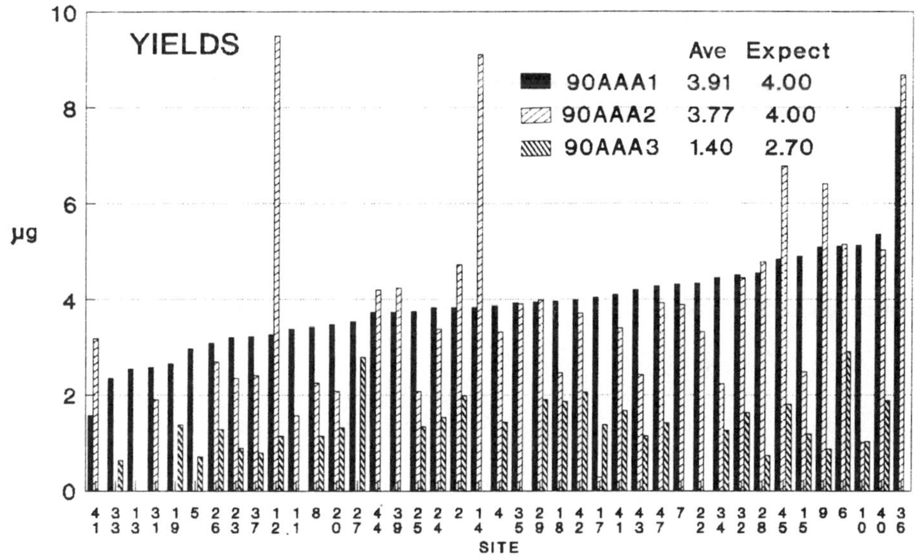

Fig. 2. Yields calculated from amino acid recovery data.

Table II. Selected amino acid analyses from 1990 ABRF study

	True	Preclm PITC	Preclm OPA	Post/OPA Salty	Postclm Ninhydrin	Pre/PITC PVDF
Ala	15	14.8	14.5	15.1	14.8	15.5
Arg	2	2.4	2.2	2.1	2.0	2.3
Asx	10	10.5	10.4	10.7	10.6	10.8
Glx	19	19.0	20.9	18.8	19.8	18.9
Gly	15	15.0	14.8	15.5	15.4	15.5
His	11	12.8	11.2	10.5	11.0	9.3
Ile	9	9.0	7.8	7.7	8.2	7.5
Leu	17	17.1	n.d.	17.1	17.6	17.0
Lys	19	18.2	18.4	18.2	17.9	17.1
Met	2	2.0	2.1	2.0	1.6	1.2
Phe	7	6.8	6.8	6.9	6.9	7.1
Pro	4	4.0	4.5	4.3	4.0	3.1
Ser	5	4.7	4.8	5.3	4.8	5.0
Thr	7	6.7	6.5	6.9	7.2	7.2
Trp	2	n.d.	n.d.	n.d.	n.d.	n.d.
Tyr	2	2.0	2.0	2.0	2.1	2.2
Val	7	6.8	6.4	6.7	6.8	6.8
% Error		4.12	5.76	3.93	4.29	9.52

Fig. 1. AAA by site (read vertically), arranged low to high by % error, distinguished according to pre-and post-column methods.

precolumn methods were evenly represented in the better analyses, but the worst ones tended to be precolumn. Precision is correlated with accuracy weakly in 90AAA1, strongly in the other two cases. Errors increase from control sample to salty sample to sample on PVDF. At the same time, total μg recovered from the 3 samples per expected recovery tend to fall (Fig. 2). While 44% of the sites recovered from 90AAA1 the expected amount ± 10%, only 26% did so from 90AAA2 and hardly any did from 90AAA3. The yields from 90AAA1 and 90AAA2 were well correlated. Quite accurate analyses could be obtained with any of the major methods: Table II shows the best analysis in several categories.

C. Coping with salty samples

Most of the facilities that utilized precolumn methods desalted 90AAA2 prior to hydrolysis (see Table III), mainly by RP-HPLC and precipitation. Desalting gave average % errors comparable to those of 90AAA1 by any method and to 90AAA2 analyzed by postcolumn methods, but frequently lower recoveries. Precolumn methods applied sans desalting, in sharp contrast to postcolumn methods, averaged no better in yield and gave distinctly worse accuracy. Apparently postcolumn methods are suitable for both accurate compositions and quantitative yields of samples with some salt contamination (see Table II for a best example). Precolumn methods give good compositions after desalting, but without special care quantification is compromised by potential and variable losses in the desalting step.

SD and error for each amino acid from 90AAA1 and 90AAA2 are summarized in Table IV. Both pre- and postcolumn methods showed high variation for Arg and Met, which is partly explained by their low abundance in myoglobin (see Data reduction). Tyr generally shared this problem. Also contributing to the Arg error for postcolumn methods is the difficulty of measurement of a broad peak on an elevated baseline, while oxidation of Met is a potential hazard with all methods. Pro was more of a problem for postcolumn methods. Low values for Ile likely reflect incomplete hydrolysis. While both Gly and Ser are common contaminants, only Ser showed high variability and error, suggesting instead a failure to correct for hydrolysis losses.

D. Samples on PVDF

Errors and SD's for the 30 sites whose data passed the filtering criteria were much larger with PVDF than for controls (see Fig. 1). At all performance levels the % error (and % SD) was about 2-fold greater. Most sites used either 1) vapor-phase hydrolysis with 6N HCl followed by extraction of the amino acids with an organic solvent-containing solution, for example, 30% methanol, or 2) liquid-phase hydrolysis with 6N HCl (Table V).

Table III. Desalting procedures, % error and yield for 90AAA2

Method	Sites	Ave % error	Range	Ave yield, µg	Range
Precolumn					
RP-HPLC	7	20.2	7.3 - 61	1.9	0.3 - 2.7
PPT	5	24.4	10.1 - 64	3.6	1.6 - 6.4
None	4	32.5	9.1 - 66	2.5	1.0 - 4.8
Postcolumn					
None	15	14.4	5.0 - 27	4.4	2.3 - 9.5
PPT	4	22.4	15.0 - 38	3.6	2.1 - 5.1

PPT: precipitation with acetone, ethanol, methanol, TCA, etc.

Table IV. Accuracy/precision 90AAA1 & 90AAA2, pre- & postcolumn

90AAA1				90AAA2			
Pre-		Post-		Pre-		Post-	
% Error	% Std dev	% E	% SD	% E	% SD	% E	% SD
A -1.5	V 8.0	F -0.5	E 6.0	Y 3.2	T 16.0	F -1.2	D 7.7
E 2.9	F 9.2	A 1.3	G 6.6	K -3.8	I 17.4	T -2.0	A 8.8
F -3.4	A 9.8	L -1.9	L 7.0	F -5.7	L 19.3	A -2.6	I 9.1
T -3.4	G 10.9	E 2.9	I 8.4	D -6.6	A 19.8	V -2.7	L 9.7
G 4.1	I 11.9	Y 3.3	K 8.6	G 7.6	F 20.4	L -3.2	V 11.0
V -4.8	E 13.4	V -3.8	A 8.7	T -8.3	G 21.6	D 4.4	K 11.1
L -4.9	K 14.1	D 4.1	H 9.6	A -8.8	H 27.3	H -5.7	T 12.1
P 5.5	S 15.5	T -5.1	T 10.8	P 9.5	E 31.5	K -7.5	F 15.7
H -7.4	T 15.8	K -6.0	F 11.5	E -12.2	S 32.1	E -8.9	E 17.0
K -10.0	H 20.1	H -6.2	V 13.0	V 14.4	P 35.9	P 12.0	G 19.7
D 10.6	L 21.9	G 7.0	D 15.7	L -14.5	D 38.5	G 12.8	S 23.7
Y 10.6	D 22.1	P 12.4	S 21.7	H -16.4	R 42.1	I -13.4	Y 25.3
I -13.0	P 22.1	I -13.2	Y 22.9	I -17.0	M 45.5	M -17.4	P 26.7
M -13.9	R 22.4	S 13.7	P 24.3	M -24.7	Y 45.8	Y 17.8	R 28.2
S 15.4	Y 31.0	R 24.2	R 29.3	S 35.4	V 56.4	S 22.2	M 33.9
R 29.9	M 38.4	M -28.3	M 32.1	R 47.7	K 56.8	R 32.5	H 45.7

Residues (single letter code) are arranged in each column low to high by absolute value of % E (% error) or % SD (% standard deviation).

The extract or acid was then dried for analysis. Both methods gave similar mean yields and errors, however, the broader range in yields for the first suggests variability in extraction procedures, which therefore merits optimization and extra care.

Table V. Methods and results for analysis of 90AAA3 (PVDF)

Method	No. sites	Calculated yield (µg)		% Error	
		average	range	average	range
Vapor-phase hydrolysis Extract amino acids with organic solution; e.g., 30% MeOH (3 sites filtered out)	13	1.52	0.78-2.90	25.7	13.2-41.1
Vapor-phase hydrolysis Extract amino acids with aqueous solution; e.g., citrate buffer (3 sites filtered out)	1	0.71		27.7	
Vapor-phase hydrolysis Derivatize with PITC without prior extraction of amino acids	2	1.07	0.86-1.28	44.0	38.6-49.5
Vapor-phase hydrolysis Extract amino acids (no conditions reported)	3	1.69	1.32-1.88	23.2	14.6-39.7
Liquid-phase hydrolysis Remove acid, dry, and analyze	8	1.39	0.89-1.97	26.4	14.8-41.1
Liquid-phase hydrolysis Dry acid, extract amino acids with organic solution	2	1.26	0.46-2.06	31.8	31.5-32.1
Auto-hydrolysis, derivatization, and analysis (2 sites filtered out)	1	0.73		15.7	

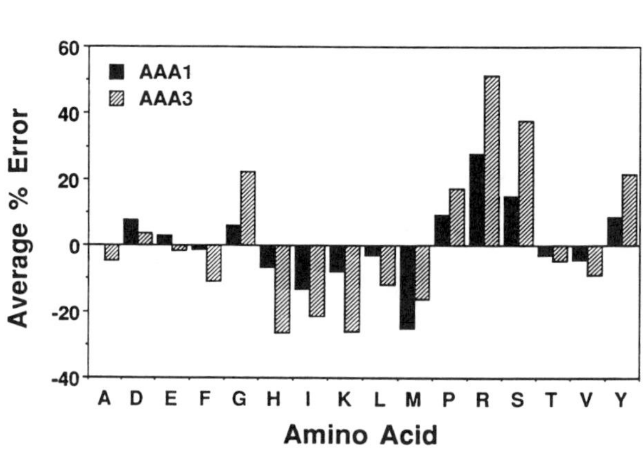

Fig. 3. Comparison of average % error for 90AAA1 and 90AAA3. For a comparison of best analyses see Table II.

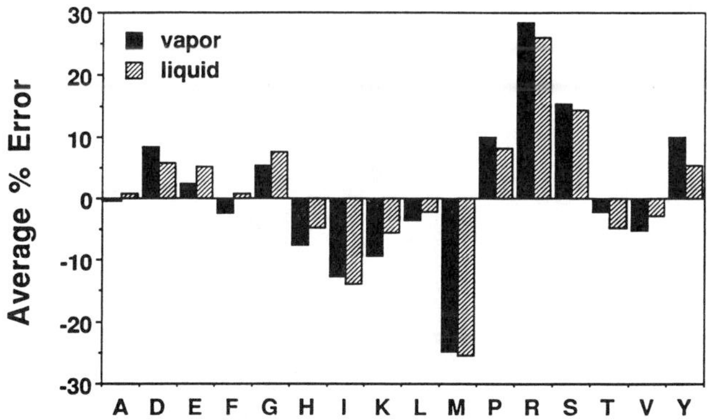

Fig. 4. Average % error with vapor & liquid-phase hydrolysis.

90AAA3 was compared to 90AAA1 according to average % error for individual amino acids (Fig. 3). For 13 of 16 amino acids, greater % errors were seen for 90AAA3. The nearly 3-fold greater errors for Gly and Ser suggests contamination during sample handling. High errors for other residues require other explanations (see above for Met and Arg). Apparently the greater errors observed for PVDF samples were due to overall difficulty in those analyses and not to a few problem residues.

E. Hydrolysis methods

Of the 41 sites in the 90AAA1 filtered data set, 13 sites used liquid-phase and 28 vapor-phase methods. Of the former, 12 hydrolyzed with 6N HCl and one with 4N methanesulfonic acid (MSA), 11 under vacuum, 2 under inert gas, all for 18-24 hr at 108-115°C. The MSA site identified 1 of 2 Trp residues. Antioxidants included phenol (4 sites), mercaptoethanol (4), thioglycolic acid (2), and both phenol and a thiol (2). All vapor-phase sites used 6N HCl, 6 for 1-2 hr at 150-170°C, 19 for 18-24 hr at 105-115°C, and 3 short auto-hydrolysis methods; 15 hydrolyzed under vacuum, 13 under gas; 16 used phenol, 2 mercaptoethanol, 2 both. There was no significant difference in accuracy with liquid- and vapor-phase methods (Fig. 4).

F. Calibration

Calibration as a source of error was evaluated for 9 sites submitting chromatograms with both integrated areas and pmol for PTC analyses. Response factors were calculated and compared with among sites and with factors derived from extinction coefficients (4). Errors due to differences in eluting % acetonitrile, wavelength (254 nm to the best of our knowledge)

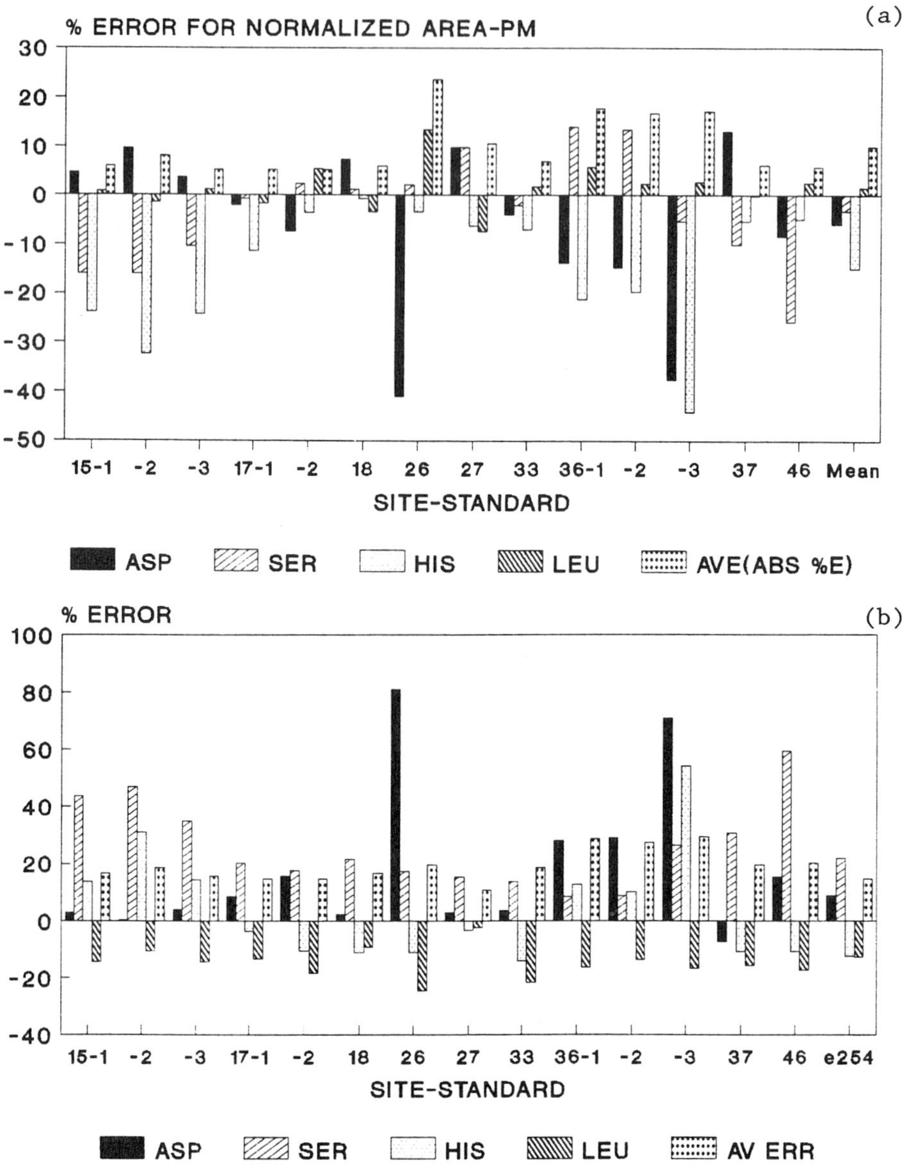

Fig. 5. Calibration by some sites employing PITC. (a) % error of calibration: factors and extinction coefficients (254) were normalized to the easily recovered residues Ala, Leu & Phe; values plotted = 100 x (area/pmol found - area/pmol 254)/ 254. (b) % error of composition: recalculation of one 90AAA1 analysis from site 26 using response factors from that and other sites.

and wavelength window in the detector are assumed to be negligible. Response values in area/pmol, expressed as % error relative to "absolute" values, are shown in Fig. 5a. Clearly many sites did not hydrolyze the calibration standard, because the Ser value should be distinctly negative to compensate for hydrolysis losses. Reanalysis of 90AAA1 samples from sites 15, 17 and 36 showed considerable variation in individual residues, even comparing different calibrations in the same facility, though the overall error was rarely strikingly different. Site 15 compensated well for their low recovery of His with all 3 of their standards, with only site 36 being close, but that consistent low recovery flags a problem. Facility 26 reveal1ed severe problems with its calibration as well as with its analysis method, one reflection being a vast overcompensation for its low recovery of Asp (Fig. 5b). Low values of Asp have been a chronic problem with precolumn methods, presumably because of complexation with divalent cations from the sample, solvents, or hydrolysis tube material. While some using PITC show this problem (large negative values in Fig. 5a), most do not, and the average results for Asp are normal (see Table IV).

IV. SUMMARY AND CONCLUSIONS

Excellent results were apparent with a variety of methods and instruments (Table I) and 11 sites were able to correctly identify the unknown 90AAA protein as horse myoglobin. To aid core facility self-evaluation, individual performance (Fig. 1) was grouped by thirds based upon accuracy into above average, average, and below average and replotted with the corresponding precision (Fig. 6). Sites performing best obtained about 91% accuracy and good precision (3-4% SD) on both control (90AAA1) and salty (90AAA2) samples. In constrast, sites in the lower 33% obtained 81% accuracy on 90AAA1 and 66% on 90AAA2 with increased variability (8.1% and 15.7% SD). Although pre- and post-column methods shared the best category equally, precolumn methods dominated troubled analyses (Fig. 1). Method of hydrolysis did not seem important (Fig. 4).

Analyses on PVDF were the most problematic, with the top 33% of participants obtaining only about 83% accuracy and 7.3% SD on 90AAA3. The low recovery from the PVDF (average 52%) illustrates the difficulties associated with the hydrolysis, extraction and analysis of this type of sample. No PVDF-specific problematic residues were identified, nor was a superior methodology evident, and good analyses were possible.

The present results demonstrate that greater than 90% compositional accuracy can be reproducibly achieved with triplicate hydrolyses/analyses from 4 μg protein in the presence or absence of salt. For now, lower accuracy is expected from

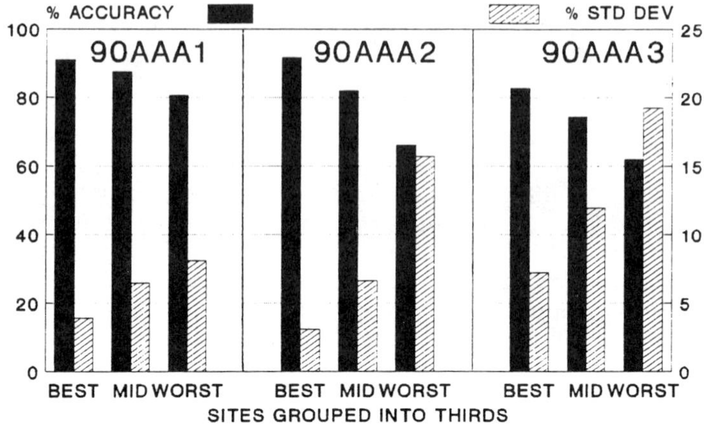

Fig. 6. Amino acid analysis ABRF-1990, grouping sites into three performance levels. See Fig. 1 for more detail.

amino acid analysis on PVDF, though current methods seem adequate for (e.g.) determining initial sequence yields.

ACKNOWLEGEMENTS

We thank the site personnel who participated in this study, especially those submitting information beyond the questionaire. Special gratitude to Janice Lamers for data reduction and Karen West for expert assistance in preparation of the PVDF samples. This work was supported in part by NSF grant DIR 8903251.

REFERENCES

1. Niece,R.L., K.R. Williams, C.L. Wadsworth, J. Elliot, K.L. Stone, W.J. McMurray, A. Fowler, D. Atherton, R. Kutny and A. Smith (1989) In: Techniques in Protein Chemistry (T.E. Hugli, ed) Academic Press, San Diego, pp. 89-101.

2. Crabb, J.W., L. Ericsson, D. Atherton, A.J. Smith and R. Kutny (1990) In: Current Research in Protein Chemistry (J.J. Villafranca, ed) Academic Press, San Diego, pp. 49-61.

3. Ericsson, L.H., D. Atherton, R. Kutny, A.J. Smith and J.W. Crabb (1991) In: Methods in Protein Sequence Analysis 1990 (H. Jornvall and J.-O. Hoog, eds) Birkhauser Verlag, Basel

4. Tarr, G.E. (1986) In: Methods of Protein Microcharacterization (J.E. Shively, ed) Humana Press, Clifton NJ, pp. 155-194.

14

Protein Sequencing from Polyvinylidenedifluoride Membranes: Design and Characterization of a Test Sample (ABRF-90SEQ) and Evaluation of Results

K. Ümit Yüksel[1], Gregory A. Grant[2], Liane M. Mende-Mueller[3], Ronald L. Niece[4], Kenneth R. Williams[5], and David W. Speicher[6]

[1]Biopolymer Analysis Laboratory, Texas College of Osteopathic Medicine, Fort Worth TX 76107.

[2]Department of Biochemistry and Molecular Biophysics, Washington University School of Medicine; St. Louis, MO 63110

[3]Protein / Nucleic Acid Shared Facility, Medical College of Wisconsin; Milwaukee, WI 53226

[4]Protein/DNA Sequence/Synthesis Facility, University of Wisconsin Biotechnology Center; Madison, WI 53705

[5]Protein and Nucleic Acid Chemistry Facility, Yale University School of Medicine; New Haven, CT 06510

[6]Protein Microchemistry Facility, Wistar Institute; Philadelphia, PA 19014

I. Introduction

Technological developments such as two dimensional electrophoresis [1], electroblotting of proteins [2], the introduction of the gas phase sequencer [3], and the introduction of polyvinylidenedifluoride (PVDF) membranes mark major advances of the last decade in protein characterization and isolation. The ability to sequence proteins that have been blotted onto PVDF membranes following their separation by polyacrylamide gel electrophoresis [4] now makes it possible to obtain amino acid sequence data on minute amounts of protein without lengthy purification procedures. Of the 49 core facilities that responded to a survey that

Techniques in Protein Chemistry II

was conducted in 1989 by the Association of Biomolecular Resource Facilities (ABRF), 92% used PVDF membranes as sequencing media, and an average 30% of their sequences (range 1-80) were run on these media. Sixty-five percent of these facilities performed electroblotting [5]. Thus, this medium is of considerable interest to those sequencing peptides and proteins. While this approach has been successfully utilized in determining 10 or more residues of sequence from 50 pmol or less of PVDF blotted protein [4], so far no quantitative study has been published that addresses the more important questions of accuracy of the resulting sequence and the "average" number of residues of sequence that might be expected to be obtained from a given amount of an "unknown" sample. To address these questions and also in part to fulfill the request of protein sequencing facilities to have an immobilized "unknown" sample, the ABRF distributed a custom designed peptide immobilized onto PVDF as this years protein sequencing test sample (ABRF-90SEQ) to 132 core facilities world-wide. The following report provides a summary and analysis of the first 56 sets of data that had been returned by 23.July.1990.

II. Materials and Methods

A. *Design of the Test Sample*

ABRF-90SEQ was designed to mimic a large electroblotted protein (ca. 100 kDa) on a PVDF membrane. A synthetic peptide was covalently attached to a protein, adsorbed to a PVDF membrane and stained with Ponceu S. The peptide whose sequence is shown in Fig. 1 was 29 residues long, contained all twenty naturally occurring amino acids, including a C-terminal cysteine for directional coupling. For ease of synthesis adjacent ß-branched amino acids were minimized and only one tryptophan was incorporated. Regular repeats of amino acids were included for repetitive yield calculations, a proline was placed between two of the repeats to evaluate its effect on lag and sequence assignment, and amino acids with traditionally low yields (e.g., His and Ser) were positioned late in the sequence to provide both a challenge and a measure for identifying these. Furthermore, an Asp-Asn sequence was included to determine the ability to assign sequence in the event some deamidation occurs, and blocks of amino acids

5 10 15 20 25
Y-L-D-N-D-W-T-S-A-L-R-V-M-P-S-I-L-K-V-Q-G-R-E-V-S-H-F-K-C-(protein)

Figure 1. Amino acid sequence of ABRF-90SEQ.

that elute next to each other (e.g., Ile-Leu-Lys or Arg-Val-Met-Pro) were included to evaluate the resolving power of chromatographs and sequence assignment abilities of the software and/or operators. Aspartate was also used to evaluate its resolution from early eluting peaks such as dithiothreitol (DTT) or ammonia derivatives, and tryptophan was included to evaluate interference from diphenylurea (DPU).

1. Synthesis and characterization of the peptide

The peptide was synthesized in the Protein Chemistry Laboratory at the Washington University School of Medicine on an Applied Biosystems Model 430A peptide synthesizer using t-BOC chemistry and PAM resins. All residues were double coupled with capping. The peptide was released from the resin by standard HF cleavage, extracted and lyophilized. Preparative scale HPLC purification was carried out on a Vydac C-18 column (22x250 mm) with a linear gradient of acetonitrile in 0.1% TFA. The sequence of the purified peptide was verified by automated Edman degradation by the authors. The peptide was judged to be pure by amino acid sequencing with an initial yield of 67%. FAB/mass spectroscopy studies that were carried out at the Comprehensive Cancer Center Facility at the Yale University School of Medicine indicated that two species with m/z of 3394.6 (major component, 66%) and 3044.6 were present. The major component had the mass expected from the sequence. The minor component was determined to be acetylated at its amino terminus but missing the three amino terminal residues. Since this component did not interfere with the sequencing process, no further purification was attempted.

2. Conjugation of the peptide to carrier protein

Human transferrin (75.2 kDa) was used as the carrier protein. Transferrin (Sigma) was activated by incubation with *m*-maleimidobenzoyl-N-hydroxysulfo-succinimide ester (Pierce; 200-fold molar excess over transferrin buffer A [50 mM sodium phosphate, pH 7.0] at room temperature for 30 min). Unreacted cross-linker was removed by size exclusion chromatography (Bio-Rad TSK-125 SW, 7.5x300 mm, preequilibrated with buffer B [buffer A containing 1 mM EDTA], 1 ml/min flow rate). The void volume peak, which contained the activated protein, was collected and purged with argon for 30 seconds. The synthetic 29 residue peptide was dissolved in buffer B at a final ratio of 15 nmol peptide/ nmol transferrin. After purging the mixture for 1 min, the reaction proceeded for 3 h. The conjugate was purified by size exclusion HPLC as described above.

3. Adsorption to PVDF membrane

The conjugate (230 μg in 1.5 ml buffer B) was treated with 2-mercaptoethanol (100 mM final concentration) for 60 min at room temperature. Then 344 μg yeast glucose-6-phosphate isomerase (EC 5.3.1.9; 2x63 kDa; naturally blocked N-terminus) was added to the conjugate to further increase the background mimicking a 100 kDa protein during the later sequencing cycles. The solution was brought to a final volume of 7.0 ml by adding buffer B. The sample was adsorbed on one side of a sheet of prewetted PVDF membrane (Immobilon-P, Millipore) by sandwiching the PVDF membrane between a solid sheet of plexiglas and an upper sheet of plexiglas containing 7 slots. Each slot was 3 mm wide by 160 mm long, and received 950 μl of sample. Prior to insertion of the membrane, the plexiglas was precoated for 30 min at room temperature with 0.1% bovine serum albumin in buffer B. After addition of the sample to the unit, it was incubated at room temperature for 2 h on a reciprocating shaker at high speed, followed by further incubation at 4°C overnight. The amount of unbound protein in the supernatant solution of each slot was quantitated by amino acid analysis. The remaining solution was removed and the slots were rinsed extensively with Milli-Q water. The PVDF membrane was removed and stained with 0.5% Ponceu S in 1% acetic acid solution for 30 seconds, and destained by several rinses with water. The membrane was air dried and stored at -20°C until distribution. Conjugation and adsorption were carried out at the Protein Microchemistry Facility, Wistar Institute. The adsorbed sample was characterized by automated Edman degradation and amino acid composition by four of the authors. The amino terminus of transferrin was >90% blocked during the coupling reaction (data not shown). In the final sample the transferrin sequence (V-P-D-K-T-V-R-W-C-A-V-S-E-H-E-...) was present at about 5% the level of the conjugated peptide sequence. While Coomassie Blue is a more commonly used stain than Ponceu S, the latter stain was chosen to avoid exposing the PVDF-adsorbed protein to organic solvents that might desorb some of the conjugated sample.

4. The test sample

The final sample mailed to 132 core facilities was a 3x5 mm strip of PVDF membrane containing 4 μg of total protein. Samples were distributed by first class mail in plastic bags and no precautions were taken to keep the samples moist or cold. Recipients were asked to load the entire sample on their instruments, and were requested to submit all data, including failed sequences, so that the state of the sequencing service in general could be evaluated. In addition, several questions were asked regarding the equipment and actual sequencing cycle used for analysis of the sample. To assure confidentiality, the data were returned to a third party who removed the postmarks and other possible identifiers. Anonymous data were then forwarded to the authors for analysis.

III. Results and Discussion

A. The Description of the Operation of the Participating Facilities

ABRF-90SEQ data (55 of the 56 respondents filled out the survey sheet with various omissions) on the number of sequencers in ABRF core facilities (1.52, range 1-6) is in good agreement with the previous data [5,6]. During the past year a total of 17,030 samples were sequenced by the responding members of ABRF (average 210, range 20-500/ instrument; average 340, range 40-3000/ facility), less than one third of these samples actually originated within the facilities (28.8%, range 0-100), the rest were sequenced as service to other principal investigators.

The amount of PTH-amino acid standards loaded onto the sequencers might provide one indication of the level of sensitivity at which sequencing is being conducted. This year the average amount of standard applied was 42.5 pmol (range 5-100). The majority (75.9%) of the facilities routinely quantify their sequencing cycles. Most facilities inject 50μl of PTH-derivatives onto the PTH-analyzer (range 20-150μl) corresponding to 41.7% (range 33.3-100%) of the PTH-derivatives. Indeed 54.5% of the facilities inject less than 42% of the PTH derivatives and only 13.2% inject ≥80%. It would appear that an increase in the injected amount to above 80% would nearly double the sensitivity of detection for most facilities. ABRF-90SEQ data (not shown) suggest that this is indeed the case.

Since ABRF-90SEQ was designed to simulate an electroblotted sample, several questions were asked about electroblotting and sequencing from PVDF membranes. Fifty-six percent of the facilities indicated that they electroblot onto PVDF membranes in their laboratories, and 26.4% of them offer electroblotting as a service compared to 10% last year. Regardless of whether they perform electroblotting in house, in the past year the facilities analyzed 39.4% of their samples on PVDF membranes, 55.7% on glass fiber filters and the remaining 4.9% on other supports. When PVDF was the sequencing medium, 38.2% of the facilities utilized alternate sequencing cycles, and 78.2% of them used a polybrene coated glass fiber filter. In 89.5% of these cases the coated filter was placed below the PVDF membrane.

B. Results of the ABRF-90SEQ

The results of the sequence data reported by 55 facilities are plotted in Figure 2. Analyses were run on sequencers produced by ABI (51), Porton (3) and Milligen (1), with the average sequencer age being 4.4 years (range <1-8). A total of 1095 (100%) sequencing cycles were reported with an average of 20 cycles / report (range 5-30). There were 849 (77.5%) positive and 124 (11.3%) tentative assignments, with 122 (11.1%) cycles remaining unassigned ("holes"). Accuracy was 82.0% for positive assignments (696/849) and 59.7% (74/124) for tentative assignments. Thus, 770 (70.3%) of the 1095 cycles were correctly identified. Half of the facilities reported assignments for the first 21 residues, the number of assignments dropped rapidly thereafter. It may therefore be more informative to examine the results in two parts, i.e. the first 21 residues and the remaining 7. Leu-2 was the most correctly identified residue overall (51 correct & 1 wrong positive; 1 correct & 1 wrong tentative; 1 hole). Asn-4 was the second best identified residue (49 correct & 4 wrong positive; 1 correct & 1 wrong tentative). Within this group, facilities had difficulty in identifying Trp-6 (13 positive & 4 tentative correct) and Arg-11 (12 positive & 5 tentative correct). Trp-6 was also the most misidentified cycle (19 positive & 9 tentative wrong). Valine was predominantly assigned to this cycle (11 out of 25). His-26 was the least identified residue in the second group and overall (1 correct & 4 wrong positive; 2 correct & 1 wrong tentative). In contrast, Val-24 and Phe-27 were rather readily identified despite their location in the sequence.

The best three responses contained the following data. One response (analysis run on Porton 2090E) positively and correctly identified the first 27 residues, had an incorrect tentative assignment for position 28, and positions 29 & 30 were overassigned. Another response (analysis run on ABI 477A) identified 28 amino acids correctly (includes three tentative assignments), and overassigned positions 29 & 30. A third response (analysis run on ABI 470A) stopped correctly at cycle 28, but this sequence contained one tentative wrong and three tentative correct assignments.

As discussed above, this peptide was designed to evaluate several performance criteria. Facilities had no difficulty in correctly identifying residues that elute next to each other on the HPLC (i.e., R-V-M-P & I-L-K). They also

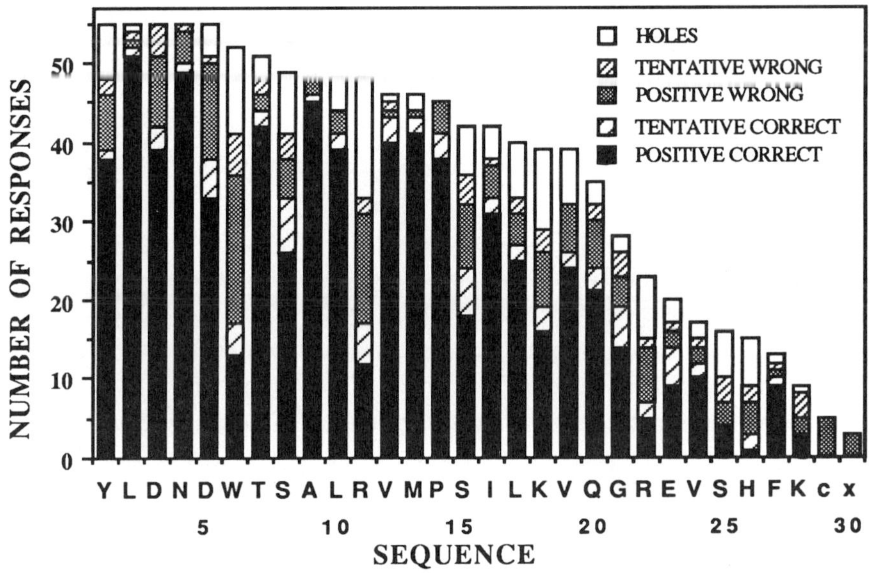

Figure 2. Results of sequence assignments for ABRF-90SEQ.

Data reported on the Sequencing Data Report Form was taken to be operator assigned sequence unless otherwise indicated by the respondent. The following example for Trp-6 shows the convention used to interpret the data: W/V = tentative correct, V/W = tentative wrong, W/(V) = positive correct, (W/V) = tentative correct, (V/W) = tentative wrong. An "X" or "?" between assigned cycles was considered as a "hole". Since no assignment after cycle 28 was expected, cycles 29 and 30 represent over assignments and were considered to be wrong.

performed well in resolving aspartate from early eluting by-products. However, identification of tryptophan was a problem. PTH-Trp is frequently not resolved from DPU, and is especially difficult to detect at low levels. Presence of a secondary sequence can complicate the situation further. The low level (<5%) human transferrin carrier protein sequence which has valine at this position, was apparently the reason why valine was frequently misassigned to this position. It is interesting that in ABRF-89SEQ, a 20% secondary sequence did not lead to any misassignments [5], while a lower molar ratio of the transferrin sequence in ABRF-90SEQ did lead to misassignments. The difference probably reflects the added complexity of sequencing of low level samples on PVDF membranes where cycle-to-cycle variations in yield can be a substantial problem, and the presence of high background from a large protein. This interpretation is further supported by a similar frequent misassignment of threonine to cycle 5 (7 Thr & 3 Asp/Thr assignments); as indicated above threonine occurs at this position in transferrin. The generally low recovery of aspartate from PVDF coupled with cycle-to-cycle variations in overall yield, and lag from the two previous cycles probably resulted in more difficulty in distinguishing between the major and minor sequence at this point. No apparent additional misassignments related to the transferrin sequence were made beyond cycle 6 presumably because the 1-2 pmol level transferrin sequence disappeared in the rapidly increasing background. Misidentification of one low yield residue (i.e. S, T, H, R) for another was not a frequent problem (4/193).

Data were further analyzed for repetitive yield and accuracy, and were broken down by instrument type (Table I). The effect of proline on lag could not be determined due to the low number of quantitative data for Leu-17 and Val-19 & 24. Although the repetitive yields and accuracy of the individual groups appear to exhibit large differences, caution should be exercised before drawing any

Table I. Sequencing results for ABRF-90SEQ [(a)]

Instrument Model [(b)]	# [(c)]	Repetitive Yield [(d)]	Number of Positive Correct	Accuracy of Positive Calls [(e)]
470/120	23 [10]	88±2.5	12±6.4	86±17.7
		(86-92)	(1-24)	(50-100)
475/120	8 [5]	88±3.2	11±6.3	83±25.0
		(83-91)	(4-23)	(42-100)
477/120	18 [9]	91±4.6	12±5.9	77±19.1
		(85-95)	(5-25)	(36-100)
473/120	1 [1]	88	13	72
2090E/1090E	3 [2]	86±6.2	21±5.1	98±4.0
		(82-90)	(17-27)	(93-100)
6600/600	1 [1]	93	17	74
All Instruments	54 [27]	89±3.7	13±6.4	82±19.1
		(82-98)	(1-27)	(36-100)

(a) Values are expressed as mean ± standard deviation, with ranges in parentheses.

(b) 2090E/1090E is a Porton sequencer with a Hewlett Packard 1090 HPLC; 6600/600 is a Milligen sequencer with Waters 600E HPLC; and all other instruments are by Applied Biosystems.

(c) Number of instruments for which data was reported; one return did not identify the instrument type. Numbers in brackets are those returns used for repetitive yield calculations. Twelve returns did not report quantification, 10 returns had incorrect calls for cycles used for calculation, and 7 did not report sequence far enough for calculation.

(d) Expressed as percent. Based on background corrected yield of Leu-2 & 10.

(e) Expressed as percent. Based on correct calls for positively identified residues only. Cycles that were left blank or had tentative assignments were not scored.

conclusions since each group has a very wide range, and for three groups very few data points were available. There was no apparent correlation between the number of positive correct assignments and the instrument age, amount of standard used or use of specialized sequencing cycles with PVDF samples.

Twenty-two of the responses contained sequence assignments made by both the operators and software allowing comparison of the accuracy of the operator *vs.* sequencing software made assignments (Table II). As was the case with the last two ABRF sequence samples [5,7], operator made assignments were more accurate than software assignments.

Although, 8 of the respondents (14.5%) encountered sequencer malfunction during the analysis of ABRF-90SEQ, the overall amount of lost data was low. Three cases involved a clogged injector needle, causing the loss of the entire sequencing run in one case, and a combined loss of 6 sequencing cycles in the other two. Two other cases involved mechanical problems (1 cycle lost) while

Table II. Accuracy of software and operator made sequence assignments [(a)]

Software assigned:

Pmole ratio[(b)]	Assigned	Correct	Wrong	% correct
0-5	88	43	45	48.9
5-10	91	45	46	49.5
10-20	104	66	38	63.5
>20	107	72	35	67.3
total	390	226	164	58.0
average / facility	17.7	10.3	7.5	

Operator assigned:

	Assigned	Correct	Wrong	% correct
positive	355	285	70	80.3
tentative	35	15	20	42.9
total	390	300	90	76.9
average / facility	17.7	13.6	4.1	

(a) Data are for 22 facilities which reported both software and operator made assignments. Some cycles in these reports were not assigned by either the operator or the software. Those cycles have not been included in the calculations.

(b) Pmol ratio is software defined and indicates the quantity of the identified amino acid relative to its background level.

the remaining three were electronic/software errors of unspecified origin (one premature termination; one cycle loss; no data loss). This year, all problems were instrument related as opposed to electrical failure which was a frequent problem last year [5].

IV. Conclusions

ABRF-90SEQ was designed to address new aspects of sequence analysis and was therefore more challenging than previous ABRF sequence analysis samples (see Table III). Reflecting this, the number of correct residues and the average number of correct positive calls fell sharply, while the ranges within these categories increased dramatically. Some obvious characteristics of ABRF-90SEQ which probably account for this trend are: (a) the decreased amount of sample available for sequencing, (b) the presence of a rapidly increasing background characteristic of a large protein, and (c) the use of PVDF as a support, which may affect certain individual residues more than others.

In contrast to these changes, the average repetitive yield of ABRF-90SEQ is comparable to STD-1 and ABRF-89SEQ. A relatively constant repetitive yield at ca. 90% may suggest that (a) the PVDF membrane does not affect this parameter, or (b) repetitive yield by itself may be a poor predictor of actual sequencer performance. In this respect it is noteworthy that, while the repetitive yield of the Porton instrument is actually lower than the other instruments, a significantly higher number of correct calls and total residues were reported (Table I). A large fraction (≥75%) of the PTH-amino acids being injected may have been a factor in longer sequence assignment. However, since only three facilities reported results from Porton instruments, this conclusion must be tempered by the possibility that these three facilities were particularly careful in their analysis since they were using a relatively new instrument. This might emphasize the importance of operators in sequence assignment. It was certainly possible to do well with other instruments (*vide supra*) since the best three responses were generated on theree different models, and overall, while 10 facilities had 5 or less positive correct calls, 8 reported 20 or more positive correct calls.

In general, this study together with the two previous sequence studies [5,7] show that as the amount of the material decreases, both the accuracy and extent of sequencing decrease substantially. As expected, low level sequence analysis in the presence of a high background is especially difficult. It is also noteworthy that the

Table III. Comparison of ABRF-90SEQ to Previous Test Samples [a]

Sample [b]	Description	Amount (pmol)	Repetitive Yield (%)	Number of Correct	Accuracy (%) [c]
STD-1 (1988)	single soluble peptide, 40 aa	100	88±3.9 (76-93)	24±7.1 (9-37)	95±7.0 (65-100)
ABRF-89SEQ	soluble peptides (5:1), 40 & 43 aa	240	89±2.4 (77-99)	30±6.5 (16-38)	95±6.6 (80-100)
ABRF-90SEQ	PVDF blot-mimic 100 kDa protein, 29 aa	30	89±3.7 (82-95)	13±6.4 [d] (1-27)	82±19.1 (36-100)

(a) Values are expressed as mean ± standard deviation, with ranges in parentheses.

(b) Data for STD-1 and ABRF-89SEQ are from references [5] and [7], respectively.

(c) Percentage of positive sequence assignments that were correct.

(d) Data from Table I. The difference between the values in Tables I & II stems from the number of data sets used (54 and 22 sets, respectively).

types of problems that lead to misassignments are sometimes quite different between a high level peptide dried from solution (ABRF-89SEQ) and a relatively small amount of large protein loaded onto PVDF membranes. In the latter case, it is much more difficult to distinguish a very minor contaminant presumably due to variable cycle-to-cycle yields of amino acids leading to difficulty in quantitation. Nonetheless, an average facility should expect to be able to identify 13 residues of a 100 kDa protein from a 30 pmol sample sequenced on a PVDF membrane. This compares very favorably with a separate survey where respondents indicated that on the average 84 pmol of sample is required to determine 15 amino acid residues of sequence information from the N-terminus of a 30 kDa protein [6].

Acknowledgements

This work was partially supported by NSF grant DIR 8903251 to K.R. Williams on behalf of ABRF. The cooperation of all the anonymous ABRF member facilities that participated in this survey by analyzing the samples and providing the data are gratefully acknowledged. The assistance of Dr. Clayton Buck and Marie Lennon (Wistar Institute) in providing the anonymity of the respondents is appreciated. The authors are especially grateful to: Mark Frazier and Ellen Gither (Washington University School of Medicine) for synthesizing the peptide; Dr. Walter McMurray (Yale University) for performing the mass spectral analysis of the peptide; Dr. Dwun-Hou Chen (Wistar Institute) for conjugation and adsorption of the peptide; Tony M. Jacobson (Texas College of Osteopathic Medicine) for amino acid composition analyses of the final sample; and Rodney Prell (Medical College of Wisconsin) for data entry.

References

1. O'Farrell, P.H. (1975) *J. Biol. Chem.*, **250,** 4007.
2. Towbin, H., Staehelin, T., and Gordon, J. (1979) *Proc. Natl. Acad. Sci. USA,* **76,** 4350.
3. Hewick R.M., Hunkapiller, M.W., Hood, L.E., and Dreyer, W.J. (1981) *J. Biol. Chem.*, **256:** 7990.
4. Matsudaira, P. (1987) *J. Biol. Chem.*, **262,** 10035-10038.
5. Speicher, D.W., Grant, G.A., Niece, R.L., Blacher, R.W., Fowler, A.V., and Williams, K.R. (1990) *in* Current Research in Protein Chemistry, J.J. Villafranca, ed., pp 159-166 Academic Press, San Diego.
6. Niece, R.L., Beach, C.M., Cook, R.F., Hathaway, G.M. & Williams, K.R.(1990) *FASEB J. (submitted for publication).*
7. Niece, R.L., Williams, K.R., Wadsworth, C.L., Elliott, J., Stone, K.L., McMurray, W.J., Fowler, A., Atherton, D., Kutny, R., & Smith, A.J. (1989) *in* Techniques in Protein Chemistry, T.E. Hugli, ed., pp. 89-101 Academic Press, San Diego.

15

EVALUATION OF SPECIFIC PROTEASES FOR DIGESTION OF MEMBRANE-BOUND PROTEINS

Wolfgang H. Fischer, Minkyu Park, Laura Lucero & Wylie Vale

The Salk Institute, 10010 N. Torrey Pines Road, La Jolla, CA 92037

I. INTRODUCTION

There are at least two different ways to obtain primary structure data from proteins purified by gel electrophoresis and electrotransferred to protein binding membranes. Proteins bound to polyvinylidene difluoride (PVDF) membranes can be subjected to Edman degradation directly (1). Difficulties are encountered with N-terminally blocked proteins or protein samples exhibiting a partially deleted N-terminus. Even in favorable cases it will be difficult to identify more than 30 residues. Alternatively, proteins transferred to nitrocellulose can be treated with protease *in situ* (2). Fragments can be resolved by reversed phase HPLC and subjected to Edman degradation. The proteolytic enzyme used most commonly is trypsin. In this laboratory, several novel proteins have been identified by combination of sequence analysis of tryptic fragments with the polymerase chain reaction (3) and cDNA cloning (4). In some cases the use of enzymes with a specificity different from that of trypsin is desirable. We have therefore evaluated the specific proteases Endoproteinase Glu-C, Endoproteinase Asp-N, and Endoproteinase Lys-C for use with this protocol. Four

Techniques in Protein Chemistry II

proteins were used for this evaluation: bovine serum albumin (BSA), ovalbumin, carbonic anhydrase, and β-lactoglobulin.

II. EXPERIMENTAL PROCEDURES

A. *Materials*

Sequencing grade enzymes were purchased from Boehringer Mannheim. BSA was from Pentex, ovalbumin, carbonic anhydrase and β-lactoglobulin were from Sigma. Electrophoresis apparatus and chemicals were from Bio-Rad. All other chemicals were high purity grade and purchased from Sigma. Nitrocellulose (pore size 0.45 μm) was obtained from Schleicher & Schuell.

B. *Instrumentation*

A Hewlett-Packard 1090L Liquid Chromatograph fitted with a Vydac C-18 column (particle size, 5 μm; pore size 300Å; dimensions, 2.1 x 150 mm) was used for HPLC separations.

Applied Biosystems 470A and 477A Protein Sequencers equipped with on-line PTH Analyzers (120A) were used to determine peptide sequences.

C. *Gel Electrophoresis and Electrotransfer*

Gel electrophoresis was carried out according to Laemmli (5), electrotransfer followed the procedure of Towbin et al. (6). Proteins were dissolved in SDS sample buffer containing 5% mercaptoethanol. Per lane, 100 pmol of protein was run. The proteins were diluted from a 1 mg/ml (based on dry weight) stock solution. On the transfer membrane, proteins were visualized with Amido Black 10B. A detailed description of these procedures can be found elsewhere (7).

D. Proteolysis and HPLC of Fragments

Excised protein bands were treated with protease employing a protocol developed for trypsin (7) based on a procedure by Aebersold et al. (2).
Briefly, membrane pieces were treated with PVP-40, rinsed with water and cut into 1 mm^2 pieces. Enzymes were added in 30 µl of digestion buffer (100 mM TES (N-tris[Hydroxymethyl]methyl-2-aminoethanesulfonic acid), pH8, 5% acetonitrile; 2mM Ca for trypsin digests). Incubation temperature was 37 C for all enzymes except Endoproteinase Glu-C which was used at room temperature. Supernatants were removed and membrane pieces were rinsed with 30 µl 0.5 M acetic acid twice. Pooled supernatant and washes were injected into the HPLC either immediately or stored at -20 C. HPLC separation was performed at a flow rate of 0.2 ml/min. Solvent A was 0.055% trifluoroacetic acid (TFA) in water, solvent B was 0.050% TFA in 90% acetonitrile and 10% water. The sample (85 µl) was applied at 100% solvent A. After 5 min at 100% solvent A, solvent B was increased to 45% within 60 min in a linear fashion. This was followed by a column wash at 100 % solvent B. Fractions were collected manually based on absorbance at 210 nm. As a control, a blank strip of nitrocellulose was treated the same way as the membrane-bound protein. This "blank digest" was also run on HPLC to identify autolysis and other artifacts. Using the HPLC conditions described above, an artifact of unknown identity elutes at approximately 30 min.

III. RESULTS

Four proteins (β-lactoglobulin, carbonic anhydrase, ovalbumin, and bovine serum albumin) ranging in molecular weight from 18,000 to 67,000 dal were used in this study. Varying numbers of fragments were obtained with the specific proteases (Endoproteinase Glu-C, Endoproteinase Asp-N, and Endoproteinase Lys-C). The proteins were also digested with trypsin for comparison.

The amount of enzyme was varied between 0.1 and 1 µg. Incubation times were varied between 2 and 16 hours.

For β-lactoglobulin, all significant peaks were sequenced. For other proteins, representative peaks were analyzed.

The optimal enzyme amount was determined to be 1 μg (using 100 pmol membrane-bound protein and 30 μl digestion buffer). The use of less enzyme results in lower yields of protein fragments. Larger amounts of enzyme do not improve yields but give rise to extraneous autolysis products.

After a reaction time of 8 hours, no further increase in protein fragment yields was observed. However, digestions were carried out overnight (16 h) as no deleterious effect was detected after this prolonged reaction time (with the exception of Endoproteinase Asp-N, for which fragment yields were slightly lower after 16 hours).

For all proteins employed in this study, sequenceable peptide fragments were obtained after proteolysis with the enzymes listed above. Fig. 1 shows the HPLC elution patterns of β-lactoglobulin (100 pmol) run out on an SDS polyacrylamide gel, electroblotted to nitrocellulose and digested with the four different enzymes used in this study. In the trypsin digest, nine fractions yielded unambiguous sequences and one fraction was a mixture of two sequences. For the other enzymes, between 4 and 6 fragments could be sequenced. The same general pattern was observed using the larger proteins. Trypsinization, giving the largest number of fragments, also resulted in the highest number of peptides overlapping in the HPLC separation. Starting with 100 pmol of each protein, initial sequencing yields varied between 30 and 10 pmol for the isolated fragments. Yields did not vary significantly between the different proteins or enzymes used. Larger fragments (>15 residues) showed lower initial yields than short peptides (<10 residues).

Due to the high enzyme concentrations employed and the long incubation times, unusual cleavage sites were observed for trypsin and for Endoproteinase Asp-N. Trypsin cleaved after a tyrosine in β-lactoglobulin (Fig. 1, this study). Previously we had observed cleavage after glutamine (8) and after methionine (9). Endoproteinase Asp-N did not only cleave N-terminal of aspartic acid but also N-terminal of glutamic acid. This cleavage pattern has been described as a side reaction for solution digests at low enzyme to substrate ratios (10).

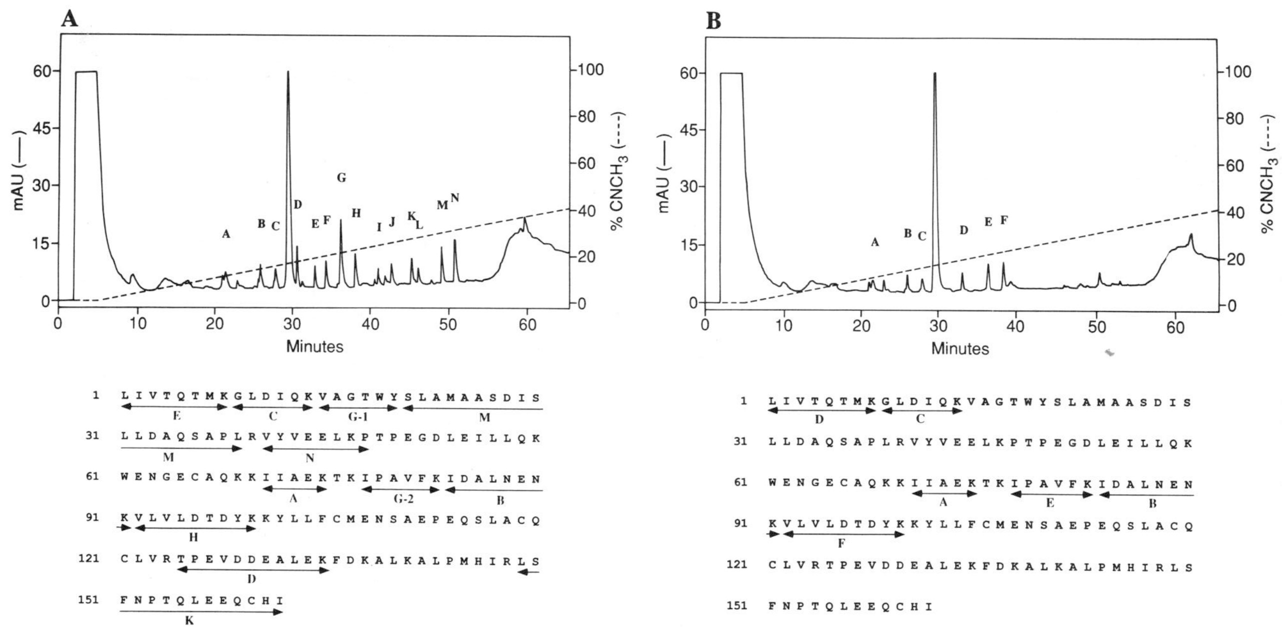

Figure 1a

Reversed phase HPLC of beta lactoglobulin digested with protease *in situ*. The proteases used were:
Panel A - Trypsin
Panel B - Endoproteinase Lys-C
The protein (100 pmol) was electrophoresed through a 12% acrylamide gel, electrophoretically transferred to nitrocellulose and digested as described in Experimental Procedures. The sequence data obtained is represented by arrows aligned to the published sequence of beta lactoglobulin (G. Preaux, G. Braunitzer, B. Schrank, and A. Stangl, Hoppe-Seyler's Z. Physiol. Chem. 360, 1595 (1979)). Absorbance was monitored at 210 nm.

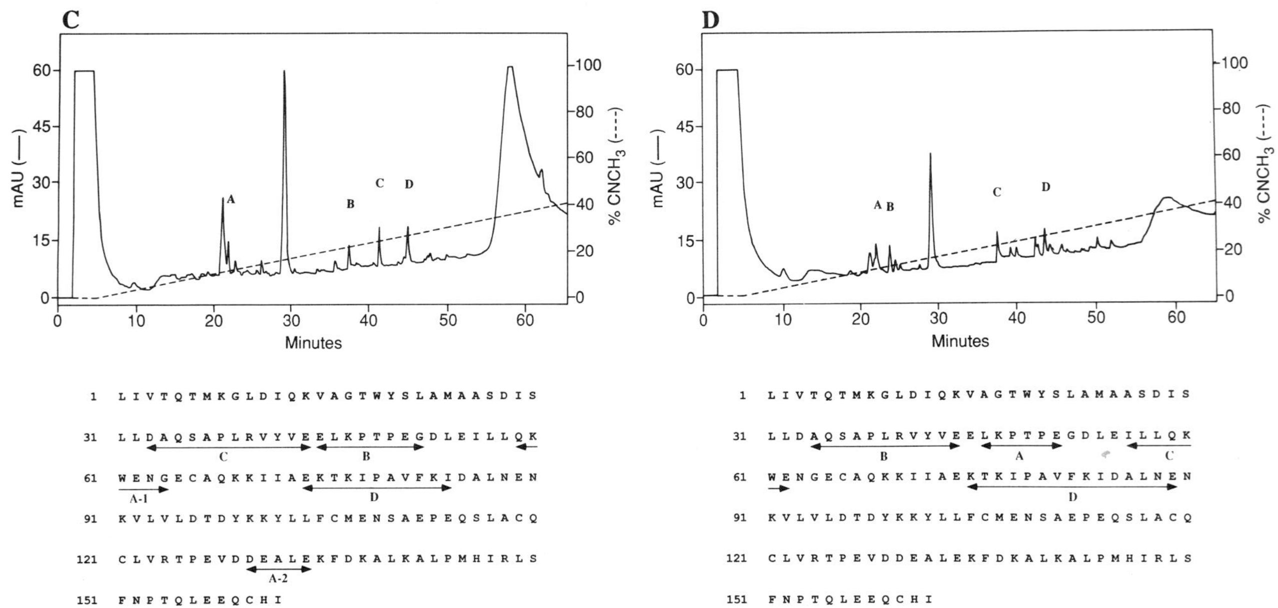

Figure 1b

Reversed phase HPLC of beta lactoglobulin digested with protease *in situ*. The proteases used were:
Panel C - Endoproteinase Asp-N
Panel D - Endoproteinase Glu-C
For conditions, see Figure 1a.

IV. DISCUSSION

All proteolytic enzymes tested in this study are compatible with *in situ* digestion procedures. It was shown that the TES buffer system developed for *in situ* digestions with trypsin can be used as a standard buffer with all enzymes tested. In general, fewer fragments were isolated when using specific Endoproteinases other than trypsin. For larger proteins containing many basic residues, trypsin digestion results in a number of fragments that is difficult to resolve by HPLC. Nevertheless, we conclude that trypsin is the enzyme most suitable for *in situ* digestions with two exceptions. For large proteins, the use of Endoproteinase Lys-C offers a good alternative as the number of cleavage sites is lower. In other cases, it might be desirable to analyze basic stretches of the protein. As these are destroyed by trypsin or Endoproteinase Lys-C, the use of Endoproteinase Asp-N or Endoproteinase Glu-C is recommended.

Not all possible fragments are observed after HPLC separation of *in situ* digested proteins. It is speculated that large fragments are either not released from the membrane in these digestions or they are not recovered from the reversed phase column. Small hydrophilic fragments are likely to be found in the void volume of the HPLC separation.

It is the current strategy in this laboratory to determine the partial protein sequence from a minimal amount of partially purified protein. This information is then used to determine the entire sequence by molecular biological techniques. The most powerful approach requires the knowledge of at least two stretches of the protein sequence. These sequences are used to design oligonucleotide primers to be employed in a polymerase chain reaction (4). In most cases, a long non-degenerate DNA will be obtained which can be used to obtain a full length cDNA coding for the protein precursor. The *in situ* procedures described in this study allow determination of the sequence of multiple protein fragments and can therefore be combined with the polymerase chain reaction based approach. Novel proteins characterized employing the procedures outlined above (using trypsin as protease) include a nuclear transcription factor (8), a circulating hormone-binding protein (11), a nerve cell survival protein (9), and a cytokine receptor (12).

V. ACKNOWLEDGEMENTS

This work was supported by NIH grants DK 26741 and HD 13527. This work was in part conducted by the Clayton Foundation for Research, California Division. W.V. is a Clayton Foundation Investigator.

VI. LITERATURE

1. P. Matsudaira, J. Biol. Chem. 262,10035 (1987)

2. R.H. Aebersold, J. Leavitt, R.R. Saavedra, L.E. Hood, S.B. Kent, Proc. Natl. Acad. Sci. USA 84, 6970 (1987).

3. R.K. Saiki et al., Science 230, 1350 (1985)

4. W.H. Fischer, G. Gonzalez, D. Behan, E. Potter, M. Montminy, W. Vale, J. Prot. Chem., 9, 341 (1990)

5. U.K. Laemmli, Nature 227, 680 (1970)

6. H. Towbin, T. Staehelin, J. Gordon, Proc. Natl. Acad. Sci. USA 76, 4350 (1979)

7. W.H. Fischer, D. Karr, B. Jackson, M. Park, W. Vale. Methods in Neuroscience, in press

8. G.A. Gonzalez, K.K. Yamamoto, W.H. Fischer, D. Karr, P. Menzel, W. Biggs, III, W.W. Vale , M.R. Montminy, Nature 337, 749 (1989)

9. D. Schubert, H. Kimura, M . LaCorbiere, J. Vaughan, D. Karr, and W.H. Fischer, Nature 344, 868-870 (1990)

10. U. Geuss, M. Schaeffer, J. Tschakert, S. Fischer, G.-B. Kresse. J. Prot. Chem. 9, 299 (1990)

11. D. Behan, E. Potter, W.H. Fischer, W. Vale et al., in prep.

12. R.A. Heller, K. Song, M. Onasch, W.H. Fischer, D. Chang, G. Ringold. Proc. Natl. Acad. Sci. USA, 87, 6151 (1990)

16

STUDY AND APPLICATIONS OF THE EFFECTS OF DETERGENTS AND CHAOTROPES ON ENZYMATIC PROTEOLYSIS

Lise R. Riviere, Mark Fleming, Christopher Elicone and Paul Tempst

Howard Hughes Medical Institute
Department of Genetics
Harvard Medical School
Boston, MA 02115

I. INTRODUCTION

Peptide mapping, in combination with chemical sequencing and mass spectrometric analysis, is a valuable tool for the study of post-translational modifications, characterization of overexpressed and mutant proteins, and for localization of artificial (e.g. chemical) modifications in the polypeptide chain. In addition, protein digestion and fragment separation is frequently required to overcome the problem of blocked N-termini during sequence analysis. Moreover, availability of internal sequences facilitates conventional and polymerase chain reaction (PCR)-based cloning techniques [1].

Efficient proteolytic cleavage is often complicated or even totally obviated due to insolubility or to protease resistance of the substrates. Drastic chemical modification prior to digestion has been the proven remedy for these problems. Manipulations, however, are tedious and frequently result in dramatic losses. In addition, certain labile chemical groups and bonds get destroyed in the process and cannot be further studied.

Supplementing the digest buffer with denaturing and/or solubility promoting agents is therefore an attractive alternative. The addition of 2 M urea is quite general [2,3]; a comprehensive study on the effects of organic co-solvents has been published [4]. Although potentially very suitable, detergents and guanidine hydrochloride (Gu.HCl) have not found widespread use for this purpose. In the absence of any ready protocol, we investigated the activities of several endoproteases in the presence of increasing concentrations of urea, Gu.HCl, CHAPS and SDS. Given here is a preliminary account of this study and a discussion of specific applications for which the use of the new methods proved to be essential.

Techniques in Protein Chemistry II

II. MATERIALS AND METHODS

Optimized digest conditions for five common proteases are listed in Table I. Digests were always carried out at 37°C with enzyme/substrate ratios of 1/25 (w/w). Cytochrome c, ß-lactoglobulin (BLG), bovine serum albumin (BSA), chymotrypsin (C) and subtilisin (S) were obtained from Sigma. Trypsin (T) and *Staph. aureus* V8 endoproteinase Glu-C (V8) were sequencing grade enzymes from Boehringer-Mannheim; *Achromobacter lyticus* M497-1 protease I (KC) was from Wako chemicals. Solutions of Gu.HCl (sequencing grade, Pierce) and acetonitrile (Burdick & Jackson) were always prepared fresh before use; CHAPS (Pierce) and SDS (Bio-Rad) were kept as 20% stock solutions and 10 M urea (ultrapure reagent, American Bioanalytical) was stored over mixed bed AG 501-X8 resin (Bio-Rad) at room temperature.

Typically, a protein pellet is dissolved in a minimal volume of 6 M Gu.HCl, heated at 50°C for 30 min, followed by the addition of two volumes of appropriate digest buffer. The enzyme of choice is then added and the mixture is incubated at 37°C. Alternatively 8 M urea can be used and the protein is warmed at 37°C for 30 min. Endoproteinase Lys-C or subtilisin are then added or the urea solution is diluted for use with other enzymes. Substrates in 1% SDS can also be preheated before digest to promote solubilization and denaturization. It is preferable to have substrate concentrations > 0.1 µg/µl with Gu.HCl and 1% SDS and > 0.05 µg/µl with urea. After digest, SDS containing samples were mixed with an equal volume of 1 M Gu.HCl, and the formed guanidine dodecyl sulfate (GuDS) precipitates were removed by a 5 min spin in an Eppendorf centrifuge before injection onto a reverse phase column. After digest, acetonitrile concentrations were diluted to 5% by adding an appropriate volume of 0.1% TFA to the sample. All other digest mixtures were injected as they were.

Peptide pattern analysis was done on Vydac C4 (214TP54) and C18 (218TP54) columns (The Separations Group) using a model 150A (Applied Biosystems) HPLC system and Turbochrom 2700 (PE Nelson chromatography software. An acetonitrile gradient in 0.1% TFA was always used unless indicated otherwise.

III. RESULTS AND DISCUSSION

A. *Systematic study of endoproteolysis*

Endoproteases were systematically tested in the presence of increasing concentrations of urea, Gu.HCl, SDS, CHAPS and acetonitrile using cytochrome c, ß-lactoglobulin (BLG) and bovine serum albumin as model substrates. Analysis of the digests was done by SDS-PAGE (except for Gu.HCl samples) and by RP-HPLC. The highest concentrations of additives that still permitted a fairly complete digest are listed in Table I.

It should be noted that the conditions listed in Table I, for urea and Gu.HCl containing samples, are the result of optimization experiments and usually represent a compromise between complete digest and fewest partials. It is not advisable to change them without prior testing and conditions should definitely be kept constant during peptide mapping experiments. The buffer and pH requirements for Lys-C and subtilisin are in general more flexible with respect

TABLE I: ENDOPROTEOLYSIS IN THE PRESENCE OF CHAOTROPES, DETERGENTS AND ACETONITRILE.

ENZYME	T	C	KC	V8	S
NATIVE					
CONDITIONS:	0.1M AB	0.1M AB	0.1M AB	0.05M AB	0.1M AB
(time)	2 h	2 h	2 h	5 h	1 h
UREA	4M	2M	8M	2M(R)	8M(R)
(buffer)	+5mM Ca^{2+}		0.2M TC		0.1M TC
(time)	15 h	5 h	5 h	5 h	1 h
Gu.HCl	(1-2)M	2M	2M	1M(R)	2M(R)
(buffer)	0.2M TC	0.2M AB	0.2M TC		
(time)	24 h	24 h	5 h	18 h	2 h
SDS	<0.1%	0.1%(R)	1%	0.1%(R)	1%
CHAPS	2%	1%	2%	2%(R)	2%(R)
ACETONITRILE	40%	30%	40%	20%(R)	40%

Buffers and times are as under native conditions, except were indicated. The highest concentrations of additives, permitting adequate proteolysis, are listed; concentrations higher than 2% CHAPS and 40% acetonitrile were not tested. For more details see Methods.

Abbreviations: T (trypsin), C (chymotrypsin), KC (*Achromobacter* protease I, a.k.a. endoproteinase Lys-C), V8 (*Staph. aureus* endoproteinase Glu-C), S (subtilisin), AB (ammonium bicarbonate), TC (Tris.HCl pH 8.5), R (restricted digest).

to obtaining complete digests than to pattern reproducibility. Gu.HCl solutions must be made fresh daily or the digests will be adversely affected. Urea solutions can be stored over mixed-bed resins at room temperature for several weeks; artificial blocking of the newly generated N-termini during digests was not evident (as determined by sequencing).

Proteases active in 2 M Gu.HCl (C,KC,S) and in 8 M urea (KC,S) provide us with efficient tools to digest compact, stable substrates whose physical properties prevent normal enzymatic degradation. A quick study of the literature and some in-house experiments indicated that, among many others, RNAse, NADase, lysozyme, amylase, superoxide dismutase, triose phosphate isomerase, xylose isomerase and pancreatic trypsin inhibitor are such proteins. After heating in 6 M Gu.HCl and subsequent dilution to 2 M, they apparently do not refold to a compact structure, native or other, rendering them amenable to digestion; alternatively, 8 M urea can be used (see Methods). All the "protease resistant" substrates listed above could be digested by one or the other method, but not always by both (results not shown). From our experience, Gu.HCl in combination with KC was the most successful method. When the Gu.HCl and urea concentrations are lowered to respectively 1 M and 4 M, the digests no longer proceed. This excludes, for instance, the use of trypsin and V8 for these purposes.

Although enzymatic digestion in the presence of SDS is not new [5], it has neither systematically been explored nor found widespread use in combination with HPLC, primarily because of interference of the dodecyl sulfate with the separation on reverse phases [6,7]. We followed a suggestion by Shively [6] to remove excess SDS by Gu.HCl precipitation prior to RP-HPLC. Chymotrypsin and V8 retain activity in 0.1% SDS. While some undigested substrate usually remains after 18 h incubation with V8, chymotrypsin gives complete cleavage albeit rather restricted (see below), which, depending on the purpose of the digest, may be an advantage. KC and subtilisin degrade almost any protein in 1% SDS after preliminary heating of the substrate at 95^{0}C. This undoubtedly has practical advantages for fragmentation of membrane proteins and unusually stable substrates. However, the recovery of peptides after GuDS precipitation is highly variable, with the worst yields for larger and/or hydrophobic peptides. This problem is nearly nonexisting with SDS concentrations of 0.1% or less.

With few exception, endoproteolysis works rather well in detergents (other than SDS) and in acetonitrile, and this practice should find widespread applications. We found comparable results for CHAPS, octylglycoside and NP-40; Triton X-100 gives a lot of background peaks during RP-HPLC and is

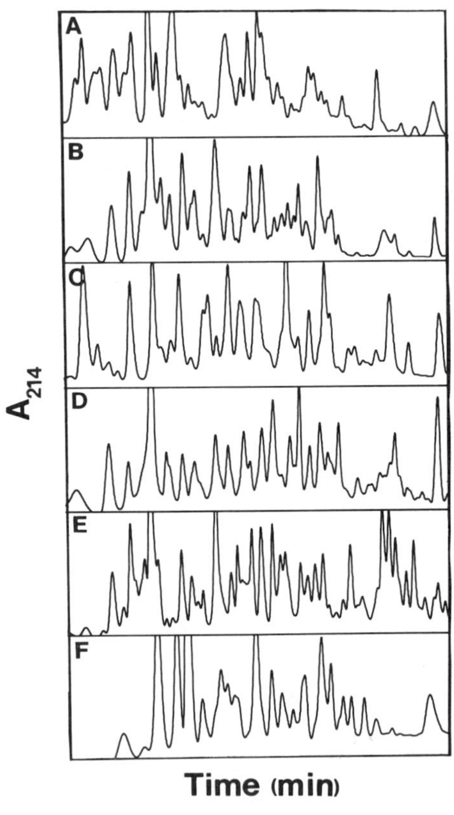

FIG. 1: RP-HPLC of BSA digested with subtilisin under standard conditions (A), or supplemented with 4 M urea (B), 2 M Gu.HCl (C), 40% MeCN (D), 1% SDS (E) and 2% CHAPS (F). Details are under Methods and in Table I.

therefore not useful. Since 2% of these nonionic and zwitterionic detergents is usually enough to keep membrane proteins solubilized, we didn't test any higher concentrations. When heated, several proteins precipitate in acetonitrile concentrations over 40% and no higher concentrations were therefore tested.

Although not the major focus of our study, we observed that under certain conditions digests were restricted to various extents; i.e. there were fewer peaks but in good yields. Enzyme/additive combinations where this was the case are marked with (R) in Table I. As illustrated in Fig. 1 for subtilisin digests of BSA, there is some serious variability of the peptide patterns under different conditions. However, the fairly "messy" profile (many small peaks and shoulders, no real prominent later eluting peaks) of the native digest is cleaned up in the presence of CHAPS and acetonitrile. Experiments are under way in our laboratory to further explore and enhance these effects. With no exceptions, even with enzymes of low specificity (e.g. subtilisin, thermolysin, pepsin, ...), peptide patterns are very reproducible under a strictly controlled set of conditions. All enzymes and additives can therefore be used for peptide mapping experiments.

B. *Sequencing, cDNA cloning and surface topography analysis of the ryanodine receptor/junctional channel complex (JCC)*

The JCC, involved in muscle excitation-contraction and calcium release, consists of 4 identical subunits with a Mr of 600,000 dalton each. The complex is not soluble without the presence of detergents, not even in 8 M urea. Final preparations of purified JCC, obtained in either 0.5% SDS or 1% CHAPS, were digested with KC. Resulting peptides were separated by two rounds of RP-HPLC (Fig. 2, panels A and B), sequenced and allowed partial cloning of the gene by PCR experiments [8]. Northern blots indicated a 16 Kb message, consistent with the extreme large size of the monomers.

Since CHAPS did not appear to disrupt the tetrameric structural integrity of the JCC or its ability to function as a calcium channel in a reconstituted system [9], we subsequently used selective proteolysis in the presence of this detergent to identify surface exposed regions. Time course experiments with several proteases and at varying temperatures indicated that complete cleavages were obtained with KC and V8 at 37°C after 6 h and 4 h incubation time respectively; prolonged incubations did not alter the electrophoretic profiles. Peptide mixtures were separated on a C4 column (Fig. 2, panels B and C), analyzed and positioned in the complete JCC primary structure (5,037 amino acids) translated from the gene sequence. The KC and V8 peptides were overlapping and highly clustered within 5 discrete regions, comprising only 5% of the total sequence. The results, combined with computer assisted analysis of the JCC sequence using predictive algorithms, allowed assignment of surface exposed areas and monomer interfaces [9].

C. *Identification of the single active site histidine in xylose isomerase from Streptomyces*

D-Xylose isomerase (XI) is a tetrameric enzyme that catalyzes the conversion of xylose and glucose to xylulose and fructose, respectively. Group specific chemical modification of XI with diethyl pyrocarbonate (DEPC) indicated that complete loss of activity is fully correlated with the acylation of a

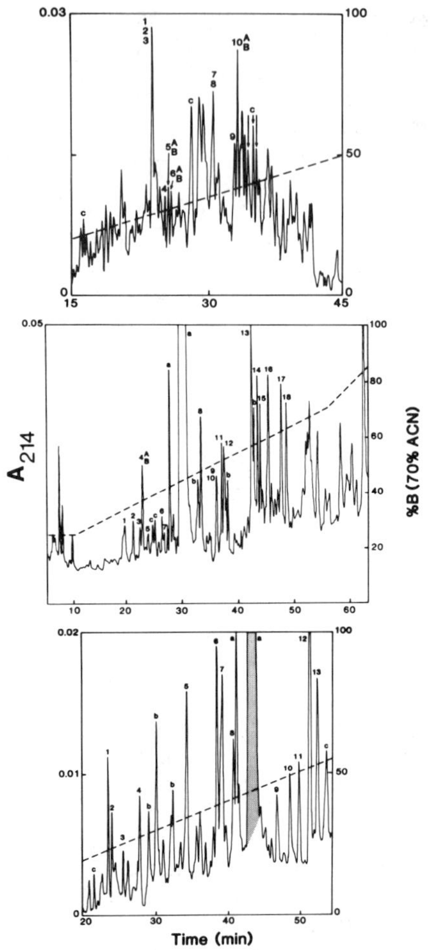

FIG. 2: RP-HPLC of the ryanodine receptor/junctional channel complex (JCC) digested with KC (A,B) and V8 (C) in the presence of 0.5% SDS (A) or 1% CHAPS (B,C). From [9], details are in the text.

single histidine; active site protection with xylitol, a competitive inhibitor, completely prevented both this modification reaction and inactivation. One approach to locate this histidine residue in the primary structure is by comparative peptide mapping between fully active and inactive forms of DEPC-derivatized XI and then isolate and sequence the differential peptide(s). The problem is that, first, XI is an extremely protease resistant enzyme (even against proteinase K [10]) and second, carbethoxy histidine derivatives are temperature sensitive and only stable strictly between pH 5 and 7.

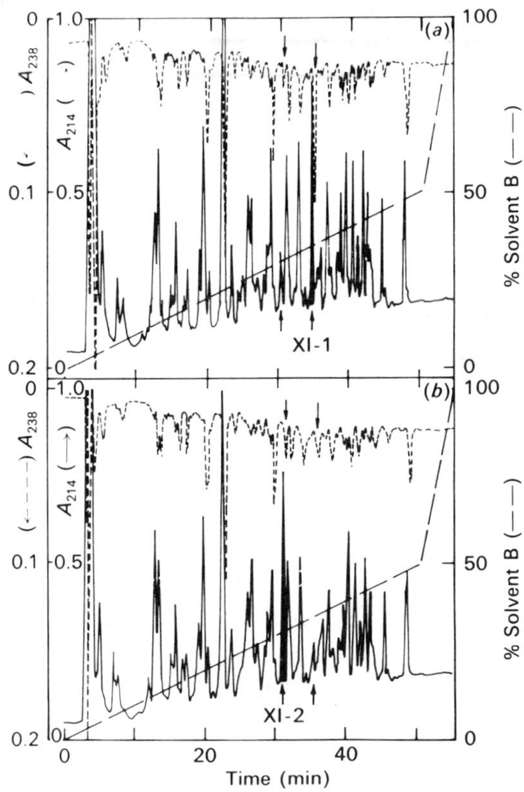

FIG. 3: RP-HPLC of xylose isomerase, derivatized with DEPC in the absence (A) or presence of xylitol, and digested with subtilisin in the presence of 2 M Gu.HCl. From [10], for more details, see text.

The following protocol was developed: modified XI's (in phosphate buffer pH 6) were denatured by adding 3 volumes 8 M Gu.HCl in 50 mM phosphate buffer pH 7 and subsequent incubation for 10 min at 30°C. The solution was then diluted three fold with buffer, subtilisin (1/20 by weight) was added and the digest allowed to proceed for 2 h at 30°C. Peptide maps were obtained on a C4 column with an acetonitrile gradient in 12 mM phosphate buffer pH 6.0 (Fig. 3); the profile of inactive XI is shown in panel A, the one of active XI in panel B. The profiles are nearly identical, except for one major peak that has shifted considerably (designated XI-1 in A and XI-2 in B), presumably because of the two additional carbon atoms (carbethoxy function) on peptide XI-1. Both peaks were sequenced and indeed found to be identical peptides. They contain a single histidine that could easily be positioned as residue 54 in the complete sequence [10]. Interestingly, this region is highly conserved among all bacterial xylose isomerase whereas most of the molecule is not.

D. Sequencing, cDNA cloning and disulfide array assignment of an NADase

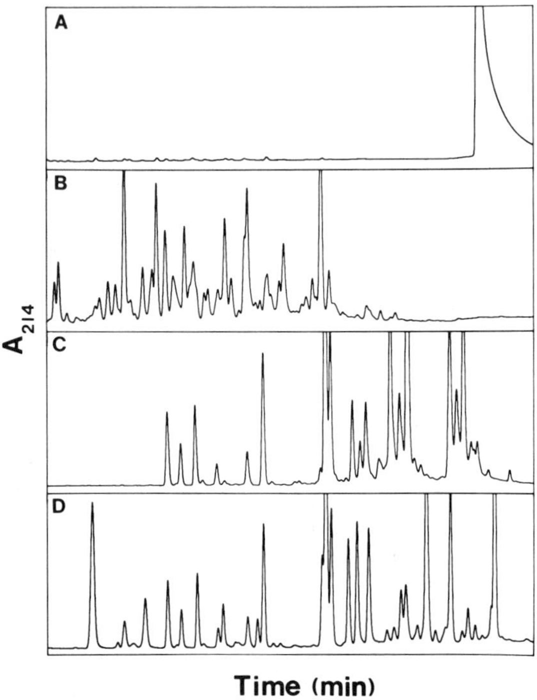

FIG 4: Narrow-bore RP-HPLC of NADase treated with subtilisin in the absence (A) or presence (B) of 2 M Gu.HCl and with KC/2 M Gu.HCl without (C) or with (D) post-digest reduction.

An ovotestis specific NAD glycohydrolase (29 KDa) was isolated from *Aplysia*. Amino terminal sequencing yielded sequences that were unsuited for probe design and lengthy incubations of the enzyme with subtilisin did not result in any noticeable fragmentation (Fig. 4A). However, peptide maps were obtained by digesting with subtilisin (Fig. 4B) and KC (Fig. 4C) in the presence of 2 M Gu.HCl, after prior denaturation of the substrate in 6 M Gu.HCl. Sequencing of several KC-peptides yielded sufficient useful data to allow cloning of the corresponding gene (Glick et al., in preparation).

Since our proteolysis techniques do not require prior reduction and alkylation of the substrate, disulfide bonds remain intact and covalently linked peptides elute as single peaks. Although this may be perceived as a disadvantage it can actually be exploited to obtain additional structural data. If preferred, there is no difficulty reducing and even alkylating the peptide mixtures after digest and before injection; this, not surprisingly, yields a larger number of peaks (Fig. 4D). More interesting is to purify the peptides to homogeneity, including disulfide linked pairs, and then perform reduction and alkylation followed by renewed chromatography of those pairs. Peaks corresponding to disulfide linked peptides are easily located by comparison of reduced versus unreduced peptide maps, as is for instance shown in Fig. 4 (profile D versus profile C). By doing so, we were able not only to provide sequences to assist in gene cloning experiments, but also to locate all five disulfide bonds in the protein (M. Fleming and P. Tempst, unpublished). Indeed, the primary structure, translated from the gene sequence, showed the presence of 10 cysteines. Pilot and micro-preparative digests, sequencing of 163 out of a total 255 amino acids and assignment of the complete disulfide array was done with a total of 530 pmol starting material.

REFERENCES

1. Tempst, P., Link, A.J., Riviere, L.R., Fleming, M. and Elicone, C. (1990) Electrophoresis **11**, 537-553.
2. Tempst, P., Woo, D., Teplow, D., Aebersold, R., Hood, L. and Kent, S. (1986) J. Chromatogr. **359**, 403-412.
3. Stone, K., LoPresti, M., Williams, N., Crawford, J., DeAngelis, R. and Williams, K. (1989) In: Techniques in Protein Chemistry (Hugli, T., Ed.) pp.377-391. Academic Press, San Diego, CA.
4. Welinder, K.G. (1988). Anal. Biochem. **174**, 54-64.
5. Cleveland, D.W., Fisher, S.G., Kirschner, M.W. and Laemmli, U.K. (1977) J. Biol. Chem. **252**, 1102-1106.
6. Shively, J.E. (1986) In: Methods of Protein Microcharacterization (Shively, J., Ed.) pp.41-84. Humana Press, Clifton, NY.
7. Tempst, P., Hood, L.E. and Kent, S.B.H. (1987) In: Modern Methods of Plant Analysis (Linskens, H. and Jackson, F., Eds.) **Vol.5**, 170-208. Springer-Verlag, Heidelberg, FRG.
8. Marks, A.R., Tempst, P., Hwang, K., Taubman, M., Chadwick, C., Fleisher, S. and Nadal-Ginard, B. (1989) Proc. Natl., Acad. Sci. USA **86**, 8683-8687.
9. Marks, A.R., Fleischer, S. and Tempst, P. (1990) J. Biol. Chem. **265**, 13143-13149.
10. Vangrysperre, W., Ampe, C., Kersters-Hilderson, H. and Tempst, P. (1989) Biochem. J. **263**, 195-199.

17

Effect of Cleavage Conditions on Edman Degradation of Proline-rich Proteins: Application to Wheat Storage Proteins

William H. Vensel
Donald D. Kasarda
Western Regional Research Center, U.S. Department of Agriculture,
Agricultural Research Service, Albany CA

I. Introduction

This study was initiated in an attempt to determine practical guidelines that would allow reliable, extended sequencing by the Edman degradation (1) of the proline-rich wheat storage (gluten) proteins. It has long been recognized that the occurrence of proline in a sequence can initiate overlap or "lag" during Edman degradation (2). The factors that influence the length of a degradation by the Edman method have been described (3-6, and references therein). The yield of an amino acid at each cycle of the degradation, which can be as high as 96% of the yield at the previous cycle, is a function of the combined efficiencies of the coupling reaction and the cleavage reaction. Failure of the coupling or the cleavage reaction to proceed to completion can result in the buildup of lag and displacement of the actual sequence position by one or more cycles. Proline residues especially contribute to this lag--with the problem arising from the cleavage reaction rather than from the coupling reaction (2). Partial cleavage of phenylthiocarbamyl-proline (PTC-Pro) when it is followed by basic amino acids has been reported (7) and differences in the kinetics of the proline cleavage reaction that are attributable to the following amino acid residue have been noted (2,8). Cleavage rates for selected peptides have been shown to vary by several orders of magnitude--with peptides having proline in the first position being the slowest to cleave (8).

The wheat storage proteins are notable for their high proportions of proline and glutamine, approximately 15% and 35%, respectively, and for repeating sequences that include recurring combinations of -Pro-X- where X is frequently glutamine, but sometimes other amino acids, especially tyrosine, phenylalanine, or glycine (9,10). The frequent occurrence of proline in the sequences of wheat gluten proteins contributes strongly to the development of lag and eventually high background values for all amino acids. The frequent occurrence of glutamine residues (often in series) contributes to the development of glutamine background.

The position of proline residues in a protein will not always be known at the start of a sequencing experiment and the frequent occurrence of proline can cause a rapid decrease in the repetitive efficiency as a result of partial cleavage at proline. The resulting lag contributes to and eventually becomes indistinguishable from background; this limits the number of cycles that can be obtained. In order to obtain extended sequence information from proline-rich proteins, the slow rate of cleavage at proline residues must be effectively countered. We have attempted to develop an improved strategy for the sequencing of proline-rich plant storage proteins, first when the proline positions are unknown and then when they have been identified. Two factors that must be balanced in order to improve the sequencing of proline-rich proteins are: 1) the effect of partial cleavage at proline to increase the out-of-phase signal and the concomitant decrease in the in-phase signal; and 2) the effect of prolonged cleavage times on the fragmentation of the protein at acid labile peptide bonds (11) with a subsequent increase in background as the result of newly-generated fragments coupling with phenylisothiocyanate.

II. Materials and Methods

The reference protein for development of sequencing strategies was A-gliadin, an α-type gliadin found in the bread wheat cultivar 'Scout 66' for which the complete sequence was available (9). The N-terminal sequence of A-gliadin through 60 residues is shown in Fig. 1.

1 11 20
V-R-V-**P**-V-**P**-Q-L-Q-**P**-Q-N-**P**-S-Q-Q-Q-**P**-Q-E-

21 31 40
Q-V-**P**-L-V-Q-Q-Q-Q-F-L-G-Q-Q-Q-**P**-F-**P**-**P**-Q-

41 51 60
Q-**P**-Y-**P**-Q-**P**-Q-**P**-F-**P**-S-Q-Q-Q-**P**-Y-L-Q-L-Q-

Figure 1. N-terminal amino acid sequence of A-gliadin.

It can be noted that this sequence includes 15 prolines. A different α-type gliadin, known to be coded on chromosome 6D, was used to test the best conditions. The protein, prepared from the cultivar 'Chinese Spring' (which does not have A-gliadin), was a gift of Dr. A. I. Rybalka. It was expected to have a similar, but not necessarily identical, sequence to A-gliadin.

Sequencing was carried out with an Applied Biosystems model 477A automatic amino acid sequencer equipped with the on-line 120A HPLC system for identification of phenylthiohydantoin (PTH) amino acids. Data analysis was carried out with software version 1.61 from Applied Biosystems. All sequencing was carried out with norleucine as an internal standard for each cycle and the data were corrected for norleucine recovery. Three different uniform cleavage times, 5 min, 10 min, and 15 min, were tested at 48 °C, and two different uniform cleavage times, 10 min and 15 min, at 53 °C. These two temperatures were chosen because 48 °C is the recommended operating temperature, and it was hoped that a moderate increase in temperature would preferentially increase proline cleavage without unduly increasing chain cleavage at acid sensitive residues and/or evaporation of trifluoroacetic acid from the protein support. The reaction chamber program for the 48 °C experiments was Normal-1 and for the 53 °C experiments, Fast-1. Coupling in Fast-1 is carried out at 53 °C, which permits shorter analysis time. Cleavage times were set according to the above scheme. After completion of sequence analysis, the data from the TABULATE.RPT file were imported into Lotus 1-2-3 for subsequent analysis. The raw data, background subtracted data, and the lag corrected data (background subtracted data corrected for lag)(6) were used to prepare Figs. 3, 4 and 5.

III. Results and Discussion

We determined the time required to cleave completely the first PTC-Pro encountered (at position 4) in the sequence of A-gliadin by subjecting it to a single coupling followed by five cycles of cleavage (5 min each) and extraction at each cycle. The data shown in Fig. 2 are for cleavages carried out at 48 °C and 53 °C; the points shown represent averages of duplicate experiments. The data demonstrate that with 5 min of cleavage at 48 °C only 60% of the proline in the PTC-Pro-Val bond was released, increasing to 82% at 53 °C. Thus, the results shown in Fig. 2 indicate that an increase in lag of about 40% would be expected when PTC-prolines are cleaved for 5 min at 48 °C. When the PTC-Pro at position 6 was cleaved at 48 °C for 5 min, however, about 70% of the proline was released. The difference presumably results mainly from the difference in cleavage behavior of PTC-Pro-Val- vs. PTC-Pro-Gln-. As shown in Fig. 2, cleavage for 10 min

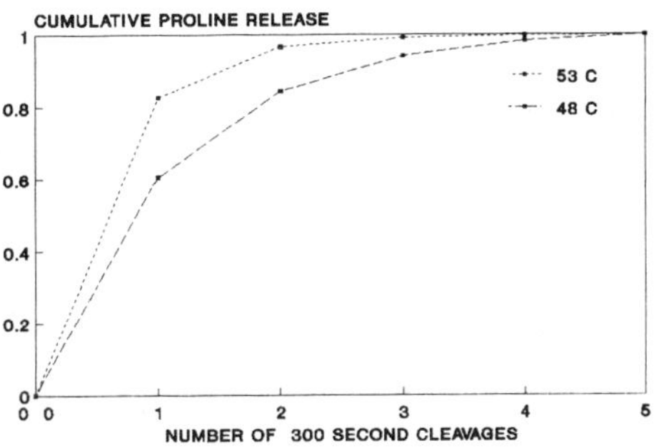

Figure 2. Cumulative proline release for proline at position 4 of A-gliadin.

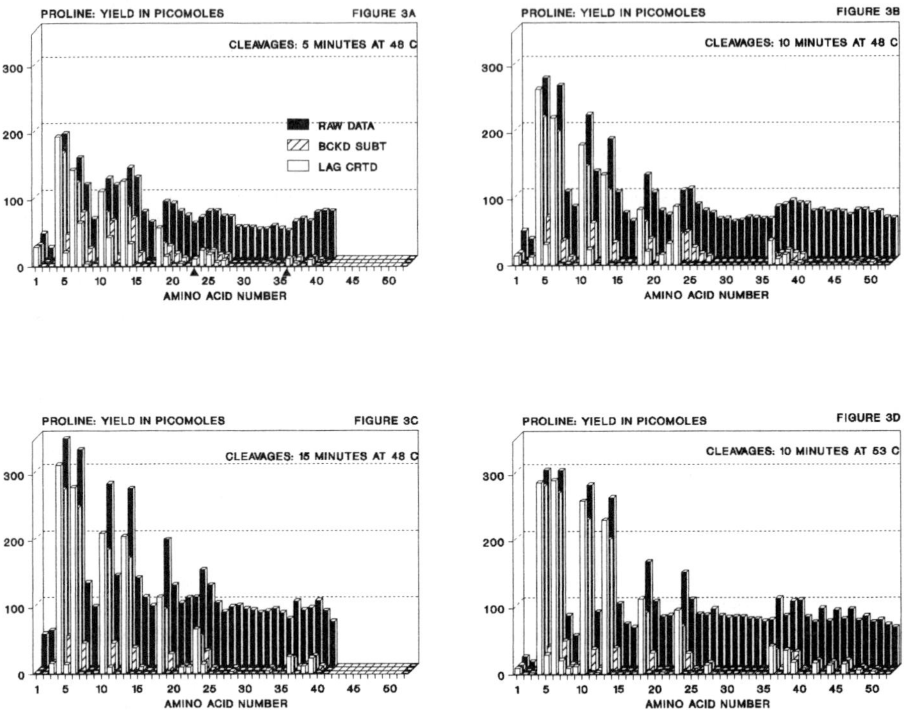

Figure 3. The proline data are the raw data (RAW DATA), background subtracted data (BCKD SUBT), and the background subtracted data that has been corrected for lag (LAG CRTD). Cleavage conditions are as described in each figure panel.

(two 5 min cleavages) at 53 °C increased the percent of the PTC-Pro cleaved to about 95%. In other experiments (data not shown), however, we found that a single 10 min cleavage at 53 °C resulted in a lower yield (90%) of PTH-Pro than two 5-min cleavages. In contrast to the cleavage rates at PTH-Pro residues, cleavages involving other PTH-amino acids were quite rapid. For example, the N-terminal valine was 90% cleaved in 1 min at 48 °C (data not shown).

Further investigation involved studies of the effects of various conditions on the length of identifiable sequence obtained for A-gliadin. Various preliminary experiments were carried out before the final data set was obtained. The precision of the single experiments that comprise the final data set was increased by the use of an internal standard for every cycle. The effect of three different uniform cleavage times at 48 °C on extended sequencing of A-gliadin is shown in Fig. 3(-A, B, and C) along with the uniform condition of 10 min cleavage at 53 °C (Fig. 3D). A uniform condition of 5 min at 53 °C was also studied, but that data is not shown. In all cases 600 picomoles of protein were loaded. Sequencer runs were terminated after the sequence could no longer by recognized by visual examination of the chromatograms. The 5 min and 15 min

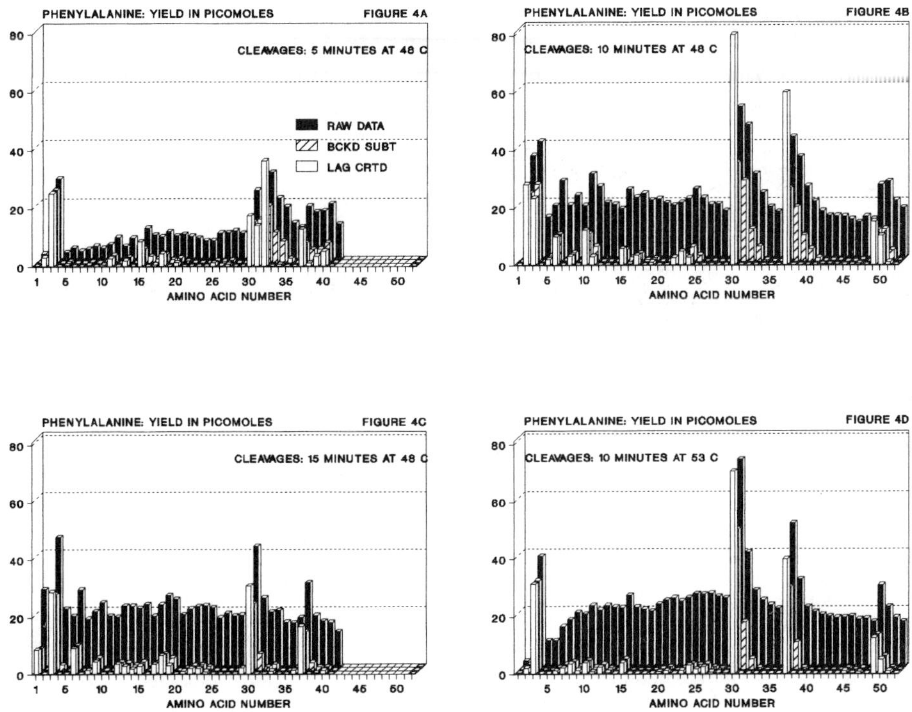

Figure 4. Cleavage conditions for the phenylalanine data are as described in each figure panel. Abbreviations are as described in the Fig. 3 legend.

experiments were terminated after 41 cycles of Edman degradation and the 10 min experiments after 52 cycles. The initial yields of valine were 395, 352, and 325 picomoles for the experiments conducted with cleavages at 48 °C for 5, 10, or 15 min, respectively, whereas for the experiment carried out at 53 °C (Fig. 3D), the initial yield of valine was 350 picomoles Although similar initial yields were obtained for all four conditions, the proline signal was most attenuated for the 5 min cleavage at 48 °C (Fig. 3A) and there was considerable lag even after correction for lag by the data analysis program. These latter conditions allowed PTC-Pro identification through position 36. When cleavages were for 10 min at 48 °C (Fig. 3B), the lag was considerably reduced and PTC-Pro could be detected above background at position 38. Increasing the cleavage time to 15 min at 48 °C (Fig. 3C) decreased the lag even further, but despite the decreased lag, the PTC-Pro recovery was less for later prolines (positions 24 and 36) than with other conditions. Examination of these results indicated to us that a 10 minute cleavage time appeared to be the best of the three different cleavage times at 48 °C. The 10 min cleavage at 53 °C (Fig. 3D) allowed clear identification of PTC-Pro above background through the proline at position 46,

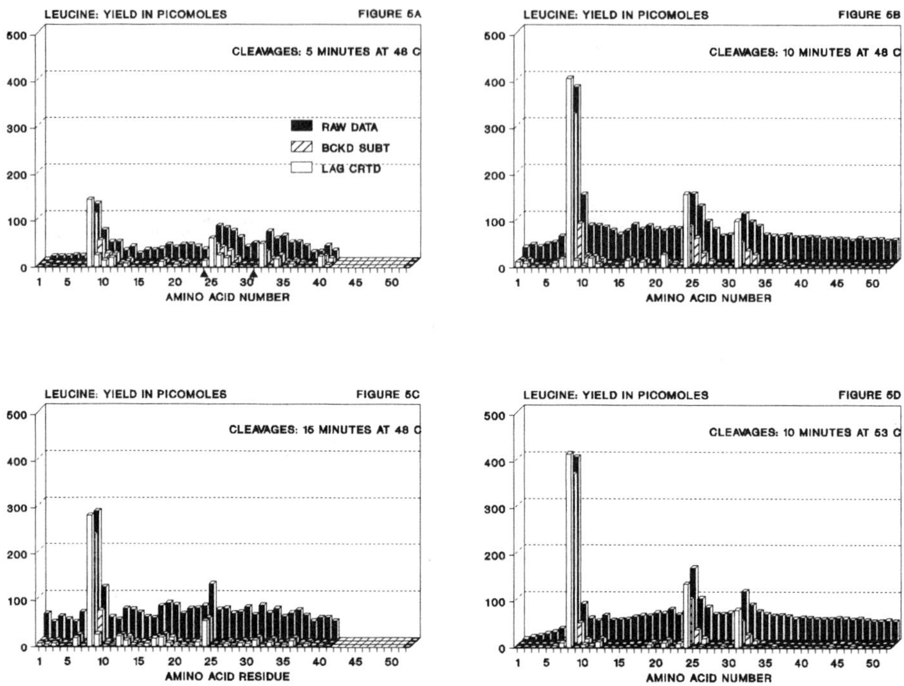

Figure 5. Cleavage conditions for the leucine data are as described in each figure panel. Abbreviations are as described in the Fig. 3 legend.

and reasonably confident identification at positions 48 and 50. Increasing cleavage temperature to 53 °C from 48 °C appears to increase proline cleavage with only minimal effect on background (Figs. 3C and 3D).

The effect of the various cleavage conditions on background and yield of other amino acid derivatives is shown for two amino acids, phenylalanine and leucine. The phenylalanine data (Figs. 4A-D) indicated that the background was lowest when all cleavages were for 5 min. Despite considerable lag in the in-phase signal (Fig. 4A), it was possible to correctly identify the occurrence of phenylalanine at positions 30 and 37. Note that the lag is considerable by position 30 (Fig. 4A) so that the maximum in the signal actually occurred at position 33. Increasing the cleavage time (Figs. 4B, 4C) reduced the lag, but the use of uniform cleavages of 15 min at 48 °C and 10 min at 53 °C resulted in a decrease in the lag-corrected signal when compared to cleavages carried out for 10 min at 48 °C. The phenylalanine peak at position 3 results from a minor variant sequence of A-gliadin.

When the leucine data (Fig. 5) were examined, it was possible to observe the presence of leucine at positions 8, 24, and 31 for cleavage times longer than 5 min (Figs. 5B, 5C, 5D), but when the cleavage time was 5 min, the maxima in the leucine signals were shifted to position 25 from position 24 and to

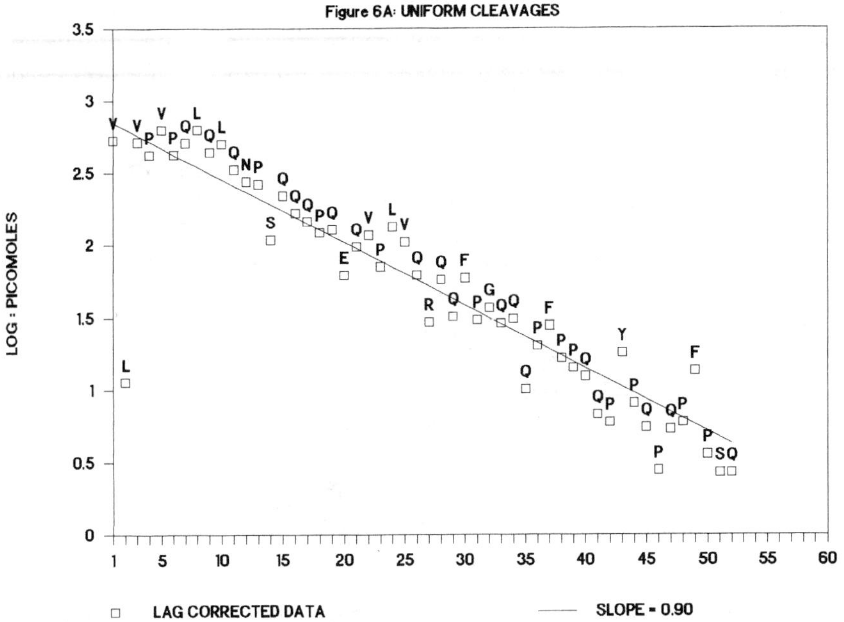

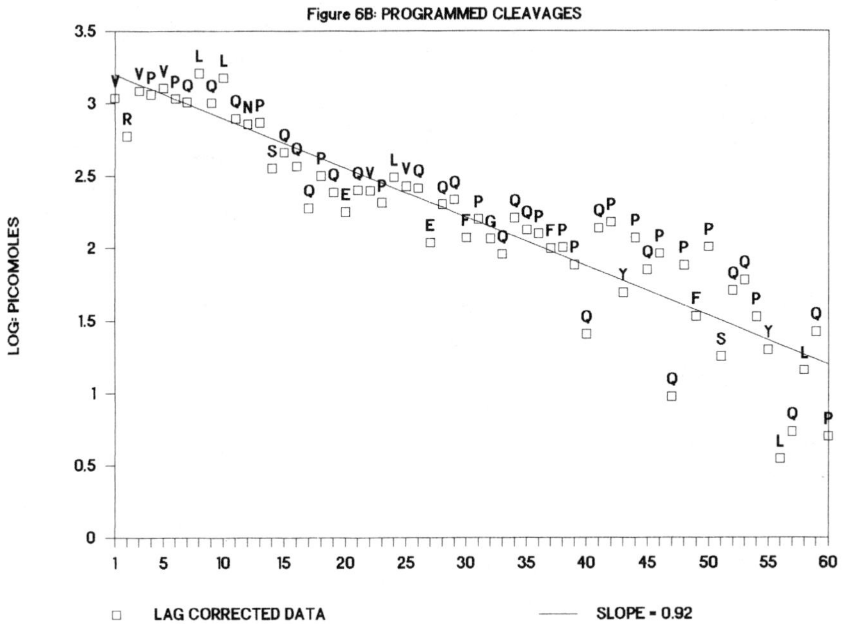

Figure 6. Maximum yield of the lag corrected, background subtracted data for the 6-D coded α-gliadin at each cycle. In panel 6A uniform cleavages were used and in 6B, programmed cleavages.

position 32 from position 31, one entire position out-of-phase. Other conditions than 5 min at 48 °C were satisfactory for recognizing the position of leucine, but the highest lag-corrected yields appeared to be at cleavage times of 10 min at 48 °C (Fig. 5B), although yields at 10 min at 53 °C (Fig. 5D) were not greatly different.

When we examined all the data for uniform cleavage conditions at different times and temperatures, we found that 10 min cleavages at 53 °C gave the best results allowing the sequence of A-gliadin to be correctly identified through the first 50 residues, which included 14 prolines. As indicated, all other conditions tested were less satisfactory.

Next, we used the uniform cleavage condition of 10 min at 53 °C on another α-gliadin, coded by a gene on chromosome 6D (6D-α-gliadin), which was known to be structurally similar to A-gliadin, but which was expected to have some substitutions in its sequence. The log of the yield of the most abundant amino acid at each cycle (background-subtracted, lag-corrected data) is plotted against cycle number in Fig. 6A. The slope of the regression line, which represents repetitive yield, was 0.90 and the r^2 value for the fit of the regression line was 0.83.

After having determined the expected positions of proline residues, we sequenced the protein using different cleavage conditions for proline and all other amino acids. Residues other than proline were programmed for 5 min cleavages at 48 °C. Positions known to contain proline were cleaved and extracted twice at 53 °C. The first cleavage at 53 °C was for 10 min and was followed by extraction and a second cleavage of 5 min and extraction at the same temperature. The PTC-Pro residues were cleaved so as to obtain about 98% total cleavage. The first cleavage would be expected to cleave PTC-Pro-Val- bonds 90%, and the second cleavage would remove another 80% of remaining PTC-Pro-Val-. It is not possible to schedule the delivery of more than one extraction of the reaction cartridge into the conversion flask. Accordingly, the product from the first cleavage was transferred to the conversion flask for conversion and analysis, and the product from the second cleavage was routed to waste. A series of three 5 min cleavages and extractions could have been used, but calculations indicated that more proline would be recovered from the combination of a 10 min cleavage followed by a 5 min cleavage. The protein was subjected to a total of 72 cycles of degradation--farther than we expected to identify the sequence. After the 52nd cycle, we assumed an unknown sequence and used 10 min at 53 °C for all cycles. After the analysis, the log of the amount of the major amino acid at each cycle indicated by the data analysis software was plotted against cycle number as shown in Fig. 6B. The slope of the regression line for the first 60 amino acids was 0.92, higher than for the unprogrammed analysis. The r^2 value was 0.85.

Comparison of the data for this 6D-α-gliadin sequenced with uniform cleavages with the data for the same protein sequenced with programmed cleavages showed that for most cycles the amino acids identified were the same. The notable exceptions were cycle 2 with leucine indicated when cleavages were uniform and arginine when the programmed cleavages were used, and cycle 27 with arginine indicated when cleavages were uniform and glutamic acid indicated when cleavages were programmed. The correct assignment at cycle 2 is likely to be arginine, which extracts better at 48 °C than at 53 °C (Model 477A Users Manual). In this regard, we noted a persistent arginine background after position 2 in the unprogrammed experiment. The correct assignment at cycle 27 is almost certain to be glutamic acid, which was recovered in tenfold excess over arginine in the programmed experiment.

The sequence data from the programmed experiment appeared solid up to cycle 55 except for cycle 47 where glutamine, leucine, methionine, and valine were of nearly equal abundance. We assigned the predominant amino acid, glutamine, to this cycle (Fig. 6B). On this basis, the sequence differed from the A-gliadin sequence at positions 10 (leucine for proline), 27 (glutamic acid for glutamine), and 31 (proline for leucine). Equivalent substitutions have been reported for protein sequences of α-type gliadins derived from cDNA sequences (12). One reported sequence (A212) (12) differs in its first 55 residues from our 6D-α-gliadin sequence only in having glutamine substituted for proline at position 31.

IV. Conclusions

Our investigation of different cleavage conditions for uniform cleavage of proline-rich wheat gluten proteins has indicated that 10 min cleavages at 53 °C are likely to provide the maximum number of proline residues in an unknown sequence. When these proline positions are known, a second experiment in which proline cleavages are programmed for 10 min at 53 °C followed by an additional 5 min cleavage at the same temperature, while cleavages for all other amino acids are kept to 5 min at 48 °C, can extend the number of correctly identified amino acids. Using these modifications, we have been able to obtain lengths of sequence for proline-rich gliadins that approach those obtained for proteins with few or no proline residues

Disclaimer

Reference to a company and/or product named by the Department is only for purposes of information and does not imply approval or recommendation of the product to the exclusion of others which may also be suitable.

References

1 Edman, P. (1956), Acta Chem. Scand. 10, 761.

2 Brandt, W. F.,Edman,P., Henschen, A. and von Holt, C. (1976), Hoppe-Seyler's Z. Physiol. Chem. 357,1505.

3 Edman, P., and Begg, A. (1967). Eur. J. Biochem. 1, 80.

4 Smithies, O., Gibson, D., Fanning, E. M., Goodfliesh, R. M., Gilman, J. G., and Ballantyne, D. L. (1971). Biochemistry 10, 4912.

5 Laursen, R. A., in "The Peptides" (E. Gross and J. Meienhofer, ed.), p. 261. Academic Press, New York (1981).

6 Hunkapiller, M.in "Methods in Protein Sequence Analysis" (K. A. Walsh, ed.) p. 140. Humana Press, Clifton, N.J. (1987).

7 Brandt, W. F. and von Holt, C. (1974), Eur. J. Biochem. 46, 419.

8 Tarr, G. E. (1977), Meth. Enzymol. 47, 335.

9 Kasarda, D.D., Okita, T.W., Bernardin, J.E.,Baeckaer, P.A.,Nimmo, C.C., Lew, E. J-L., Dietler, M.D. and Greene, F.C. (1984) Proc.Natl. Acad. Sci. USA 81, 4712.

10 Shewry, P.R., Halford, N.G.,Tatham, A.S. (1989) Oxford Surveys of Plant Molecular and Cell Biology 6, 163.

11 Brandt, W. F., Henschen, A., and von Holt, C. (1980), Hoppe-Seyler's Z. Physiol. Chem. 361, 943.

12 Okita, T. W., Cheesbrough, V., and Reeves, C. D., J. Biol. Chem. 260:8203-8213, 1985.

18

Glycopeptidase F Treatment and Amino Acid Sequence Determination of Glycoproteins Immobilized on PVDF and Its Derivatives

Brian M. Martin and William K. Eliason

Section on Molecular Neurogenetics, Clinical Neuroscience Branch
NIMH, Bethesda, Maryland

I. Introduction

The importance of protein sequence analysis has grown over the past 10 years. With the increasing interest in proteins (of low abundance) isolated from human/animal sources, the need for primary structure information has required development of new isolation procedures and more sophisticated methods for sequence analysis. Proteins of low abundance are now routinely isolated by 2-dimensional electrophoresis, and transferred to PVDF for amino-terminal sequence analysis. Internal amino acid sequence from these proteins has proven extremely difficult to obtain. Therefore, the determination of the amino-terminal sequence of these proteins (needed prior to probe design for cDNA isolation) requires the maximum number of identified residues. One of the main difficulties involved is post-translationally modified amino acids such as asparagine linked to carbohydrate and cysteine involved in disulfide bridges. The identification of these post-translational modifications remains in the realm of the protein chemist and is the focus of this report.

The two proteins chosen for this study are glycoproteins with CHO attachment sites within the first 20 residues. Carboxypeptidase Y has four Asn-linked sites with each site bearing a high mannose chain (man_{5-7})(Trimble et al, 1983). Human glucocerebrosidase from placenta also has four Asn-linked sites, with approximately 80% being complex bi- and tri- antennary and the remaining 20% high mannose (Takasaki et al, 1984). In addition, glucocerebrosidase has three half-cystines within the first twenty residues (Martin et al, 1989). Attempts to sequence either protein through 20 residues gave poor results, with large overlap/low yield after the carbohydrate in carboxypeptidase and similar problems with the half-cystines in glucocerebrosidase (Martin et al, 1982, Martin et al, 1989). Because of the typical difficulties seen in sequencing these types of proteins and the need to determine continuous strings of amino acid sequence to design probes for cloning projects, we have attempted to remove the holes in the N-terminal region of these proteins.

II. Materials and Methods

A. Materials

Carboxypeptidase Y was a gift from Carlsberg Laboratories and was kindly provided by Dr. Jack Johansen. Human glucocerebrosidase was isolated from placenta as described by Takasaki et al (1983) and run on reverse-phase prior to use (Eliason, unpublished). Sequelon™-AA and DITC kits were purchased from Milligen Biosearch. Problott and Immobilion-P (PVDF) were products of Applied Biosystems and Millipore respectively. Sequencer reagents and solvents were obtained from the manufacturers.

B. Blotting

Both Carboxypeptidase Y (Martin et al, 1982) and human glucocerebrosidase (Martin et al, 1989) were run on pre-cast mini gels (ISS-12.5% or Novex 12.0%) under denaturing conditions. After electrophoresis, the proteins were transferred to PVDF/Problott (Matsudaira, 1987) using 25mM TRIS, 192mM glycine, 20% MEOH as buffer. Initially, the transfer was done at 25mA, twenty minutes later jumped to 70mA for fifteen minutes and left overnight at 100mA. Following transfer, the membrane was removed from the Hoffer Mighty Small unit, washed extensively with deionized water and stained with Ponceau S(0.1% in 1% acetic acid). The membrane was destained with water washes and air dried. The bands of interest were excised using a clean razor blade and stored in a screw cap eppendorf tube under nitrogen.

C. Sequelon™-AA and DITC

The proteins were covalently attached to each sequelon membrane following the manufacturer's instructions. These instructions and all the necessary reagents/buffers are supplied in kit form (Milligen/Biosearch). The covalently attached protein discs were placed in a 1.5 ml. eppendorf until needed.

D. Reduction and Alkylation

Complete reduction of the immobilized proteins under denaturing conditions was accomplished by treating the band or disc in 800 μl 100 mM TRIS, 6 M guanidine pH 8.6 with DTT (10mM) at 55 degrees Celsius for a period of 60-90 minutes, with occasional vortexing. The solution was then cooled to room temperature (22 degrees Celsius) and the protein alkylated with either iodoacetic acid (Crestfield et al, 1963) or 4-vinyl pyridine (Friedman et al, 1970). Alkylation with iodoacetic acid was accomplished by treating the reduced protein with excess iodoacetate (4mg) in 320 ul 1M TRIS, pH 7.5. The solution was flushed with nitrogen, the tube recapped and covered with foil. Alkylation with 4-vinyl pyridine was accomplished in a similar manner using 4 μl of neat reagent. After thirty minutes, the alkylated protein was freed of excess reagents/salts by extensive washing of the PVDF pieces/discs. Typically, the solution was removed from the tube and the membrane/discs washed in the tube with 1 ml water twice, 1 ml MEOH twice and a final 2 times 1 ml water. The membrane was then dried in a Speed-Vac (Savant) in the screw cap tube after punching several holes in the cap prior to amino acid sequencing or analysis or rinsed with deglycosylation buffer prior to the carbohydrate removal.

E. Deglycosylation

A wet membrane (either PVDF or Sequelon) was submerged in 500 ul 100mM potassium phosphate pH 8.3, containing 0.01% NP-40. Glycopeptidase F (Boehringer Mannheim) was added to a nominal concentration of 6 units/ml, the sample vortexed several times, capped and incubated overnight at 37 degrees Celsius. Prior to sequence analysis, the solution was removed and saved for carbohydrate analysis. The membrane(s) was then washed and dried as described above.

F. Sequence Determination

Edman degradation of Immobilon P samples were performed on both an ABI 470A gas-phase sequencer with a 120A on-line PTH analyzer and a Porton 2090E gas phase sequencer. Problott samples of carboxypeptidase Y were run on the 470A and Sequelon™-AA and DITC samples were analyzed on a Milligen 6600 Prosequencer. Sequenet™ (Milligen/Biosearch) samples of glucocerebrosidase were analyzed only for carboxymethylated material using the Milligen/Biosearch 6600, following the manufacturer's instructions.

III. Results and Discussion

Due to the excessive number of residues which could not be identified in glucocerebrosidase and other proteins recovered from gels by transblotting we initially directed our attention to half-cystine. Since our samples were run on reducing gels we attempted to alkylate the sulfhydryls after boiling the samples in reducing buffer. This approach worked well for alkylation with iodoacetic acid, however, it suggested the possibility of a change in mobility for proteins run on 2-dimensional gels. It was evident that removal of proteins in a quantitative fashion from PVDF was extremely difficult and therefore we decided to reduce and alkylate the proteins directly on PVDF. We tried a number of buffers/denaturants and found guanidine to be best as denaturant. It would appear that any alkylating agent may be effective as both iodoacetic acid and 4-vinyl pyridine have been successfully employed. The amino-terminal amino acid sequence of glucocerebrosidase and carboxypeptidase Y are presented are in Table 1 (Martin et al, 1989, Martin et al, 1982).

Figure 1 illustrates the utility of this method for identification of the half-cystines reduced and pyridylethylated from human glucocerebrosidase immobilized on Sequelon-AA. The results were similar for samples transblotted to PVDF/Problott after gel electrophoresis. Initial yields for electroblotted protein samples before and after reduction/alkylation were within 5% showing that glycoproteins with apparent molecular weights in the range of 60 KDa are neither removed from the PVF nor blocked by this procedure. (data not shown)

In addition to obtaining a respectable repetitive yield from the Edman degradation, the removal of carbohydrate from asparagine would provide a positive identification of the attachment site. We felt that the hydrophobic nature of PVDF based membranes would encourage the carbohydrate to be

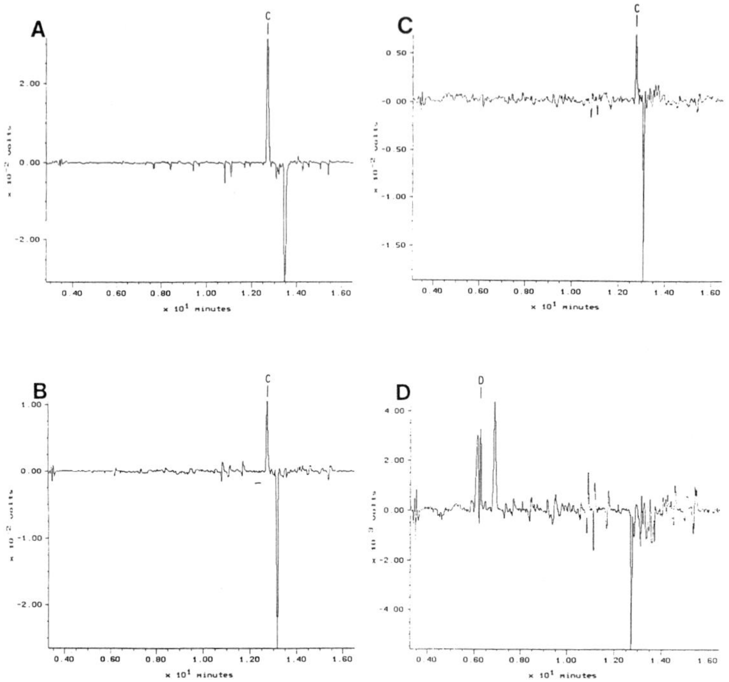

Figure 1. Human glucocerebrosidase PTH-amino acids from cycles 4 (Panel A), cycle 16 (Panel B) cycle 18 (Panel C) and cycle 19 (Panel D). Pyridylethyl cysteine (labelled C) can be seen in Panels A-C and aspartic acid (labelled D) in Panel D. The glycoprotein was covalently attached to Sequelon-AA and then reduced and alkylated with 4-vinyl pyridine prior to deglycosylation with Glycopeptidase F (see text). The chromatograms were generated on the Milligen/Biosearch 6600 Prosequencer after dynamic subtraction to remove the background from the previous cycle.

TABLE I. Amino-terminal sequence of Carboxypeptidase Y and Human Glucocerebrosidase.

CHO (attached at N, position 13)

K I K D P K I L G I D P N V T Q Y T G Y

CARBOXYPEPTIDASE Y

CHO (attached at N, position 19)

A R P C I P K S F G V S S V V C V C N A

HUMAN GLUCOCEREBROSIDASE

accessible in an aqueous environment. Initial attempts to use glycopeptidase F in phosphate buffer alone were unsuccessful as the enzyme preparation bound to the PVDF. Subsequently, a number of reaction mixtures were employed which gave unsuccessful results, due to partial removal of the protein of interest from the membrane. The inclusion of NP-40 in the phosphate buffer provided maximum release of carbohydrate and minimum effect on sequencable protein. In Figure 2, the Asn linked to carbohydrate in position 13 is recovered as the Asp after glycopeptidase F treatment of Carboxypeptidase Y covalently immobilized to Sequelon-DITC. Panel D of Figure 1 shows the Asp recovered in position 19 of human glucocerebrosidase after deglycosylation. In both cases,

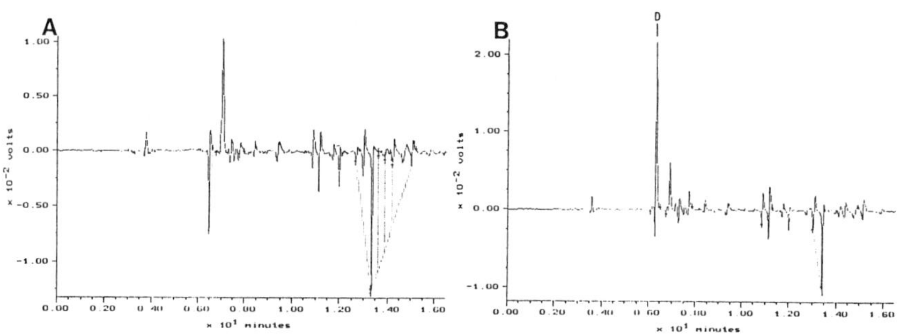

Figure 2. Carboxypeptidase Y PTH-amino acids from cycle 13 (Panel A) of native enzyme and cycle 13 (Panel B) of reduced carboxymethyl-enzyme after deglycosylation of enzyme covalently attached to Sequelon-DITC. The appearance of the aspartic acid (labelled D) in Glycopeptidase F treated sample (Panel B) is evident. The chromatograms were generated after dynamic subtraction as in Figure 1.

the deglycosylation reaction was conducted on reduced alkylated samples. Similar deglycosylation results were obtained when the Glycopeptidase F treatment was done on unreduced samples.

The reaction of Glycopeptidase F on carbohydrate chains using this method does not appear to be dependent on carbohydrate composition as re-electrophoresis of deglycosylated glucocerebrosidase gave a band corresponding to the molecular weight of the nascent protein. (data not shown)

We have demonstrated the utility of these procedures for protein samples transblotted to Immobilon-P/Problott and found that the amino-terminal sequence analysis can be accomplished on either the 470A or PI2090E-gas phase sequencer as well as the Milligen/Biosearch 6600 after sequenet chemistry has been performed. By taking advantage of the covalent attachment chemistry from Sequelon-AA and DITC, we extended the utility of the method to samples obtained from gels by electroelection, HPLC or traditional chromatography which can provide extended sequence data (>40 cycles). The method should allow more straight forward design of probes for cloning projects, in particular those involving proteins recovered from 2-dimensional gel electrophoresis. In addition, the method should prove useful in the positive identification of N-linked carbohydrate attachment sites.

In conclusion, we have developed a buffer system for successful removal of N-linked carbohydrate from glycoproteins which does not interfere with subsequent amino acid sequence analysis of samples immobilized on PVDF or its derivatives. Recovery of sequenceable protein approaches 100% and reduction/alkylation can be done before or after deglycosylation to provide the maximum possibility of amino acid identification.

Acknowledgments

We thank the National Gaucher Foundation for partial support of this work and Elizabeth Alzona for preparing the manuscript.

References

Crestfield, A.M., Moore, S., and Stein, W.H. (1963). J. Biol. Chem. **238,** 622-627.

Friedman, M., Krull, L.H., and Cavins, J.F. (1970). J. Biol. Chem. **245,** 3868-3871.

Martin, B.M., Svendsen, I., Viswanatha, T., and Johansen, J.T. (1982). Carlsberg Res. Comm. **47,** 1-13.

Martin, B.M., Sidransky, E., and Ginns, E.I. (1989). In "Adv. Pediatr." (Barness, L.A., ed.), Vol.36, p. 277. Year Book Medical Publishers, Chicago.

Matsudaira, P. (1987). J. Biol. Chem. **262,** 10035-10038.

Takasaki, S., Murray, G.J., Furbish, F.S., Brady, R.O., Barranger, J.A., and Kobata, A. (1984). J. Biol. Chem. **259,** 10112-10117.

Trimble, R.B., Maley, F., and Chu, F.K. (1983). J. Biol. Chem. **258,** 2562-2567.

19

The Complete Primary Amino Acid Sequence of Soluble Bovine Dopamine ß-Hydroxylase: Confirmation of N-Terminal, C-Terminal, and Internal Unassigned Regions of the Protein

James G. Robertson, Alok Kumar, Parimal Desai[1], and Joseph J. Villafranca
The Pennsylvania State University, University Park, PA

1. Introduction

Dopamine ß-hydroxylase catalyzes the physiologically important conversion of dopamine to norepinephrine within the chromaffin granules of the adrenal medulla (1, 2). The enzyme is a copper containing tetrameric glycoprotein of M_r 290,000 (1, 2). Both soluble and membrane-bound forms of the enzyme have been purified (3, 4). Extensive studies of the bovine enzyme have characterized both the kinetic and chemical mechanisms of the enzyme (5-7), and physical studies have shown that there are 4 histidine ligands equatorially coordinated to each copper atom in a square planar environment (8).

The enzyme's involvement in the catecholamine pathway has made it an attractive target for drug design, and numerous mechanism based inhibitors of the enzyme have been synthesized (7). Moreover, the medical community has become increasingly interested in the enzyme since the discovery of several patients with deficiencies in dopamine ß-hydroxylase (9). In addition, the report that dopamine ß-hydroxylase contains covalently bound pyrroloquinoline quinone has prompted close scrutiny of the protein (10, 11). Therefore, the determination of the primary sequence of the enzyme has been an important goal for the last several years.

Early sequencing studies of the bovine enzyme indicated that the protein contains at least two N-termini (12). The sequences of five internal peptides from the bovine enzyme provided some additional structural information (13), and the available sequence information was used by several groups in attempts to clone the bovine enzyme, but with little success. However, a clone of the human enzyme was obtained in 1987, and the deduced amino acid sequence of the human enzyme provided the first complete elucidation of the protein's primary structure (14). Nevertheless, the cDNA predicted sequence of the human enzyme was not completely homologous with the known bovine peptide sequences. Therefore it remained important to obtain the complete sequence for the bovine enzyme.

[1]Present address: American Cyanamid Company, Medical Research Division, Pearl River, New York, 10965

Techniques in Protein Chemistry II

Simultaneously, protein sequencing in our laboratory and cloning efforts in two other laboratories succeeded in determining the sequence of the bovine enzyme (15-17). Very soon after these results were published, a third laboratory reported another clone of the bovine enzyme and demonstrated that 93% of the predicted sequence could be verified by protein sequencing and mass spectrometry (18). However, in our original studies, several small peptides were not identified and there were several positions where the peptide sequencing was uncertain. Several of these regions have remained unconfirmed even after the recent protein and mass spectrometry studies (18). In addition, the four reported sequences contain several discrepancies between the peptide sequences and between the peptide and cDNA predicted sequences. Therefore, we have conducted additional peptide sequencing to resolve these differences and present here a complete primary amino acid sequence based on peptide sequencing.

2. Methods

Dopamine ß-hydroxylase was purified as previously described (3). Cysteine residues were modified by S-ß-(4-pyridylethylation) essentially as described (19). An enzyme sample containing 2.7 mg in 2 ml was dialyzed into 250 mM Tris, pH 8.5, 7 M guanidine-HCl, and was transferred to a micro reaction flask (from Kontes). The sample was evacuated, flushed with argon, and sealed under an argon atmosphere. A diluted solution of ß-mercaptoethanol was prepared by mixing 5 µl of ß-mercaptoethanol with 50 µl distilled water. The ß-mercaptoethanol solution then was injected into the reaction flask through a rubber septum and left to react at 37 °C for 6 hours. After reduction, an emulsion of 4-vinylpyridine was prepared by vortexing 15 µl of 4-vinylpyridine with 50 µl distilled water, and the emulsion was injected into the reaction vessel. After derivatization for 1 hour at 37 °C, the sample was dialyzed extensively against 10 mM NH_4HCO_3, pH 7.8, to remove ß-mercaptoethanol and 4-vinylpyridine.

For digestion with *Staph. aureus* V8 protease, the enzyme sample was dialyzed into 50 mM NH_4HCO_3 containing 4 M urea at a pH of 7.8. For digestion with trypsin, the enzyme sample was dialyzed into 50 mM KH_2PO_4 containing 4 M urea at a pH of 6.5. Protease was added at a 1:50 ratio (mg protease/mg dopamine ß-hydroxylase) and digestion was allowed to proceed for 18 hours at room temperature.

Peptides were separated by HPLC on a Vydac C_{18} column. Buffer A contained 0.1% trifluoroacetic acid in HPLC grade water, and buffer B contained 0.1% trifluoroacetic acid in HPLC grade acetonitrile. The elution gradient started at 90% A-10% B and ran to 50% A-50% B over 80 minutes, which corresponds to a 0.5% change in solvent per minute. Peptide elution was monitored by diode array detection. Individual peptide peaks were collected in Nunc cyrotubes and were concentrated by lyophilization in a Speed Vac. Concentrated peptides were rechromatographed in the same solvents at 0.33% change per minute to assess peptide purity and to separate small amounts of contaminating peptides that remained from overlapping peaks in the original chromatograph. Repurified peptides were concentrated and sequenced in the DNA and Protein Chemistry Laboratory at The Penn State Biotechnology Institute.

Peptides were sequenced on an Applied Biosystems Model 477 gas phase sequencer. Sequence analysis of the intact N-terminus of the protein was performed on a Milligen Prosequencer model 6600. Enzyme was dialyzed into water before sequencing. Two filters, one containing dopamine ß-hydroxylase covalently linked to a Sequelon AA membrane (carboxyl linkage) and one covalently linked to a Sequelon DITC membrane (lysine linkage), were sequenced at once. This permitted sequencing up to 40 residues in the mature protein with yields in the last cycle greater than 10 pmol.

3. Results

Previous peptide sequencing in our laboratory identified 553 amino acids out of a possible 597 amino acids predicted by the cDNA of the bovine gene of dopamine ß-hydroxylase (15). The amino acids were numbered according to the sequence predicted by the human cDNA in order to maintain consistency with the first published dopamine ß-hydroxylase sequence. Thus, the predicted leader sequence of human dopamine ß-hydroxylase contains a methionine as amino acid 1. According to this system, the N-terminus of the mature bovine protein begins with Ser 26, due to cleavage of the leader sequence.

In the N-terminal region of the protein, our peptide sequencing data were uncertain over residues 49-51 and suggested three changes from the bovine gene. A change also was suggested a position 67. In addition, peptide mapping with trypsin and *Staph. aureus* V8 protease failed to identify a peptide spanning residues 57-65, and these amino acids remained unassigned. A second study also failed to confirm residues 41-65 by either peptide sequencing or mass spectrometry (18). Over the internal region of the protein, our initial sequence contained small gaps at residues 435-437, 459-460, and 557-558. At the C-terminal end of the protein, the cDNA predicted the final three residues as G-K-G. An isolated tryptic peptide confirmed the C-terminus up to the penultimate K, but trypsin presumably cleaved the final G and G was not detected in the sequence data. Therefore, the final residue of the protein remained unverified, and was not verified in later work by others (18).

Analysis of the published protein and cDNA derived sequences indicated that all of our uncertain and unsequenced regions, including the C-terminus, could be isolated as *Staph. aureus* V8 protease or tryptic peptides containing either cysteine or tryptophan. Digestion with *Staph. aureus* V8 protease would not cleave the final K-G peptide bond and therefore would permit confirmation of the C-terminal G. Consequently, our strategy was to identify cysteine and tryptophan containing peptides in our peptide map and to resequence these peptides. Improved methods for extended N-terminal protein sequencing also made it possible to attempt confirmation of the N-terminal gap in our sequence by direct protein sequencing.

Protein derivatization with 4-vinylpyridine produces an S-ß-(4-pyridylethylation) derivative of cysteine with an absorption maximum at 254 nm, and dopamine ß-hydroxylase was modified with 4-vinylpyridine to take advantage of this chromophor. Tryptophan has a natural absorption maximum at 280 nm, well separated from the S-ß-(4-pyridylethylation) chromophor, and therefore it was possible to monitor the presence of cysteine and tryptophan simultaneously. These chromophors were monitored in our peptide map by diode array detection, and Figure 1 shows representative peak scans of peptides containing cysteine, tryptophan, or both cysteine and tryptophan.

Digestion by *Staph. aureus* V8 protease predicts one peptide containing tryptophan and cysteine, and one peptide containing one tryptophan and two cysteines. It also predicts two peptides containing two cysteines. By monitoring the relative peak heights of tryptophan and cysteine in the peak scans, it was possible to identify these four peptides in the HPLC chromatogram. This left only seven peaks containing one cysteine and four peaks containing exclusively tryptophan. By these comparisons it was possible to focus the sequencing efforts on only a few relevant peaks in the chromatogram.

Extended N-terminal protein sequencing on the Milligen Prosequencer with covalent protein attachment and simultaneous multiple filters permitted determination of the first 40 amino acids of the protein (Ser 26 - Arg 65). Table 1 shows the N-terminal sequence for 40 residues. This result spans a previously unsequenced region from residues 57-65. Residues 47-56 were confirmed by sequencing a tryptophan containing peptide shown as peptide #1 in Table 1.

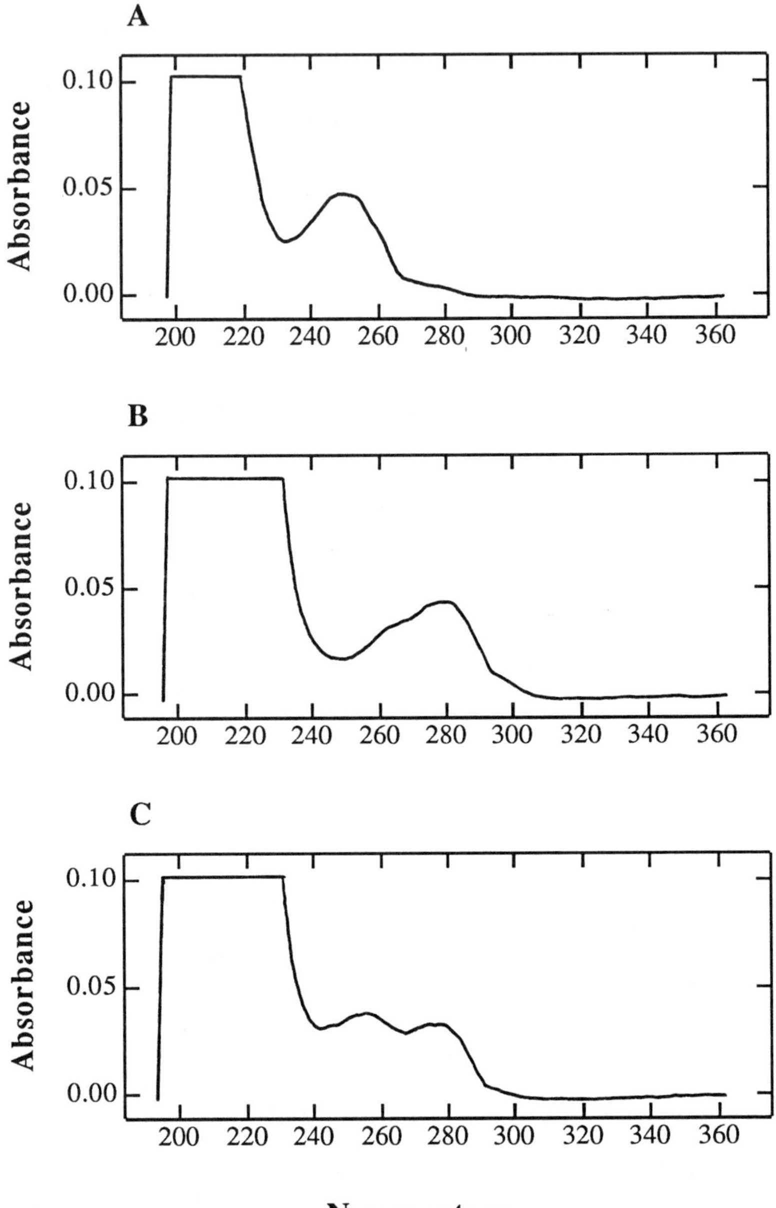

Figure 1 UV-Visible Profiles of Peptides Containing Cysteine and Tryptophan Dopamine ß-hydroxylase was modified with 4-vinylpyridine and digested as described under Methods. Panel A shows the absorbance profile of a peptide that has been pyridylethylated on cysteine, Panel B the absorbance profile of a peptide containing tryptophan, and Panel C the absorbance profile of a peptide containing both tryptophan and a modified cysteine.

Table 1

		N-Terminal Region	
NT	26	SAPAESPFPFHIPLDPEGTLELSWXISYAQETIYFQLLVR	65
#1	47	LSWXISYAQE	56
		Internal Regions	
#2	433	IRMLKKVVSVQPGDVLITSCTYNTE	457
#3	459	RRLATVGGFGILEEMCVNYVHYYPQTQLELCK	490
#4	537	VLKALYGFAPISMHCNRSSAVRFQGE	562
		C-Terminus	
#5	571	IVSRLEEPTPHCPASQAQSPAGPTVLNISGGKG	603

Table 1 Newly Obtained Protein N-Terminal, Internal, and C-Terminal Sequences. Amino acid residues have been numbered according to the human cDNA derived sequence in order to maintain consistency with a previous publication (15). Thus, the N-terminus of the mature protein starts at residue 26 due to cleavage of the leader sequence. NT represents the N-terminus. Numbers 1-5 indicate new peptides that have been sequenced. X represents a blank in the sequencing data.

Peptide 1 contained a blank in cycle 4. This corresponds to a predicted asparagine at a consensus site for glycosylation, and therefore supports the idea that position 50 in the sequence is a glycosylated asparagine. These two results clarify the original uncertainty in residues in 49-51 and demonstrate that this sequence is W-N-I rather than the original tentative assignment as R-Y-V.

Peptide #2 in Table 1 was isolated as a cysteine containing peptide by *Staph. aureus* V8 protease digestion and covered the region from amino acids 433-457. The beginning of this peptide confirms the sequence M-L-K and fills the gap in our original sequence from residues 435-437.

Peptide #3 in Table 1 was obtained as a cysteine containing peptide by trypsin digestion and confirmed residues 459-460 as R-R. In fact, two sequences were obtained for peptide #3, one beginning R-L-A-T and one beginning L-A-T. This result indicates the presence of two arginines and a staggered cleavage by trypsin.

Peptide #4 in Table 1 was obtained as a cysteine containing peptide by *Staph. aureus* V8 protease digestion and identified residues 557 and 558 as V-R, which assigns a previously unfilled gap in the sequence. This peptide also clarifies the identity of residue 553. Residue 551 is a cysteine and residue 552 is a glycosylated asparagine. The original data for this peptide yielded a blank cycle for the glycosylated asparagine and demonstrated significant carry over from the cysteine at position 551. In fact, the yield of cysteine in position 553 doubled from the previous cycle and appeared to represent a true cysteine at 553. Therefore the original sequence tentatively assigned it as such. However, peptide #4 in Table 1 clearly demonstrated arginine in high yield at position 553, and arginine agrees with the cDNA predicted sequence. Moreover, a tryptic peptide following residue 553 has been confirmed by mass spectrometry (18), which indicates that position 553 must be either a lysine or an arginine. Therefore, position 553 can be assigned as arginine.

The C-terminal peptide was identified as a cysteine containing peptide in a *Staph. aureus* V8 protease digest. Peptide #5 in Table 1 shows the confirmed sequence of the C-terminus, including the terminal glycine.

Table 2

cDNA Predicted DßH N-Terminal Leader Sequences

```
-7 MQVPSPSVREAASMYGTAVAVFLVILVAALQG   25 Bovine  (16)
1         MREAASMYGTAVAVFLVILVAALQG   25 Human   (14)
+7              MYGTAVAVFLVILVAALQG   25 Bovine  (17)
```

Protein Sequence of Mature Soluble DßH

```
1      -------------------------SAPAESPFPFHIPLD 40
41     PEGTLELSWNISYAQETIYFQLLVREFKAGVLFGMSDRGE 80
81     LENADLVVLWTDRDGAYFGDAWSDQKGQVHLDSQQDYQLL 120
121    RAQRTPEGLYLLFKRPFGTCDPNDYLIEDGTVHLVYGFLE 160
161    EPLRSLESINTSGLHTGLQRVQLLKPSIPKPALPADTRTM 200
201    EIRAPDVLIPGQQTTYWCYVTELPDGFPRHHIVMYEPIVT 240
241    EGNEALVHHMEVFQCAAEFETIPHFSGPCDSKMKPQRLNF 280
281    CRHVLAAWALGAKAFYYPEEAGLAFGGPGSSRFLRLEVHY 320
321    HNPLVITGRRDSSGIRLYYTAALRRFDAGIMELGLAYTPV 360
361    MAIPPQETAFVLTGYCTDKCTQLALPASGIHIFASQLHTH 400
401    LTGRKVVTVLARDGRETEIVNRDNHYSPHFQEIRMLKKVV 440
441    SVQPGDVLITSCTYNTEDRRLATVGGFGILEEMCVNYVHY 480
481    YPQTQLELCKSAVDPGFLHKYFRLVNRFNSEEVCTCPQAS 520
521    VPEQFASVPWNSFNREVLKALYGFAPISMHCNRSSAVRFQ 560
561    GEWNRQPLPEIVSRLEEPTPHCPASQAQSPAGPTVLNISG 600
601    GKG                                      603
```

Table 2 Leader Sequences and Complete Primary Amino Acid Sequence of Bovine Dopamine ß-Hydroxylase. Leader sequences have been taken from references (14, 16-17). Amino acid residues have been numbered according to the human cDNA derived sequence in order to maintain consistency with a previous publication (15). Thus, residue 1 is the starting methionine of the human leader sequence. According to this scheme, the mature bovine protein N-terminus begins at residue 26. The bovine leader sequence in reference 16 is longer than either the human leader sequence or the bovine leader sequence in reference 17, and hence starts at residue -7 relative to the human sequence. N-terminal sequencing of the membrane bound form of the enzyme has confirmed 22 of 27 positions predicted by the cDNA leader sequence (16).

---

4. Discussion

Protein and peptide sequencing have now provided direct confirmation of all 578 amino acids in mature soluble bovine dopamine ß-hydroxylase. From the combined data, several comparisons may now be made between the four sequences reported for the bovine enzyme (this work, 16-18).

All three cDNA derived sequences predict the sequence of residues 49-52 as W-N-I-S, and hence predict a consensus site for glycosylation. Chemical tests did not identify a carbohydrate modified peptide containing the sequence W-N-I-S (15). Also, peptide chromatography on concanavalin A failed to isolate a glycosylated peptide containing the sequence W-N-I-S (18). However, cycle #4 in peptide #1 in Table 1 was obtained as a blank and indicates glycosylation at this site. Failure to detect glycosylation at Asn 50 by chemical testing or by concanavalin A binding is consistent with previous studies demonstrating that dopamine ß-hydroxylase contains both high mannose oligosaccharides and biantennary complex oligosaccharides (20). The complex oligosaccharides in

Table 3

Apparent Polymorphisms

	Residue	A.A.	Sequencing	Reference
#1	67	Leu	cDNA, peptide,	(16-18)
	67	Phe	peptide	(15)
#2	461	Glu	peptide, MS	(18)
	461	Arg	cDNA, peptide	(15-17)

dopamine ß-hydroxylase consist of sialic acid, galactose, n-acetyl glucosamine, and limited amounts of mannose. Consequently, they would not be expected to bind to concanavalin A or to provide reducing ends detectable by chemical tests. Therefore, there is good evidence that position 50 in the sequence corresponds to a site modified by N-linked complex oligosaccharide.

Also, all three cDNA derived sequences predict a threonine at position 57. However, Figure 4 in (18) indicates that position 57 is a tyrosine, although there is no peptide to confirm this assignment. This appears to be a discrepancy between Figures 3 and 4 in (18). Sequencing data in Table 1 demonstrate that position 57 is a threonine, in agreement with the cDNA.

Similarly, all three cDNA derived sequences predict a leucine at position 67. Peptide sequencing and mass spectrometry confirmed this leucine (18). However, previous peptide sequencing (15) indicated that position 67 is a phenylalanine. This position may represent a polymorphism because the conflicting data appear to be strong in each of the separate cases.

One of the bovine cDNA derived sequences predicts a cysteine in position 198 (16). However, the other two cDNA derived sequences predict an arginine (17,18), and both peptide sequencing and mass spectrometry confirm an arginine at position 198 (15, 18). A cysteine at position 198 would give a total of 15 cysteines in the protein. It is known that all the cysteines in dopamine ß-hydroxylase are paired as disulfides, so an odd number of cysteines would indicate a free sulfhydryl, which could only be accounted for as an intermolecular disulfide between subunits. However, analysis of disulfides in the protein has not detected an intermolecular disulfide at this position (unpublished results). A single base change can account for this discrepancy, so the weight of evidence in the bovine protein would indicate that position 198 is an arginine. However, it should be noted that the human cDNA predicts a cysteine at this position.

The largest discrepancy between bovine sequences occurs at positions 260-262. Here, one cDNA sequence predicts R-D-H (16), whereas the other two cDNA sequences predict E-T-I (17, 18). In addition, both peptide sequences agree with the assignment E-T-I (15, 18). Evidence for this region would indicate that E-T-I is the correct sequence, and it is not clear why one cDNA predicted sequence would deviate from all the other sequences to such an extent.

Peptide #2 in Table 1 confirms the sequence from 435-437 as M-L-K, in agreement with the peptide sequencing and mass spectrometry data (18), and therefore we have listed this in Table 2 as the confirmed sequence where our original data contained a small gap (15).

At position 461 the two peptide sequences differ (15, 18). In one report, peptide sequencing and mass spectrometry indicate at glutamate at this position (18). However, peptide sequencing and cDNA sequencing in three reports indicate an arginine at position 461 (15, 16, 17), and the cDNA in the same report (18) indicates arginine. This may represent a polymorphism.

All three cDNA predicted sequences contain an arginine at position 553, and the peptide and mass spectrometry data support this assignment. The new data presented here agree with arginine at position 553, and this is a correction from our previous sequence.

In one other position, His 581, two cDNA sequences and two peptide sequences agree on histidine (15, 17, 18), whereas the third cDNA sequence predicts glutamine (16). At this position, a single base change in the codon would interconvert the two amino acids, and it is possible that this is a DNA sequencing error. The weight of evidence indicates that histidine is the correct amino acid in this position.

Based on the analysis of the four reported sequences (15-18) and the new data presented here, the sequence of mature soluble bovine dopamine ß-hydroxylase in Table 2 may be proposed with a high degree of certainty. All 578 amino acids have been verified by protein and peptide sequencing, and a significant number of peptide sequences have been confirmed by mass spectrometry (18). In the few cases where there is disagreement among the sequences, we have proposed a "correct" amino acid if three of the four sequences agree, or listed a polymorphism in Table 3 if one set of data appeared to be very strong.

The sequences confirmed by peptide sequencing and mass spectrometry agree that there are two glycosylation sites, one at Asn 170 and one at Asn 553 (15, 18). There also is consistent evidence for complex oligosaccharide glycosylation at Asn 50. Also, the complete peptide sequencing of the protein provides very convincing evidence that dopamine ß-hydroxylase does not contain covalently bound pyrroloquinoline quinone. Based on the sequence in Table 2, the calculated molecular weight of the protein is 65,218. Finally, using the calculated molecular weight and the identification of 7 tryptophans and 24 tyrosines in the protein, the extinction coefficient may be calculated as described previously (21). Calculation from the sequence in Table 2 and using a molecular weight of 65,218 yields the value $\varepsilon_{280}{}^{1\%} = 10.8$.

References

1. Villafranca, J. J., Ash, D. E., Colombo, G., Fitzpatrick, P. F., Papadopoulos, N., and Rajashekhar, B. (1985) in *Molecular Architecture of Proteins and Enzymes* (Bradshaw, R. A., and Tang, J., eds.) pp. 31-50, Academic Press, Orlando, FL
2. Stewart, L. S. and Klinman, J. P. (1988) *Annu. Rev. Biochem.* **57**, 551-592.
3. Colombo, G., Papadopoulos, N. J., Ash, D. E., and Villafranca, J. J. (1987) *Arch. Biochem. Biophys.* **252**, 71-80.
4. Saxena, A. and Fleming, P. J. (1983) *J. Biol. Chem.* **258**, 4147-4152.
5. Fitzpatrick, P. F., Harpel, M. R., and Villafranca, J. J. (1986) *Arch. Biochem. Biophys.* **249**, 70-75.
6. Ahn, N. and Klinman, J. P. (1983) *Biochemistry* **22**, 3096-3106.
7. Fitzpatrick, P. F. and Villafranca, J. J. (1987) *Arch. Biochem. Biophys.* **257**, 231-250.
8. McCracken, J., Desai, P. R., Papadopoulos, N. J., Villafranca, J. J. and Peisach, J. (1988) *Biochemistry* **27**, 4133-4137.
9. Robertson, D., Hollister, A. S., and Biaggioni, I. (1990) in *Hypertension: Pathophysiology, Diagnosis, and Management* (Laragh, J. H. and Brenner, B. M. eds.) pp. 749-759, Raven Press, Ltd., New York, NY
10. van der Meer, R. A., Jongejan, J. A., and Duine, J. A. (1988) *FEBS Lett.* **231**, 303-307.
11. Robertson, J. G., Kumar, A., Mancewicz, J. A., and Villafranca, J. J. (1989) *J. Biol. Chem.* **264**, 19916-19921.
12. Skotland, T., Ljones, T., Flatmark, T., and Sletten, K. (1977) *Biochem. Biophys. Res. Commun.* **74**, 1483-1489.

13. McCafferty, B. and Angeletti, R. H. (1987) *J. Neurosci. Res.* **18**, 289-292.
14. Lamouroux, A., Vigny, A., Faucon Biguet, N., Darmon, M. C., Franck, R., Henry, J.-P., and Mallet, J. (1987) *EMBO J.* **6**, 3931-3937.
15. Robertson, J. G., Desai, P. R., Kumar, A., Farrington, G. K., Fitzpatrick, P. F., and Villafranca, J. J. (1990) *J. Biol. Chem.* **265**, 1029-1035.
16. Taljanidisz, J., Stewart, L., Smith, A. J., and Klinman, J. P. (1989) *Biochemistry* **28**, 10054-10061.
17. Lewis, E. J., Allison, S., Fader, D., Claflin, V. and Baizer, L. (1990) *J. Biol. Chem.* **265**, 1021-1028.
18. Wang, N., Southan, C., DeWolf, W. E., Wells, T. N. C., Kruse, L. I., and Leatherbarrow, R. J. (1990) *Biochemistry* **29**, 6466-6474.
19. Fullmer, C. S. (1984) *Anal. Biochem.* **142**, 336-339
20. Margolis, R. K., Finne, J., Krusius, T., and Margolis, R. U. (1984) *Arch. Biochem. Biophys.* **228**, 443-449
21. Edelhoch, H. (1967) *Biochemistry* **6**, 1948-1954.

SECTION III

Peptide Synthesis Techniques

20

COMPARISON OF SEVERAL METHODS FOR SYNTHESIS OF A DIFFICULT TEST PEPTIDE BY BOC STRATEGY

Richard L. Stevens, Kirk L. Stevens, and Janis D. Young

Peptide Synthesis Facility, Department of Biological Chemistry
UCLA School of Medicine, Los Angeles, CA 90024

I. INTRODUCTION

One of the more common methods employed in the solid phase peptide synthesis (SPPS)* of synthetic peptides in the last decade utilizes amino acids N-terminally blocked with t-butyloxycarbonyl (BOC) (1). With the advent of improved starting resins for SPPS, the availability of more solvent systems for the coupling reaction, the use of pure BOC-amino acids, and the flexibility of reaction protocols by automated procedures, coupling efficiencies greater than 99% are frequently obtained and extraordinarily pure peptides are produced (2). However, not all peptide bonds are made with equal facility nor are all peptides synthesized with equal ease.

Specific amino acids which contribute to hard-to-synthesize peptide bonds were incorporated into a peptide designed by us to investigate various aspects of synthetic peptide chemistry. We recently submitted for publication data on more than 5000 peptide couplings which were analyzed and characterized as either complete, incomplete, or highly incomplete (3,4). These single couplings were made on a Beckman 990B synthesizer, an Applied Biosystems 430A synthesizer, or by manual synthesis. Incomplete couplings (less than 99% coupling) occurred in one of every four peptide bonds; high incomplete couplings (less than 98% coupling) occurred with a frequency of about 8% of the time. This data was used to devise the 21 amino acid peptide that contained the peptide bonds made with the greatest difficulty. The sequence of this Difficult Test Peptide is:

R-V-K-A-Q-E-T-L-R-R-Q-V-I-F-R-V-K-A-Q-Q-I.

---

* Abbreviations Used: ABI: Applied Biosystems Inc., BOC: t-butyloxycarbonyl , BOP: benzotriazolyl-oxy-tris(dimethylamino)-phosphonium hexafluorophosphate, Bzl: benzyl, Cl-Z: 2-chlorobenzyloxycarbonyl, DCC: dicyclohexylcarbodiimide, DCM: dichloromethane, DIEA: diisopropylethyl amine, DMF: dimethylformamide, DMSO: dimethyl sulfoxide, FMOC: 9-fluorenylmethyloxycarbonyl, HF: hydrofluoric acid, HOBt: 1-hydroxybenzotriazole, NMP: N-methyl pyrrolidone, OBzl: benzyl ester, Pam: phenylacetamidomethyl, SPPS: solid phase peptide synthesis, TEA: triethyl amine, TBTU: 2-(1H-benzotriazolyl)-1,1,3,3-tetramethyluronium tetrafluoroborate, TFA: trifluoroacetic acid.

Techniques in Protein Chemistry II

In addition to the activating reagents, solvation of the resin-peptide is an important aspect of peptide synthesis. Solvation of the resin-peptide is dependent on the type of solvent, the nature of the resin, and the specific protected peptide (5,6). In this investigation we have kept constant the resin type (Pam) and the protected peptide (Difficult Peptide) but have varied the solvation system (DCM,DMF, and NMP solvents) and the amino acid activation reagents (DCC, HOBt, and BOP).

The efficiency of synthesis of each peptide bond in the 21 amino acid peptide was quantitatively determined for several different methods. These methodologies, all using BOC-protected amino acids, were as follows: 1) NORMAL-ABI regular cycles, 2) RAPID-Kent version of method 1, 3) NMP/HOBT-ABI cycles, 4) BOP I, and 5) BOP II - a higher pH version of method 4.

II. PROCEDURE

The general scheme for the five methods in the synthesis of the Difficult Test Peptide is outlined in Table I. Additional details of the syntheses are described here.

NORMAL. ABI version 1.4 cycles were used in the synthesis identified as NORMAL (14). Each residue was coupled three times. Side chain protection of the BOC-L-amino acids was Cl-Z for lysine, Tos for arginine, Bzl for threonine and OBzl for glutamic acid. The mole ratio of amino acid to DCC for each symmetric anhydride coupling was 2:1 and 1:1 for the HOBt-DCC couplings. A 4 molar excess of amino acid (2 mmoles) was used for each coupling in the synthesis of 0.5 mmole of peptide. Deprotections were with 33% to 50% TFA (containing 1 mg/ml indole) in DCM for a total time of 18 min. Each coupling was done in triplicate to insure that every coupling step was driven toward completion.

Table I. Outline Of The Five Methods.*

METHOD	MACHINE	RESIN	ACTIVATION REAGENTS	REACT. TIME	CAP	CYCLE TIME
NORMAL	ABI 430A	PAM	S.A., DCC; HOBt-DCC (N,Q,R)	30	NO	60
RAPID	ABI 430A	PAM	S.A., DCC; HOBt-DCC (N,Q,R)	12	NO	25
NMP/HOBt	ABI 430A	PAM	HOBt-DCC in NMP	55	YES	105
BOP I	SHAKER	PAM	BOP:HOBt:NMM (1:1:1) in NMP	60	NO	110
BOP II	SHAKER	PAM	BOP:HOBt:DIEA (1:1:2.1) in NMP	30	NO	60

* *Basic features of the five methods used to synthesize the Difficult Test Peptide.*
Details of each method are described in the Procedure section.

RAPID. A rapid version of ABI 1.4 cycles was devised by S.B.H. Kent (8); in principle it was similar to the NORMAL method of synthesis but was more than twice as fast and used concentrated reagents and flow through steps. The deprotection time was 5 min with 100% TFA; the activation time was 13 min; the reaction time was 12 min; and a shorter (20 ml) reaction vessel was used. Each amino acid was coupled in triplicate to produce the best possible peptide.

NMP/HOBt. ABI 1.4 HOBt-NMP cycles were used for the NMP/HOBt synthesis of the Difficult Test Peptide (9). Mole ratios and amounts of reactants were identical to the NORMAL synthesis protocol. The solvent for the reaction was NMP throughout and all couplings were HOBt-DCC activations. Couplings were performed for 30 minutes in NMP and then 16 minutes in a 20:80 solution of DMSO/NMP. Finally 1.9 mmol DIEA was added and the coupling continued for 7 minutes. The deprotection time was 18 min with 75% TFA; activation time was 36 min; the reaction time was 53 min; and the 20 ml reaction vessel was used. After the third coupling each residue was capped using acetic anhydride.

BOP I. The BOP I strategy of synthesizing the Difficult Test Peptide was performed using a manual shaker and a 15 x 100 mm glass reaction vessel (Milligen, Medford, MA). The amounts of reagents were the same as the NORMAL synthesis with the following exceptions. Deprotections were 30 min with 25% TFA in DCM; neutralizations were with 10% TEA in DCM; NMP was used as the solvent for the HOBt-BOP activation. The ratio of the reactants was 1:1:1 of BOP:amino acid:N-methyl morpholine. Activation times were 2-4 min and reaction times were 60 min. The efficiency of coupling was also measured by the quantitative ninhydrin test at time intervals of 10 and 30 min. If the coupling was less than 98% after 60 minutes, a recoupling step was used to force the synthesis closer to completion.

BOP II. Synthesis conditions for BOP II were adapted from that used by Briand et al. (10). The dry ingredients of BOC-amino acid (2 mmole), HOBt (2 mmole) and BOP (2 mmole) were dissolved in 10 ml of 25% NMP in DCM for approximately 3 min. Four mmoles of DIEA was added to the mixture. The solution was added to the the swollen resin (0.5 mmole) and shaken for 15 min. Then 6.2 ml of NMP was added and the mixture was shaken for an additional 15 min. The mole ratio of the reactants was 1:4:4:4:8 for resin-amino acid:BOC-amino acid:HOBt:BOP:DIEA. Because the first four residues were not as well coupled as expected, the procedure was modified for the last 16 residues. The solvent was changed to only 10 ml NMP with no DCM for a coupling time of 30 min and the amount of DIEA was increased to a mole ratio of 8.5. Deprotection was accomplished throughout the synthesis with 60% TFA in DCM for 15 min and neutralization by 10% DIEA in DCM for 3 min. No recouplings were attempted.

NINHYDRIN MONITORING. Each coupling or specific time period of coupling was monitored by a quantitative ninhydrin reaction (11).

CLEAVAGES. All resin-peptides were cleaved for one hour at 0 °C with anhydrous HF containing 2.5% p-cresol and 2.5% thiocresol.

AMINO ACID ANALYSIS. The amino acid composition of the dried peptide after HF cleavage was determined by an 18 hr acid hydrolysis at 110 °C and analysis of the phenylthiocarbamoyl derivatives (12).

CAPILLARY ELECTROPHORESIS. After HF cleavage the peptide was analyzed using an HPE 100 high performance electrophoresis system (Bio-Rad Laboratories) fitted with a coated capillary cartridge of 20 cm x 25 μm. Peptides from

each of the 5 synthetic methods was that obtained after the final re-coupling step of each residue. Ten μl of each peptide solution (43-53 μg/ ml) was applied of which 2 μl was loaded onto the column in an 8 sec period at 8 kV with a variable amperage of 3-8 μA. A 15 min running time followed with a voltage drop of 8 kV in a running buffer of 0.1 M sodium phosphate, pH 2.5 and peptides were detected at 200 nm.

III. RESULTS and DISCUSSION

In the synthesis of the Difficult Test Peptide by five methods (Figure 1), the method employing NMP/HOBt was clearly superior in coupling efficiency with an average per cent coupling of 99.5. The peptide was also synthesized in an efficient manner (an average coupling of 98.4% excluding residues 1-4) by the BOP II procedure. The RAPID, NORMAL, and BOP I synthesis methods followed in order of coupling efficiency (Figure 1) with 97.3, 95.4, and 94.3 percent, respectively. These differences are significant and important. In order to produce high quality peptides, it is necessary to have high overall coupling efficiency. This is especially true for long peptides.

In general as the peptide increased in length the degree of difficulty in peptide synthesis increased with each of the five methods (Figure 2). This is in keeping with previous observations by us with 350 peptides from 10-20 residues in length (3). A comparison of the coupling efficiencies at the beginning and the end of the peptide was made possible by the repeating sequence QAKVR. Thus, counting from the C-terminal end of the peptide, residues 4-7 and 18-21 were coupled under identical conditions. Residues 4-7 were

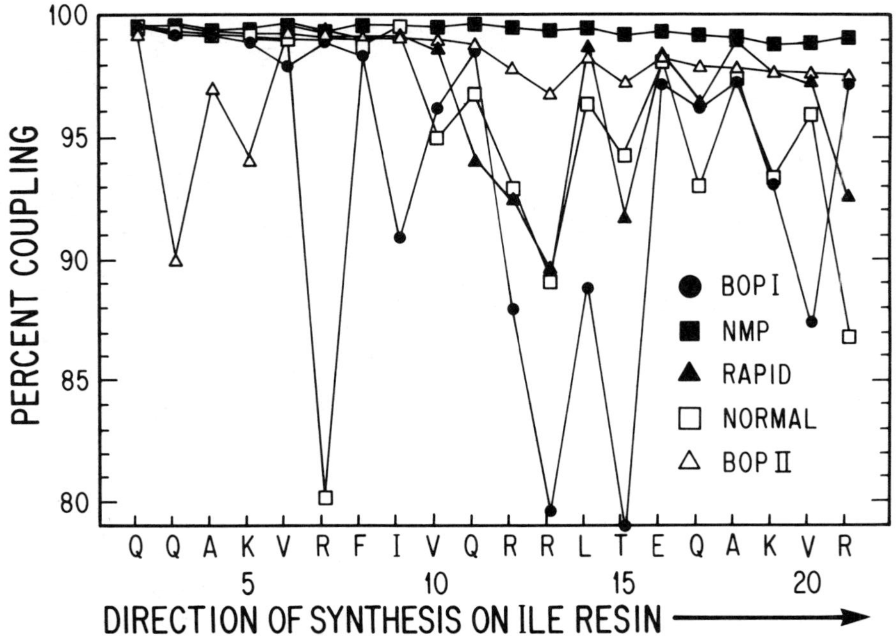

Figure 2. A comparison of the efficiency of each coupling step in the synthesis of the Difficult Test Peptide. The per cent coupling of each residue of each of the five methods was measured by a quantitative ninhydrin test after the first coupling step.

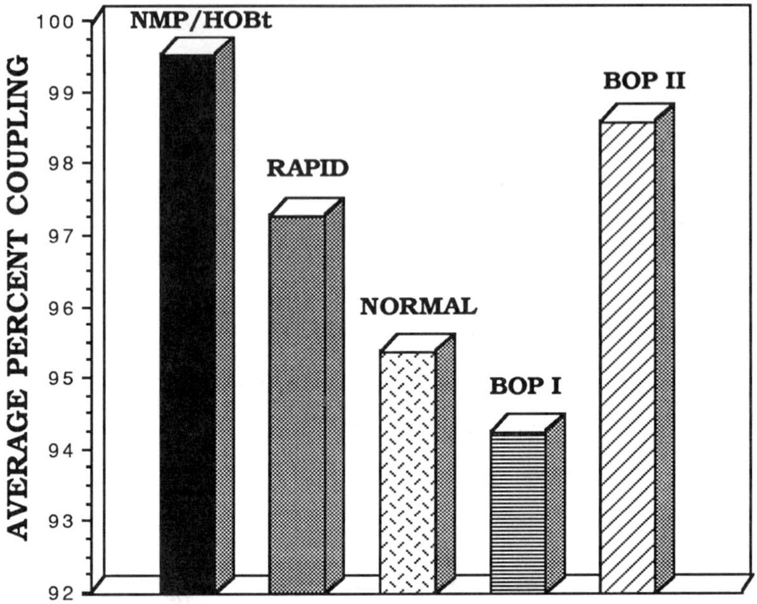

Figure 1. A comparison of the average per cent coupling of the five coupling methods in the synthesis of the Difficult Test Peptide. This data was collected after single couplings of each residue.

generally coupled (average of the five methods) at efficiencies greater than 99% whereas the identical sequence of residues 18-21 were coupled at an average of about 97%. This increase in difficulty of coupling efficiency as the peptide increased in length has also been observed by others for several other peptides (13).

In addition to this data, of possible greater significance was the observation that in the middle of the peptide residues 11-15 were inefficiently synthesized by several of the methods (Figure 2). This data is not incompatible with that of some investigators who have advanced the idea of certain peptides having sequence-dependent coupling difficulties in the range of 8-16 residues from the resin (14-16). The possibility of beta-sheet formation at certain lengths of peptide by intra- as well as inter-chain interactions can lead to peptide aggregation. This has been invoked by several investigators in the synthesis of peptides with problem, difficult or sequence-dependent couplings (14-20). Aggregation of the peptide chain by any means will result in lower solvation and thus lower availability for reaction of the covalently attached peptide.

The inaccessibility of the resin attached N-terminal amino acid of the resin-peptide to react with the free BOC-amino acid will be exacerbated in poor solvents such as DCM. DMF has been a commonly used solvent to maintain solvation and reactivity of the peptide chain. Our data supports the idea of NMP as being an excellent solvent for peptide coupling. With the NMP/HOBt method the peptide was efficiently synthesized throughout its entire length even in the range of 8-16 residues. It appears that the NMP solvent is able to make the N-terminal end of the resin-peptide more accessible for coupling by possibly minimizing intra- and/or inter-chain hydrogen-bonding. It should be noted that the NMP/HOBt method used by us did contain 20 % DMSO for a portion of the reaction time. Others have used DMSO as a structure-disrupting solvent in the commonly used DCM system to effect a more favorable coupling environment (20). More recently Klis and Stewart (21) have suggested using chaotropic salts (e.g., KSCN or $NaClO_4$) to improve SPPS coupling reactions. Their results with a 0.4 M salt solution in a 1:1 DCM:DMF solvent system increased the efficiency of synthesis of difficult coupling reactions. A rapid quantitative reaction necessary for

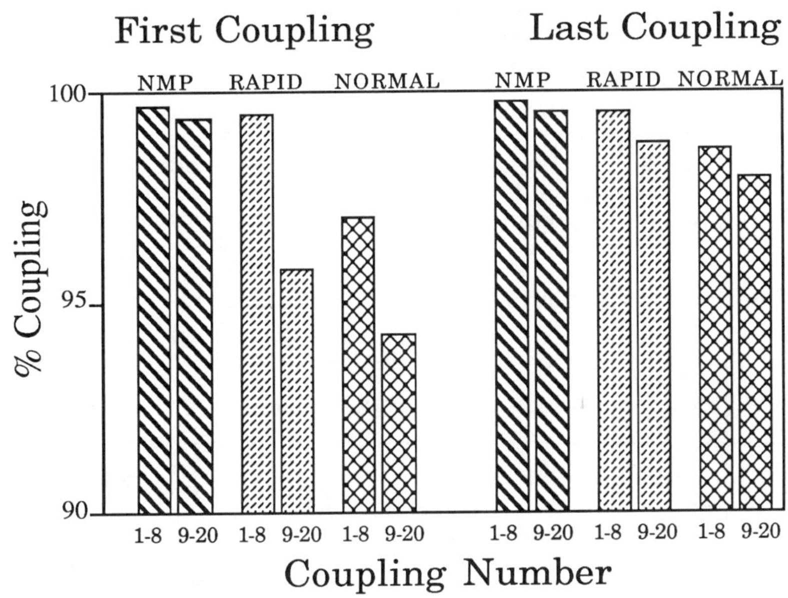

Figure 3. Comparison of three methods for efficiency of coupling in the first half of the sunthesis (residues 1-8) and the last half of the synthesis (residues 9-20). In each instance the last coupling represents the third coupling of the three couplings made of each residue in forcing to near completion.

a stepwise SPPS synthesis will depend on the ability of the covalently attached peptide to be solvated at high concentration as determined by the nature of the resin, the permeating solvent, and the peptide sequence with its protective groups.

In the first three methods of synthesis three couplings at each step were made with every residue (Figure 3). This insured the best possible intermediate product for synthesis to be continued. The final average per cent couplings were NORMAL method 98.2%, RAPID method 99.1%, and NMP/HOBt method 99.6%. The final average per cent coupling for BOP I and BOP II was 97.5 and 98.1%, respectively. Amino acid analysis and capillary electrophoresis were performed on the final peptide product. There was no noticeable difference in amino acid composition or migration of the major product by capillary electrophoresis between peptides made by the five methods (Table II and Figure 4). By capillary electrophoresis the peptide preparation by NMP/HOBt appears to have less contaminants than peptide synthesized by the other methods.

Since the chemistry used in the NORMAL and RAPID methods is the same and the RAPID method is twice as fast, it has the advantage of routinely using double coupling to keep the coupling efficiency near 99% (about one third of all couplings with this protocol require a double couple). With the RAPID method a 21 amino acid peptide is easily synthesized overnight.

From this data the method of choice would be NMP/HOBt synthesis. With only single coupling an average of 99.5% coupling was achieved. An excellent product at minimum cost for amino acids was obtained. Since this is the only method to employ capping after each residue, another

Table II. Amino Acid Analysis Of Difficult Test Peptide.

	MOLE RATIO					
AMINO ACID	EXPECTED	NORMAL	RAPID	NMP-HOBt	BOP I	BOP II
GLU (E)	5	5.4	5.1	4.9	5.6	5.2
ARG (R)	4	4.3	3.9	4.2	4.0	3.7
THR (T)	1	1.0	1.0	1.0	1.0	1.0
ALA (A)	2	2.0	2.0	2.1	2.1	2.1
VAL (V)	3	2.6	2.5	2.7	2.6	2.9
ILE (I)	2	1.7	2.0	1.9	2.2	2.2
LEU (L)	1	1.0	1.2	1.2	0.8	1.0
PHE (F)	1	1.0	1.3	1.1	1.0	1.1
LYS (K)	2	2.0	2.0	1.8	1.7	1.9

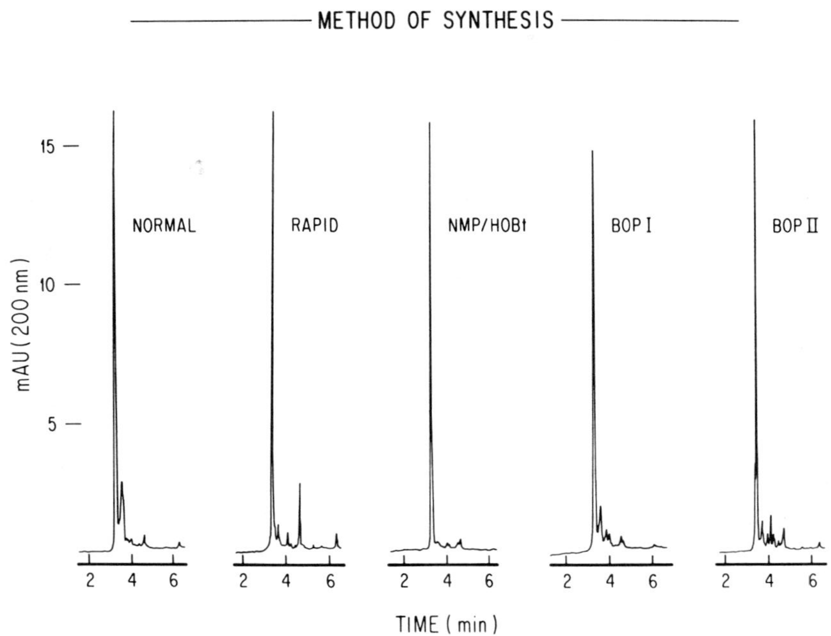

Figure 4. Capillary electrophoresis profiles of the Difficult Test Peptide. About 0.1 µg of peptide was electrophoresed at 8kV. All peptides were driven to near completion in the synthesis by multiple couplings.

advantage would be a cleaner product on capillary electrophoresis or HPLC. However, if time were a factor, it is obvious that not as many peptides could be made per week by this method.

BOP was introduced fifteen years ago as a peptide coupling agent by Castro (22) and that same group has recently published a paper on the present status and new developements of BOP and its congeners (23). It has been used for several years in solid phase peptide synthesis with the BOC/TFA strategy (24, 25) and more recently with the FMOC/morpholine system (26). In the synthesis of the Difficult Test Peptide by the BOP I procedure one equivalent of base was used. Under these conditions coupling proceeded in a rapid fashion and to near completion for the first third of the synthesis, but the remaining residues of the 21 amino acid peptide were coupled poorly (approximately 90%). With only one equivalent of DIEA the basicity of the reaction mixture was not at the level for favorable amine reaction. This method clearly emphasized the difficulty of synthesis of this peptide in the region of 8-16 residues (Figure 2). However, even under these adverse pH conditions synthesis proceeded well in the earlier phase of the peptide and residues 1-8 could be forced with extra couplings to greater than 99% coupling efficiency. The latter part of the peptide was poorly synthesized with couplings as low as 80%.

BOP II synthesis was carried out with 2.2 equivalents of base per one equivalent each of BOC-amino acid and HOBt; the pH of the coupling reaction was between 8 and 9. This method of synthesis was very efficient and successful and resulted in an overall coupling efficiency 1% less than that achieved by the NMP/HOBt method (see Figures 1 and 2). Our findings corroborate those of experienced BOP investigators (17, 22, and 23) who concluded that BOP in the presence of HOBt provides efficient coupling in solid phase synthesis. It is interesting to note that the automated procedure by Briand et al. (17) using BOP with a DCM/NMP solvent did not work well in a manual mode employed by us (BOP II method). The first four couplings averaged 95% whereas in all other procedures the values were over 99%. It was assumed that the presence of DCM was responsible for the poor coupling efficiency. Therefore, for the remainder of the synthesis DCM was replaced by NMP in the reaction mixture. Couplings 5-20 of the BOP II method, performed in only the NMP solvent, were easily accomplished with 98.4% efficiency.

The Difficult Test Peptide appears to have the requirements necessary to be used to evaluate new synthesis protocols. It is of sufficient length and varied composition to form secondary structural features in certain solvent systems and the BOC chemistry methodology employed here. It would be of interest to determine the efficiency of synthesis of this peptide by other methodologies and chemistries. Promising preliminary results have been achieved for the synthesis of the

Difficult Test Peptide by FMOC chemistry with activators TBTU/HOBt in the solvent DMF (26). Recently a predictive method for difficult sequences in SPPS has appeared using as a basis optimized Chou and Fasman type coil conformational parameters (27). Applying this test to the Difficult Test Peptide did result in a prediction that difficult couplings would occur at a high frequency at every segment from residues 5-21.

REFERENCES

1.Barany, G., N. Kneib-Cordonier and D.G. Mullen. (1987) *Int. J. Pept. Prot. Res.*, **30**:705-739.
2.Kent, S.B.H. (1988) *Ann. Rev. Biochem.*, **57**:957-989.
3.Young, J.D., A.S. Huang, J Bruins and D. Ng (1988) *J. Cell Biol.*, **107**:630a.
4.Young, J.D., A.S. Huang, N. Ariel, J. Bruins, D. Ng, and R.L. Stevens. submitted to *Peptide Research*.
5.Merrifield, R.B. (1963) *J. Am. Chem. Soc.*, **85**:2149
6.Sarin, V.K., S.B.H. Kent and R.B. Merrifield (1980) *J. Am. Chem. Soc.*, **102**:5463.
7.Applied Biosystems Users Manual for the ABI 430A Peptide Synthesizer, Version 1.3B July 15, 1989, section 6-8, Applied Biosystems, Inc., Foster City, CA.
8.Schneider, J. and S.B.H. Kent. (1988) *Cell*, 54:363-368.
9.Applied Biosystems Users Manual for the ABI 430A Peptide Synthesizer, Version 1.3B July 15, 1989, section 6-48, Applied Biopsystems, Inc., Foster City, CA.
10.Briand, J.P., A. Van Dorsselaer, B. Raboy and S. Muller. (1989) *Peptide Research*, **6**:381-388.
11.Sarin, V.K., S.B.H. Kent, J.P. Tam, and R.B. Merrifield. (1981) *Anal. Biochem.* **117**:147-157.
12.Bidlinger, B.A., S.A. Cohen, and T.L. Tarvin. (1984) *J. Chromatogr.* **336**:93-104.
13.Applied Biosystems Newsletter, May 1988. Synthesis of magainin-1, magainin-2, human growth hormone releasing factor, and rat atrial natriuretic peptide.
14.Deber, C.M., M.K. Lutek, E.P. Heimer and A.M. Felix. (1989) *Peptide Research*, **2**:184-188.
15.Meister, S.M. and S.B.H. Kent. (1985) Peptides, Structure and Function. Proc. 9th Amer. Pept. Symp., Eds., C.M. Deber, V.J. Hruby and K.D. Kopple, Pierce Chem. Co., Rockford, IL., pp. 103-107.
16.Kent, S.B.H. and I. Clarke-Lewis. (1985) Synthetic Peptides in Biology and Medicine, Eds., K. Alitalo, P. Partnen and A. Vaheri, Elsevier Science Publishers, B.V. Biomedical Division, pp. 29-56.
17.Kent, S.B.H. (1985) Peptides, Structure and Function. Proc. 9th Amer. Pept. Symp., Eds., C.M. Deber, V.J. Hruby and K.D. Kopple, Pierce Chem. Co., Rockford, IL., pp. 407-414.
18.Atherton, E. and R.C. Sheppard. (1985) Peptides, Structure and Function. Proc. 9th Amer. Pept. Symp., Eds., C.M. Deber, V.J. Hruby and K.D. Kopple, Pierce Chem. Co., Rockford, IL., pp. 415-418.
19.Mutter, M., K.-H. Altmann, D. Bellof, A. Florsheimer, J. Herbert, M. Huber, B. Klein, L. Strauch, T. Vorherr and H.-U. Gremlich. (1985) Peptides, Structure and Function. Proc. 9th Amer. Pept. Symp., Eds., C.M. Deber, V.J. Hruby and K.D. Kopple, Pierce Chem. Co., Rockford, IL., pp. 397-405.
20.Toniolo, C., G.M. Bonora, V. Moretto and M. Bodansky. (1985) Peptides, Structure and Function. Proc. 9th Amer. Pept. Symp., Eds., C.M. Deber, V.J. Hruby and K.D. Kopple, Pierce Chem. Co., Rockford, IL., pp. 419-422.
21.Klis, W. and J.M. Stewart. (1990) Peptides, Chemistry, Structure and Biology. Proc. 11th Amer. Pept. Symp., Eds., J.E. Rivier and G.R. Marshall, Escom, Leiden, p. 904-906.

22. Castro, B., J.R. Dormoy, G. Evin and C. Selve. (1975) *Tetrahedron Letters*, **14**:1219-1222.
23. Coste, J., D. Le-Nguyen and B. Castro. (1990) *Tetrahedron Letters*, **2**:205-208.
24. Hudson, D. (1988) *J. Org. Chem.*, **53**:617-624.
25. Fournier, A., C.-T. Wang and A. Felix. (1988) *Int. J. Pept. Prot. Res.*, **31**:86-97.
26. Personal communication from Cynthia G. Fields of Applied Biosystems, Inc, Foster City, CA.
27. Milton, R.C. de L., S.C.F. Milton and P.A. Adams. (1990) *J. Am. Chem. Soc.*, **112**:6039-6046.

21

PEPTIDE CYCLIZATION ON OXIME RESIN (PCOR)*

George Ösapay, Marlène Bouvier and John W. Taylor

The Rockefeller University
Laboratory of Bioorganic Chemistry and Biochemistry,
New York, NY 10021

1. INTRODUCTION

Side-chain and main-chain cyclized peptides are usually synthesized for the purpose of investigating the relationship between chiroptical properties or biological activity and conformation of peptides. Cyclization *via* amide bond formation is achieved by the intramolecular condensation of an N-terminal or side-chain amine with an activated (active ester, anhydride, azide, chloride) C-terminal or side-chain carboxylic acid.[1] To minimize undesired intermolecular condensations, cyclizations are generally carried out in solution phase at inconveniently high dilutions.[2] Even so, significant amounts of linear and cyclic oligopeptides are generally formed during these reactions, reducing the yields and increasing the difficulty of isolating the desired product.

Cyclization reactions have also been performed on high molecular weight polymer supports. For example, the selective deprotection of lysine and aspartic acid side-chain functional groups during the linear assembly of growth hormone-releasing factor analogues by the standard Merrifield approach, allowed their efficient condensation to be achieved on the solid support using the BOP coupling method.[3] Even when two such lactam bridges were built into the same peptide chain by this method, the intrachain-cyclized monomeric peptides could be isolated from the polymeric products of

*Dedicated to the memory of Professor Emil Thomas Kaiser.

Techniques in Protein Chemistry II

interchain condensation reactions after completion of the synthesis and deprotection and cleavage from the solid support.

An alternative solid-phase strategy that should, in principle, eliminate the problem of isolating the desired monomeric peptide from polymeric side products is to link the carboxyl component of the condensation reaction to a polymer support as an active ester intermediate.[4-6] Intrachain cyclization then releases the monomeric peptide into solution, whereas the polymeric products of interchain reactions remain attached to the solid support.[4,6] This approach has generally been explored through the coupling of protected peptides that had been assembled by solution-phase methods to appropriate polymer functionalities. Although high yields were obtained for the cyclization step, coupling of the linear precursors to the supports was rather unsatisfactory.

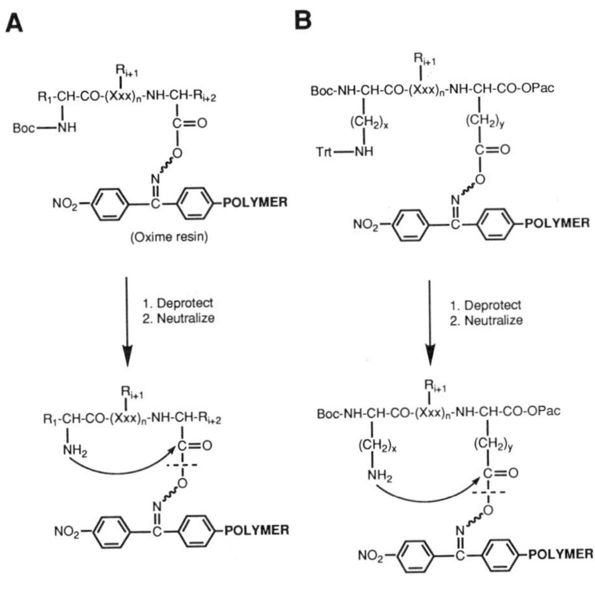

Fig. 1. Use of *p*-nitrobenzophenone oxime resin for the preparation of protected cyclic peptides. A: synthesis of main-chain cyclized peptides; B: synthesis of side-chain cyclized peptides suitable for incorporation into larger peptide structures by segment-condensation synthesis.

We have recently extended this approach through the use of the *p*-nitrobenzophenone oxime polymer (oxime resin) developed by Kaiser and coworkers.[7,8] This oxime resin was originally developed as a polymer support that allows peptides to be assembled by solid-phase methods and then released into solution in a fully protected form through nucleophilic cleavage of the peptidyl oxime ester linkage.[7-10] When applied to the synthesis of protected cyclic peptides, oxime resin allows both the solid-phase assembly of the peptide chain and the subsequent intrachain cyclization and concomitant cleavage of the peptidyl-resin linkage (Figure 1) to be performed in consecutive steps on the same support. As we now describe, this approach of peptide assembly and cyclization on oxime resin (PCOR) allows the efficient synthesis of side-chain and main-chain cyclized protected peptides, including ring sizes of up to 10 amino-acid residues, in high yield and purity.

2. SYNTHESIS OF CYCLIC PEPTIDES

2.1 Synthesis of Cyclic Dipeptides (Diketopiperazines)

It is well known that significant peptide cleavage (in the form of diketopiperazines) occurs during coupling of the third amino acid to a neutralized dipeptide that is attached to a resin by an ester linkage.[7,11] Our initial investigations of the utility of oxime resin for synthesizing cyclic peptides were based on this usually undesired reaction. Boc-LeuLeu-oxime resin was synthesized according to the general procedures described previously[10] and three different protocols were then used to deprotect, neutralize and cyclize this dipeptide: A, (1) 25% TFA/DCM, (2) DCM and Pr^iOH washes, (3) 5% DIEA/DCM, (4) DCM washes, (5) reaction in DCM; B, (1) 25% TFA/DCM, (2) DCM and Pr^iOH washes, (3) reaction in DCM with 2.2 eq. DIEA ; C, (1) 25% TFA/DCM, (2) DCM and Pr^iOH washes, (3) reaction in DCM with 2.2 eq. DIEA and 10 eq. AcOH. Procedure A was based on the methodology for normal stepwise synthesis on oxime resin.[10] Since AcOH has been

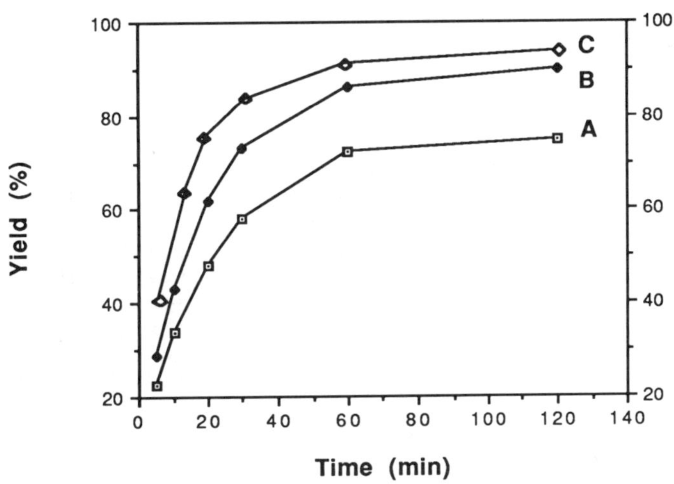

Fig. 2. Kinetics of *cyclo*-(LeuLeu) formation from Boc-LeuLeu-oxime resin in methylene chloride by procedures A, B and C (see text). The cyclization reaction was followed by RP-HPLC (analytical Vydac C_4 column; linear gradient 50-70% CH_3CN/H_2O in 20 min.; flow rate 1.2 mL/min).

reported to catalyze the aminolysis of oxime esters[10,12] and the formation of diketopiperazines[11,13] in solution and on solid supports, procedures B and C were developed to determine whether or not the same catalytic effect of the residual 1 eq. TFA salt or excess AcOH could be observed during solid-phase cyclization on oxime resin.

Figure 2 shows the time course of the cyclization reaction for each of the procedures A, B and C, as followed by RP-HPLC. These results demonstrate that cyclization by all three procedures was rapid, with a 58-84% yield of the diketopiperazine achieved at 30 min and nearly complete reaction after 2 hours. In each case, the desired cyclic product was obtained in high purity directly from the reaction mixture. Procedure C gave slightly superior results in terms of both reaction rate and final yield. Similar results were obtained for the preparation of other diketopiperazines by this PCOR method (Table I, peptides **1**-**3**) using, in most cases, procedure C.

Table I

Summary of PCOR Reaction Conditions and Yields

Peptide[a]	Oxime Groups on Resin (meq/g)	Peptide Substitution Level (meq/g)[b]	Capping Reagent	Added AcOH	Reaction Time (h)	Purification Method	Yield (%)
1	0.67	0.428	Ac_2O	-	4	cryst.	75.7
1	0.67	0.428	Ac_2O	+	4	cryst.	94.4
2	0.67	0.091	Ac_2O	-	4	cryst.	95
3	0.67	0.339	Ac_2O	+	48	cryst.	88
4	0.67	0.308	Ac_2O	+	48	HPLC/ $cryst^n$	84.7
5	0.54	0.114	Ac_2O	-	24	HPLC	55
6	0.67	0.041	Ac_2O	+	72	silica gel/ $cryst^n$	61
7	0.67	0.120	Ac_2O	+	72	silica gel/ $cryst^n$	54
8	0.54	0.300	$Tmac_2O$	+	72	silica gel/ $cryst^n$	68

[a]*Peptide products of PCOR reactions were as follows:* ***1****, cyclo-(LeuLeu);* ***2****, cyclo-(GlyGly);* ***3****, cyclo-(ProVal);* ***4****, cyclo-(ProValProVal);* ***5****, cyclo-[D-PheProPhe-D-PheAsnGlnTyr(Br*$_2$*-Z)ValOrn(Z)Leu];* ***6****, Boc-Lys(2Cl-Z)Leu-cyclo-[LysGlu(OBzl)Leu-Lys(2Cl-Z)Glu]-OPac;* ***7****, Boc-Lys(2Cl-Z)Leu-cyclo-[LysGlu(OBzl)Leu-Lys(2Cl-Z)Asp]-OPac;* ***8****, Boc-cyclo-[OrnGlu(OBzl)LeuLys(2Cl-Z)Glu]-OPac.*

[b]*Refers to the substitution level of the amine participating in the cyclization reaction.*

2.2 Synthesis of Main-Chain Cyclized Peptides

The scope of the PCOR method for preparing main-chain cyclized peptides larger than diketopiperazines has been investigated. The cyclic tetrapeptide, *cyclo*-(ProValProVal) (Table I, peptide **4**), was successfully prepared in high yield by assembly of the linear peptide from Boc-Val-oxime resin, followed by cyclization in the presence of

AcOH over two days. In this reaction, *cyclo*-(ProVal) appeared as a significant impurity during the first 5 min. of the reaction, but could be separated from the desired tetrapeptide simply by washing the resin at this time. (Presumably, the dipeptide resulted from peptide losses from the oxime resin through diketopiperazine formation during coupling of the third residue.) The high yield of this cyclic tetrapeptide contrasts favorably with the low yield (15%) reported for this cyclization reaction by solution-phase methods.[14]

The largest cyclic peptide prepared by the PCOR method to date is tyrocidine A, a 30-member ring antibiotic peptide (Table I, peptide **5**).[15] The stepwise assembly of the linear decapeptide precursor on the oxime resin was designed such that cyclization took place between the Leu-oxime ester bond and the N-terminal amino group of D-Phe. This choice was based on our observation that the Leu-oxime ester bond is relatively susceptible to aminolysis and other evidence that D-amino acids are preferred at the N-terminal position for solution-phase cyclization. In this case, the cyclization reaction proceeded without addition of AcOH as catalyst and the product was obtained in high yield and purity (Table I).

2.3 Synthesis of Side-Chain Cyclized Peptides

Peptides incorporating side-chain to side-chain and side-chain to main-chain amide bridges have frequently been useful in exploring the functional conformations of biologically active peptides. With this goal in mind, we investigated the utility of the PCOR method for the synthesis of side-chain bridged, fully protected peptides that may subsequently be incorporated into larger peptide structures by segment-condensation methods. A series of protected peptides (Table I, peptides **6**-**8**) having the general structure shown in Figure 1B have been assembled on oxime resin and allowed to cyclize as indicated, after selective acidolytic deprotection and neutralization of the trityl-protected side-chain amine. In these peptides, lactam bridges compatible with the α-helical conformation were formed by reacting the ε-amino group of Lys^{i} with the side-chain oxime ester resin linkage of Glu^{i+4} or Asp^{i+4}, and the δ-amino group of Orn^{i} with the side-chain oxime ester resin linkage of Glu^{i+4}. Addition of AcOH was shown to be necessary to obtain a significant rate for the Lys^{i}/Glu^{i+4} reaction. Subsequently, AcOH was added to all of the cyclization mixtures in this series. Similar high yields (Table I) were obtained in each case, and the purity of the products was such that flash chromatography on silica gel generated HPLC-pure products (Figure 3).

The protecting group stategy used in the synthesis of these side-chain cyclized peptides is compatible with their subsequent incorporation into larger peptide structures by segment-condensation methods since the N-terminal amine and C-terminal carboxylic acid may be selectively deprotected using TFA and Zn/AcOH, respectively. The utility of this approach in synthesizing multicyclic peptides has recently been demonstrated[16] through the preparation of a 21-residue, amphiphilic α-helical peptide comprising three repeats of peptide **6**.

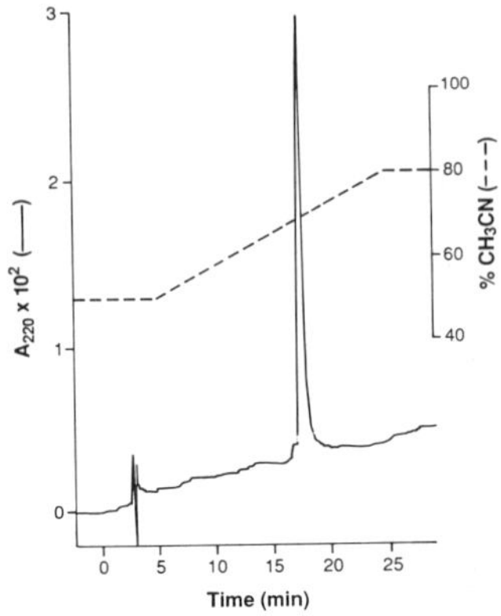

Fig. 3. RP-HPLC chromatogram of cyclic model peptide **5** (Table I) as obtained after purification by flash chromatography. (Analytical Vydac C_4 column [0.45 X 25 cm]; eluant A, 0.1% AcOH in H_2O, eluant B, 0.1% AcOH in 90% CH_3CN/H_2O; linear gradient 50-80% B in 20 min; flow rate 1.2 mL/min).

3. BLOCKING OF UNSUBSTITUTED OXIME GROUPS

Undesired transacylation reactions involving the acetyl groups normally used to block excess oximes are known to be a major source of premature peptide chain termination (and the subsequent initiation of new peptide chains as additional side products) during peptide assembly on oxime resin. The blocking acetyl groups are also susceptible to reactions with nucleophiles, including amino acid and peptide esters, used during peptide cleavage from oxime resin. Likewise, in the case of PCOR, analysis of the polymer support[17] remaining after the cyclization/cleavage of protected peptide **6** identified the N^{ε}-acetylated peptide as a significant side product.

One approach to reducing the frequency of side reactions involving acetylated oxime groups is to reduce the number of oxime groups on the resin during its preparation so that this matches more closely the desired peptide substitution level and fewer excess oxime groups need to be blocked. In a separate study,[18] this approach has been applied to the PCOR synthesis of another Lys^i, Asp^{i+4}, lactam-bridged protected pentapeptide. Unfortunately, however, the condition of nearly complete peptide substitution of the oxime groups on the solid support strongly promoted <u>inter</u>chain aminolysis reactions, even under conditions of low oxime substitution (0.1 meq/g) that were expected to favor the <u>intra</u>chain reaction. Consequently, the side-chain cyclized pentapeptide product of this PCOR reaction was contaminated with a significant amount of the double-length cyclic peptide (a 40-membered ring!) and was difficult to purify. (The larger peptide presumably resulted from one interchain aminolysis reaction followed by intrachain cyclization of that product with concomitant cleavage from the solid support.) Since double-length peptides were not observed as side products during the syntheses of peptides **6** and **7** (Table I), where low peptide substitution levels in the same range (0.04 to 0.12 meq/g resin) on oxime resin functionalized at a much higher level were employed, it is apparent from this experience that the distribution of products resulting from PCOR depends in a complex fashion on both the peptide substitution level on the oxime resin and the substitution level of the excess, blocked oxime groups.

An alternative approach to the problem of reducing the transacylation side reactions that might not promote interchain aminolysis reactions during PCOR is to search for a more stable blocking group. This led us to investigate the use of a bulkier and less electrophilic group, the trimethylacetyl (Tmac) group, as a substitute for the acetyl (Ac) group. Acetic anhydride (Ac_2O) and trimethylacetic anhydride ($Tmac_2O$) were initially compared for their abilities to quantitatively react with oxime groups on the resin. The oxime ester bonds formed by the acetyl and trimethyacetyl groups were then investigated and compared for their stabilities towards (a) acidolysis in the 25% TFA/DCM mixture normally used for N^{α}-Boc deprotection during peptide synthesis on oxime resin, and (b) aminolysis by H-Leu-OBu^t·HCl.

3.1 Stability of oxime ester bonds towards acidolysis

The relative abilities of Ac_2O and $Tmac_2O$ to quantitatively react with free oxime groups on the resin (0.544 meq oxime groups/g resin) were investigated as described in Scheme I A. In both cases, less than 1% of the total oxime groups reacted with the preformed symmetric anhydride of Boc-Gly-OH after acylation of the oxime resin with either Ac_2O or $Tmac_2O$ for 24 h (5% reaction with Boc-Gly-OH anhydride was found after acylation of the oxime resin with $Tmac_2O$ for only 4 hours). This result suggests that the bulkier $Tmac_2O$ reagent couples equally well to the oxime resin, provided that a 24 h reaction period is used.

The stabilities of the oxime ester bond formed by the acetyl and trimethylacetyl groups towards 25% TFA/DCM were then studied according to Scheme I B. The two types of oxime ester bond were found to be equally stable in this acidic medium after 4 h exposure, since less than 1% of the total oxime groups on each capped resin could subsequently react with the symmetric anhydride of Boc-Leu-OH. This acid treatment corresponds to an exposure time equivalent to the synthesis of a nonapeptide.

STABILITY OF OXIME ESTER BOND

TOWARD ACIDOLYSIS

oxime resin

↓ 40 eq. Ac_2O (or $Tmac_2O$) in DCM, 24 hr

A

↓ 10 eq. sym. anh. BocGlyOH in DCM, overnight

1

↓ 1.) 25 % TFA/DCM 4 hr
2.) 5 % DIEA/DCM 3x1 min

B

↓ 3 eq. sym. anh. BocLeuOH in DCM, 2 hr

2

SCHEME I

TOWARD AMINOLYSIS

oxime resin

↓ 0.55 eq. BocAlaOH
0.55 eq. DCC
in DCM, overnight

↓ 40 eq. Ac_2O (or $Tmac_2O$) in DCM, 24 hr

1

↓ 10 eq. sym. anh. BocGlyOH in DCM, overnight

2

↓ 3.0 eq. H-Leu-OtBu.HCl
3.0 eq. DIEA
1.5 eq. HOAc
in DCM, 24 hr

3

↓ 5.0 eq. BocAlaOH
5.0 eq. DCC
in DCM, overnight

4

SCHEME II

3.2 Stability of oxime ester bonds towards aminolysis

The stabilities of the acetyl and trimethylacetyl oxime ester bonds towards aminolysis were studied according to Scheme II. Initially, Boc-Ala-OH (0.55 eq.) was coupled to oxime resin (0.544 meq oxime groups/g resin), which was then acylated with either Ac_2O or $Tmac_2O$ for 24 h, yielding resins **1** (0.222 meq Boc-Ala/g resin). Subsequent coupling of Boc-Gly-OH to resins **2** modified less than 1% of the total oxime groups, demonstrating that quantitative acylation had been obtained. Aminolysis of the oxime esters in resins **2** using H-Leu-OBu^t and AcOH resulted in the crude product mixtures shown in the RP-HPLC chromatograms in Figure 4. Amino acid analysis and FAB mass spectrometry were then used to identify each component of these mixtures. Product peak 1 was identified as excess H-Leu-OBu^t. The major additional product, peak 2, in the chromatogram of the Ac_2O experiment (Figure 4A) was identified as Ac-Leu-OBu^t. In marked contrast, no trace of Tmac-Leu-OBu^t (peak4) in the chromatograph of the $Tmac_2O$ experiment (Figure 4B) was found at the expected elution time of this compound indicated by the arrow. The yield of Boc-Ala-Leu-OBu^t (peak 3) was found to be 98% and 100% for the Ac_2O and $Tmac_2O$ experiments, respectively. Repeat coupling of Boc-Ala-OH to the regenerated oxime resins **3** allowed us to determine, based on the resulting Boc-Ala substitution level, that 72% of the acetyl groups had been cleaved by Leu-OBu^t, whereas 0% of the trimethylacetyl groups had reacted.

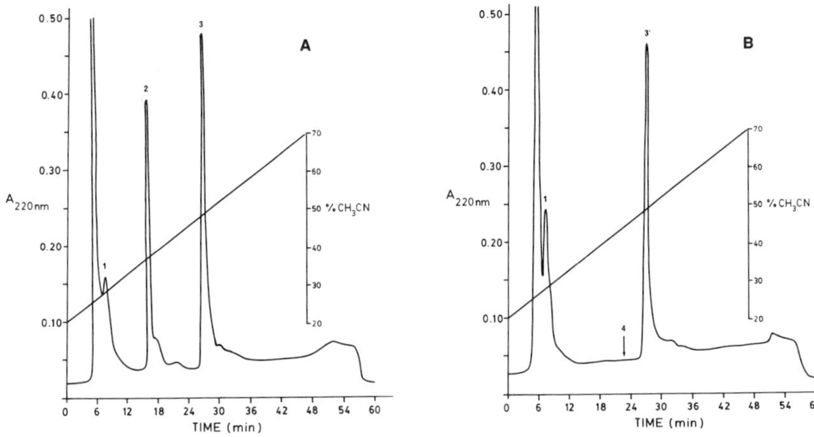

Fig. 4. RP-HPLC chromatograms of the crude reaction mixtures from the aminolysis experiment described in Scheme II, using (**A**) Ac_2O and (**B**) $Tmac_2O$ as the blocking reagent. (Semi-preparative Vydac C_4 column [1.0 X 25 cm]; eluant A, 0.1% TFA in H_2O; eluant B, 0.1% TFA in 90% CH_3CN/H_2O; linear gradient 20-70% B in 45 min; flow rate 3.0 mL/min).

From this study, it is clear that $Tmac_2O$ is superior to Ac_2O as a blocking reagent for unsubstituted oxime groups owing to the increased stability of the oxime ester product towards aminolysis. Blocking with $Tmac_2O$ was subsequently employed in the synthesis by PCOR of the Orn^i, Glu^{i+4} lactam-bridged peptide **8** (Table I) with satisfactory results in terms of both yield and purity of the desired product.

4. GENERAL PROCEDURE FOR THE PCOR METHOD

Direct comparisons of the different protocols used to date to perform peptide cyclizations on oxime resin mostly still need to be performed. Furthermore, it is clear (see Table I) from the various requirements of these reactions for acetic acid as a catalyst, and from the dependence of the yield and purity of the desired cyclic products on both the peptide and oxime substitution levels, that each PCOR reaction should be optimized on an individual basis, as necessary. Nevertheless, the following stepwise protocol is recommended:

(a) protected peptide chains should be assembled on oxime resin at a peptide substitution level of 0.2 meq/g or less, using an oxime resin having 0.5 meq oxime groups/g or more;
(b) standard protocols for peptide chain assembly on oxime resin[10] should be followed, except that excess oxime groups should be capped by reaction with $Tmac_2O$ for 24 h and not Ac_2O;
(c) after selective acidolytic deprotection of the appropriate amino group (N-terminal or side chain) and extensive washing, the cyclization reaction is initiated in DCM or DCM/DMF by neutralization with DIEA and addition of 10 eq. AcOH;
(d) cyclization is allowed to proceed for 2-72 hours and the released cyclic peptide is obtained by filtering the reaction mixture and washing the resin extensively with DCM and EtOH or MeOH;
(e) the combined filtrates are evaporated, redissolved in DCM, washed (if the desired product is water insoluble), dried and finally purified by crystallization or chromatography.

One of the most attractive features of this new method is that cyclization takes place directly on the resin with concomitant release of the cyclized peptide into solution, free of polymeric side-products. Other useful advantages of this method include its general applicability to any peptide sequence and to both main-chain and side-chain cyclizations, rapid product isolation in high yield and purity, and adaptability to multicyclic peptide syntheses and large scale syntheses. The use of $Tmac_2O$ as a blocking reagent for unsubstituted oxime groups has also been shown to eliminate transacylation side-reactions associated with the PCOR method in particular, and peptide chain assembly on oxime resin in general. This reagent should replace Ac_2O as the blocking reagent of choice in these procedures.

ACKNOWLEDGMENTS

The authors thank the laboratory of Dr. Brian Chait, at The Rockefeller University, for performing mass spectrometry analyses. This research was supported by PHS grant DA 04197.

REFERENCES

1. Kopple, K.D. (1972) J. Pharm. Sci. **61**, 1345.
2. Ziegler, K. (1955). In "Methoden der Organischen Chemie" (Houben-Weyl, ed.) Vol. 4/2, p. 738. Georg Thieme Verlag, Stuttgart.
3. Felix, A.M., Wang, C.-T., Heimer, E.P., Campbell, R.M., Madison, V.S., Fry, D., Toome, V., Downs, T.R. and Frohman, L.A. (1990). *In* "Proc. 11th Amer. Peptide Symp." (Rivier, J.E., Marshall, G.R., eds.), p. 226. ESCOM Science Publishers B.V., Leiden, The Netherlands.
4. Fridkin, M., Patchornik, A. and Katchalski, E. (1965) J. Am. Chem. Soc. **87**, 4646.

5. Fridkin, M., Patchornik, A. and Katchalski, E. (1972) Biochemistry **11**, 466.
6. Flanagan, E. and Marshall, G.R. (1970) Tetrah. Lett. 2403.
7. DeGrado, W.F. and Kaiser, E.T.(1980) J. Org. Chem. **45**, 1295.
8. DeGrado, W.F. and Kaiser, E.T. (1982) J. Org. Chem. **47**, 3258.
9. Lansbury, P.T., Jr. Hendrix, J.C. and Coffman, A. (1989) Tetrah. Lett. **30**, 4915.
10. Nakagawa, S.H. and Kaiser, E.T. (1983) J. Org. Chem. **48**, 678.
11. Gisin, B.F. and Merrifield, R.B. (1972) J. Am. Chem. Soc. **94**, 3102.
12. Fujino, M. and Nishimura, O. (1969) Chem. Pharm. Bull. **17**, 1937.
13. Suzuki, K., Sasaki, Y., Endo, N. and Mihara, Y. (1981) Chem. Pharm. Bull. **29**, 233.
14. Ueda, T., Sada, I., Kato, T. and Izumiya, N. (1985) Int. J. Peptide Protein Res. **25**, 475.
15. Ösapay, G. and Taylor, J.W. (1990) Tetrah. Lett. (in print).
16. Ösapay, G. and Taylor, J.W. (1990) J. Am. Chem. Soc. **112**, 6046.
17. Peptides were released from the resin by reaction with excess N-hydroxypiperidine and identified by mass spectrometry.
18. Bouvier, M. and Taylor, J.W., unpublished results.

22

Chemical Synthesis of Human Transforming Growth Factor-α

Kerry L. Spear* and Mark X. Sliwkowski

Triton Biosciences Inc.
Department of Protein Chemistry
Alameda, CA 94501

I. Introduction

One superfamily of growth factors consists of a group of secreted polypeptides having shared homology, most notably a unique arrangement of 6 half-cystine residues. These peptides elicit related biological responses which potentiate cell growth and differentiation. Members of this group include transforming growth factor-α (TGF-α) (1), epidermal growth factor (EGF) (2), amphiregulin (3), vaccinia virus growth factor (4), Shope fibroma growth factor (5) and myxoma growth factor (6).

TGF-α consists of a linear chain of 50 amino acids and includes 3 disulfide bonds (see Figure 1). It was originally isolated from a retrovirus-transformed mouse cell line (7), although it is now known (8) to be secreted from other types of cells, including tumor cells and embryo cells. This peptide has been associated

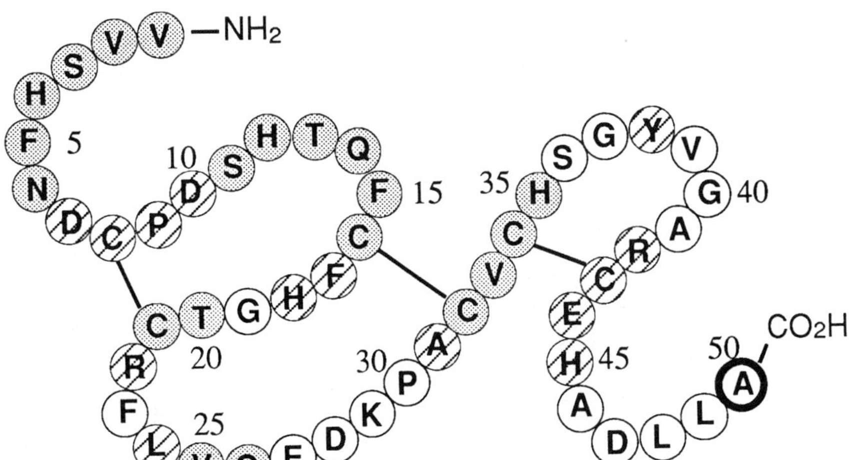

Figure 1. Sequence of TGF-α showing disulfide bonds. A putative degree of difficulty is assigned for each coupling (- easy; - moderately difficult; - difficult).

Techniques in Protein Chemistry II

with a variety of physiological functions such as wound healing, angiogenesis and malignant transformation (1, 9). The biological activities appear to be initiated by binding to the same cell surface receptor as EGF. Since the EGF receptor is over-expressed in several classes of malignant tumors and since the level of TGF-α is elevated in these cases, this growth factor has been implicated in the onset and development of some types of cancer (9).

Our own interest in TGF-α has focused on establishing a structure/activity relationship (SAR) which, when taken together with conformational data and molecular modelling, will provide insights into the receptor-bound conformation. It is hoped that such information might prove useful in designing TGF-α antagonists as novel therapeutic agents for the treatment of several types of human malignancies, most notably some forms of breast and colorectal cancers. Several SAR studies of full length TGF-α employing either site-directed mutagenesis (10, 11, 12) or chemical synthesis (13) have been reported to date. In addition, a synthetic peptide which includes residues 34-43 of TGF-α was reported to be a weak antagonist (14).

A useful prelude to developing an SAR for TGF-α is to synthesize the natural product. Indeed, several such syntheses via solid phase peptide synthesis (SPPS) techniques have been reported. These include the preparation of 3 forms of this growth factor, rat (15), human (16, 17) and bovine (18), using tert.-butoxycarbonyl (Boc) chemistry. Human TGF-α (hTGF-α) was also reported to have been prepared via a fluoren-9-ylmethoxycarbonyl (Fmoc) strategy, although the yield was low (19). Our experience with SPPS and Fmoc chemistry led us to believe that this method could be competitive with a Boc strategy. We describe here the Fmoc-based chemical synthesis of TGF-α in good overall yield.

II. Materials and Methods

Fmoc amino acids were purchased from MilliGen/Biosearch or Bachem with the following sidechain protecting groups: Asp, Glu, Ser, Thr and Tyr: tert.-butyl (tBu); Cys and His: trityl (Trt); Asn and Gln: 2,4,6-trimethoxybenzyl (Tmob); Lys: Boc; and Arg: 2,2,5,7,8-pentamethylchroman-6-sulfonyl (Pmc). Dimethylformamide (A.C.S. grade) was purchased from Baker and treated with alumina (activated, acidic, Brockmann 1, standard grade, 150 mesh) and filtered prior to use. Dichloromethane (HPLC grade), methanol (HPLC grade) and toluene (resi-analyzed grade) were also obtained from Baker. Acetonitrile (ACN; HPLC grade) was purchased from Burdick and Jackson. Prior to use, water was purified using a Millipore Milli-Q Water System. Trifluoroacetic acid (TFA) used for cleavage reactions was purchased from Baker (HPLC grade). TFA used as a mobile phase modifier for HPLC solvents was purchased in ampules from Pierce (spectrograde). Piperidine, anisole, thioanisole, 1,2-ethanedithiol, Na_2SO_3, 1,3-diisopropylcarbodiimide (DIPCDI) and 1-hydroxybenzotriazole (HOBT) were all purchased from Aldrich. Urea and Tris base were purchased from BRL. $Na_2S_4O_6$ was obtained from Pierce. Radiolabelled (^{125}I) murine EGF (mEGF) was purchased from Amersham. Recombinant TGF-α (rTGF-α) was prepared at Triton Biosciences Inc.

A. *Peptide Synthesis*

Automated peptide synthesis was performed on a Milligen/Biosearch Model 9500 peptide synthesizer using the standard Fmoc-DIPCDI/HOBT protocols except that, in several instances, the coupling time for a specific residue was modified. The reaction was performed on a 0.4 mmol scale and the initial residue (alanine) was derivatized through a hydroxymethylphenoxyacetic acid linker to a methyl benzhydrylamine polystyrene resin. This Fmoc-alanine-PAC support (0.33 mmol/g loading) was purchased from MilliGen/Biosearch.

B. *Cleavage*

Dried resin-bound peptide was taken up in a solution of 90:5:3:2 (v/v)/ TFA:thioanisole:1,2-ethanedithiol:anisole (1.0 g/20 mL). The resulting slurry was gently shaken at room temperature for 3.5 h before being filtered and washed with TFA (4x5 mL). The combined filtrates were evaporated in vacuo. The residue was redissolved in TFA (7 mL) and diethyl ether (~200 mL) was added dropwise with stirring. The resulting white ppt. was vigorously stirred for 15 min and filtered. The filter cake was washed with ether (4x), dissolved in 50% aq. ACN and lyophilized to afford a white powder.

C. *HPLC*

Analytical HPLC (Vydac; C_{18}, 0.46 x 15 cm, 5 mm, 300Å) was carried out on a Waters system consisting of 2 Model 510 pumps connected to a Model 680 automated gradient controller. Column effluent was monitored at 220 nm by a Model 490 spectrophotometer and peak area was measured with a Model 745B Data Module. Preparative HPLC (Vydac; C_{18}, 5 x 25 cm, 15-20 mm, 300Å) was carried out on a Gilson system consisting of 2 Model 303 pumps equipped with extended flow heads (permitting a maximum flow rate of 99 mL/min) and connected to a Model 803C manometer module and a Model 811 mixer. Column effluent was monitored at 220 nm by a Holochrome Model spectrophotometer. Gradients were run using a binary solvent system consisting of water (A) and ACN (B), both containing 0.05% TFA.

D. *Folding*

Synthetic TGF(1-50) [sTGF(1-50)] was folded to its native disulfide bond configuration (see Figure 1) by converting the cysteines in the peptide to their corresponding S-sulfonates (20). In a typical reaction, 5 mg of reduced lyophilized sTGF(1-50) was dissolved in 159 µL of water and 81 µL of 1.5 M Tris HCl, pH 8.8. The cysteines in the peptide were spontaneously oxidized to disulfides by allowing the solution to stand at room temperature for 90 min. Sulfitolysis was performed by adding 960 µL of 10 M urea to this solution followed by 150 µL of 1 M Na_2SO_3 and then 150 µL of 0.2 M $Na_2S_4O_6$. The reaction mixture was kept at room temperature for 2 h, after which the solution was dialyzed versus 1 M urea, 20 mM Tris HCl, pH 7.1. Initially this derivatized sTGF(1-50) was chromatographed on an anion exchange column (Mono Q HR 10/10, Pharmacia) in an effort to assess the extent of derivatization and also to detect minor impurities. In subsequent preparations, this step was found to be unnecessary and reagents used in the sulfitolysis reaction were removed by desalting over G-25 or dialysis. S-sulfonated sTGF(1-50) at a concentration of 0.2-3 mg/mL was adjusted to pH 8.5-8.8. EDTA was added to a final concentration of 1 mM and folding was initiated by the addition of cysteine to a final concentration of 3 mM. Analytical reverse phase HPLC was used to monitor the folding efficiency. Generally, the reaction was allowed to proceed from 16 h to 24 h. Prior to the final reverse phase chromatography step, the solution was acidified by the addition of one-tenth volume of 10% TFA.

E. Characterization

Compositional analysis data were collected from 6 M HCl hydrolysates (vapor phase, 150° C, 1 h) by o-phthaldehyde (OPA)/Fmoc-based analysis performed on a Hewlett-Packard Model 1090 Amino Quant high-performance analyzer. Automated Edman degradation chemistry was used to determine the peptide sequence analysis data. An Applied Biosystems, Inc. (ABI) Model 475A gas-phase sequencer equipped with an on-line ABI Model 120 HPLC was employed for the degradations.

F. Radioreceptor Assay

The ^{125}I-mEGF competitive binding assay was performed on subconfluent monolayers of formalin-fixed A431 human carcinoma cells using the method of Marquardt and Todaro with minor modifications (21). The monolayer of cells was rinsed with Hank's balanced salts and Dulbecco's phosphate buffered saline (DPBS) before fixing. The fixed cells were then washed with DPBS, Hank's balanced salts and finally binding buffer (Dulbecco's modified Eagle's medium containing 0.1% bovine serum albumin and 50 mM HEPES solution, pH= 6.8). Competitions were initiated by the addition of 80 μL of binding buffer, a solution of binding buffer (10 μL) with or without the factor (synthetic TGF-α or rTGF-α), and finally a solution (10 μL) of binding buffer containing ^{125}I-mEGF (0.65 mM). After incubation for 2 h at 37° C, the cells were rinsed 3 times with binding buffer, treated with 1 N NaOH and harvested on cellulose acetate wicks. Specifically bound ^{125}I-mEGF was then measured.

III. Results and Discussion

Recent advances in the utilization of Fmoc derivatized amino acids in SPPS make this synthesis strategy an attractive alternative to Boc chemistry (22). Several syntheses of large polypeptides using an Fmoc approach have been reported, including that of the 86 residue HIV TAT protein (23). These successes led us to conclude that the total synthesis of TGF-α by this methodology would compare favorably to the previously reported syntheses of this peptide via Boc chemistry.

Sidechain protected TGF(1-50) was assembled in a stepwise fashion from residue 50 (alanine). Coupling time per residue varied from 1 h to 4 h and depended on a perceived degree of difficulty for each step (see Figure 1). Those couplings which were predicted to be easy were allowed to proceed for one hour.

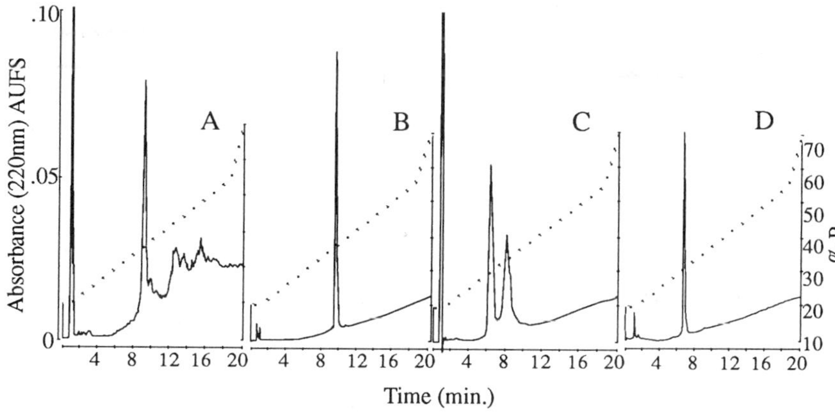

Figure 2. HPLC profile of (A) crude product after cleavage and lyophilization; (B) acyclic TGF(1-50) after purification; (C) crude product after cyclization of S-sulfonated-TGF(1-50); and (D) purified sTGF-α using a Vydac C_{18} analytical column (4.6 mm I.D. x 5 cm; 5mm; 300A). Conditions: isocratic at 20% B for 2 min followed by 20% B to 55% B over 17.5 min. A: water (+ 0.05% TFA); B: ACN (+0.05% TFA). Flow rate: 2.0 mL/min.

Couplings which were expected to be of medium difficulty were reacted for 2 h and "difficult" couplings were continued for 4 h. The degree of difficulty was estimated by considering possible sidechain-sidechain interactions (including any protecting groups) between the N-terminal residue and the amino acid to which it will couple. The proximity of steric bulk to the peptide backbone (i.e., sidechain-backbone interactions) might also affect coupling rates. For example, coupling of Fmoc-Glu(tBu) to a peptide would be expected, a priori, to be somewhat more facile than coupling of Fmoc-Asp(tBu) to the same peptide chain. Finally, amino acids or N-terminal residues containing a β-disubstituted sidechain (Thr and Val in this synthesis) would be expected to couple at a slower rate than residues in which the sidechain was monosubstituted at the β-position, again for reasons of steric crowding near the reaction site. Since qualitative ninhydrin analyses of the most difficult couplings were negative after the prescribed coupling times, double-coupling was not performed on any of the residues in this synthesis.

After completion of the synthesis and removal of the N-terminal Fmoc protecting group, 3.98 g of resin-bound peptide was obtained (theoretical = 5.06 g). Cleavage of the peptide from its resin support also effectively removed all protecting groups to afford 1.91 g of crude peptide (theoretical = 2.63 g assuming 9 trifluoroacetate counter-ions). Analytical HPLC (see Figure 2A) confirmed the presence of a predominant peak, presumably acyclic TGF(1-50). Treatment with dithiothreitol (DTT) resulted in only minimal changes in the HPLC profile, suggesting that few, if any, disulfide bonds were present. Purification via preparative HPLC afforded a product (246 mg having a single symmetrical peak and >98% homogeneity by integration of the HPLC chromatogram from each of 2 different gradients; see Figure 2B) which was positive by Ellman's colorimetric test (24) and remained unchanged upon treatment with DTT.

Since air oxidation of the putative acyclic TGF(1-50) afforded only a poor yield (<10%) of sTGF-α, this peptide was cyclized to its native conformation after conversion of the cysteines to their corresponding S-sulfonates (25). Disulfide exchange was initiated by the addition of a thiol. Cysteine at a final concentration of 3 mM was found to afford the optimal yield of correctly folded

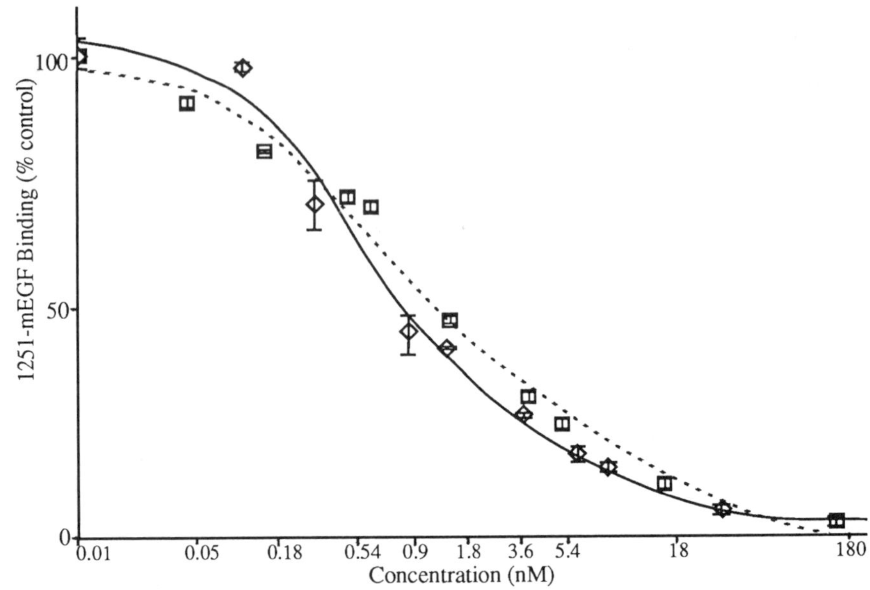

Figure 3. Competition for binding to fixed A431 cells. Human carcinoma cells (A431) were incubated with ^{125}I-mEGF and increasing concentrations of unlabelled rTGF-α (□), or sTGF-α (♦). Values shown are the mean +/- standard error (n=4 for each sample).

synthetic TGF-α (sTGF-α; see Figure 2C). After final purification, sTGF-α (128 mg having a single symmetrical peak and >98% homogeneity; see Figure 2D) and recombinantly derived TGF-α (rTGF-α) were found to have identical HPLC retention times by co-injection. The nature of the improperly folded TGF-α was not explicitly determined.

Amino acid analysis of the purified sTGF-α was in good agreement[1] with the expected amino acid composition.(theoretical values in parentheses): Asx 4.9 (5), Glx 4.1 (4), Ser 2.7 (3), His 4.0 (5), Gly 3.1 (3), Thr 1.5 (2), Ala 4.3 (4), Arg 2.3 (2), Tyr 0.9 (1), Val 3.8 (5), Phe 3.6 (4), Leu 3.2 (3), Lys 1.0 (1), Pro 2.0 (2). In addition, the identity of the first 25 residues was confirmed by sequence analysis. Examination of the Edman degradation data also indicated that the overall purity was much greater than 95% with respect to deletion peptides.

Receptor binding of sTGF-α was evaluated in a radio-receptor assay (RRA) involving competitive displacement of ^{125}I-mEGF from a preparation of human EGF receptors expressed on fixed whole cells. Since the activity of sTGF-α (EC_{50} = 0.87 nM) was found to be equivalent to that of rTGF-α (EC_{50} = 1.15 nM; see Figure 3), this provides additional evidence that the synthetic variant of TGF-α is correctly folded and structurally identical to rTGF-α.

[1] For comparison, amino acid analysis for rTGF-α is also given (average of 3 replicates): Asx 5.1 (5), Glx 4.2 (4), Ser 2.9 (3), His 5.0 (5), Gly 3.1 (3), Thr 1.9 (2), Ala 4.1 (4), Arg 2.0 (2), Tyr 1.0 (1), Val 4.0 (5), Phe 4.2 (4), Leu 3.2 (3), Lys 1.0 (1), Pro 2.1 (2).

IV. Conclusions

The SPPS of hTGF-α is readily achieved using standard Fmoc chemistry in which the coupling is activated by a carbodiimide (DIPCDI) and HOBT. The isolated yield of acyclic TGF(1-50) having no protected sidechain functional groups was 9.4%. Under the conditions described here, oxidation of the 6 cysteine residues to cystine moities proceeded smoothly to afford a product which appears to be structurally and functionally identical to recombinantly-derived hTGF-α. The overall yield based on Fmoc-Ala-Pac resin is 4.9%.

Acknowledgements

We thank R. N. Harkins, T. K. Slattery and J. R. Speckart for sequence analysis and rTGF-α standard; J. A. Kassel and A. Telang for the RRA data; and P. M. Johnson for excellent technical assistance. We also gratefully acknowledge D. Hudson and P. M. Lester for insightful discussions.

References

1. Derynck, R. *J. Cell. Biochem.* **1986**, 32, 293 and references cited within.
2. Carpenter, G.; Cohen, S. *Ann. Rev. Biochem.* **1979**, 48, 193; Savage, C. R., Jr.; Hash, J. H.; Cohen, S. *J. Biol. Chem.* **1973**, 248, 7669.
3. Shoyab, M.; Plowman, G. D.; McDonald, V. L.; Bradley, J. G.; Todaro, G. J. *Science* **1989**, 243, 1074.
4. Stoobant, P.; Rice, A. P.; Gullick, W. J.; Cheng, D. J.; Kerr, I. M.; Waterfield, M. D. *Cell* **1985**, 42, 383.
5. Chang, W.; Upton, C.; Hu, S., Purchio, A. F.; McFadden, G. *Mol. Cell. Biol.* **1987**, 7, 535.
6. Upton, C.; Macen, J. L.; McFadden, G. *J. Virol.* **1987**, 61, 1271.
7. DeLarco, J. E.; Todaro, G. J.; *Proc. Natl. Acad. Sci. USA* **1978**, 75, 4001.
8. Todaro, G. J.; Fryling, C.; DeLarco, J. E. Proc. Natl. Acad. Sci. USA **1980**, 105, 163.
9. Derynck, R. *Cell* **1988**, 54, 593 and references cited within.
10. Lazar, E.; Watanabe, S.; Dalton, S.; Sporn, M. B. *Mol. Cell. Biol.* **1988**, 8, 1247; Lazar, E.; Vicenzi, E.; Van Obberghen-Schilling, E.; Wolff, B.; Dalton, S.; Watanabe, S.; Sporn, M. B. *Mol. Cell. Biol.* **1989**, 9, 860.
11. Defeo-Jones, D.; Tai, J. Y.; Wegrzyn, R. J.; Vuocolo, G. A.; Baker, A. E.; Payne, L. S.; Garsky, V. M.; Oliff, A.; Riemen, M. W. *Mol. Cell. Biol.* **1988**, 8, 2999.
12. Defeo-Jones, D.; Tai, J. Y.; Vuocolo, G. A.; Wegrzyn, R. J.; Schofield, T. L.; Riemen, M. W.; Oliff, A. *Mol. Cell. Biol.* **1989**, 9, 4083.
13. Tam, J. P.; Sheikh, M.; Harris, P.; Pileggi, N.; Tsai, L.; Hampton, R. *Proc. West. Pharmacol. Soc.* **1986**, 29, 471; Tam, J. P. in Peptides: Chemistry and Biology, Proceedings of the Tenth Am. Peptide Symposium, G. R. Marshall, Ed.; ESCOM Science Publishers B. V., Leiden, The Netherlands; 1988; pp. 561-565; Tam, J. P.; Lin, Y.-Z.; Wu, C.-R.; Shen, Z.-Y.; Galantino, M.; Liu, W.; Ke, X.-H. in Peptides: Chemistry, Structure and Biology, Proceedings of the Eleventh Am. Peptide Symposium, J. E. Rivier and G. R. Marshall, Eds.; ESCOM Science Publishers B. V., Leiden, The Netherlands; 1990; pp. 75-77.
14. Nestor, J. J.; Newman, S. R.; DeLustro, B. M.; Schreiber, A. B. in Peptides: Structure and Function, Proceedings of the Ninth Am. Peptide Symposium, C. M. Deber, V. J. Hruby, K. D. Kopple, Eds; Pierce Chemical Co., Rockford, IL; 1985; pp. 39-42; Eppstein, D. A.; Marsh, Y, V.; Schryver, B. B.; Bertics, P. J. *J. Cell. Physiol.* **1989**, 141, 420.

15. Tam, J. P. *Int. J. Pept. Prot. Res.* **1987**, 29, 421.
16. Tam, J, P.; Sheikh, M. A.; Solomon, D. S.; Ossowski, L. *Proc. Nati. Acad. Sci. USA* **1986**, 83, 8082.
17. Woo, D. D. L.; Clark-Lewis, I.; Chait, B. T.; Kent, S. B. H. *Prot. Engineering* **1989**, 3, 29.
18. Tou, J. S.; McGrath, M. F.; Zupec, M. E.; Byatt, J. C.; Violand, B. N.; Kaempfe, L. A.; Vineyard, B. D. *Biochem. Biophys. Res. Commun.* **1990**, 167, 484.
19. Scanlon, D. B.; Eefting, M. A.; Lloyd, C. J.; Burgess, A. W.; Simpson, R. J. *J. Chem. Soc. Chem. Commun* **1987**, 516.
20. Bailey, J. L.; Cole, R. D. *J. Biol. Chem.* **1959**, 234, 1733; Frank, B.H., Pettee, J. M., Zimmerman, R. E., Burck, P. J. in Peptides: Synthesis, Structure and Function, Proceedings of the Seventh Am. Peptide Symposium, D. H. Rich , E. Gross, Eds.; Pierce Chemical Co., Rockford, IL; 1981; pp. 729-739; Furman, T. C., Epp, J., Hsiung, H. M., Hoskins, J., Long, G. L., Mendelsohn, L. G., Schoner, B. Smith, D. P., Smith, M. C. *Bio/Technology* **1987**, 5, 1047; Sliwkowski, M. X.; Spear, K. L. Fourth Symposium of The Protein Society, 1990, abstract M11.
21. Marquardt, H.; Todaro, G.J. *J. Biol. Chem.* **1982**, 257, 5220.
22. Fields, G. B.; Noble, R. L. *Int. J. Pept. Prot. Res.* **1990**, 35, 161.
23. Cook, R. M.; Hudson, D.; Tsou, D.; Teplow, D. B.; Wong, H.; Zou, A. Q.; Wickstrom, E. in Peptides 1988, G. Jung and E. Bayer, Eds.; Walter de Gruyter and Co., Berlin; 1989; pp. 187-189.
24. Ellman, G. L. *Arch. of Biochem. and Biophys.* **1959**, 87, 70; Deakin, H.; Ord, M. G.; Stocken, L. A. *Biochem. J.* **1963**, 89, 296.
25. Sliwkowski, M. X.; Spear, K. L. unpublished results.

23

ENHANCEMENT OF COUPLING REACTION IN PEPTIDE SYNTHESIS BY MICROWAVE IRRADIATION

Kung-Tsung Wang, Shui-Tein Chen & Shyh-Horng Chiou

Institute of Biochemical Sciences, National Taiwan University and Institute of Biological Chemistry, Academia Sinica, Taipei, Taiwan R.O.C.

I. INTRODUCTION

Microwaves are electromagnetic radiation of wavelength between 0.1 cm and 100 cm, i.e. microwave energy lies in the range of the radio and infrared region of the electromagnetic spectrum. Recently there have been growing interests in applying microwave heating to the rapid thermal digestion prior to element and chemical analysis of inorganic and biological samples (1-4). The rapid heating capability of the microwave oven leads to a considerable saving in dissolution time, which may eventually replace some of the conventional flame and hot-plate heating protocols. During the last few years, we have introduced a rapid method of microwave heating for the facile preparation of protein and peptide hydrolysates prior to amino acid analysis (5-7). Microwave irradiation has also been applied to several organic reactions in a continuous-flow process by which preparative-scale samples (> 20 g) may be synthesized with ease using a kitchen microwave oven (8). We report here a novel application of microwave technology to enhance the coupling reactions in peptide synthesis. Both liquid-phase and solid-phase synthesis were tried and the reaction rates increase at least 2-4 fold under the comparable conditions and similar ambient temperature (25-30 ^{o}C).

---

[1] **Abbreviations used**: Moz-Val, *p*-methoxybenzyloxycarbonyl valine; Val-OMe, L-valine methyl ester; Boc-(Br-Z)-Tyr, α-(tert-butyloxycarbonyl)-O-(2-bromobenzyloxycarbonyl)-L-tyrosine.

Techniques in Protein Chemistry II

II. MATERIALS AND METHODS

A. Reagents and amino acid derivatives

Amino acids of L-configuration were purchased from Kyowa Hakko Co. Japan. The Boc, Moz[1] protecting groups for the α-amino group were introduced by the oxime carbonate method [reaction of the amino acid with α-(tert-butyloxycarbonyloxyimino)- or α-(benzyloxycarbonyloxy)-phenylacetonitrile/base]. The high-purity solvents were obtained from Alps Chem. Co. Taipei, Taiwan.

B. Synthesis procedures:

1. Liquid-phase synthesis of Moz-Val-Val-OMe.

Moz-Val (0.504 g, 2.0 mmol) dissolved in 1,4-dioxane (15 mL) was reacted with dicyclohexylcarbodiimide (0.424 g, 2.2 mmol) and N-hydroxybenzotriazole (0.324 g, 2.4 mmol) for an hour. The mixture was then added Val-OMe (0.165 g, 1 mmol) to start the reaction. The coupling reaction was conducted via: (a) microwave irradiation for a known period, (b) stirring at room temperature and (c) reaction at room temperature with stirring. The reactions under different conditions were quenched by adding 0.1 N HCl (100 mL) and the Moz-Val-Val-OMe was isolated by extraction with ethyl acetate as the routine procedure. The yield of each reaction was calculated based on Val-OMe (1 mmol).

2. Solid-phase synthesis of Tyr-Ile.

The solid-phase synthesis was performed via symmetric anhydride method. In a general reaction, two equivalents of Boc-(Br-Z)-Tyr anhydride dissolved in methylene chloride (10 mL) was added to Ile-resin (2.34 g, 1 mmol) to start the reaction. The reaction was conducted via: (a) microwave irradiation for known periods or (b) stirring at room temperature in a common solid-phase reaction vessel. The reaction was stopped by filtering off the reaction solution, and the yields of each reaction were calculated.

3. Solid-phase synthesis of Leu-Ala-Gly-Val.

The Leu-Ala-Gly-Val was synthesized via symmetric anhydride method. In a general reaction, two equivalents of Boc-amino acid anhydride dissolved in methylene chloride (10 mL) was added to Val-resin (1,86 g, 1 mmol) to start the reaction. The reaction was conducted under microwave irradiation for 20 min in a reaction vessel (Fig. 1) and stopped by filtering off the reaction solution. The coupling yield was calculated by measuring OD of the reaction product from ninhydrin test. After deblocking the Boc-protecting group, the next amino acid residue was incorporated onto the peptide using the same procedure. The final product was cleaved from the resin by treatment with liq. HF at 0 ^{o}C. The purity of product was analyzed by amino acid analyzer and HPLC and the results are shown on Table I & Table II.

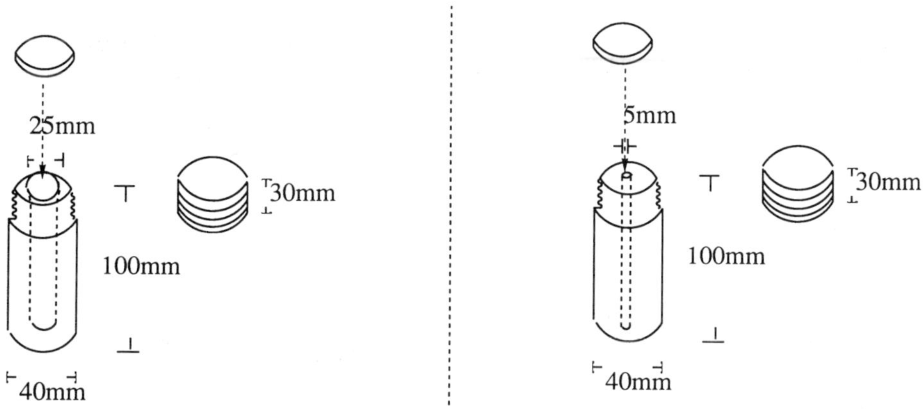

Fig. 1. The design of custom-made 50 mL/ 2 mL Teflon reaction vials

C. Apparatus and reaction vessels

The microwave oven used was a commercially available cooking apparatus without any modification (Tatung microwave oven TMO-1110, Tatung Co. Taipei, Taiwan). The total power of microwaves was 650 W with 9 power setting starting from 72 W. The temperature indicator is not available, so the temperatures for different setting need to be measured separately using several compounds with known melting points (5,6). For each measurement a melting-point standard encapsulated in a capillary tube immersed with 15 mL of the solvent in a Teflon reaction vessel was heated in the microwave oven for 1-5 min. At the end of irradiation period, the temperature was estimated from these standards with known melting points. The custom-made Teflon reaction vessels are shown in Fig. 1.

III. RESULTS AND DISCUSSION

The present report further extends the novel heating method of microwave irradiation to the reactions of peptide synthesis which would seem to be a major advantage over the conventional procedure both with respect to the speed and ease in the application of this rapid heating technique.

Table I & II show the enhancement of the reaction rate and the coupling yields in the synthesis of tetrapeptide Leu-Ala-Gly-Val (LAGV). It is clearly shown that by microwave irradiation in the three steps of amino-acid coupling the yields are in general over 98.5 %, and side products of the final reaction are less than 0.5 %. The great advantages lie both in the speed of the heating protocol and also in the final purity of peptide products.

Fig. 2 shows the comparison of the yields in the solid-phase synthesis of Boc-(Br-Z)-Tyr-Ile-resin in dichloromethane using microwave irradiation and conventional room-temperature stirring. It is noteworthy that microwave irradiation increases the yield of at least 20 % for each reaction period and the reaction is complete in about 5 min as compared to 30 min in the conventional stirring procedure.

TABLE I. Yields of various products in the synthesis of tetrapeptide Leu-Ala-Gly-Val (LAGV) by microwave irradiation

	LAGV (%)[a]	AGV(%)	LAV(%)	LGV(%)	AV+GV(%)
#1	98.8	0.19	0.27	0.33	0.41
#2	99.5	0.12	0.13	0.10	0.15
#3	98.9	0.19	0.27	0.36	0.28

a. Crude products from HF cleavage of the peptide resins were directly injected into the amino acid analyzer for the identification of various peptides (in triplicate exp.).

TABLE II. The yields of each coupling step in peptide synthesis by microwave irradiation

	#1	#2	#3
Boc-Gly + Val-resin	99.1[a]	98.7	99.5
Boc-Ala + Gly-Val-resin	98.3	99.6	98.7
Boc-Leu + Ala-Gly-Val-resin	99.4	99.1	99.2

a. The percentage yields for each step (in triplicate exp.) were calculated from the OD values of the products of reactions between each N-free peptide resin and ninhydrin reagent.

Fig.3 shows the comparison of peptide coupling reactions of Moz-Val and Val-OMe in 1,4-dioxane. The coupling reaction between these two refractory and sterically hindrant ß-branched amino-acid derivatives to form the protected dipeptide usually takes longer time than most other amino acid derivatives. The reaction was slow if the coupling was performed at the ambient temperature either with or without mechanical stirring. In case of microwave irradiation using a commercial microwave oven (operating at a frequency of 2450 MHz and having a full-power level of 650 W) set at the lowest power input (72 watts), the reaction rate increases at least 2-4 fold under the comparable conditions and similar ambient temperature (25-30 ^{o}C). The coupling reactions can be completed in 4 min with microwave heating whereas stirring the reaction mixture for 10 min only reaches at the level of 75% completion. This rate enhancement does not seem to be accounted for by the temperature effect of microwave heating.

We have measured the temperature of the reaction vessels containing different organic solvents usually used in the peptide coupling. For some microwave-absorbable solvents such as methylene chloride and acetonitrile the vessel temperatures increase with the irradiation time of microwaves (Fig. 4). It is noteworthy that for nonpolar solvent such as 1,4-dioxane without dipole moment, the microwave irradiation does not elevate the temperature of

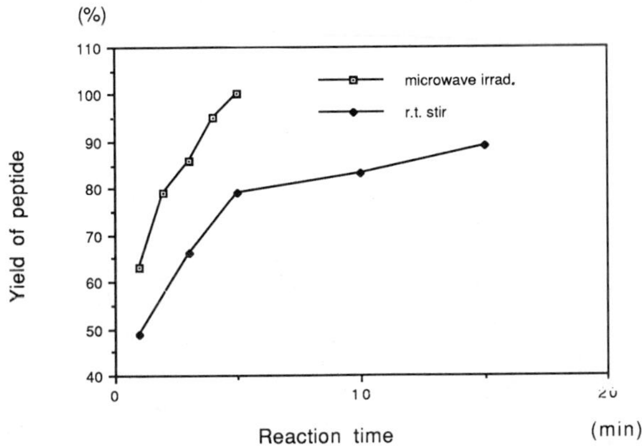

Fig. 2. The comparison of the yields in the solid-phase synthesis of Boc-(Br-Z)-Tyr-Ile-resin in dichloromethane using microwave irradiation and conventional room-temperature stirring.

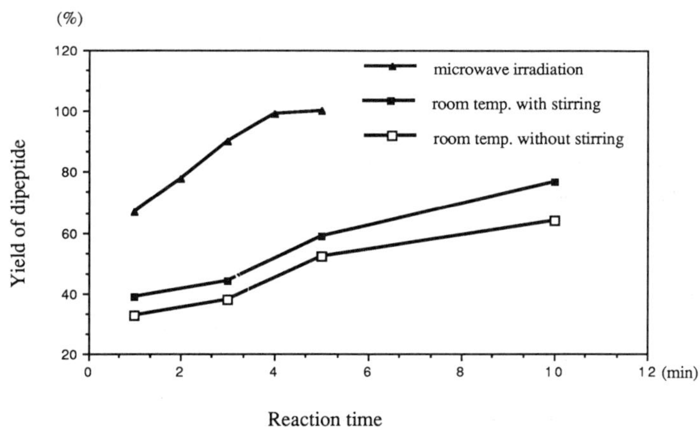

Fig. 3. Comparison of coupling reactions of Moz-Val and Val-OMe in 1,4-dioxane by microwave irradiation and mechanical stirring at ambient temperature. The dipeptides under different conditions were isolated and yields calculated on the basis of the initial weight of Val-OMe. Note that ambient temperature with stirring shows a slightly higher coupling yield than that without stirring. The microwave power was at the lowest setting (72 W).

solvent media appreciably and the vessel temperature remains relatively constant even with long-time irradiation. Although internal heating is always considered a key element in most microwave processes, the results shown above does not conform to the common concept of enhancement of reaction rate purely on the basis of temperature effect arising from microwaves. It is

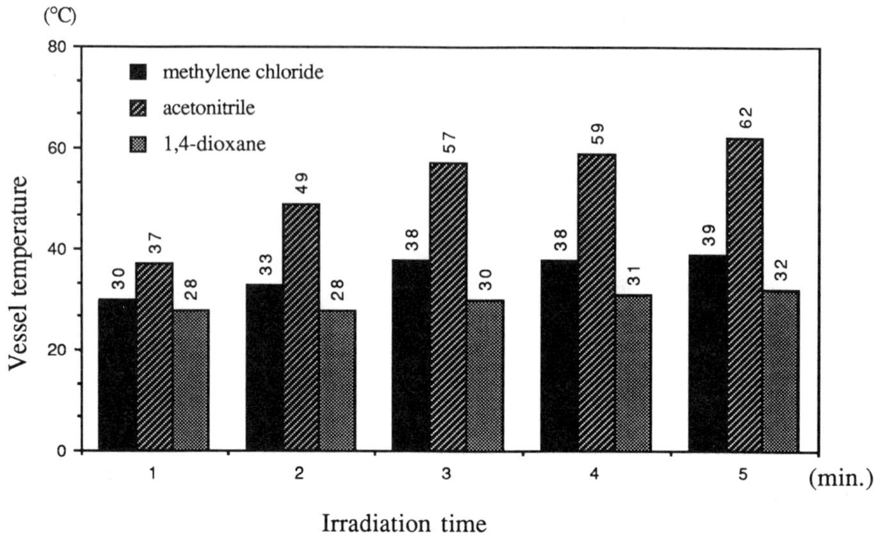

Fig. 4. Temperature measurements of three organic solvents after different periods of microwave irradiation. For each measurement a melting-point standard encapsulated in a capillary tube immersed with 15 mL of the solvent in a Teflon reaction vessel was heated in a microwave oven for 1-5 min. At the end of irradiation period, the temperatures were estimated from these standards with known melting points. Each value above the bar columns represents the mean of triplicate measurements with a standard error of less than 1 °C. The temperature of the vessel with 1,4-dioxane reached 32 °C after 5 min irradiation whereas that of acetonitrile reached 62 °C.

well known microwaves will agitate the solvent molecules in the reaction mixture at about 2.5 billion times per sec (2450 MHz) and the friction between agitated molecules is probably the cause of the rapid heating effect of microwaves. In the case of microwave transparent solvents or media, it is conceivable that the frictional energy dissipated to the bulk solvent molecules induced by microwave agitation is relatively small and there is only a small temperature increment of the solvent media. The enhancement of reaction rates in the microwave transparent media such as 1,4-dioxane shown here probably arises from the nonthermal microwave effect of some other causes. We have studied the microwave-accelerated coupling reactions in different solvents and at different reaction temperatures by control of energy input in microwave power. It is found that no racemization side reactions have occurred to a significant degree in the dipeptide products from coupling reactions between different amino-acid derivatives by microwave heating. This new heating protocol has been applied to the coupling reactions between various N-protected Boc-amino acids and amino-acid esters or immobilized amino-acid Merrifield resin in the routine liquid- and solid-phase peptide synthesis with high efficiency.

In conclusion, microwaves are low electromagnetic energy which induces molecular perturbation by stimulation of ionic diffusion and enhancement of dipole rotation without causing rearrangement of molecular structures. Microwave energy irradiated on the samples in various liquid media is lost or absorbed to the samples by the following two mechanisms: ionic conduction and dipole rotation. These two effects take place simultaneously to account for the phenomenon of rapid heating associated with many practical application of microwave irradiation (9,10). In contrast to conventional heating, the salient feature of "dipole rotation" constitutes one efficient form of "molecular agitation or stirring" which can be explored for many aspects in chemical reactions. We believe that this novel type of "molecular agitation" is independent of the regular thermal heating induced by microwaves and open a new avenue in exploring the potentials of applying microwave irradiation in place of conventional heating to some of important processes of biochemical analysis and organic synthesis.

IV. ACKNOWLEDGEMENT

We thank the generous support of Academia Sinica and the National Science Council, Taipei, Taiwan.

V. REFERENCES

1. Nadkarni, R. (1984) Anal. Chem. 56, 2233-2237 .
2. Fischer, L.B. (1986) Anal. Chem. 58, 261-263 .
3. Fernando, L.A., Heavner, W.D. & Gabrielli, C.C. (1986) Anal. Chem. 58, 511-512.
4. Kingston, H.M. & Jassie, L.B. (1986) Anal. Chem. 58, 2534-2541.
5. Chen, S.-T., Chiou, S.-H., Chu, Y.-H. & Wang, K.-T. (1987) Int. J. Peptide Protein Res. 30, 572-576.
6. Yu, H.-M., Chen, S.-T., Chiou, S.-H. & Wang, K.-T. (1988) J. Chromatogr. 456, 357-362.
7. Chiou, S.-H. & Wang, K.-T. (1989) J. Chromatogr. 491, 424-431.
8. Chen, S.-T., Chiou, S.-H. & Wang, K.-T. (1990) J. Chem. Soc. Chem. Commun. 807-809 .
9. Jassie, L.B. & Kingston, H.M. (1988) in *Introduction to Microwave Sample Preparation: Theory and Practice* (Eds. Jassie, L.B. & Kingston, H.M.) pp. 1-6, American Chemical Society, Washington, D.C.
10. Neas, E.D. & Collins, M.J. (1988) in *Introduction to Microwave Sample Preparation: Theory and Practice* (Eds. Jassie, L.B. & Kingston, H.M.) pp. 7-32, American Chemical Society, Washington, D.C.

SECTION IV

Protein Structure/ Function Analysis

24

STRUCTURE - FUNCTION RELATIONSHIPS IN EPIDERMAL GROWTH FACTOR

T.J. DUDGEON[1], U. HOMMEL[1], R.M. COOKE[1], A. FALLON[2] & I.D. CAMPBELL[1]

[1] Department of Biochemistry, South Parks Road, Oxford, OX1 3QU, UK
[2] British Biotechnology Ltd, Watlington Road, Cowley, Oxford, OX4 5LY, UK.

I. INTRODUCTION

Epidermal growth factor (EGF) is a 53 amino acid protein which is mitogenic for epithelial and endothelial cell lines. It and acts by binding to and activating the intrinsic tyrosine kinase activity of a cell surface protein, the EGF receptor. There is much interest at present in understanding the binding of EGF to its receptor. We present here a methodology for investigating this interaction, based on rational selection of amino acid targets for a site-directed mutagenesis program, the expression and purification of mutant EGFs, the measurement of their receptor affinities and investigation of their structural integrity by nuclear magnetic resonance (NMR) spectroscopy.

EGF is a member of a large family of proteins defined by amino acid sequence homology. Transforming growth factor alpha (TGFα), which is found associated with several transformed cell lines, competes with EGF for the EGF receptor and demonstrates a range of biological activities similar to that of EGF. Some pox viruses have been found to produce similar proteins. Many other extracellular proteins of diverse biological functions also contain regions of homology to EGF. These include blood clotting factors, fibrinolytic proteins, components of the complement system, cell surface receptors and cell adhesion molecules. These proteins are mosaic in nature, that is, they are composed of units, defined by amino acid homology, which fold, and probably function, as discrete units, or domains, of which EGF homologous (or GF) domains are just one (see Doolittle,

Techniques in Protein Chemistry II

1985 for a review of mosaic proteins). The sequences of GF domains from a number of these proteins are shown in Figure 1. The understanding of the functions of the GF and other domains from these mosaic proteins lies at the heart of understanding the functions of these proteins as a whole. We expect that the methodologies presented here can be extended to these larger and more complex proteins.

```
GROWTH FACTORS
                           1       2       3                     4   5              6
EGF-human          NSDSE   C PLSHDG Y C LH D G V C  MYIEALDKYA   C N C VV GY I G E R C Q YR D L KWWELR
EGF-Mouse          NSYPG   C PSSYDG Y C LN G G V C  MHIESLDSYT   C N C VI GY S G D R C Q TR D L RWWELR
EGF-Rat            NSNTG   C PPSYDG Y C LN G G V C  MYVESVDRYV   C N C VI GY I G E R C Q HR D L R
EGF-Guinea pig     QDAPG   C PPSHDG Y C LH G G V C  MHIESLNTYA   C N C VI GY V G E R C E HQ D L DLWE
TGF-Human          VVSHFND C PDSHTQ F C FH   G T C  RFLVQEDKPA   C V C HS GY V G A R C E HA D L LA
TGF-Rat            VVSHFNK C PDSHTQ Y C FH   G T C  RFLVQEEKPA   C V C HS GY V G V R C E HA D L LA
VGF                DIPAIRL C GPEGDG Y C LH   G D C  IHARDIDGMY   C R C SH GY T G I R C Q HV V L VDYQRS
MGF                        C NDDYKN Y C LN N G T C FTVALNNVSLNPF C A C HI NY V G S R C Q FI N L ITIK
SFGF                       C NHDYEN Y C LN N G T C FTIALDNVSITPF C V C RI NY E G S R C Q FI N L VTY

PLASMINOGEN ACTIVATORS
                1      2       3                4   5           6
tPA-Human       C SEPR C FNG G T C  QQALYFSDFV  C Q C PE GF A G KC C
uPA-Human       C  D   C LNG G T C VSNKYFSNIHW  C N C PK KF G G QH C
uPA-Pig         C  G   C LNG G K C VSYKYFSNIQR  C S C PK KF Q G EH C
uPA-Mouse       C  G   C QNG G V C VSYKYFSRIRR  C S C PR KF Q G EH C

CLOTTING FACTORS
                                 1         2         3            4   5                 6
Human Protein S (1)        IPDQ C  SPL    C NED G YMS C KBGKASFT C T C KP GW   Q     G EK C E
Human Protein S (3)        DVDE C SLKPSI  C     G TAV C KBILGDFE C E C PE GY RYNLK  S KS C E
Bovine Protein S (1)       ISDQ C  NPLP   C NED G FMT C KBGQATFT C I C KS GW   Q     G EG C E
Bovine Protein S (3)       DVDE C VLKPSI  C     G TAV C KBIPGDFE C E C AE GY KYNPV  S KS C D
Human Factor IX (1)        DGDQ C  ESNP   C LNG G  S  C KDDINSYE C W C PF GF   E     G KN C E
Bovine Factor IX (1)       DGDQ C  ESNP   C LNG G  M  C KDDINSYE C W C QA GF   E     G TN C E
Human Factor X (1)         DGDQ C  ETSP   C QNQ G  K  C KDGLGEYT C T C LE GF   E     G KN C E
Bovine Factor X (1)        DGDQ C  EGHP   C LNQ G  H  C KDGIGDYT C T C AE GF   E     G KN C E
Human factor VII (1)       DGDQ C  ASSP   C QNG G  S  C KDQLGSYI C F C LP AF   E'    G RN C ET
Human Protein C (1)        LEHP C  ASL    C CGH G  T  C IDGIGSFS C D C RS GW   E     G RF C QR
Bovine Protein C (1)       SGSP C  DLP    C CGR G  K  C IDGLGGFR C D C AE GW   E     G RF C LH
Bovine Protein Z (1)       GGSP C  ASQP   C LNN G  S  C QDSIRGYA C T C AP GY   E     G PN C A

CELL ADHESION MOLECULES
                 1      2         3            4   5              6
ELAM-1         A C TNTS C SGH G E C VETINNYT C K C DP GF S G LK  C E
GMP-140        S C QDMS C SKQ G E C LETIGNYT C S C YP GF Y G PER C EYV
Leu-8          S C QPWS C SGH G E C VEIINNYT C N C DV GY Y G PQ  C QFV
MEL-14           C QPGS C NGR G E C VETINNHT C I C DA GY Y G PQ  C QY
```

FIGURE 1. Amino acid sequences of a few of the GF domains. Highlighted in solid boxes are the residues which are conserved between all GF domains and are thought to form the structural core. The six cysteines are numbered. The overall degree of homology excluding these residues is low, though within the classes distinct sequence homologies can be seen, such as the residues in the growth factors which are enclosed by non-solid boxes. EGF = epidermal growth factor, TGF = transforming growth factor alpha, VGF = Vaccinia virus growth factor, MGF = Myxoma virus growth factor, SFGF = Shope Fibroma virus growth factor.

II. METHODS

A. *IDENTIFYING TARGETS FOR MUTAGENESIS*

The amino acid sequence homology of GF domains is highly characteristic and can give an important insight into funtional aspects of these molecules (Campbell *et al.*, 1990). Although the overall degree of homology is not very high, even within the EGFs and TGFαs, a few amino acids are almost completely conserved. These include the cysteines involved in the three disulphide bonds in EGF, three glycines and an aromatic residue corresponding to tyrosine 37 in EGF. These highly conserved residues are illustrated in Figure 1 by solid boxes, and shown in Figure 2 as non-shaded circles. It is these amino acids that define a GF domain.

The conservation of these residues, combined with knowledge of the structure of EGF and TGFα determined by NMR spectroscopy allows us to predict that these residues form the structural basis of GF domains. As many of the proteins containing GF domains do not possess EGF activity these residues might not play

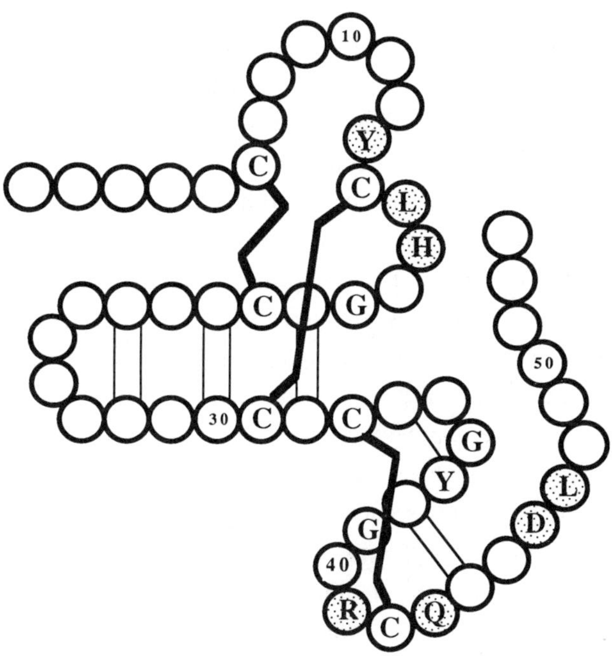

FIGURE 2. Schematic representation of the EGF structural motif showing the position of the three disulphides (thick lines) and the hydrogen bonds (thin lines). Residues implicated as forming the structural core are shown as non shaded circles and those conserved only or predominantly in the proteins which bind to the EGF receptor are shaded.

a major part in the interaction of EGF with its receptor. A further study of the sequences in Figure 1 reveals that there are some residues which are conserved in the EGFs, TGFαs and pox virus growth factors, but not in the GF domains which do not bind to the EGF receptor. Arginine 41 and leucine 47 are present in all these proteins which bind to the EGF receptor. An aromatic ring (tyrosine or phenylalanine) is similarly always present at the residue corresponding to tyrosine 13 in EGF. These three residues are not conserved in other GF domains. This identifies these as potential sites of interaction with the EGF receptor and so are key targets for mutagenesis. Leucine 47 had already been implicated in receptor binding by enzymatic trimming from the C-terminus of the purified protein (Gregory *et al.*, 1988, Burgess *et al.*, 1988). There are other candidates for residues involved in receptor binding, though none as clearly implicated as tyrosine 13, arginine 41 and leucine 47. Aspartate 46 is conserved in all EGFs and TGFαs, but not in the pox virus proteins. Leucine 15, histidine 16 and glutamine 43 have conservative substitutions in the proteins which bind to the EGF receptor, but similar homology is also found in some other GF domains. These three residues, as well as tyrosine 13 and arginine 41, are located at the interface between two somewhat distinct regions in the molecule, so substitutions to these residues may affect the structure of the protein. In testing the involvement of these residues in receptor binding by mutagenesis, a structural characterisation would therefore be required to be able to distinguish residues that formed direct contacts with the receptor from those that altered the structure of the protein.

A residue making interactions with the receptor would be expected to be located on the surface of the protein. The model of the EGF structure reveals that the residues identified as possible sites of contact with the receptor are indeed generally located on the surface, though leucine 15 and histidine 16 are somewhat less exposed than the others. These residues are located in two distinct regions on the surface of the protein, one containing tyrosine 13, leucine 15, histidine 16, arginine 41 and glutamine 43 and the other containing aspartate 46 and leucine 47. This indicates that there may be more than one region of interaction with the receptor and that a region involved in receptor interactions would probably involve residues distant in the primary sequence of the protein located in close proximity due to the tertiary fold of the protein. The highly conserved cysteines, glycines and the aromatic residue corresponding to tyrosine 37 conversely are generally buried within the protein, perhaps further implicating these residues as playing a structural role in the GF domain rather than being involved in the interaction with the EGF receptor

B. EXPRESSION OF EGF IN YEAST

A synthetic human EGF gene has been expressed in *Saccharomyces cerevisiae* using an alpha factor secretion system with plasmids based on the 2-micron origin of replication and *Leu-2d* selectable marker. Expression is driven from the yeast phosphoglycerate kinase promoters and terminators. A typical plasmid is illustrated in Figure 3. The alpha factor leader sequence directs EGF into the cell supernatant where it accumulates to levels of around 2 milligrams per litre of a 2 - 3 day culture grown in 1% glucose and 0.67% yeast nitrogen base w/o amino acids (Difco).

C. CONSTRUCTION OF EGF MUTANTS

A number of the potential receptor binding residues identified above, as well as some additional residues, were chosen for mutagenesis. This was performed using standard techniques. M13mp18 and M13mp19 templates containing the PstI - BamHI fragment of the alpha factor - EGF fusion were used. Oligonucleotide

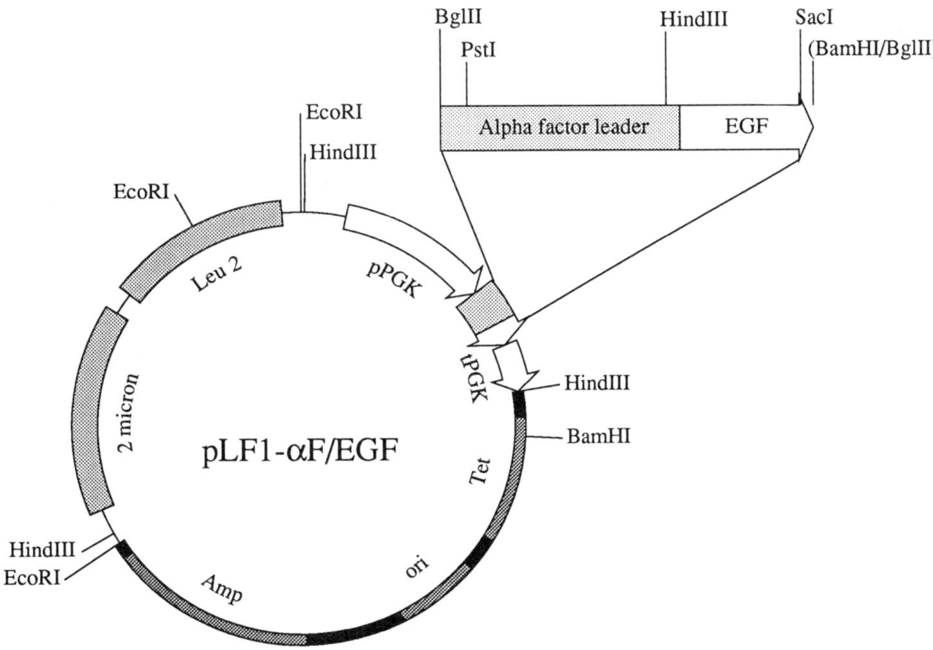

FIGURE 3. Plasmid map of a typical yeast expression vector containing the alpha factor - EGF fusion. Some restriction sites are shown. pPGK = yeast phosphoglycerate kinase promoter. tPGK = yeast phosphoglycerate kinase terminator. Amp Ori Tet = *E. coli* origin of replication and selectable markers. 2 micron = yeast 2 micron origin of replication. Leu 2 = yeast *Leu-2d* selectable marker.

primers were synthesised which had some degeneracy at the required position so that multiple changes could be made at a site with a single primer. Mutants were identified by sequencing and the fragments containing the mutant EGF gene cloned into the expression vector.

D. *PURIFICATION OF EGF MUTANTS*

EGF and the EGF mutants were purified to homogeneity by a three-stage purification scheme. EGF was first adsorbed from a cell free supernatant at pH 3.5 using a Zetaprep 60-SP cation exchange cartridge (Cuno Inc.) and eluted at pH 6.5. The EGF was further purified by anion exchange followed by reversed phase HPLC and then lyophilised. The purified protein is virtually homogeneous as determined by reversed phase HPLC and SDS-PAGE. As had previously been shown (George-Nascimento *et al.* 1988) the form of EGF secreted by yeast corresponds to the 1-52 form of the molecule with the C-terminal arginine being removed during secretion. The concentration of purified EGF was determined spectrometrically using an extinction coefficient of 2.89 calculated from the contributions of tryptophans and tyrosines.

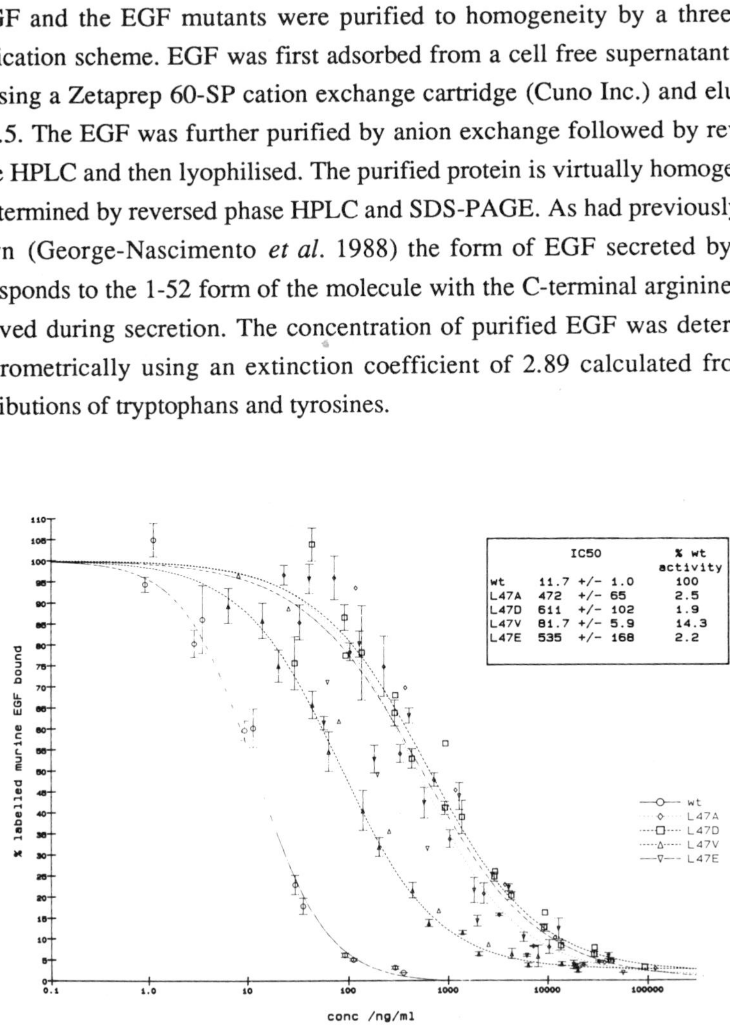

FIGURE 4. Receptor binding assay of wild-type EGF and four leucine 47 mutants using a monkey Vero cell line and ^{125}I labelled murine EGF. The curves of the leucine 47 mutants are displaced to the right of that of wild-type, indicating that larger amounts of the mutant proteins are required to displace the labelled EGF, and hence that the mutant proteins have a reduced affinity for the receptor. INSET. IC_{50} values of the receptor binding curves obtained from the concentration of protein required for 50% displacement of the labelled EGF.

E. DETERMINATION OF EGF RECEPTOR AFFINITIES

The relative receptor affinities of EGF and the mutants were determined by a competition binding assay using ^{125}I labelled murine EGF and a monkey Vero cell line which expresses the EGF receptor. A reduction in receptor affinity of a mutant EGF will be reflected by larger concentrations required to displace 50% of the labelled EGF.

F. STRUCTURAL CHARACTERISATION BY NMR

Standard 1-D and 2-D COSY and NOESY NMR experiments were performed on purified protein in D_2O at pH 2.9 (uncorrected meter reading) using 500 MHz and 600 MHz Bruker spectrometers. Satisfactory 1-D spectra could be obtained using as little as 0.3mM protein whereas 1.5mM was required for 2-D spectra.

III. RESULTS

Although EGF is produced at about 2 mg/litre levels, the expression of some of the mutants was found to be considerably less. For instance the expression of non-conservative substitutions at leucine 47 and methionine 21 is generally similar to that obtained with wild-type EGF, whilst most histidine 16, tyrosine 37 and arginine 41 mutants have greatly reduced expression levels. This indicates that the substitution of these residues alters some properties of these proteins. In general the mutants which express poorly are those that, by the arguments above, might play a part in the structural integrity of the protein, and this suggests that some property such as protein folding or stability has been affected by substitution of these residues. However, other possibilities such as effects on translational efficiency cannot be ruled out.

As leucine 47 mutants were expressed at high levels, these were the first to be investigated and will be discussed here. Receptor binding assays, shown in Figure 4, show that the receptor affinities of the leucine 47 mutants are considerably reduced compared with that of wild-type EGF. The ID_{50} of the valine mutant is reduced some sevenfold while the ID_{50}'s of the alanine, aspartate and glutamate mutants are reduced a further sevenfold. This is broadly consistent with results which have appeared in the literature (Gregory *et al.* 1988, Burgess *et al.* 1988, Lazar *et al.* 1988, Engler *et al.* 1988, Ray *et al.* 1988) who have all shown a considerable drop in activity on removal or substitution of leucine 47. What is not clear from these publications is whether the drop in receptor affinity is a result of the change of a residue which forms a direct contact with the receptor, or whether it is due to a structural modification which causes changes in the relative orientations of other residues that do form such contacts with the receptor. A structural characterisation by NMR was undertaken to answer this question.

The NMR spectra obtained from the leucine 47 mutants were found to be very similar to that of the wild-type protein. This is illustrated in the downfield region of 1-D spectra which contain the aromatic and amide resonances (Figure 5), though the similarities also extend to 2-D COSY and NOESY spectra. These 1-D and 2-D spectra indicate that very similar chemical shifts are seen for most resonances in the mutant proteins as are found in the wild-type. Also seen in Figure 5 are the backbone amide protons which are slow to exchange with D_2O. As the same backbone amide protons are seen in the spectra of the leucine 47 mutants this indicates that the same solvent accessibility is seen for these mutants.

IV. DISCUSSION

The approach described above illustrates how structural studies can be used with site-directed mutagenesis to investigate the areas of contact between two proteins. It is important to be able to target certain residues for mutagenesis, as a random mutagenesis scheme would not be practical, even on a small protein such as EGF. The approach of combining sequence homologies and knowledge of the EGF structure has allowed us to target a limited number of residues for mutagenesis. Several of these residues chosen for mutagenesis may have an important influence on the structure of the protein and it might be expected that for a small protein such as EGF, the effect of a single residue in stabilising or destabilising the structure would be relatively greater than for larger proteins. The interpretation of changes in biological function of a protein upon substitution of an amino acid therefore cannot be made with confidence unless it is known whether the substitution has caused structural changes in the protein.

Complete structures can be calculated from sufficient restraints observed in 2-D NMR spectra (Wuthrich, 1986), although such a study requires a complete spectral assignment and careful quantification of the data. Qualitative information can however be obtained from the 1-D spectrum of a protein. The chemical shift of a resonance is very dependent on conformation, so a structural change in a mutant protein would be expected to produce changes in the chemical shifts of some resonances. The very similar chemical shifts seen for the leucine 47 mutants and the wild-type protein argues for there being no significant structural changes

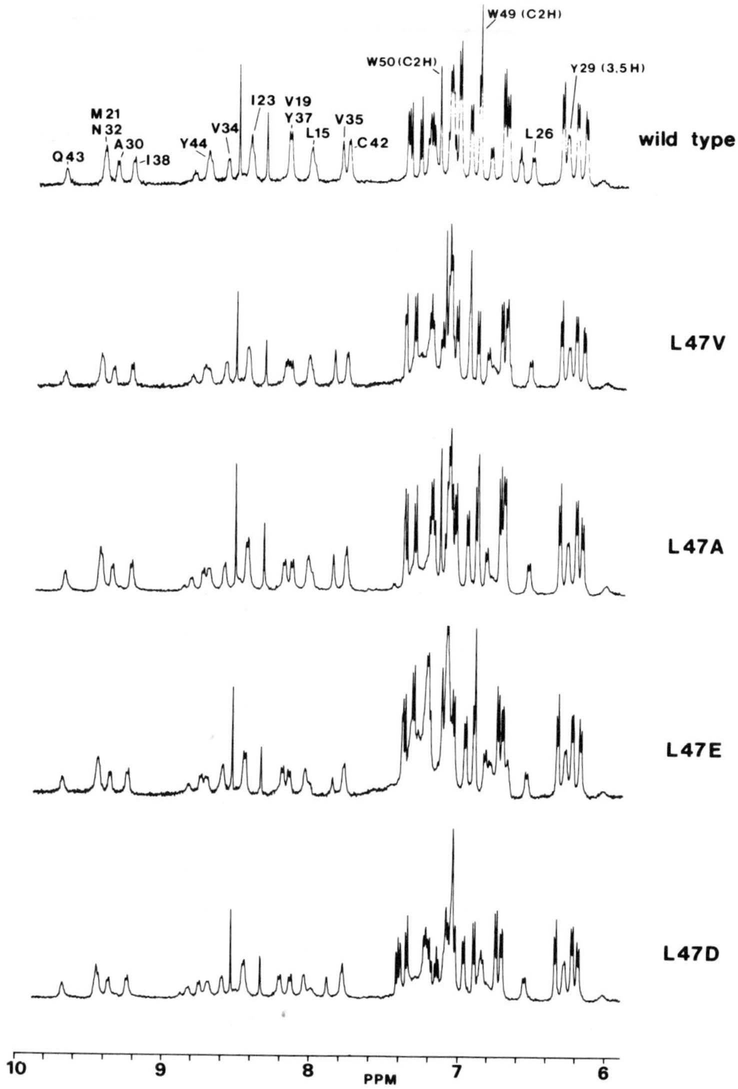

FIGURE 5. Downfield regions of the 1-D NMR spectra in D_2O of the 1-52 form of human EGF and the four leucine 47 mutants. The resonances of some aromatic resonances and slowly exchanging amide protons for wild-type EGF are shown.

on substitution of leucine 47. The spectral changes that can be seen are small and can be interpreted as local changes around position 47.

The argument for similar structures is further strengthened by the patterns of slowly exchanging amide protons and of cross-peaks in the NOESY spectra. The rate of exchange of amide protons is dependent on the accessibility to D_2O, so similarities in the patterns of exchange are evidence that factors influencing exchange, such as secondary structure and flexibility, have not changed in the mutant proteins. Cross peaks in a NOESY spectrum give information about distances between protons, and a similar pattern of cross-peaks is seen for the mutant and wild-type proteins. In the case of the glutamate mutant, assignment of the spectrum was carried out to enable specific identification of the NOEs that are most charactaristic of the tertiary fold of the protein.

Several lines of evidence therefore argue that the structures of the leucine 47 mutants are very similar to that of the wild-type protein (Dudgeon *et al.* 1990). This is consistent with the results of Moy *et al.* 1989, who have studied the structure of murine EGF with leucine 47 substituted for a serine. It can thus be concluded that the reduction in receptor affinity of leucine 47 mutants is an indication that leucine 47 is a residue which binds directly to the receptor.

Whereas the similarity of the 1-D spectra of the leucine 47 mutants is strong evidence on its own for unaltered structures, the 1-D spectra of some other mutants is not so easy to interpret. We have prepared a number of other mutants which show significant changes in chemical shifts of some resonances. For instance some of the aromatic ring proton resonances have altered chemical shifts in spectra of tyrosine 13, methionine 21 and arginine 41 mutants. These changes in chemical shift cannot be immediately interpreted as evidence for a perturbed structure for a number of reasons. The chemical shift of a proton resonance is a parameter which is extremely sensitive to the local environment of the proton. However, the effect of conformation on chemical shift cannot be accurately predicted and so it is not possible to interpret chemical shift changes of mutants in terms of changes in conformation. Similar complications occur if the amino acid substitution involves an aromatic residue, as ring currents can cause large chemical shifts. Thus, where the 1-D spectrum of a mutant is similar to that of the wild-type protein, it can reasonably be inferred that the structures are very similar, but when significant spectral differences are seen, more complete NMR studies are required to complement the information available from 1-D spectra. In particular, this will include the observation and quantification of NOEs, which give information on distances between protons and hence the tertiary fold of the protein, and measurement of amide proton exchange rates which can give information on the dynamic stability of the protein.

ACNOWLEDGEMENTS

This is a contribution from the Oxford Centre for Molecular Science, which is suported by the SERC and MRC. We also thank EMBO and Monsanto for financial support.

REFERENCES

Burgess, A.W., Lloyd, C.J., Smith, S., Stanley, E., Walker, F., Fabri, L., Simpson, R.J. & Nice, E.C. Biochemistry 27, (1988), 4977-4989

Campbell, I.D., Baron, M., Cooke, R.M., Dudgeon, T.J., Fallon, A.F., Harvey, T.S. & Tappin, M.J. Biochenical Pharmacology 40, (1990), 35-40

Doolittle. R.F. Trends in Biological Science 10, (1985), 223-237

Dudgeon, T.J., Cooke, R.M., Baron, M., Campbell, I.D., Edwards, R.M. & Fallon, A.F. FEBS Lett. **261**, (1990), 392-396

Engler, D.A., Matsunami, S.R., Campion, S.R., Stringer, C.D., Stevens, A. & Niyogi, S.K. J. Biol. Chem. 263, (1988), 12384-12390

George-Nascimento, C., Gynes, A., Halloran, M., Merryweather, J., Vanezuela, P., Steimer, K.S., Masiarz, F.R. & Randolph, A. Biochemistry 27, (1988), 797-802

Gregory, H., Thomas, C.E., Young, J.A., Willshire, I.A. & Garner, A. Regul. Pept. 22, (1988), 217-226

Lazar, E., Watanabe, S., Dalton, S. & Sporn, M.B. Moll. Cell. Biol. 8, (1988), 1247-1252

Moy, J.M., Scheraga, H.A., Liu, J-F., Wu, R. & Montelione, G.T. PNAS 86, (1989), 9836-9840

Ray, P.R., Moy, F.J., Montelione, G.T., Liu, J-F., Narang, S.A., Scheraga, H.A. & Wu, R. Biochemistry, 27, (1988), 7289-7295

Wuthrich, K. NMR of Proteins and Nucleic Acids. John Wiley & Sons, New York

25

MULTIPLE CONFORMATIONS OF STAPHYLOCOCCAL NUCLEASE[1]

Roger A. Kautz[2]
Robert O. Fox[2,3]

Yale University, New Haven, Connecticut 06511

I. INTRODUCTION

Nuclear Magnetic Resonance is a powerful tool in probing the structure and dynamics of proteins under native solution conditions. Recently NMR has also been applied to kinetic studies of protein folding[1]. Here these kinetic techniques are applied to illustrate the folding and isomerization of staphylococcal nuclease using magnetization transfer (MT) NMR techniques. Nuclease exhibits multiple folded conformations in equilibrium, which can be utilized to probe the forces stabilizing the folded forms of a protein. In addition, variant proteins produced by site-directed mutagenesis will be discussed.

---

[1] This work was supported in part by the NIH grant RO1 AI23923 to ROF.
[2] Department of Molecular Biophysics and Biochemistry
[3] Howard Hughes Medical Institute

Techniques in Protein Chemistry II

Staphylococcal nuclease (deoxyribonucleate 3' nucleotidohydrolase, EC 3.1.4.4) is a small (16.8 kD) globular protein of 149 amino acid residues with no disulfide bonds, cysteines or prosthetic groups. Staphylococcal nuclease (nuclease) requires Ca^{2+} to hydrolyze DNA and RNA into 3' phospho mono- and dinucleotides. It binds with high affinity and is inhibited by 3',5'-diphosphonucleotides, particularly pdTp. Nuclease has been the subject of extensive biophysical and biochemical studies[2,3,4,5]. The crystal structure of nuclease has been solved to high resolution in both its unliganded[4,6] form and in complex with Ca^{2+} and pdTp[7,8].

Nuclease NMR Spectrum

The downfield aromatic region of the ^1H NMR spectrum of wild type nuclease is shown in Figure 1. This sample has been unfolded in the presence of D_2O to exchange backbone amide protons for deuterons, simplifying this region of the spectrum by leaving only the resonances of histidine $H^{\varepsilon 1}$ protons. Histidine $H^{\varepsilon 1}$ protons are particularly well suited for studying protein folding since their resonances are well resolved at pH below 6, and they show a very broad chemical shift dispersion which is exquisitely sensitive to their environments in folded protein: their chemical shift changes by about 1 ppm depending on the protonation state of the imidazole ring, and the pK_a of histidine in protein is extremely sensitive to its environment.

The resonances of the four histidine $H^{\varepsilon 1}$ protons of nuclease are labelled in Figure 1. These assignments were made by comparing spectra of variant proteins in which one of the histidines was replaced with another amino acid. In each case the spectrum was relatively unperturbed except for the absence of the resonances of one of the four histidines[9,10].

When the protein is thermally unfolded, the histidines' environments are no longer well ordered and their resonances are at a chemical shift similar to histidine free in solution. The cluster of the four histidine resonances of unfolded protein is labelled U in Figure 1.

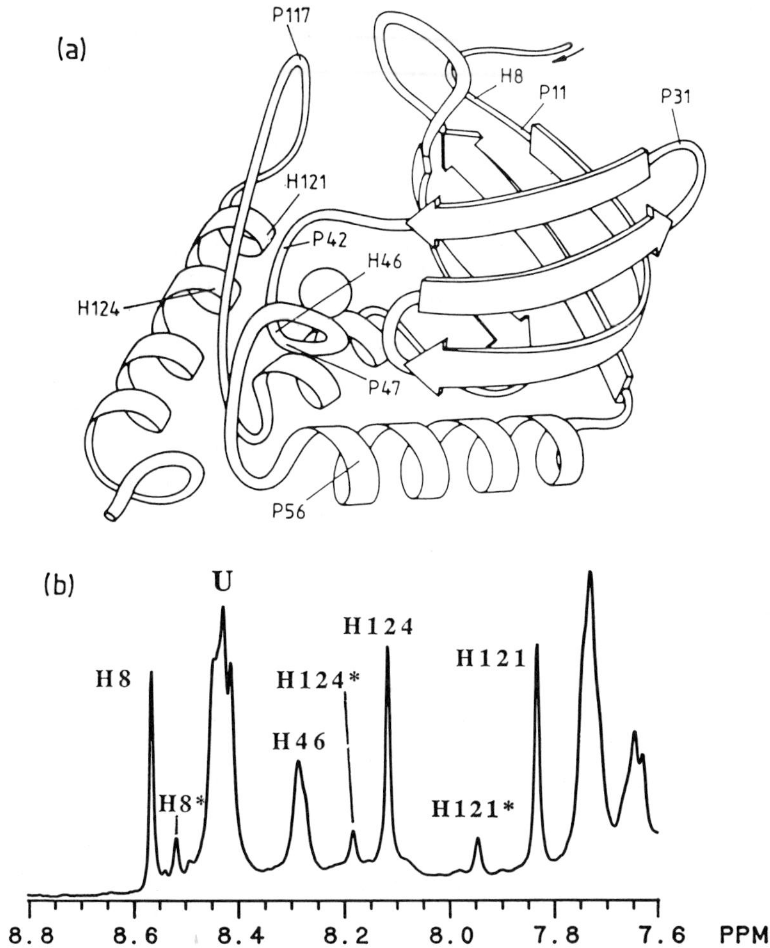

Figure 1: Staphylococcal nuclease. (a) Ribbon diagram of the staphylococcal nuclease backbone[18] with histidines and prolines noted. The circle represents the Ca^{2+} ion seen in the crystal structure. (b) Downfield aromatic region of 500 MHz ^1H NMR spectrum of nuclease at 48° C showing resonances of nuclease's 4 histidine $H^{\varepsilon 1}$ in folded and unfolded protein.

II. THE THREE-STATE MODEL

The resonances of His-8, His-121, and His-14 in the spectrum of Figure 1 are accompanied by minor resonances H8*, H121* and H124*. The identity and properties of these minor peaks is the subject of this work. These minor resonances were proposed to be due to an alternative folded conformation (N*) which is native-like but distinct from the major form (N)[11].

Several lines of evidence lead to the multiple conformations hypothesis: Each minor resonance has a chemical shift comparable to its associated major peak, and follows it closely through the entire range of a pH titration[9]. In the histidine substitution mutants used to assign the histidine resonances, each minor peak is absent together with its associated major peak[9,10]. These results suggest that the minor peaks are due to one or more alternative folded forms which are native-like but distinct from the major form. The relative intensities of the major and minor peaks varies with temperature, suggesting their relationship as two states of the same molecule in equilibrium. The ratios H124*/H124, H121*/H121, and H8*/H8 are equal to each other, vary together over a range of temperatures[12], and also diminish in concert as Ca^{2+} and the inhibitor 3',5'-diphosphothymidine (pdTp) are added[13,14]. These results suggest that these three minor resonances are reporting the same conformer. It was suggested that the minor form may represent an dimerized or aggregated form of nuclease but the amount of the minor form is independent of concentration over the range 0.2 mM to 5.5 mM[12,13], ruling out this possibility. An additional minor resonance H8** sometimes be seen between H8 and H8* has been found to be concentration dependent and is believed to report an aggregated state[14,15]. Another unrelated minor resonance is partially resolved from H46 at pH 5.3-5.5, and may monitor a *cis/trans* equilibrium of the His-46 — Pro-47 peptide bond.

The strongest evidence for multiple conformations comes from magnetization transfer experiments[16]. Magnetization transfer can be used to demonstrate that separate resonances arise from the same proton in different environments in chemical exchange, such as a molecule interconverting between two isomers.

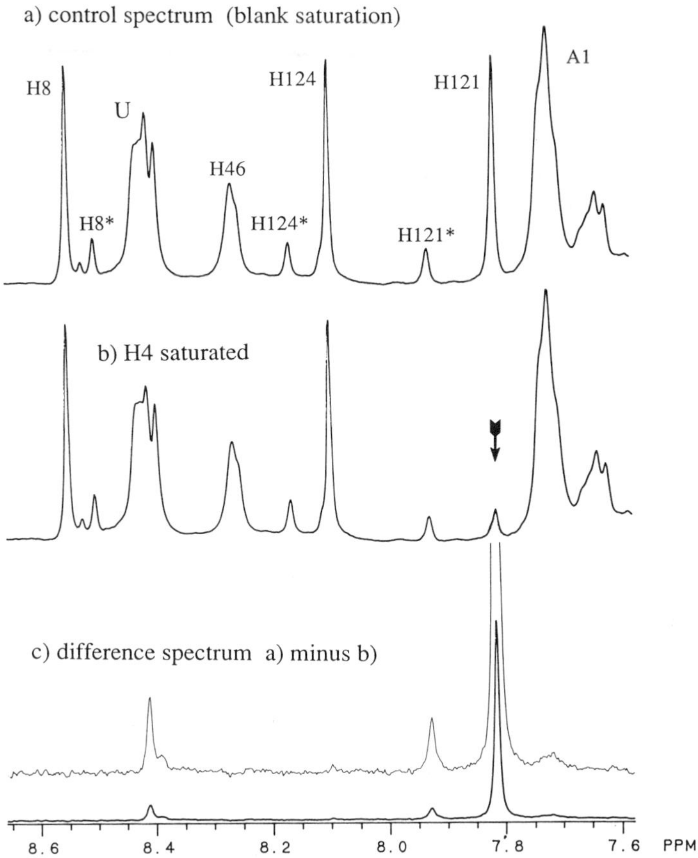

Figure 2: Magnetization Transfer experiment of nuclease at pH 5.5, 47° C. Spectrum (a) is an unperturbed spectrum in which the decoupler was applied to a blank region. Spectrum (b) shows the effect of a selective saturation pulse applied to H121. To highlight effects on other regions of the spectrum, spectrum (b) has been subtracted from the spectrum (a) to produce a difference spectrum (c).

Figure 2 shows a magnetization transfer experiment demonstrating the conformational exchange between two folded states and the unfolded state[11]. The difference spectrum shows that perturbation of H121 has caused a diminution of the intensities of resonances H121* and of a resonance in the unfolded cluster. These experiments revealed exchange between at least three states of nuclease:

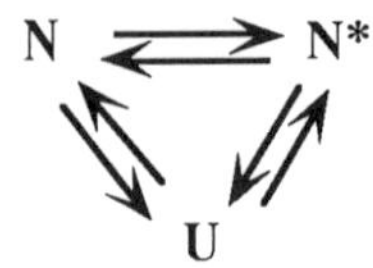

Double saturation experiments were also performed which demonstrated that the conversion between all 3 states is direct: N* is not a required folding intermediate (U↔N*↔N); nor do N* and N need to unfold to interconvert[11] (N*↔U↔N).

The two folded states differ in the cis/trans *isomerization state of Proline 117.*

A number of lines of evidence lead to a hypothesis that this structural heterogeneity is due to *cis/trans* isomerization about the Lys-116 — Pro-117 peptide bond[25]. The rates observed for isomerization between the major and minor forms are similar to the known rates for proline isomerization in model peptides[17]. The crystal structure of a Nuclease-Ca^{2+}-pdTp complex[8] shows a single *cis*-proline at residue 117. This is quite close to histidines 121 and 124, which show the greatest chemical shift difference between the major and minor resonances. Pro-117 is in a type VI_a turn[18] on an isolated loop near the cleft of the active site. Comparison of crystal structures of liganded and apo-nuclease show dramatic differences in local backbone conformation in this region, suggesting that loop conformers containing both *cis* and *trans* peptide bonds could be accommodated.

The hypothesis that Pro-117 is involved in the heterogeneity was tested by the generation of variants of nuclease A in which Pro-117 was replaced with another amino acid. Spectra of mutants P117G (Pro 117 → Gly)[13] and P117T (Pro 117 → Thr) (Figure 3) show a pattern of four major histidine resonances similar to wild type, indicating that the folded structures of the variant proteins are comparable to that of wild type. The spectra of the proline substitution variants are strikingly different in that no trace of minor resonances can be seen near resonances H8, H124, or H121.

This is strong evidence that the multiple folded forms of nuclease observed in the histidine $H^{\varepsilon 1}$ resonances of the ^1H NMR spectrum are due to *cis/trans* isomerization at Pro-117, although the possibility remained that the proline substitution affected the means of detecting the heterogeneity, rather than the multiplicity itself. The model has been further supported

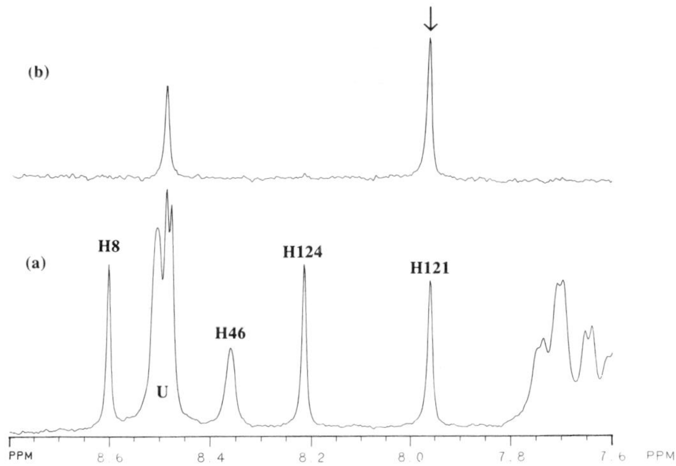

Figure 3: (a) Spectrum of nuclease variant P117T at 50° C. This variant shows no evidence of conformational heterogeneity in the folded form. (b) Magnetization transfer difference spectrum showing effects of saturating H121. This variant similarly shows no evidence of the conformational heterogeneity in the unfolded form.

by a large body of consistent experiments, discussed below. Also a direct measurement of the *cis/trans* ratio of Pro-117 in a $^{13}C4$-proline labelled variant is in progress and also supports the model[19].

The major form has Proline 117 in the cis *conformation*

The crystal structure of the nuclease-Ca^{2+}-pdTp complex shows Pro-117 to be unambiguously in the *cis* conformation, while the NMR spectrum of unliganded nuclease shows significant heterogeneity. When these ligands are added to an NMR sample the intensity of the downfield minor resonance diminishes to zero while the intensity of the upfield major resonance increases. This implies that Pro-117 is *cis* in the major form, and thus *trans* in the minor form.

The presence of 10 mM Ca^{2+} and 4 mM pdTp also increases the midpoint of thermal denaturation of a 1.5 mM nuclease A sample from 48° to 66° C. Since this very strong binding energy stabilizes the *cis* form, the *cis* form must bind inhibitor much more tightly than the *trans* form. It is yet to be determined whether the *trans*-Pro-117 form binds inhibitor, although both Pro-117 substitution mutants are enzymatically active.

III. THE FOUR STATE MODEL

In light of evidence supporting the hypothesis that the heterogeneity is due to *cis/trans* isomerization the 3-state model must be extended to a four state model with unfolded *cis* and *trans* forms U and U* as well as the folded *cis* and *trans* forms N and N*:

$$\begin{array}{ccc} N & \longleftrightarrow & N^* \\ \updownarrow & & \updownarrow \\ U & \longleftrightarrow & U^* \end{array}$$

This diagram shows the interconversion between N, N* and the two unfolded forms U and U* which has been demonstrated by magnetization transfer. Interconversion of U and U* is implied by the experiments showing a difference in the transfer of magnetization between N and N* when the unfolded states U and U* are saturated. Direct interconversion between folded *cis* and unfolded *trans*, and vice versa, is not shown: although it has been demonstrated that in some cases folding and isomerization are linked events[20], this is not believed to be the case in nuclease because all of the states are stable; interconversion is observed in both directions between all of the states; and the lifetime of each state is long compared to the time spent in the activated states of folding or isomerization processes. Thus isomerization and folding are probably discrete events.

The populations of the states in the diagram are discussed below with the measurements of the rate constants of interconversion.

Evidence of Multiple Unfolded States

If the effects on U of saturating H121 in Figure 2 are examined carefully it is apparent that there are two difference peaks in the unfolded cluster. This heterogeneity in the unfolded states could be observed only through magnetization transfer from the *cis* resonance of His-121, and not from its *trans* resonance or from either resonance of the more distant histidines His-124 and His-8. The time development shows that transfer occurs first to the smaller upfield peak U, then relays to the larger downfield peak U* at longer times[13]. This indicates that U is the unfolded *cis* resonance.

The Proline 117 Substitution Mutants Show Only a Single Unfolded Form

A magnetization transfer experiment on the P117T variant is shown in Figure 3. This variant behaves similarly to the P117G variant[25]. There is only one difference peak in the unfolded resonances upon saturation of H121. A similar result was found in the P117G variant[13]. The proline substitution abolishes not only the heterogeneity in the folded states, but heterogeneity in the unfolded states as well. This supports the hypothesis that the heterogeneity in the unfolded states is due also to *cis/trans* isomerization of Pro-117. P117G and P117T thus fit the two state model: N↔U.

IV. MEASUREMENT OF FIRST ORDER RATE CONSTANTS

The magnitudes of the MT effects observed in single and double saturation experiments can be analyzed to determine first order rate constants of the interconversion processes[16]. In nuclease it has been possible to determine all 8 first order rate constants[13], shown below (units are sec^{-1}; uncertainties are shown as minimum and maximum in parentheses):

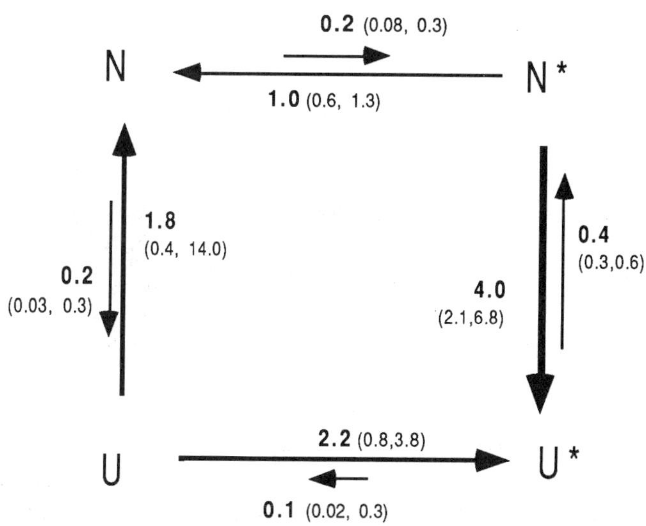

From these measured rates it can be demonstrated that the state N*, the *trans* form, has higher entropy and higher enthalpy, while N, the *cis* form, has lower enthalpy and lower entropy[13]. Higher entropy can be interpreted as greater flexibility.

A comparison of the unfolding rates of the N* and N forms, and the folding rates of U and U* lead to the conclusion that there is a strong enthalpic contribution to the stabilization of the *cis* state, and that this stabilization must occur late in the folding process, relative to a transition state[13]. This is consistent with the view that the sum of weak forces that stabilize the folded form of a protein (H-bonds, van der Waals forces, etc.) depend on specific short range interactions which can only come into effect when the protein is in a conformation close to that of the native fold.

V. APPLICATIONS

Kinetic Studies

Kinetic studies provide insight into the intermediates, not just the endpoints, of protein folding. It has been possible to monitor the transient formation of an intermediate in the refolding of RNase A at low temperatures using NMR[21]. Although the *trans* form of nuclease is not technically an intermediate because it exists in equilibrium under native conditions, it has been possible to transiently trap a larger fraction of molecules in the *trans* state by rapid refolding, then monitor the return to equilibrium[14,19]. This experiment provides information on the isomerization rates at low temperatures where they are too slow to measure by magnetization transfer; and on the refolding kinetics of slow folding phases of nuclease. Studying refolding by NMR has advantages because the relative intensities of histidine $H^{\varepsilon 1}$ resonances of folded and unfolded protein are a better quantitative measure of the folding equilibrium than fluorescence, and CD may not distinguish a collapsed state, such as a molten globule, from native protein.

Magnetization transfer is a particularly powerful tool in the kinetics of protein folding in cases where it can dissect the folding equilibrium into the separate first order rate constants for the folding and unfolding processes under equilibrium conditions[1,13]. These measurements are complicated in wild type nuclease by the heterogeneity in both folded and unfolded states, but the Pro-117 substitution mutants P117G and P117T do not show this heterogeneity so the folding and unfolding rates can be readily measured over a range of temperatures through the transition. The relation between the different thermal stabilities of these two variants and their folding or unfolding rates can reveal how the substitution has altered the stability — by affecting the free energy of the folded, the unfolded, or a transition state[14,19].

What Stabilizes a cis *proline?*

The nuclease 4-state system provides an assay for the *cis/trans* isomerization state of a proline in a native protein in solution as a function of the amino acid sequence. This enables us to ask what forces stabilize the unfavorable *cis* proline conformation. A *cis* proline is locally unfavorable, yet its presence must be thermodynamically favorable to the protein as a whole. Understanding the forces that stabilize the *cis* proline in a type VI turn is generally informative since these are the same forces that stabilize other turn types, and surface structures of proteins in general.

Crystallographic studies of nuclease variants have suggested that the primary constraint on turn type is the local geometry: the position of the backbone adjacent to the turn and steric constraints between the turn and adjacent residues[22,23,24]. In addition, the residues within the turn may have a preference for a particular turn type. A grid search of all allowed ϕ, ψ angles for a segment of backbone connecting residues 114 and 119 of nuclease's crystal structure have demonstrated that a type VI turn is particularly well suited to make this connection[14]. If residue 117 is not proline the *cis* conformation becomes extremely unfavorable. Crystal structures show residues 115-118 to have adopted a type I' turn in the P117G variant. Glycine is strongly preferred in turn types II, II', and I'; nuclease variant P117T which contains no glycines in this turn adopts a type I conformation[6,24].

We have investigated glycine substitutions within the 115-118 turn sequence which suggest that adding flexibility to the turn relaxes the constraints maintaining the *cis* proline[14,19].

Global Reflections of the cis/trans *isomerization*

Although the two forms differ primarily in the *cis/trans* isomerization state of Pro-117, the fact that discrete major and minor resonances are seen for histidines 121, 124, and 8 (Figure 1) suggest that the differences between the two forms are not localized to this turn but propagate through the molecule[25]. This contrasts with what is often seen in crystal structures of amino acid substitution mutants where local perturbations are absorbed within a few shells of atoms. Conversely, variants with substitutions at sites remote from the turn have shown effects on the equilibrium between the *cis* and *trans* forms[26].

REFERENCES

1. Dobson, C.M. and P.A. Evans. (1984) *Biochemistry* **23**:4267-4270.
2. Tucker, P.W. (1978) *Mol Cell Biochem.* **22**:67-77.
3. Tucker, P.W., E.E. Hazen Jr. and F.A. Cotton. (1979) *Mol. Cell Biochem* **23**:3-16.
4. Tucker, P.W., E.E. Hazen Jr. and F.A. Cotton. (1979) *Mol Cell Biochem* **23**:67-86.
5. Tucker, P.W., E.E. Hazen Jr. and F.A. Cotton. (1979) *Mol. Cell. Biochem.* **23**:131-141
6. Hynes, T.R. and R.O. Fox. (manuscript in preparation).
7. Cotton, F.A. and Hazen, E.E. Jr. (1971) "The Enzymes", Vol IV. 3rd. ed., p. 153 Academic Press, New York.
8. Loll, P.A. and E.E. Lattman. (1989) *Proteins* **5**:183-201.
9. Kautz, R.A., J.F. Gill and R.O. Fox. (1990) in "Protein and Pharmaceutical Engineering" (C. Craik, ed.) Wiley, Liss, Inc.
10. Alexandrescu, A.T., D.A. Mills, E.L. Ulrich, M. Chinami, and J.L. Markley. (1988) *Biochemistry* **27**:2158-2165.
11. Fox, R.O., P.A. Evans, and C.M. Dobson (1986) *Nature* **320**:192-194.
12. Evans, P.A. (1986) D. Phil Thesis, Oxford University.
13. Evans, P.A., R.A. Kautz, R.O. Fox, and C.M. Dobson. (1989) *Biochemistry* **28**:362-370.
14. Kautz, R.A. (1990) PhD thesis, Stanford University.
15. Alexandrescu, A.T., E.L. Ulrich, and J.L. Markley. (1989) *Biochemistry* **28**: 204-211.
16. Forsen, S. and R.A. Hoffman. (1964) *J. Chem Phys* **40**:1189-1196.
17. Grathwohl, C. and K. Wuthrich. (1981) *Biopolymers* **20**:2623-2633.
18. Richardson, J.S. (1981) *Adv. Protein Chem.* **34**:167-339.
19. Kautz, R.A. and R.O. Fox. (manuscript in preparation).
20. Cook, K.H., F.X. Schmid, and R.L. Baldwin. (1979) *PNAS USA* **76**:6157-6161
21. Blum, A.D., S.H. Smallcombe, and R.L. Baldwin. (1978) *J. Mol. Biol.* **118**:305-316.
22. Hynes, T.R., R.A. Kautz, M.A. Goodman, J.F. Gill and R.O. Fox. (1989) *Nature* **339**:73-76.
23. Goodman, M.A. (1990) PhD thesis, Stanford University.
24. Hynes, T.R. (1989) PhD Thesis, Stanford University.
25. Evans, P.A., C.M. Dobson, R.A. Kautz, G. Hatfull, and R.O. Fox. (1987) *Nature* **329**:266-268.
26. Alexandrescu, A.T., A.P. Hinck, and J.L. Markley. (1990) *Biochemistry* **29**: 4516-4525.

26

DETECTION AND KINETIC CHARACTERIZATION OF A NOVEL PROLINE ISOMERISM IN STAPHYLOCOCCAL NUCLEASE BY NMR SPECTROSCOPY†

Stewart N. Loh, Charles W. McNemar and John L. Markley

Department of Biochemistry, University of Wisconsin-Madison,
Madison, WI 53706

I. INTRODUCTION

Proline cis/trans isomerism makes a singular contribution to protein folding kinetics[1,2] as well as to the multiple conformations found in an increasing number of native proteins.[3,4] Nuclear magnetic resonance (NMR) spectroscopy provides an effective means for measuring rates of processes that occur slowly on the NMR timescale, yet rapidly compared to the T_1 relaxation times of the exchanging nuclei. Isomerizations about X-Pro peptide bonds in proteins often have rates within this regime of 10^{-1} to 10^3 s^{-1}. In this paper we demonstrate the use of saturation transfer and inversion transfer techniques to measure the cis/trans rate constants of a previously unreported proline isomerism in staphylococcal nuclease.

Wild-type staphylococcal nuclease (nuclease wt), first obtained from the Foggi strain of *Staphylococcus aureus*[5], and nuclease H124L, first isolated from the V8 strain and differing from wt by a leucine in place of His124[6], have proven to be convenient systems for the study of conformational heterogeneity in both the native and denatured states. These studies have primarily focused on the NMR signals from the histidine $^1H^{\epsilon1}$ resonances, which report on the conformational substates.[7-11] Doubling of the $^1H^{\epsilon1}$ signals of the imidazole of His46 was first noted by Markley et al.[12] Splitting of the peaks from the other histidines of nuclease wt (His8, His121, and His124) has since been firmly linked to cis/trans heterogeneity about the Lys116-

†This work was supported by Grant GM35976 from the National Institutes of Health and made use of the National Magnetic Resonance Facility at Madison. Equipment in the facility was purchased with funds from the University of Wisconsin, the NSF Biological Biomedical Research Technology Program (DMB-8415048), the NIH Biomedical Research Technology Program (RR2301), the NIH Shared Instrumentation Program (RR02781) and the U.S. Department of Agriculture. S.N.L. is supported by a Training Grant in Cellular and Molecular Biology (NIH GM07215).

Techniques in Protein Chemistry II

Pro117 peptide bond.[13] In agreement with what had been observed with nuclease wt, Alexandrescu et al.[9] observed, with H124L, two monomeric species in slow exchange on the NMR timescale. We have designated the dominant monomeric conformational substate (containing cis-Pro117) the a1 substate, and the minor component (containing trans-Pro117) the a2 form. Although the a1 ⇄ a2 equilibrium is affected by a number of solution conditions, such as temperature and pH, the hallmark of this process is its dependence on ligand concentration: titration with the inhibitor 3′,5′-thymidine bisphosphate (pdTp) progressively increases the a1 form at the expense of a2.

Alexandrescu[14] and Wang et al.[15] provided evidence that a conformational equilibrium (denoted b1 ⇄ b2) reported by His46 is distinct from the a1 ⇄ a2 equilibrium. Here we report preliminary kinetic analysis of the b1 ⇄ b2 interchange. This process is not linked to inhibitor binding as strongly as the a1 ⇄ a2 equilibrium and is similarly unaffected by the presence of mutations that perturb, and even invert the a1 and a2 populations. Resonances of residues distant from His46 are unaffected by the b1 ⇄ b2 transition. Thus it may reflect a local reorientation of the "disordered loop" containing His46 seen in the X-ray crystal structure[16] of the nuclease-Ca^{2+}-pdTp ternary complex. Since no doubling of the His46 $^1H^{\varepsilon 1}$ peak is seen in spectra of the mutant P47G+H124L, we attribute the origin of the splitting to cis/trans heterogeneity about the His46-Pro47 peptide bond.

II. MATERIALS AND METHODS

Protein production. The parent plasmid pTSN2cc, which contains the nuclease H124L gene under control of the T7 promoter, was a gift of Dr. D.M. LeMaster (Northwestern University). Mutagenesis was performed and mutant nucleases were overproduced from *Escherichia coli* as previously described,[17] via co-transformation with the plasmid pGP1-2.[18]

NMR spectroscopy. Samples prepared for NMR experiments consisted of 2.5-3.5 mM protein and 300 mM KCl, in ^2H_2O, pH* 5.5 (unadjusted glass electrode reading). Nuclease-pdTp-Ca^{2+} ternary complexes contained three- and six-fold excesses of pdTp and $CaCl_2$, respectively. Data were collected at 45°C on a Bruker AM500 spectrometer. Data acquisition parameters are given in figure captions. Chemical shifts are referenced to internal TSP. Peak areas were calculated from computer simulations of peak heights and linewidths by the GLINFIT program (A.D. Bain, Bruker User Group Software) as described previously.[10]

The treatment of a two-spin system presented below originates from the formalism of McConnell[19] and others.[20-23] For the process

$$b1 \underset{k^{b2}}{\overset{k^{b1}}{\rightleftharpoons}} b2 \tag{1}$$

in which the His46 $^1H^{\varepsilon 1}$ resonance interconverts between two nonequivalent environments defined by the b1 and b2 subconformations, relaxation of the two spins is described by the general solutions to the modified Bloch equations

$$I^{b1}(t) = C_1 e^{-\lambda_1 t} + C_2 e^{-\lambda_2 t} + I_o^{b1} \tag{2}$$

$$I^{b2}(t) = \frac{\lambda_1 - k_1^{b1}}{k^{b2}} C_1 e^{-\lambda_1 t} + \frac{\lambda_2 - k_1^{b1}}{k^{b2}} C_2 e^{-\lambda_2 t} + I_o^{b2} \tag{3}$$

where

$$\lambda_{1,2} = \frac{1}{2}\{(k_1^{b1} + k_1^{b2}) \pm [(k_1^{b1} - k_1^{b2})^2 + 4k^{b1}k^{b2}]^{1/2}\} \tag{4}$$

$$k_1 = k + 1/T_1 \tag{5}$$

and T_1 is the longitudinal relaxation time. C_1 and C_2 are best obtained in simplified form given the boundary conditions of a particular experiment, as will be illustrated below. Biexponential decays were fitted to Eqs. 2 and 3 by the method of Marquardt.[24]

Eqs. 2-5 provide the basis for the two methods used to determine rate parameters of the b1 ⇌ b2 interconversion. The first set of results was obtained by applying a saturation pulse from the decoupler to the His46 $^1H^{\epsilon1}$ b2 peak, and by observing saturation transfer to the b1 signal. The time dependence of this effect is described by

$$\frac{d}{dt}\ln\frac{I^{b1}(t) - I_\infty^{b1}}{I_\infty^{b1}} = -k_1^{b1} \tag{6}$$

At equilibrium, the ratio of the b1 peak intensities before and after saturation of b2 is given by

$$\frac{I_o^{b1}}{I_\infty^{b1}} = k_1^{b1}T_1^{b1} \tag{7}$$

The value k_1^{b1} is obtained from Eq. 6 and T_1^{b1} is extracted from Eq. 7. The possibility of decoupler spillover contributing to changes in intensity of the observed signal was eliminated by offsetting the decoupler frequency 55 Hz (equaling the separation between His46 $^1H^{\epsilon1}$b1 and $^1H^{\epsilon1}$b2) from the His8 $^1H^{\epsilon1}$ peak and by noting no saturation of the His8 signal.

In the second approach, either the b1 or the b2 peak was inverted by a selective 180° pulse, and the intensities of the inverted as well as the noninverted spins were followed over time. Although better results might have been obtained by a comprehensive analysis incorporating both data sets, we found it possible to deduce the full set of kinetic parameters from a fit of either the inverted or the noninverted signal to Eqs. 2 and 3, since the two exponentials could be resolved and the second spin was not significantly perturbed by the inversion.[††] In the case of b1 inverted, the initial conditions are $I^{b1}(0) = -I_o^{b1}$ and $I^{b2}(0) = I_o^{b2}$, and the inverted spin yields the following

$$k_1^{b1} = \frac{C_1\lambda_1 + C_2\lambda_2}{C_1 + C_2}, \quad k_1^{b2} = \lambda_1 + \lambda_2 - k_1^{b1}, \quad k^{b1} = [(k_1^{b1}k_1^{b2} - \lambda_1\lambda_2)/K_{eq}]^{1/2} \tag{8}$$

Alternatively, if b1 is the noninverted resonance, the initial conditions become $I^{b1}(0) = I_o^{b1}$ and $I^{b2}(0) = -I_o^{b2}$, and the rates are calculated from Eqs. 9 and 10:

$$k_1^{b1} = \frac{\lambda_1 + \lambda_2}{2} + (\lambda_1 + \lambda_2)(1 - \frac{C_2C_2^*}{I_o^{b1}I_o^{b2}})^{1/2}, \qquad k_1^{b2} = \lambda_1 + \lambda_2 - k_1^{b1} \tag{9}$$

[††]To minimize perturbation of the second spin, instead of a soft 180° pulse, we employed a $90°_x$-$1/2\Delta$-$90°_{-x}$ inversion sequence where Δ is the difference in frequency between the two peaks.

$$k^{b1} = C_2^*(\lambda_1 - \lambda_2)/2I_o^{b1}$$
$$k^{b2} = C_2(\lambda_1 - \lambda_2)/2I_o^{b2} \qquad (10)$$

where C_2^* is obtained from the decay of the noninverted b2 signal in the complementary experiment in which the inverted and noninverted spins are interchanged.

III. RESULTS AND DISCUSSION

The b1 ⇌ b2 equilibrium. In order to test the hypothesis that the b1 ⇌ b2 transition results from isomerism at a proline near His46, we mutated to glycines the two proline residues, Pro42 and Pro47, closest in sequence to His46. Both prolines are contained in a large loop that extends into the solvent and forms part of the active site. Residues within this loop are critical to binding and catalysis, although the loop itself is rather poorly defined in both the X-ray[16] and NMR[25,26] structures of nuclease. Cis/trans heterogeneity at either of these positions could conceivably perturb the loop locally, without causing a gross structural rearrangement (as does the a1 ⇌ a2 equilibrium) that would be reported by the other histidines.

Portions of the nuclease ^{1}H spectrum, containing $^1H^{\epsilon 1}$ peaks from the three histidine residues and the upfield-shifted methyl group of Val74, are shown in Fig. 1. Histidine assignments in the P42G+H124L and P47G+H124L mutants were made by reference to the H124L spectrum. A direct comparison is possible because the overall fold of the three proteins is highly similar as evidenced by two-dimensional spectra (not shown). In the absence of pdTp, doubling of each of the signals is evident (Fig. 1(a), (c), (e)).[†††] These minor forms disappear with the addition of pdTp and Ca^{2+} (Fig. 1(b), (d), (f)) and are assigned accordingly to the a2

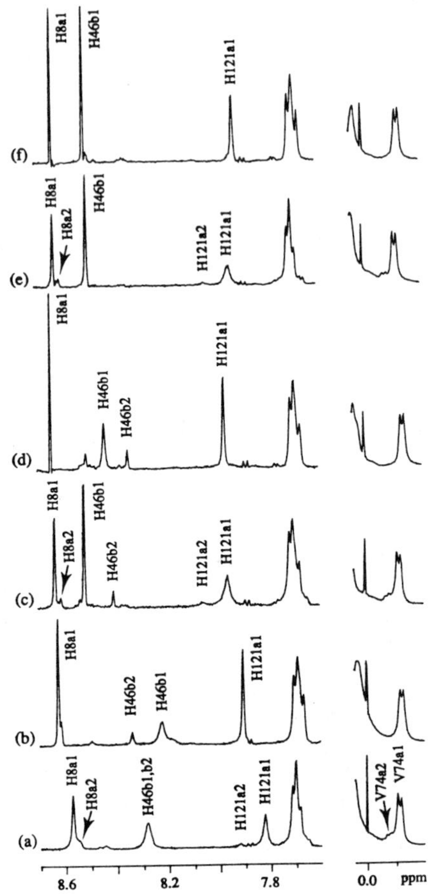

Fig. 1. 500 MHz ^{1}H spectra of nuclease H124L (a) unligated, (b) ternary complex; nuclease P42G+H124L (c) unligated, (d) ternary complex; nuclease P47G+H124L (e) unligated, (f) ternary complex. Each FID was multiplied by a π/20-shifted sine-bell filter prior to Fourier transformation. The small peak near 8.45 ppm arises from histidines in the denatured state.

[†††] Although the His46 signal of H124L does not appear to be split at pH* 5.5, doubling is clearly seen at both higher and lower pH. The major and minor peaks apparently overlap near pH* 5.5.

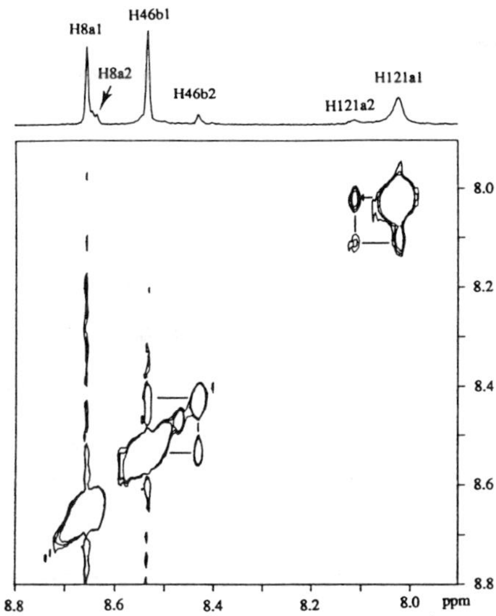

Fig. 2. 500 MHz phase-sensitive exchange-correlated spectrum of unligated nuclease P42G+H124L (histidine region). The standard NOESY pulse sequence used employed a mixing time of 1 s. 417 t1 increments of 4096 time domain data points were collected and zero filled to a final size of 2048 points. Prior to Fourier transformation, $\pi/2$-shifted sine-bell filters were applied to both dimensions.

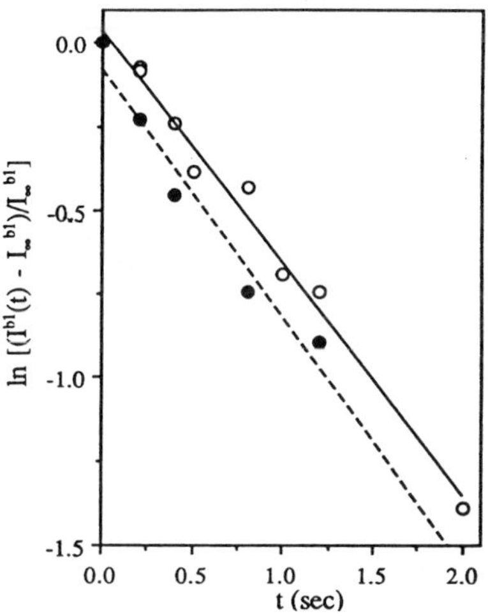

Fig. 3. Relative intensities of the nuclease P42G+H124L His46 $^1H^{\epsilon 1}$b1 signal plotted against the duration of a saturating pulse to the b2 peak. The b1 peak heights were scaled relative to that of the saturated b2 peak within the same time point, after the blank FID had been subtracted. Open (solid line) and closed (dashed line) circles are data from unligated and ligated enzymes, respectively.

conformation of the enzyme, since the a1 form has a higher affinity for the inhibitor. The most striking inconsistency is the persistence of an additional peak in the H124L and P42G+H124L ternary complex spectra. Several lines of evidence indicate that this is attributable to His46 in a minor form of the enzyme. The possibility that it is due to an impurity is discounted by the fact that it was observed in ^{13}C-edited spectra of ^{13}C-His labeled nuclease.[14] Furthermore, it does not arise from histidines in the denatured state as revealed by variable temperature studies (not shown). Finally, the two-dimensional exchange-correlated spectrum of P42G+H124L shown in Fig. 2 confirms appreciable interconversion of this peak with the major His46 $^1H^{\epsilon 1}$ peak within the one second mixing time of the experiment.

It is also apparent from Fig. 1 that P47G+H124L has but a single His46 $^1H^{\epsilon 1}$ peak, both in the presence and absence of pdTp. Thus, it is likely that doubling of this peak in H124L and P42G+H124L is caused by cis/trans heterogeneity at the His46-Pro47 peptide bond. The lone His46 $^1H^{\epsilon 1}$ signal in P47G+H124L coincides with the major His46 $^1H^{\epsilon 1}$ peak in P42G+H124L. This fact, together with Pro47 being observed as trans in the crystal structure of the nuclease wt-Ca^{++}-pdTp ternary complex, suggests that the predominant isomer has the trans peptide bond.

Saturation transfer. Poor separation and broad linewidths of the b1 and b2 resonances precluded use of nuclease H124L to measure rate constants of the cis-trans interconversion. The

Table I. Kinetic parameters for the mutant P42G+H124L.

parameter	unligated, sat. trans.	ternary complex, sat. trans.	unligated, inv. trans.	ternary complex, inv. trans.
k_1^{b1}, s^{-1}	0.7 ± 0.1[a]	0.7 ± 0.2[a]	0.9 ± 0.2[b]	1.0 ± 0.2[b]
k_1^{b2}, s^{-1}	nd	nd	1.7 ± 0.5[b]	1.9 ± 0.5[b]
k^{b1}, s^{-1}	0.05 ± 0.07[a]	0.06 ± 0.1[a]	0.2 ± 0.2[b]	0.3 ± 0.2[b]
k^{b2}, s^{-1}	0.4 ± 0.5[c]	0.3 ± 0.5[c]	1.4 ± 1[c]	1.6 ± 1[c]
T_1^{b1}, s	1.5 ± 0.1[d]	1.5 ± 0.2[d]	1.3 ± 0.5[d]	1.5 ± 0.7[d]
T_1^{b2}, s	nd	nd	4 ± 10[d]	4 ± 10[d]
K_{eq}=[b1]/[b2]	8.1 ± 0.6	5.1 ± 0.2	8.1 ± 0.6	5.1 ± 0.2

[a]From Eqs. 6 and 7. [b]From Eq. 8. [c]Calculated using K_{eq}. [d]From Eq. 5. nd, not determined.

mutant P42G+H124L, however, proved to be a tractable candidate. Fig. 3 shows the buildup of saturation of the His46 b1 resonance as a function of irradiation time of the b2 peak for this mutant. The equilibrium values I_0^{b1}/I_∞^{b1} were determined to be 1.1 (unligated P42G+H124L) and 1.1 (ternary complex), and were entered into Eq. 3 to obtain T_1^{b1}. The slopes of the lines yielded k_1^{b1}. We did not attempt to measure k_1^{b2} owing to the poor signal-to-noise ratio of the complementary experiment that resulted from the low intensity of the b2 peak. Results are summarized in Table I.

Inversion transfer. Several factors prevented us from obtaining the full data set consisting of fitted intensities for both the inverted and noninverted spins. Chief among these was the inherently slow b1 $\rightarrow$ b2 rate which approached the detectable limit set by T_1 relaxation. We were prompted to fit the intensities of the inverted peak rather than those of its noninverted partner because of the greater dynamic range covered by the relaxation of the former. In addition, only inversion of the b2 signal yielded data of sufficient quality to resolve the two exponentials in Eq. 2, presumably because of the relatively large contribution from the intense b1 peak. These effects are reflected in the large uncertainties associated with the parameters summarized in Table I. The data set generated by inversion of the His46 $^1H^{\epsilon 1}$ b2 resonance of unligated P42G+H124L is shown in Fig. 4.

The change in k^{b1} for the unligated vs ternary complexed enzymes in both the saturation transfer and inversion transfer experiments suggests that the decrease in K_{eq} for the b2 $\rightleftharpoons$ b1 equilibrium upon complexation primarily results from an increase in the trans to cis rate when pdTp is bound. The large errors, however, caution against assigning too much significance to these changes. Additional experiments need to be performed to improve the quality of the calculated parameters before further conclusions can be drawn regarding the kinetic effects of inhibitor binding.

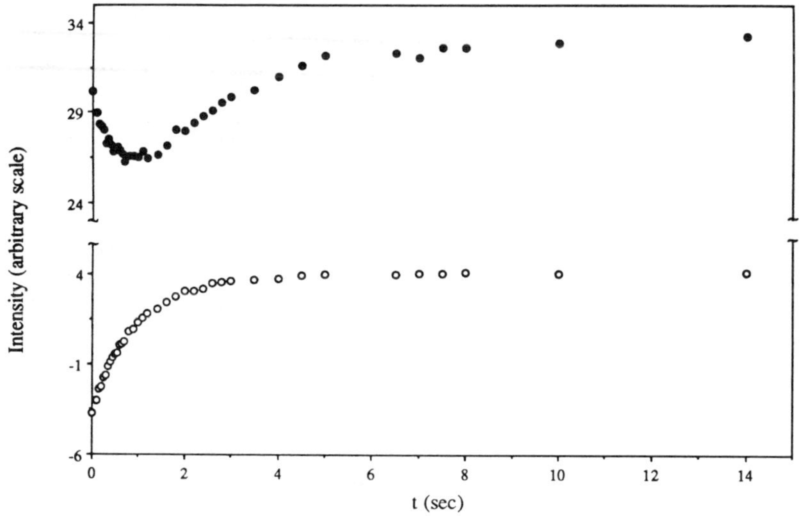

Fig. 4. Inversion transfer decay curves for unligated nuclease P42G+H124L. The open and closed circles represent the inverted b2 and noninverted b1 signals, respectively.

IV. CONCLUSIONS

A second conformational equilibrium has been observed in staphylococcal nuclease. It is distinct from the a1 ⇌ a2 equilibrium that results from isomerism at Pro117. Chemical exchange spectroscopy shows that the b1 ⇌ b2 process involves exchange between two conformations of the His46 $^{1}H^{\epsilon 1}$ proton with lifetimes on the order of 1 s. Because this equilibrium is abolished by the Pro47 → Gly mutation, we have attributed it to cis/trans heterogeneity at the His46-Pro47 peptide bond. The associated rate constants were measured by magnetization transfer NMR methods, and the slow (trans → cis) rate was found to increase upon formation of the nuclease-Ca^{++}-pdTp ternary complex.

V. ACKNOWLEDGMENTS

We thank Ross Reedstrom for oligonucleotide synthesis, Professor Richard Burgess for the pGP1-2 plasmid, and Andy Hinck for helpful discussions.

REFERENCES

1. Brandts, J.F., Halvorson, H.R. & Brennan, M. *Biochemistry 14,* 4953 (1975).
2. Cook, K.H., Schmid, F.X. & Baldwin, R.L. *Proc. Natl. Acad. Sci. U.S.A. 76,* 6157 (1979).
3. Touchette, N.A., Perry, K.M. & Matthews, C.R. *Biochemistry 25,* 5445 (1986).

4. Chazin, W.J., Kördel, J., Drakenberg, T., Thulin, E., Brodin, P., Gundström, T. & Forsén, S. *Proc. Natl. Acad. Sci. U.S.A. 86,* 2195 (1989).
5. Anfinsen, C.B., Cuatrecasas, P. & Taniuchi, H. *In* "The Enzymes" (P. Boyer, ed.), Vol. 4, pp. 177-204. Academic Press, New York, 1971.
6. Taniuchi, H. Anfinsen, C.B. & Sodja, A. *J. Biol. Chem. 242,* 4752 (1967).
7. Fox, R.O., Evans, P.A. & Dobson C.M. *Nature 320,* 192 (1986).
8. Evans, P.A., Dobson, C.M., Kautz, R.A., Hatfull, G. & Fox, R.O. *Nature 329,* 266 (1987).
9. Stanczyk, S.M., Bolton, P.H., Dell'Acqua, M., Pourmotabbed, T. & Gerlt, J.A. *J. Am. Chem. Soc. 110,* 7908 (1988).
10. Alexandrescu, A.T. Ulrich, E.L. & Markley J.L. *Biochemistry 28,* 204 (1989).
11. Alexandrescu, A.T., Hinck, A.P. & Markley, J.L. *Biochemistry 29,* 4516 (1990).
12. Markley, J.L., Williams, M.N. & Jardetzky, O. *Proc. Natl. Acad. Sci. U.S.A. 65,* 645 (1970).
13. Evans. P.A., Kautz, R.A., Fox, R.O. & Dobson, C.M. *Biochemistry 28,* 362 (1989).
14. Alexandrescu, A.T., Ph.D. Thesis, University of Wisconsin-Madison, 1989.
15. Wang, J., Hinck, A.P., Loh, S.N. & Markley, J.L. *Biochemistry 29,* 4242 (1990).
16. Loll, P.J. & Lattman, E.E. *Proteins: Struct. Funct. Genet. 5,* 183 (1989).
17. Hinck, A.P., Loh, S.N., Wang, J. & Markley, J.L. *J. Am. Chem. Soc.,* in press.
18. Tabor, S. & Richardson, C.C. *Proc. Natl. Acad. Sci. U.S.A. 82,* 1074 (1985).
19. McConnell, H. *J. Chem. Phys. 28,* 430 (1958).
20. Forsén, S. & Hoffman, R.A. *J. Chem. Phys. 39,* 2892 (1963).
21. Noggle, J.H. & Schirmer, R.E. "The Nuclear Overhauser Effect". Academic Press, New York, 1971.
22. Campbell, I.D., Dobson, C.M., Ratcliffe. R.G. & Williams, R.J.P. *J. Magn. Reson. 29*, 397 (1978).
23. Led, J.J. & Gesmar, H. *J. Magn. Reson. 49,* 444 (1982).
24. Marquardt, D.W. *J. Soc. Indust. Appl. Math. 11,* 431 (1963).
25. Torchia, D.A., Sparks, S.W. & Bax, A. *Biochemistry 28,* 5509 (1989).
26. Wang, J., LeMaster, D.M. & Markley, J.L. *Biochemistry 29,* 88 (1990).

27

Structure in a Partially Folded State of Ubiquitin as Determined by NMR

D.N. Woolfson[1,2]
M.M. Harding[1,§]
P.A. Evans[2]
D.H. Williams[1]

[1]University Chemical Laboratory, Lensfield Rd., Cambridge, UK
[2]Department of Biochemistry, Tennis Court Rd., Cambridge, UK
[§]Present Address: Department of Organic Chemistry, University of Sydney, Broadway, NSW, 2006 Australia.

I. INTRODUCTION

One approach taken in tackling the protein folding problem has been to discern structural details of folding pathways.[1,2] Such studies have been hampered by the apparent high cooperativity of the folding process which renders states other than the native and unfolded only transiently stable. Fortunately, recent studies indicate that this property of protein organisation breaks down to some degree: some small peptides do show significant secondary structure;[3,4] kinetic intermediates in folding can be observed and studied;[5,6,7,8] equilibrium intermediates — often referred to as molten globules or compact intermediates — have also been observed for some proteins.[9] The technique of choice for studying such structures has been ^1H NMR, as it gives high resolution information about local and tertiary interactions.[10]

The more detailed of these studies have come from four groups. Those of Baldwin, Roder and Fersht used a combination of hydrogen exchange, rapid mixing techniques and NMR to follow hydrogen bond formation in RNase A, cytochrome *c* and barnase during refolding.[5,6,7,8] Baum and co-workers have succeeded in identifying the slowly exchanging amide protons of an equilibrium partly folded state, the A-state of α-lactalbumin .[11] In all these cases the structure present appears to be a subset of the secondary interactions present in the folded states of the protein.

In this paper we report the structural characteristaion of a partially folded state of ubiquitin. We use a combination of NMR techniques to illustrate how structural details in such species can be determined.

Techniques in Protein Chemistry II

II METHODS

Proton NMR spectra were recorded with Bruker AM-400 and AM-500 spectrometers equipped with aspect 3000 computers. Spectra were recorded over spectral widths of 4400 or 5500 Hz (at 400-MHz and 500-MHz respectively) with quadrature detection employed throughout. Samples were typically 1.5 - 2.5 mM in concentration. Chemical shifts were measured relative to internal dioxan at 3.73ppm. Two-dimensional proton NMR spectra in D_2O were acquired in the phase sensitive mode using quadrature detection in f_2, and with time proportional phase incrementation in f_1.[12] Data sets resulting from 256 to 512 increments of t_1 were acquired and zero filled to 1024 points, with each free induction decay composed of 2048 data points. Typically 64 or 96 transients were recorded for each increment of t_1 with a recycle delay of 1.2-2.0 seconds between transients. Solvent suppression in NOESY and COSY spectra was achieved by coherent presaturation during the recyle delay. HOHAHA spectra were recorded using a 7 kHz spin locking field with an MLEV-17 sequence of 40-80ms. Solvent supression was achieved by a DANTE pulse sequence directly prior to a SCUBA preirradiation pulse to enhance the alpha resonances.[13] Baseline correction was achieved using the sine modulation method, this was useful when acquiring HOHAHA spectra from 90% H_2O/10% D_2O solutions and in the 2-D exchange spectroscopy. Data was subjected to shifted sine-bell or squared sine-bell weighting functions in f_1 and f_2.

III METHANOL-INDUCED CHANGES IN UBIQUITIN

The 1-D proton NMR spectrum of native ubiquitin at pH 5.8 is shown in Figure 1 (A). The high degree of chemical shift dispersion is indicative of ordered interactions within the protein. There are a number of resonances shifted upfield of the main methyl evelope at about 1ppm. These are the result of ring current shifts on the chemical shifts of aliphatic protons sitting above aromatic rings. There are also some C_α protons that resonate downfield of the water signal. Such resonances are typically associated with protons in ß-structure. Finally, the most down field of the resonances in the spectrum are amide protons that are stabilised from hydrogen exchange with solvent, (D_2O for this sample), through hydrogen bonding. These groups of resonances are, thus, all good probes of structure in the protein. For example, when the protein is fully unfolded using urea and heat the chemical shift dispersion and the amide resonances are lost, Figure 1 (C).

Many proteins show maximal stability at their isoelectric points, because electrostatic repulsion in the compact folded states of proteins is minimised when the net charge on them is zero.[14] Increase or decrease in the pH away from this value does generally result in a decrease in the thermal stability of proteins.[15] In the extreme some proteins undergo acid or base unfolding transitions at room temperature. Recently, some of the unfolded states have been shown to contain a high degree of structure.[11,16] Lowering the pH of a

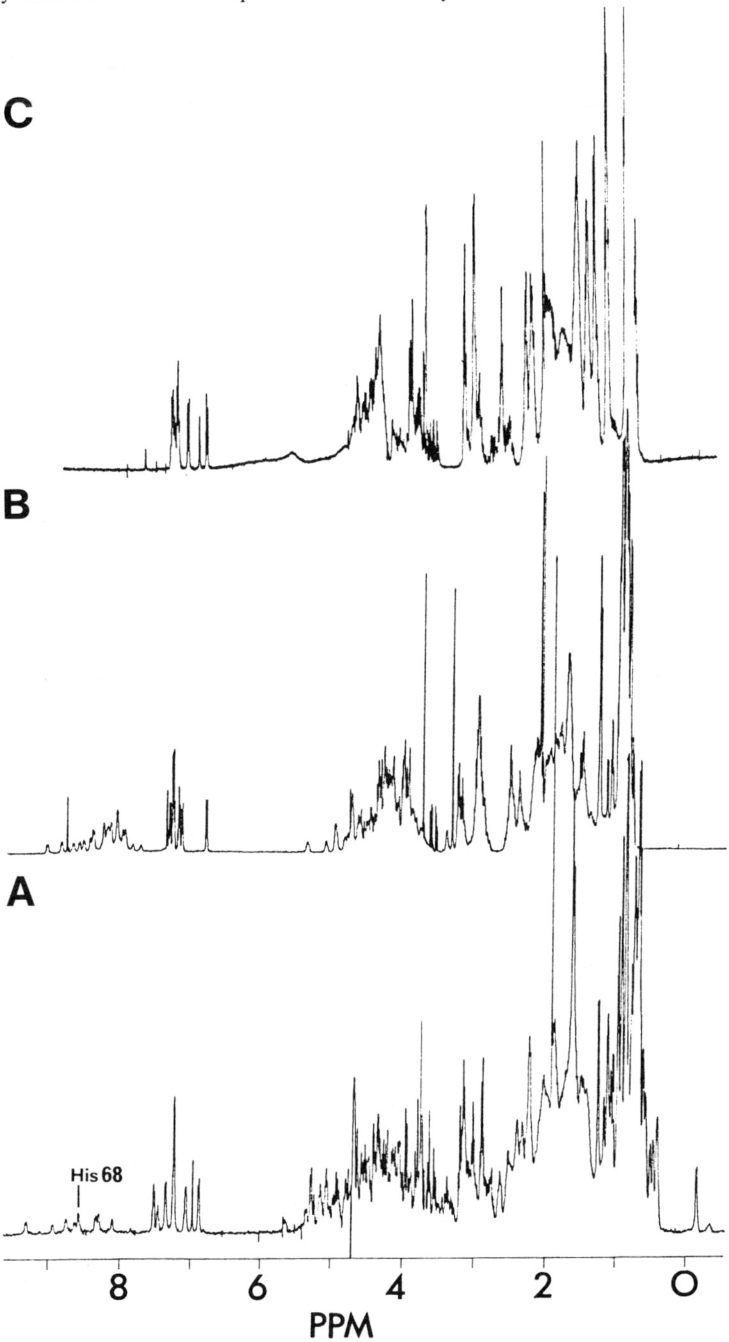

Figure 1 1-D ^{1}H NMR spectra of ubiquitin in deutero solvents: (A) in the native state, pH 5.8; (B) in the A-state, pH2 and 60% methanol; (C) in the unfolded state, 350K and 8M urea.

ubiquitin sample has no substantial effect on its ^1H NMR spectrum; the protein remains folded, although thermal stabilty is reduced compared with solutions at pH 5. Titration of methanol into samples at low pH (after Biringer & Fink[17]), however, is accompanied by marked changes in the spectrum and thus the structure of ubiquitin, Figure 1 (B), the transition is complete in 60% methanol.

Changes in the appearance of the 1-D spectrum, assignments for the N-state are known,[18,19] can be summarised as follows:

1. Much of the chemical shift dispersion associated with spectra of fully folded states is lost . This is particularly apparent in the upfield region of the spectrum where ring current shifted methyl resonances have collapsed back into the methyl envelope;

2. Many of the C_α H resonances downfield of the HDO signal have also collapsed back toward random coil positions. One resonance in this region (the C_α H of Phe-4), however, is still prominent and served well as a starting point for resonance assignments in this spectrum;

3. Approximately 15 - 20 amide resonances are still present in the methanol-induced form or A-state of ubiquitin. These signals are sharp, long-lived and the majority have $^3J_{NH\alpha}$ coupling constants of 8 Hz of more. This strongly suggests that these protons are hydrogen bonded within ß-structure.

Structure in the A-state was further investigated using 2-D NMR techniques. The first step in the study was to assign resonances to types of spin system, and thus to classes of residue, using methods which correlate resonances through scalar or J-couplings, COSY and HOHAHA Once identified the spin systems were then pieced together using the nOe, through NOESY experiments.[10] A starting point for this assignment work was the Phe-4 C_α H peak which is unperturbed at 5.2 ppm during the N- to A-state transition.

HOHAHA and NOESY spectra recorded in D_2O/d_4-methanol are presented in Figures 2 and 3. The secondary structure determined through these experiments is also shown in the second of the figures. The methanol induced structural changes in ubiquitin are summarised in Figure 4. The diagram shows that a large proportion of the hydrogen bonded structure of the molecule is lost, notably residues 36 through 65 appear disordered,(on the basis that amide exchange is relatively rapid in this part of the molecule; amide proton resonances are not observed in 2D experiments). The first three strands of what was a five strand ß-sheet are preserved in the A-state. It is interesting to note that this piece of secondary structure is not made up from a contiguous stretch of sequence; the first two strands are part of a hairpin formed by residues 1 through 18 and the third comes from the C-terminus of the protein. NOESY experiments in H2O/d4-methanol solutions (spectra not shown) identified d_{NN} type nOe's between — what were in the D2O spectra — rapidly exchanging protons. This suggests that residues Val-26, Lys-27

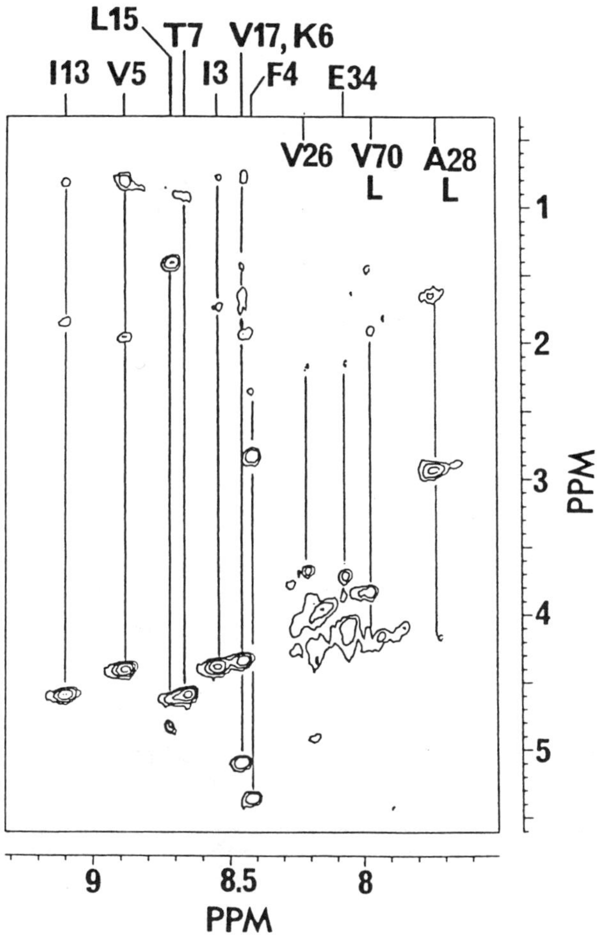

Figure 2 Identification of the spin systems of the amino acids with slowly exchanging amides in the A-state using the HOHAHA experiment (53 ms mixing time at 400 MHz).

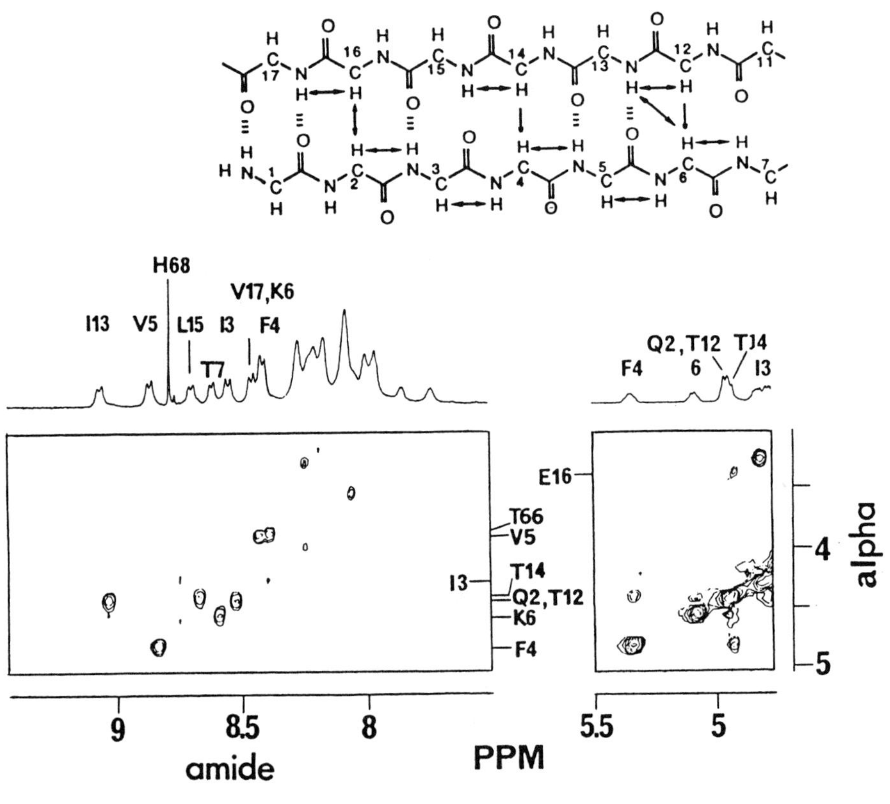

Figure 3 Sequence specific assignment of amide and CαH resonances and identification of the long-lived secondary structure in the A-state.

N-STATE

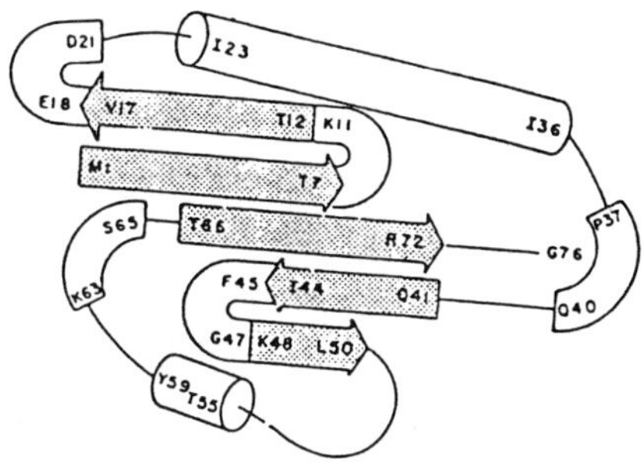

A-STATE

α-helix destabilised

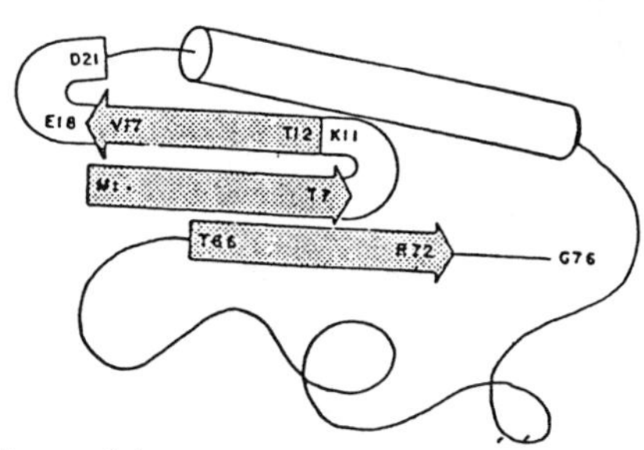

disruption of turn-rich area
essentially random coil

Figure 4 Secondary structure differences between the native and A-states of ubiquitin.

and Ala-28 are in α-helical structure in the A-state, although this element of structure is much less stable than it is the native state.

Despite the high degree of hydrogen bonded structure in the A-state of ubiquitin it appears, from the 1-D NMR experiments, that core or hydrophobic interactions in it are considerably diminished compared with those in the native state. To test this further the COSY spectra shown in Figure 5 were recorded. It is clear that there are large differences in the aliphatic regions of spectra of the totally unfolded and partly folded protein. In the latter there is greater chemical shift dispersion, several resonances are shifted upfield of the random coil positions. Assignment of the crosspeaks in the A-state spectra is not possible using conventional techniques as there is considerable overlap of resonances as a consequence of the break-down of secondary and tertiary interactions. 2-D magnetisation transfer or exchange spectroscopy, however, can be used to take advantage of the dispersion in shifts of the folded states.[20] Under certain conditions "exchange" crosspeaks in NOESY or ROESY spectra can be observed between resonances of the same proton that are in different environments. The aromatic region of a 2-D exchange spectrum of ubiquitin is shown in Figure 6. The spectrum was recorded at 310K an a sample in 40% methanol; conditions under which the N- and A-states of the protein are in slow exchange. Thus, the assignment of partly folded and unfolded protein spectra is possible provided that the N-state assignments have been made and that magnetisation transfer is observed between resonances of the folded and unfolded resonances.

IV CONCLUSIONS

A patially organised equilibrium species of ubiquitin can be prepared in water/methanol mixtures at low pH. The hydrogen bonded structure in this state has been characterised using proton NMR, and its structural outline is summarised in Figure 4. Hydrophobic interactions are reduced in the A-state compared with those in the N; this is a consequence of both the solvent and pH conditions as the protein is fully folded in high concentrations of methanol at pH 5. Upfield shifts in the aliphatic region of A-state spectra indicate that residual hydrophobic interactions are present, these are likely to be responsible, in part, for the stability of the species. The system clearly warrants further attention and we are at present engaged in 2-D magnetisation transfer studies to determine which residues participate in the hydrophobic interactions, amide exchange experiments to determine the extent and stability of any hydrogen bonding interactions not observed in the 2-D NMR studies performed to date and a series of 1-D magnetisation transfer studies to determine the kinetics of the folding transition.

V ACKNOWLEDGEMENTS

We would like to thank Dr. Tom Frenkiel for help with some of the NMR experiments, and Pfizer Central Research, UK and the SERC of the UK for financial support. MMH gratefully acknowledges the receipt of a Ramsay Memorial General British Fellowship. DHW and PAE are members of the Cambridge Centre for Molecular Recognition. Figure 4 was adapted from ref 18.

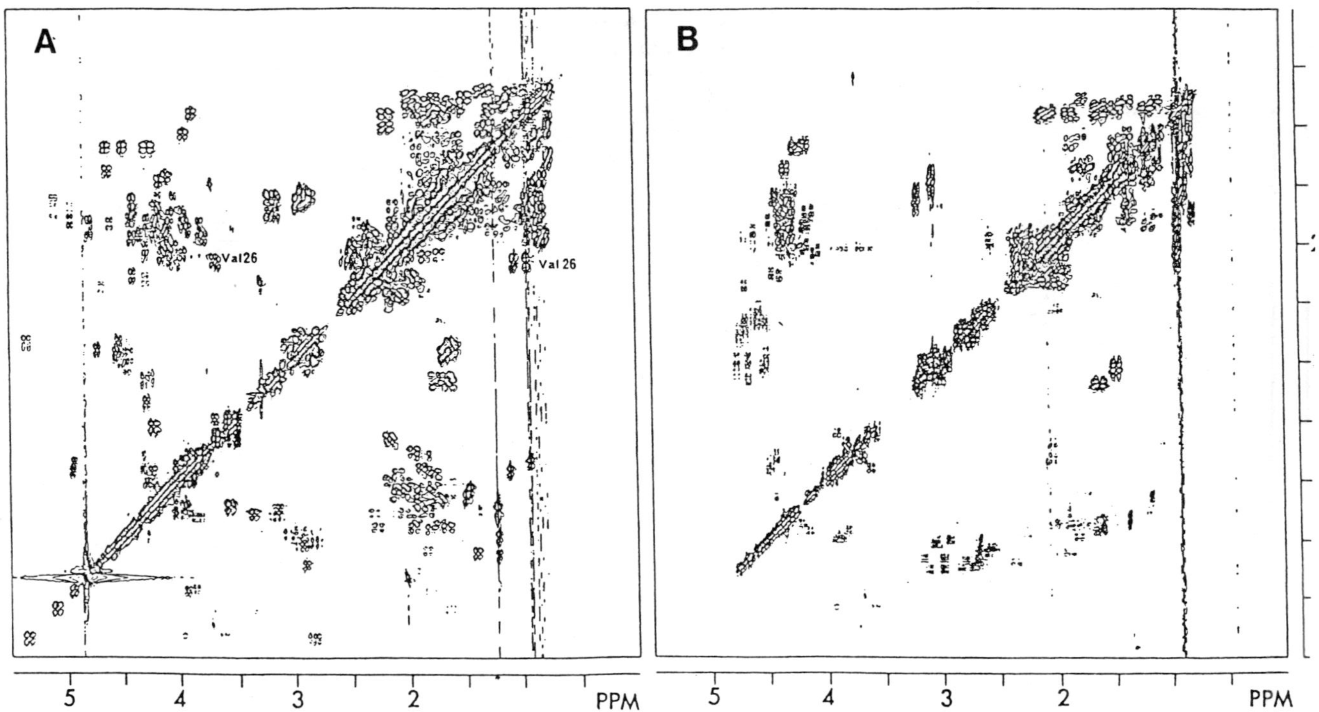

Figure 5 Aliphatic regions of 400 MHz COSY spectra of ubiquitin in the A- (A) and unfolded states (B).

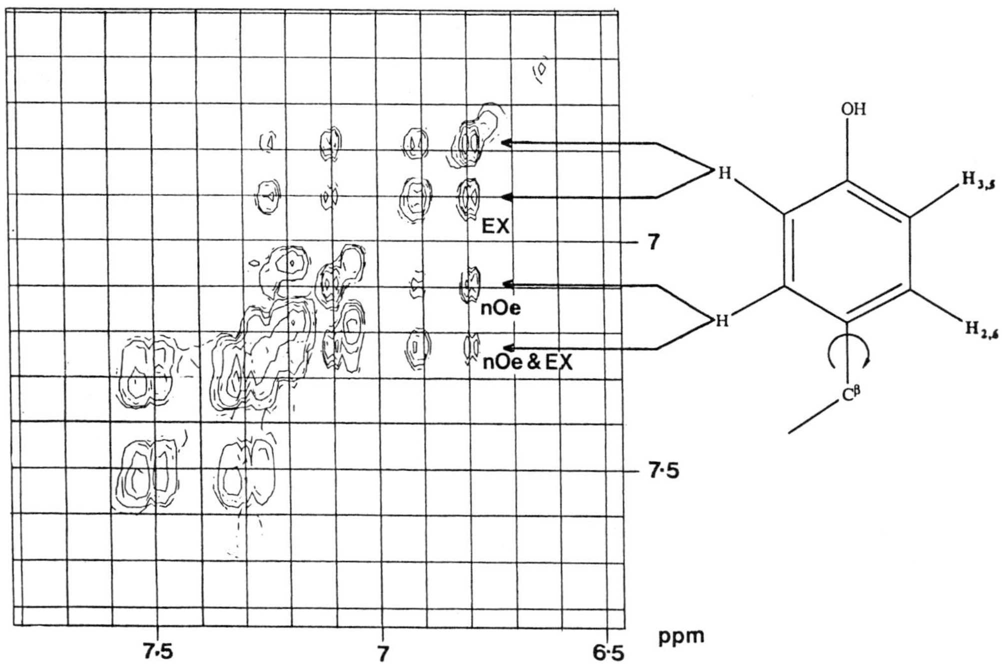

Figure 6 The aromatic region of a 2-D exchange spectrum of approximately a 50/50 mixture of the A- and N-states of ubiquitin. Highlighted are the Tyr-59 resonances. Note, in particular, the three crosspeaks to the A-state $H_{3,5}$ resonance. These arise from: exchange processes between resonances of the same proton in different conformers, (EX); nuclear Overhauser effects between the $H_{3/5}$ and $H_{2/6}$ protons in the A-state, (nOe); and a combination of nOe and exchange processes, (nOe & EX). This spectrum was recorded using a NOESY pulse-sequence with a 400 ms mixing time and on a sample in 40% methanol, pH 2 at 310K.

REFERENCES

[1]Kim, P.S. & Baldwin, R.L. (1982), Ann. Rev. Bioch. 51 459-489

[2]Kim, P.S. & Baldwin, R.L. (1990), Ann. Rev. Bioch. 59 631-660

[3]Marqusee, S. & Baldwin, R.L. (1987), Pro. Natl. Acad. Sci. USA 84 8898-8902

[4]Wright, P.E., Dyson, H.J. & Lerner, R.A. (1988), Biochemistry 27 7167-7175

[5]Udgaonker, J.B. & Baldwin, R.L. (1988), Nature 335 694-700

[6]Roder, H., Elove, G.A. & Englander, S.W. (1988), Nature 335 700-

[7]Matouschek, A., Kellis, J.T., Serrano, L., Bycroft, M. & Fersht, A.R. (1990), Nature 346 440-445

[8]Bycroft,M., Matouschek, A., Kellis, J.T., Serrano, L. & Fersht, A.R. (1990), Nature 346 440-445

[9]Kuwajima (1989), Proteins 6 87-103

[10]Wüthrich K. (1986), "NMR of Proteins and Nucleic Acids" Wiley, NewYork

[11]Baum, J., Dobson, C.M., Evans, P.A. & Hanley, C. (1989), Biochemistry 28 7-13

[12]Marion, D., & Wüthrich, K. (1983) Biochem. Biophys. Res. Commun. 113, 967.

[13]Brown, S.C., Weber, P.L., & Mueller, L. (1988) J. Magn. Reson. 17, 166.

[14]Dill, K.A. (1990), Biochemistry 29 7134-7155

[15]Privalov, P.L & Kechinasvili, N.N. (1974), J. Mol. Biol. 86 665-

[16]Ptitsyn,O.B., Pain, R.H., Semisotnev, G.V., Zerovnik, E. & Razgulyaev, O.I. (1990), FEBS Lett. 262 20-24

[17]Biringer, R. G., & Fink, A. L. (1982) Biochemistry 21, 4748

[18]D.L Di Stefano & A.J. Wand (1987), Biochemistry 26 7272-7281

[19]P.L. Weber, S.C. Brown & L. Mueller (1987), *ibid* 7282-7290

[20]Evans, P.A., Topping, K.D., Woolfson, D.N. & Dobson, C.M., Proteins in press

28

STABILIZATION OF BARNASE BY AN INTERACTION BETWEEN TWO OPPOSITELY CHARGED SIDE CHAINS

Daša Šali, Mark Bycroft and Alan R. Fersht

MRC Unit for Protein Function and Design
University Chemical Laboratory,
Cambridge CB1 2EW, UK

I. Introduction

Coulombic interactions between point charges are ubiquitous in proteins; they play an essential role in molecular recognition, binding, catalysis, assembly of macromolecular structures, cooperativity of allosteric transition and protein stability (Warshel & Russel, 1984; Perutz, 1987; Russell *et al.*, 1987; Horovitz *et al.*, 1990; Perutz & Raidt, 1975). Although the interactions between point charges are generally thought to stabilize proteins, the magnitude of this stabilization has only been estimated in a few instances. Studies of binding between tyrosyl-tRNA synthetase and its substrates revealed that a hydrogen bond involving two oppositely charged side chains contributed 3–6 kcal/mol to the specificity of binding (Fersht, 1987). Stabilization of T4 lysozyme by a half-buried salt bridge between His31 and Asp70 was found to be 3–5 kcal/mol (Anderson *et al.*, 1990). Marqusee *et al.* (1987) examined the interaction between oppositely charged pairs of residues in a synthetic polypeptide α-helix; they found that the greatest contribution to stability of the helix was obtained when the two residues were four positions apart, with a negatively charged glutamate closer to the N-terminus of the helix and a positively charged lysine closer to the C-terminus.

In order to find out whether surface salt bridges, such as those known to stabilize secondary structure of small peptides, necessarily contribute to protein stability, a new electrostatic interaction was engineered into the second α-helix of barnase.

Barnase, the extracellular ribonuclease from *Bacillus amyloliquefaciens* (Nishimura & Nomura, 1958), is a monomeric enzyme of 110 amino acid residues, with M_r of 12,382 (Hartley & Barker, 1972). Its gene has been cloned and expressed in *Escherichia coli* (Paddon & Hartley, 1986). The X-ray crystal structure of barnase has been solved at high resolution (Mauguen *et al.*, 1982), and two-dimensional NMR has been used to assign resonances for protons in all

Techniques in Protein Chemistry II

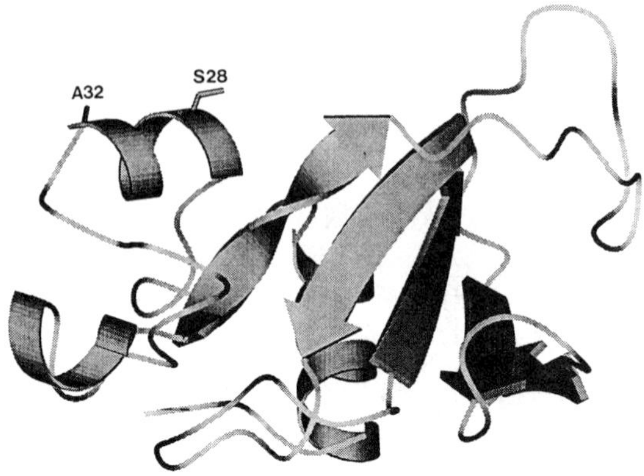

Fig. 1. Ribbon structure of barnase, showing the side chains of Ser28 and Ala32.

its residues (Bycroft *et al.*, 1990b) and to determine its structure in solution (Bycroft *et al.*, 1990c). Despite its small size, barnase has extensive secondary structure, including two α-helices and a five-stranded anti-parallel β-sheet (Figure 1). It undergoes reversible, two-state denaturation induced *in vitro* by solvent or high temperature (Hartley, 1968; Kellis *et al.*, 1989), allowing for straightforward thermodynamic measurements of protein stability. As barnase contains no cysteines, there are no complications due to presence of disulfide bridges in the folded and unfolded state.

Several studies investigating the importance of specific interactions for barnase stability and folding pathway have been carried out (Kellis *et al.*, 1988; Šali *et al.*, 1988; Kellis *et al.*, 1989; Matouschek *et al.*, 1989; Serrano & Fersht, 1989; Bycroft *et al.*, 1990a; Horovitz *et al.*, 1990; Matouschek *et al.*, 1990; Serrano *et al.*, 1990). The general approach has been to make conservative site-directed mutants of the enzyme in order to disrupt, or add, a particular interaction and then to compare thermodynamic or kinetic properties of wild-type and mutant enzymes.

The residues mutated in this study are Ser28 and Ala32 on the 26–34 α-helix of barnase (Figure 1). Since these two residues are located on the side of the helix which is fully accessible to the solvent and make no contacts with residues outside the helix, their substitution with charged residues was not expected to disrupt the structure of the enzyme. Ser28 was mutated into glutamic acid and Ala32 into lysine, mimicking the $(i + 4)$ Glu,Lys ion pairs from the experiments on model peptides by Marqusee *et al.* (1987). In addition to the double mutant, the two single mutants were also made and the contribution of the interaction between Gly28 and Lys32 to stability of the protein was determined using double mutant cycles. The mutations were shown to be non-disruptive and the orientation of the mutant side chains explored using 2D ^{1}H NMR.

II. Materials and Methods

A. Expression and mutagenesis

The enzyme was expressed in *E. coli* harboring a derivative of pTZ18U plasmid, which contains the gene for barnase fused to the alkaline phosphatase promoter and signal sequence (Serrano *et al.*, 1990). It was purified to homogeneity using ion exchange chromatography (Paddon & Hartley, 1987). Mutants were made using the double-priming Kunkel method (Kunkel *et al.*, 1987).

B. Urea denaturation

Urea denaturation was observed using fluorescence intensity as the probe, with excitation at 290 nm and emission at 315 nm. Experiments were performed at 25±0.05°C, *p*H 6.3. MES acid/base buffer was used at concentrations of 7 mM, 10 mM and 50 mM. NaCl was added for high salt experiments to the final concentration of 250 mM and 556 mM. The concentration of barnase was 1 μM. The denaturation curves were analyzed as described previously (Kellis *et al.*, 1988; Serrano & Fersht, 1989) to obtain the free energy of unfolding in water (ΔG_{H_2O}), the slope (m) and the half point of denaturation ([urea]$_{50\%}$) for each protein. The difference in free energy of unfolding between different forms of the protein is given by:

$$\Delta\Delta G_{U12} = \frac{m_1 + m_2}{2}([urea]_{50\%2} - [urea]_{50\%1})$$

C. Nuclear magnetic resonance

To make the NMR samples, pure barnase was dissolved in 0.5 ml of 90% H_2O/10% D_2O to a final concentration between 4 and 6 mM. *p*H was adjusted to 4.5. NMR spectra were acquired at 37°C on a Bruker AM 500 spectrometer with an Aspect 3000 computer. COSY (Aue *et al.*, 1976; Bax & Freeman, 1981) and NOESY (Jeener *et al.*, 1979; Anil-Kumar *et al.*, 1980) experiments were recorded with 2048 data points in t_2 and 512 t_1 increments, with a spectral width of 8000 Hz in both dimensions. A relaxation delay of 1.5 s was used. Suppression of the solvent signal was achieved by presaturation of the water resonance during the relaxation delay.

III. Results

A. Wild-type and mutant stability and determination of the free energy of interaction between Glu28 and Lys32

Table I lists the free energies of unfolding of the wild-type, the two single mutants and the double mutant under the different ionic strength conditions.

The simplest way of estimating the magnitude of the interaction between Glu28 and Lys32 would be to compare the stabilities of wild-type and the double mutant,

SE28AK32. The problem with this approach is that it does not distinguish between the contribution to stability due to the interaction between the two charges introduced by mutation and any contribution due to the interaction of either of these side chains with the rest of the protein. For example, it is clear from Table I that the single mutation from Ser28 to Glu already stabilizes the protein under most conditions, while the single mutation from Ala32 to Lys makes barnase less stable. Both these effects have to be taken into account when measuring the interaction between Glu28 and Lys32.

This problem can be overcome by using double mutant cycles (Carter *et al.*, 1984; Horovitz, 1987). Such cycles consist of the form of the enzyme in which neither of the partners in the interaction are present (in this case, the wild-type), the form in which both partners are present (SE28AK32) and the two forms in which only one of them is present (SE28 and AK32) (Figure 2). The free energy of the interaction between the two residues is determined by measuring the unfolding free energies of each of the proteins in the cycle. The difference between the change in the free energy of unfolding upon mutation of the first residue when the second residue is present and absent is the apparent free energy of the interaction between the two residues. If the two mutations are not independent, i.e., there is an interaction between the two residues in the double mutant, the following inequalities are true (using the symbols defined in Figure 2):

$$\Delta G_{\text{E}\rightarrow\text{E-X}} \neq \Delta G_{\text{E-Y}\rightarrow\text{E-XY}} \tag{1}$$

$$\Delta G_{\text{E}\rightarrow\text{E-Y}} \neq \Delta G_{\text{E-X}\rightarrow\text{E-XY}} \tag{2}$$

The free energy of interaction between X and Y is given by:

$$\Delta G_{\text{int}} = \Delta G_{\text{E}\rightarrow\text{E-X}} - \Delta G_{\text{E-Y}\rightarrow\text{E-XY}} = \Delta G_{\text{E}\rightarrow\text{E-Y}} - \Delta G_{\text{E-X}\rightarrow\text{E-XY}} \tag{3}$$

The free energy of interaction between Glu28 and Lys32 under different salt conditions, as determined by double mutant cycles, is given in the last column of Table I.

B. Comparison of 2D ¹H NMR spectra of wild-type and the mutants

Fingerprint regions of mutant 2D NMR spectra were assigned using the similarities between the spectra of wild-type barnase and the mutants as the starting point. About 75% of the cross-peaks corresponding to the main chain protons in the mutants could be assigned by analogy with the wild-type spectra, with another 15–20% of assignments obtained from sequential connectivities (Wüthrich, 1986). The plots of chemical shifts of equivalent main chain protons in the mutants and the wild-type (Figure 3) show that the only differences occured at the position of the mutation or at the immediately adjacent positions. This is consistent with the notion that most main chain protons are in the same environment in the mutants as in the wild-type. The patterns of NOE connectivities between

Protein	[MES] [mM]	[NaCl] [mM]	ΔG_{H_2O} [kcal/mol]	m	[urea]$_{50\%}$ [M]	$\Delta\Delta G_U$ [kcal/mol]	ΔG_{int} [kcal/mol]
wt	7	0	8.48	1.89	4.48		
SE28	7	0	10.21	2.16	4.73	+0.52	
AK32	7	0	9.22	2.10	4.39	−0.19	-0.15
SE28AK32	7	0	10.24	2.17	4.71	+0.48	
wt	10	0	8.96	1.97	4.54		
SE28	10	0	9.44	1.99	4.74	+0.39	
AK32	10	0	8.85	1.99	4.45	−0.18	-0.18
SE28AK32	10	0	8.95	1.90	4.71	+0.33	
wt	50	0	10.25	2.23	4.59		
SE28	50	0	10.14	2.16	4.69	+0.23	
AK32	50	0	10.69	2.36	4.53	−0.14	-0.19
SE28AK32	50	0	10.63	2.26	4.70	+0.25	
wt	50	250	9.93	2.00	4.97		
SE28	50	250	10.06	2.01	5.00	+0.06	
AK32	50	250	10.21	2.10	4.86	−0.22	-0.15
SE28AK32	50	250	9.93	1.99	4.99	+0.04	
wt	50	556	10.24	1.95	5.25		
SE28	50	556	10.73	2.05	5.23	−0.04	
AK32	50	556	10.42	2.03	5.13	−0.24	-0.14
SE28AK32	50	556	10.70	2.07	5.16	−0.18	

TABLE I. Free energies of unfolding for wild-type and the mutants at positions 28 and 32, at *p*H 6.3. $\Delta\Delta G_U$'s were calculated for each ionic strength as a difference between wild-type and each mutant. The errors were ±0.01 M for [urea]$_{50\%}$, ±0.02 kcal/mol for ΔG_{H_2O}, ±0.03 kcal/mol for $\Delta\Delta G_U$ and ±0.04 kcal/mol for ΔG_{int}.

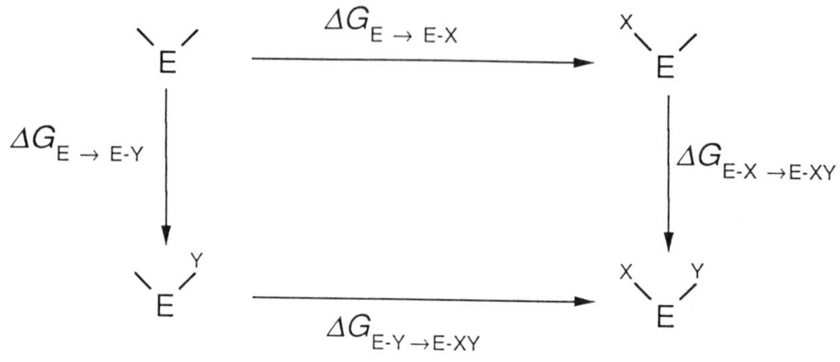

Fig. 2. The scheme for a double mutant cycle. X and Y are the two residues being introduced by mutations and $\Delta G_{E \rightarrow E\text{-}X}$, $\Delta G_{E \rightarrow E\text{-}Y}$, $\Delta G_{E\text{-}X \rightarrow E\text{-}XY}$ and $\Delta G_{E\text{-}Y \rightarrow E\text{-}XY}$ the changes in free energy of unfolding for the appropriate transitions in the cycle. ΔG_{int} is the free energy of interaction between X and Y.

backbone protons which characterize specific secondary structure elements, i.e. d_{NN}, $d_{\alpha N}(i, i+3)$ and $d_{\alpha N}(i, i+4)$ connectivities for the α-helices, and strong $d_{\alpha N}$ and inter-strand NOE contacts for the β-sheets, were identical for the wild-type and the three mutants throughout the molecule, including the connectivities of the 26–34 α-helix (data not shown). This again suggests that the overall fold of barnase was not altered by mutation; consequently, the effect on stability of the mutations at positions 28 and 32 can be safely attributed to local effects of the side chains introduced by mutagenesis.

IV. Discussion

Stabilization energies listed in Table I indicate that the SE28 mutation by itself stabilizes barnase by about as much as the double mutant (SE28AK32). The greatest increase in stability occurs in 7 mM MES and is 0.52 kcal/mol for SE28 and 0.48 kcal/mol for SE28AK32. At higher salt concentrations, the difference in stability between these two mutants and the wild-type decreases, until at 556 mM NaCl the mutants are slightly less stable than the wild-type. In contrast, the AK32 mutation slightly destabilizes the enzyme, the change in free energy of unfolding between the mutant and the wild-type being 0.04–0.24 kcal/mol. The fact that the single mutation from serine to glutamic acid at position 28 already stabilizes barnase may be due to the introduction of a favourable interaction between Glu28 and the α-helix dipole of the 26–34 α-helix.

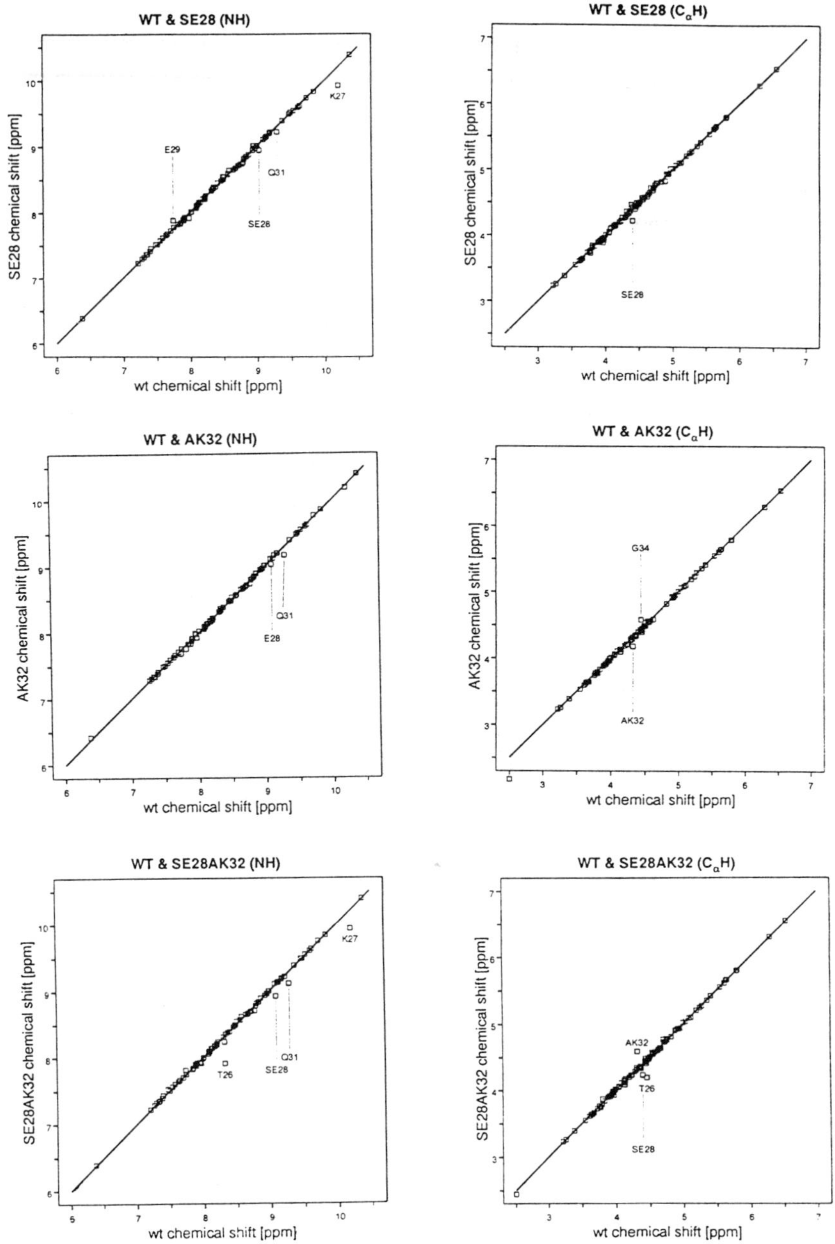

Fig. 3. Chemical shifts of amide and C_α protons in the mutants plotted against the chemical shifts in wild-type.

When stability of wild-type barnase and the mutants at positions 28 and 32 was analyzed using double mutant cycles, the interaction energy between Glu28 and Lys32 side chains was shown to be 0.16 ±0.04 kcal/mol throughout the ionic strength range examined. As the variation of the interaction energy between the two side chains with ionic strength is smaller than the error estimated for calculation of this value, this variation with ionic strength cannot be considered to be significant.

The interaction energy between two oppositely charged side chains on barnase surface measured here is an order of magnitude lower than that determined previously for interaction energies of buried, hydrogen-bonded salt bridges (Anderson *et al.*, 1990). This may be explained by the fact that the dielectric constant in the protein core is known to be much lower than in water or at the surface of the protein (Warshel & Russel, 1984), therefore the enthalpy of interaction between two buried charges would be expected to be larger than that between two charges separated by the same distance at the surface of the molecule. In addition, side chains on the surface of a protein have a greater degree of freedom than those in the centre of the protein; they would be expected to spend only a portion of their time in close proximity to oppositely charged side chains, interacting with solvent water the rest of the time.

The free energy of interaction between Glu28 and Lys32 obtained here is comparable to the interaction energy measured by Serrano *et al.* (1990) for another engineered surface ion pair in barnase. They found the interaction energy between Asp12 and Arg16 on the surface of the N-terminal α-helix in barnase to be 0.33 kcal/mol. Both these studies seem to suggest that interactions between oppositely charged side chains on protein surface do not necessarily make a significant contribution to the overall stability of the proteins.

V. Acknowledgements

This work was supported by the MRC and the British Biochemical Society.

VI. References

Anderson, D.E., Becktel, W.J.& Dahlquist, F.W. (1990) *Biochemistry* **29**, 2403–2408.

Anil-Kumar, Ernst, R.R. & Wüthrich, K. (1980) *Biochem. Biophys. Res. Comm.* **95**, 1-6.

Aue, W.P., Bartholdi, E. & Ernst, R.R. (1976) *J. Chem. Phys.* **64**, 2229–2246.

Bax, A. & Freeman, R. (1981) *J. Mag. Res.* **44**, 542–561.

Bycroft, M., Matouschek, A., Kellis, J.T., Jr., Serrano, Ll & Fersht, A.R. (1990a) *Nature* **346**, 488–490.

Bycroft, M., Shepard, R.N., Lau, F.T.-K. & Fersht, A.R. (1990b) *Biochemistry*, in press.

Bycroft, M., Poulsen, F.M. & Ludvigsen, S. (1990c), manuscript in preparation.

Carter, P.J., Winter, G., Wilkinson, A.J. & Fersht, A.R. (1984) *Cell* **38**, 835–840.

Fersht, A.R. (1987) *Trends Biochem. Sci.* **12**, 301–304.

Hartley, R.W. (1968) *Biochemistry* **7**, 2401–2408.

Horovitz, A. (1987) *J. Mol. Biol* **196**, 733–735.

Horovitz, A., Serrano, L., Avron, B., Bycroft, M. & Fersht, A.R. (1990) *J. Mol. Biol.* in press.

Jeener, J., Meier, B.H., Bachmann, P. & Ernst, R.R. *J. Chem. Phys.* **71**, 4546-4553.

Kellis, J.T.,Jr., Nyberg, K., Šali, D. & Fersht, A.R. (1988) *Nature* **333**, 784–786.

Kellis, J.T.,Jr., Nyberg, K. & Fersht, A.R. (1989) *Biochemistry* **28**, 4914–4922.

Kunkel, T.A., Roberts, J.D. & Zakour, R.A. (1987) *Methods Enzymol.* **154**, 367–382.

Marqusee, S., Robbins, V.H. & Baldwin, R.L. (1987) *Proc. Natl. Acad. Sci. USA* **86**, 5286–5298.

Matouschek, A., Kellis, J.T., Jr., Serrano, L. & Fersht, A.R. (1989) *Nature* **340**, 124–126.

Matouschek, A., Kellis, J.T., Jr., Serrano, L., Bycroft, M. & Fersht, A.R. (1990) *Nature* **346** 440–446.

Mauguen, Y., Hartley, R.W., Dodson, E.J., Dodson, G.G.,Bricogne, G., Chothia, C. & Jack, A. (1982) *Nature* **297**, 162–164.

Paddon, C.J. & Hartley, R.W. (1987) *Gene* **53**, 11–19.

Perutz, M.F. & Raidt, H. (1975) *Nature* **255**, 256–258.

Russell, A.J., Thomas, P.G. & Fersht, A.R. (1987) *J. Mol. Biol.* **193**, 803–813.

Serrano, L. & Fersht, A.R. (1989) *Nature* **342**, 296–299.

Serrano, L., Horovitz, A., Avron, B. & Fersht, A.R. (1990) *Biochemistry*, in press.

Šali, D., Bycroft, M. & Fersht, A.R. (1988) *Nature* **335**, 740–743.

Warshel, A. & Russel, S.T. (1984) *Quart. Rev. Biophys.* **17**, 283–422.

Wüthrich, K. *NMR of Proteins and Nucleic Acids* (1986) Wiley, New York.

29

Analysis of Disulfide Bonds Within The Novel Apolipoprotein J Isolated From Human Plasma

Gregory J. Kelso, Terence L. Kirley and Judith A. K. Harmony

Department of Pharmacology and Cell Biophysics, University of Cincinnati College of Medicine, Cincinnati, OH 45267

I. INTRODUCTION

A new apolipoprotein designated apolipoprotein J has recently been purified from human plasma by immunoaffinity chromatography and RP-HPLC. ApoJ exists in plasma as a unique 70 kDa protein associated with subclasses of HDL that also contain apoAI and cholesteryl ester transfer activity (1). ApoJ is a glycoprotein consisting of two subunits termed apoJα (34-36 kDa) and apoJβ (36-39 kDa), both of which share a pI range of 4.9 - 5.4. Upon deglycosylation, the apparent molecular mass of apoJα is 24 kDa and apoJβ is 28 kDa, indicating an approximate 30% increase in mass due to glycosylation (2). The primary structure of apoJ has been deduced from the cDNA; apoJ is synthesized as a 427 amino acid polypeptide that is post-translationally cleaved between Arg-205 and Ser-206 (3). ApoJ appears to be the human analog of rat sulfated glycoprotein 2 (SGP-2), present in high concentrations in the testis. ApoJ and SGP-2 show 77% sequence homology (3). Walker et al. (4) have sequenced a human serum protein termed SP-40,40 that is

Techniques in Protein Chemistry II

100% homologous to apoJ. SP-40,40 was isolated from glomerular deposits in nephrotic kidneys; it appears to inhibit complement induced cytolysis (4).

The purpose of this study was to define the location of the disulfide bonds in apoJ. Five cyst(e)ines are clustered within a 30 amino acid stretch in each subunit. Utilizing the sulfhydryl-specific reagent 4-(aminosulfonyl)-7-fluoro-2,1,3-benzoxadiazole (ABD-F) which gives rise to a fluorescent adduct upon reaction with sulfhydryls, the oxidation state of the 10 cyst(e)ine residues was determined. A comparison of ABD-F alkylation of apoJ before and after reduction indicated the absence of free sulfhydryls in native apoJ. In addition, tryptic peptides of apoJ were isolated by RP-HPLC and subjected to reduction by tributylphosphine and alkylation by ABD-F. Those peptides containing disulfide bonds were identified by their fluorescent intensity and sequenced. To date, two interchain disulfide bonds have been identified.

II. MATERIALS AND METHODS

4-(aminosulfonyl)-7-fluoro-2,1,3-benzoxadiazole (ABD-F) was obtained from Wako Chemicals, Dallas, TX. Tributylphosphine (TBP) and dimethylsulfoxide were purchased from Aldrich. TPCK-treated trypsin, β-octylglucoside, cyanogen bromide, urea, and guanidine-HCl (GdnCl) were from Sigma Chemical Co. Trifluoroacetic acid (TFA) was obtained from Pierce. PNGase-F was obtained from Genzyme, Boston MA. Boric acid, EDTA and all other buffer components and solvents were obtained from Fisher.

ApoJ was purified from human plasma according to the method of deSilva et al. (2).

ApoJ (1-1.5 mg purified protein) was deglycosylated in 20mM Na_2HPO_4, pH 7.3, 10mM EDTA and 0.6% β-octylglucoside. PNGase-F was added at 12 U/ml and allowed to react at 37 ^{0}C for 24 hr.

ApoJ (1-1.5 mg deglycosylated protein) was subjected to cyanogen bromide cleavage in 6M GdnCl, 0.2M HCl at a ratio of 100:1 [CNBr]:[methionine]. The reaction proceeded for 18 hr at 22 ^{0}C, light protected and under nitrogen (5).

HPLC-purified CNBr fragments containing intact

disulfides were digested with TPCK-treated trypsin in 2M urea, 100mM NH_4HCO_3 and 2mM $CaCl_2$ at 37 ^0C for 20 hr, at a substrate to enzyme ratio of 20:1.

All reductions and alkylations were performed at various protein concentrations on aliquots adjusted to 0.7 ml. To each 0.7 ml fraction was added 0.1 ml of a freshly mixed solution containing 1.0M $NaBO_3$, pH 8.0, 8mM EDTA, 2mM TBP and 2mM ABD-F. The fractions were vortexed and allowed to react at 60 ^0C for 10 min. Relative fluorescence of treated fractions was determined, using a Perkin-Elmer 650-10S spectrofluorometer (excitation at 385 nm, emmision at 520 nm) (7).

HPLC separations were performed on Vydac C_{18} or C_4 reverse phase columns (4.6 x 250 mm) with 0.1% TFA/H_2O (A) and 0.1% TFA/acetonitrile (B). Linear gradients of increasing (B) were used at a flow rate of 0.8 ml/min to separate peptides. The eluant was monitored for absorbance at 220 nm and relative fluorescence (excitation at 385 nm, emmision at 520 nm).

Purified peptides were sequenced on an Applied Biosystems 470A gas-phase sequencer, and the phenylthiohydantoin amino acids were analyzed on a Waters Nova-Pak C_{18} reverse phase column. Amino acids were determined with the Waters PICO-TAG system. All amino acid analyses and protein sequences were determined by Cleris Gil of the Protein Chemistry Core Facility in the Department of Pharmacology and Cell Biophysics, University of Cincinnati.

III. RESULTS AND DISCUSSION

Post-translational modification of apoJ involves the addition of carbohydrate moieties to all 6 of the N-linked glycosylation sites (6). Each cys-rich region contains one asparagine which is glycosylated, contributing to the inaccessibility of this region to cleavage reagents. Hence, the first step was to deglycosylate the protein with PNGase-F which removes the complete carbohydrate moiety. Figure 1 shows a 10% SDS-PAGE gel of apoJ before and after deglycosylation. Non-reduced apoJ (lane 3) has an apparent molecular mass of 70 kDa before deglycosylation, and 50 kDa (lane 5) after treatment with PNGase-F; reduced apoJ (lane 4) is about 35 kDa before treatment and about 24 kDa

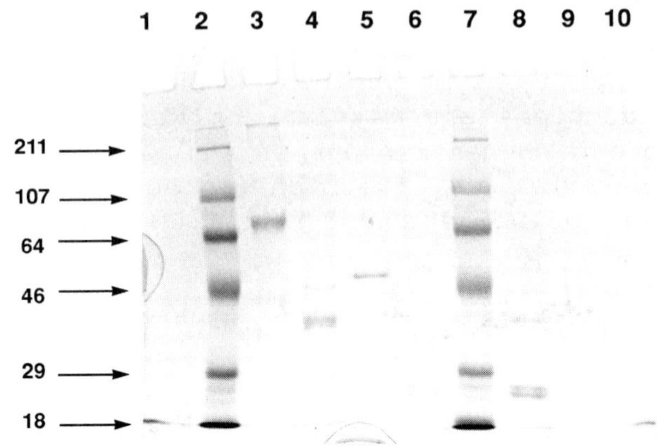

Fig. 1. SDS-PAGE (10%) analysis of apoJ before and after deglycosylation with PNGase-F. Lanes 2 & 7 are mol. wt. stds. Lane 3 is non-reduced apoJ before deglycosylation. Lane 4 is reduced apoJ before deglycosylation. Lane 5 is non-reduced apoJ after deglycosylation. Lanes 6 & 8 are reduced apoJ after deglycosylation.

(lanes 6 & 8) after treatment.

Following deglycosylation, the protein was cleaved with the disulfides intact. Cyanogen bromide was chosen as the initial cleavage reagent for two reasons. First, apoJ has trypsin cleavage sites between all of the cysteines except between Cys^{80} and Cys^{91} which contains a CNBr labile methionine. Also, CNBr treatment can act to 'crack' the protein open and allow easier access by trypsin for subsequent digestions. This cleavage reaction is typically performed in 70% formic acid, which has reducing properties and may promote disulfide interchange (5). Thus, formic acid was replaced with a solution of GdnCl which is not only an excellent protein solvent, but also does not affect the integrity of the disulfide bonds.

After CNBr treatment of deglycosylated apoJ, the peptide fragments were separated on a C_{18} reverse phase HPLC column as shown in Figure 2. Aliquots of each fraction were reduced with TBP and the cystine-containing fragments were alkylated with ABD-F. The aliquots were then assayed for relative fluorescence. Any fluorescence observed should originate from

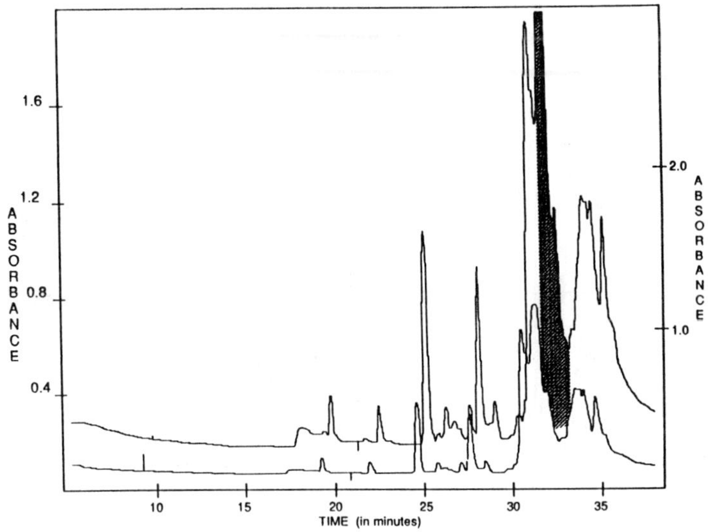

Fig. 2. HPLC separation of CNBr digested apoJ. After deglycosylation, 1.5 mg of apoJ was cleaved with CNBr and the resultant fragments isolated by injection onto a C_{18} reverse phase column (shaded peak indicates one of the disulfide-containing fragments which was further purified, and the data reported in Figures 3 and 4). Upper trace: 220 nm, 2.0 AUFS. Lower trace: 220 nm, 3.0 AUFS.

disulfides present in the CNBr fragments, thus the untreated remainder of the most fluorescent fractions (e.g. peak eluting between 32 and 33 minutes and shaded in Figure 2) were then subjected to trypsin digestion.

The peptides generated from the tryptic digest were then separated on a C_{18} reverse phase HPLC column. Figure 3 shows the HPLC chromatogram of a tryptic digest of one such cystine-containing CNBr peak (shaded peak in Figure 2 eluting at 32 min). The individual C_{18} tryptic peaks were then reduced with TBP, and the cystine-containing peptides were alkylated with ABD-F. The fractions were assayed for relative fluorescence and the most fluorescent fractions were re-chromatographed on a C_{18} column. Figure 4 shows one such chromatogram derived from a single tryptic peak (shaded peak in Figure 3 eluting at 11 minutes). Two major fluorescent peaks that coincided with two 220 nm absorbing peaks (shaded peaks in Figure 4 eluting at 18.5 and 19 min) were detected. This is the expected result if the fraction (shaded peak in Figure 3) contained one disulfide-bonded peptide since each half

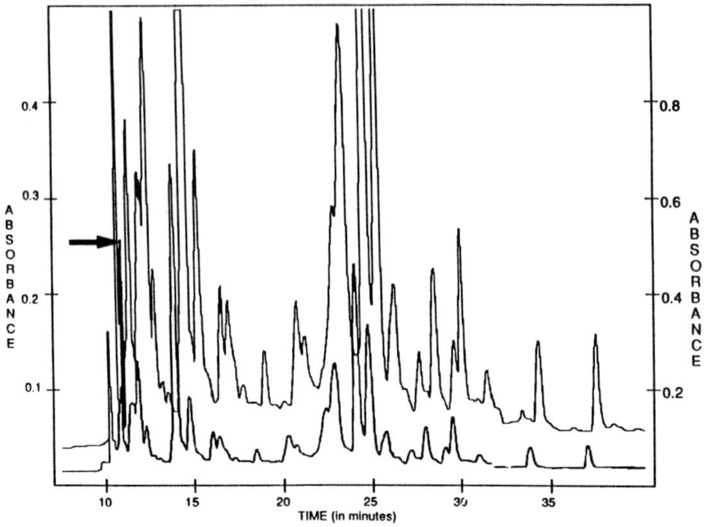

Fig. 3. HPLC separation of trypsin digested CNBr fragment (shaded peak in Figure 2). After CNBr cleavage, the ABD-F reactive fragments (determined by analysis of aliquots from CNBr fragments that were treated with TBP/ABD-F) were digested with trypsin and isolated by injection onto a C_{18} reverse phase column. Arrow points to disulfide-containing peak as determined by treatment with TBP/ABD-F. This peak was further purified after treatment with TBP/ABD-F and the results obtained are shown in Figure 4. Upper trace: 220 nm, 0.5 AUFS. Lower trace: 220 nm, 1.0 AUFS.

of the TBP reduced disulfide bond would be fluorescently labelled. The two non-fluorescent UV absorbing peaks eluting at 17-18 min in Figure 4 are peptides contained in the 11 min peak of Figure 3 which do not contain disulfide bonds. The off-scale UV peak eluting at 23 min in Figure 4 is due to excess reagent.

Finally, the fluorescent peptides were repurified on a C_4 reverse phase HPLC column and sequenced after an amino acid composition was determined using 20% of each peptide. Peptide one sequence was determined to be Val(345)-ABD-Cys(133)-Arg(148)-End while peptide two sequence was determined to be Thr(10)-Val(41)-ABD-Cys(21)-Arg(18)-End. In a similar experiment, an additional disulfide bond was demonstrated by sequence analysis of ABD-labeled peptides, linking Cys^{80} to Cys^{291} (data not shown). Thus, Figure 5 shows our current understanding of the disulfide bonding pattern in apoJ. Cys^{80} is disulfide-linked to Cys^{291} and Cys^{107}

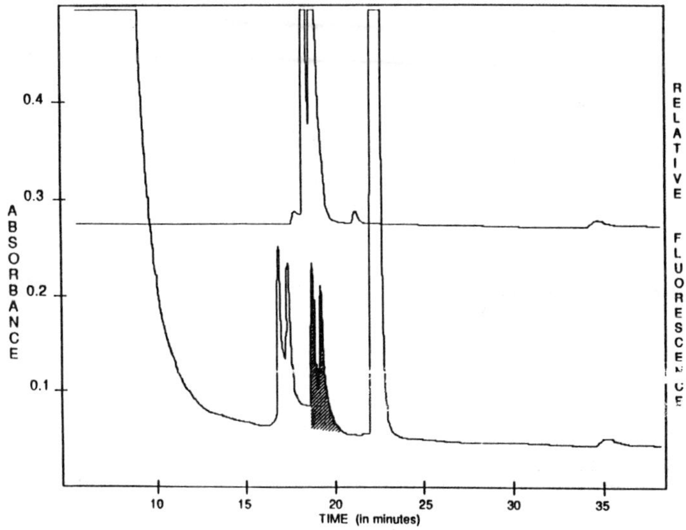

Fig. 4. HPLC purification of a single tryptic fraction after treatment with TBP/ABD-F. A single fraction (shaded peak in Figure 3) after reduction and alkylation (with TBP/ABD-F) was isolated by injection onto a C_{18} reverse phase column. Shaded 220 nm peaks correspond to the two major fluorescent peaks. Upper trace: Relative fluorescence. Lower trace: 220 nm, 0.5 AUFS.

is disulfide-linked to Cys^{263}.

This study describes a strategy which allows assignment of residues involved in disulfide bonds. The method involves separation of disulfide-linked peptides and sequencing of these peptides after reduction and labeling of the generated cysteines (7). A relatively new fluorescent reagent, ABD-F, was used which has proved to be superior to other commonly used alkylating reagents (8). Its advantages include rapid

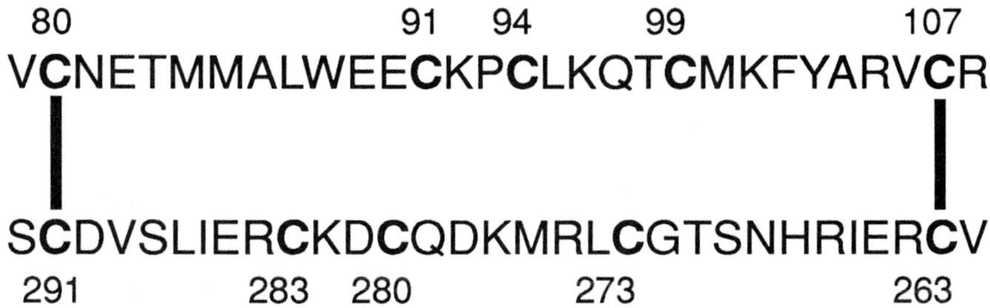

Fig. 5. Disulfide bonding pattern of apoJ determined thus far.

and selective labeling of cyst(e)ine residues that is quantifiable by absorbance, fluorescence and amino acid analysis, as well as demonstrated stability during sequence analysis (9). Future directions include locating the remaining disulfides in apoJ and determining the specific role of this cysteine-rich region in the function of apoJ.

IV. ACKNOWLEDGEMENT

This project was supported by NIH Grant HL30999.

IV. REFERENCES

1. de Silva, H., Stuart, W., Duvic, C., Wetterau, J., Ray, M., Ferguson, D., Albers, H., Smith, W., and Harmony, J.A. (1990). J. Biol. Chem. **265**, 13240-13247.
2. de Silva, H., Stuart, W., Park, Y., Mao, S., Gil, C., Wetterau, J., Busch, S., and Harmony, J.A. (1990). J. Biol. Chem. (in press).
3. de Silva, H., Harmony, J.A., Stuart, W., Gil, C.,and Robbins, J. (1990). Biochemistry **29**, 5380-5389.
4. Kirszbaum, L., Sharpe, J., Murphy, B., d'Apice, A., Classon, B., Hudson, P., and Walker, I.D. (1989). EMBO J. **8**, 711-718.
5. Villa, S., De Fazio, G., and Canosi, U. (1989). Anal. Biochem. **177**, 161-164.
6. Kirszbaum, L., Murphy, B., Walker, I.D. (1990). XIIIth Int. Complement Workshop. 354.
7. Kirley, T. (1989). J. Biol. Chem. **264**, 7185-7192.
8. Toyo'oka, T., and Imai, K. (1989). Anal. Chem. **56**, 2461-2464.
9. Kirley, T. (1989). Anal. Biochem. **180**, 231-236.

30

SURFACE RECOGNITION DOMAINS OF GLUTAMINE SYNTHETASE FOR ADENYLYLTRANSFERASE

Beth M. Cader
Jens Hemmingsen
Joseph J. Villafranca

Department of Chemistry, The Pennsylvania State University,
University Park, PA 16802

I. INTRODUCTION

Bacterial glutamine synthetase (GS) is a key enzyme in nitrogen assimilation that is subject to a variety of regulatory mechanisms (1). Expression of *gln A*, the gene that encodes GS in enteric bacteria is controlled by the Ntr regulatory system; the Ntr regulatory proteins also modulate expression of other genes involved in nitrogen utilization. In enteric bacteria, GS is also regulated by covalent modification of a specific tyrosyl residue that results in enzyme inactivation (2). The reaction in *E. coli* involves adenylylation of Tyr-397 by ATP (Eq. 1) catalyzed by ATP:GS adenylyltransferase (ATase). GS is a dodecamer (M_r = 12 x 51, 814) and all twelve subunits can be adenylylated.

$$\text{GS} + \text{nATP} \xrightarrow{\text{ATase, } Mg^{++}\text{, } P_{II} \text{ or glu}} \text{GS-(AMP)n} + \text{nPP}_i \quad (1)$$
$$n = 1 - 12$$

Techniques in Protein Chemistry II

Comparison of GS in a variety of bacteria reveals that GS is not regulated by adenylylation in all strains (3). For example, in gram-positive bacteria there is no evidence for adenylylation in *Bacillus subtilis*, but *Steptomyces coelicolor* has an adenylylating system; GS from the cyanobacterium *Anabaena 7120* is not adenylylated *in vivo*. The amino acid sequence (derived from the DNA sequence) has been reported for GS from many bacteria; the surprising finding is that a tyrosine corresponding to position 397 in *E. coli* (4) is found in many sequences (5) whether the bacteria contain ATase or not.

Several interesting experiments have been conducted to study whether GS found in organisms that do not have ATase can be adenylylated in *E. coli*. Thus, when *Anabaena 7120* GS was cloned and expressed in *E. coli*, the enzyme was not adenylylated (5). Also of interest was a report that cloned *S. coelicolor* GS was not adenylylated in *E. coli* although it is adenylylated in its natural host (3). The purpose of the present study was to investigate the specificity of *E. coli* ATase with 1) *in vitro* studies of isolated *Anabaena 7120* and *E. coli* GS and 2) computer modeling studies to identify possible ATase binding domains around of the region of Tyr-397 based on the reported crystal structure of *S. typhimurium* GS (6).

II. MATERIALS AND METHODS

The availability of the crystal structure coordinates of *S. typhimurium* GS (6) allows for an 'up close' examination of the adenylylation region centered at Tyr-397. There is 98% homology between *E. coli* and *S. typhimurium* GS and both are regulated by adenylylation *in vivo*, justifying the interchangability of the crystallographic data for our purposes. Molecular modelling in conjunction with potential energy minimization techniques was used to evaluate the region around Tyr-397 in the wild type enzyme, as well as various mutant enzymes, to potentially identify possible amino acid recognition sites for ATase. The QUANTA/CHARMm software package licensed by Polygen Corporation was used for this study and implemented on a Silicon Graphics computer system using the Iris 4D/240GTX workstation. Minimizations of wild type and mutant enzymes were identically performed using the Adopted-Basis Newton-Raphson algorithm (7). The processes were halted when the percent change of the total potential energy reached 0.1 and the root mean square atomic shift dropped below 1.0. Covalently-bound hydrogens, constraining forces, and waters of hydration were introduced prior to minimization, the details of which are described later.

The rates of adenylylation of *E. coli* GS (0.5 equivalents AMP per dodecamer, initially) and *Anabaena 7120* GS were measured by following the radioactive incorporation of ^{32}P-AMP. Initial reaction mixtures contained 20 mM Tris (pH 8.0), 30 mM $MgCl_2$, 10 mM glutamine, 5 mM α-^{32}P-ATP, GS (1 mg/ml) and ATase at 37°C. Assays for bound radioactivity were performed using the method of Krebs (8).

III. RESULTS AND DISCUSSION

Adenylylation of *E. coli* GS is a site-specific reaction occurring only at Tyr-397 (2). Our interest in this reaction centers on the specificity of the adenylylating enzyme, ATase, not only for Tyr-397 but for the region around Tyr-397. One reason for the present study was the report of Fisher et al (5) that cloned *Anabaena 7120* GS was not adenylylated *in vivo* in *E. coli*. This result is surprising since a tyrosine residue is in an analogous location in the sequence of *Anabaena 7120* GS. We decided to compare purified GS from *E. coli* and *Anabaena 7120* in an *in vitro* adenylylation experiment to test whether the *in vivo* result truly reflected 'zero' adenylylation of *Anabaena 7120* GS by *E. coli* ATase.

Figure 1 shows the result of radiolabel incorporation of α-^{32}P-ATP into GS in the presence of ATase. *E. coli* GS is fully adenylylated within minutes under the experimental conditions used. Importantly, GS from *Anabaena 7120* <u>does</u> get adenylylated, but at a much slower rate than *E. coli* GS. This result means that the recognition site on *Anabaena 7120* GS differs from that on *E. coli* GS.

Table I presents a comparison of the sequences around Tyr-397 in several bacterial strains of GS. There are several obvious differences in primary sequence between *Anabaena 7120* and *E. coli* GS, notably at positions 396, 398, and 400. This comparison suggested that molecular modelling and computational

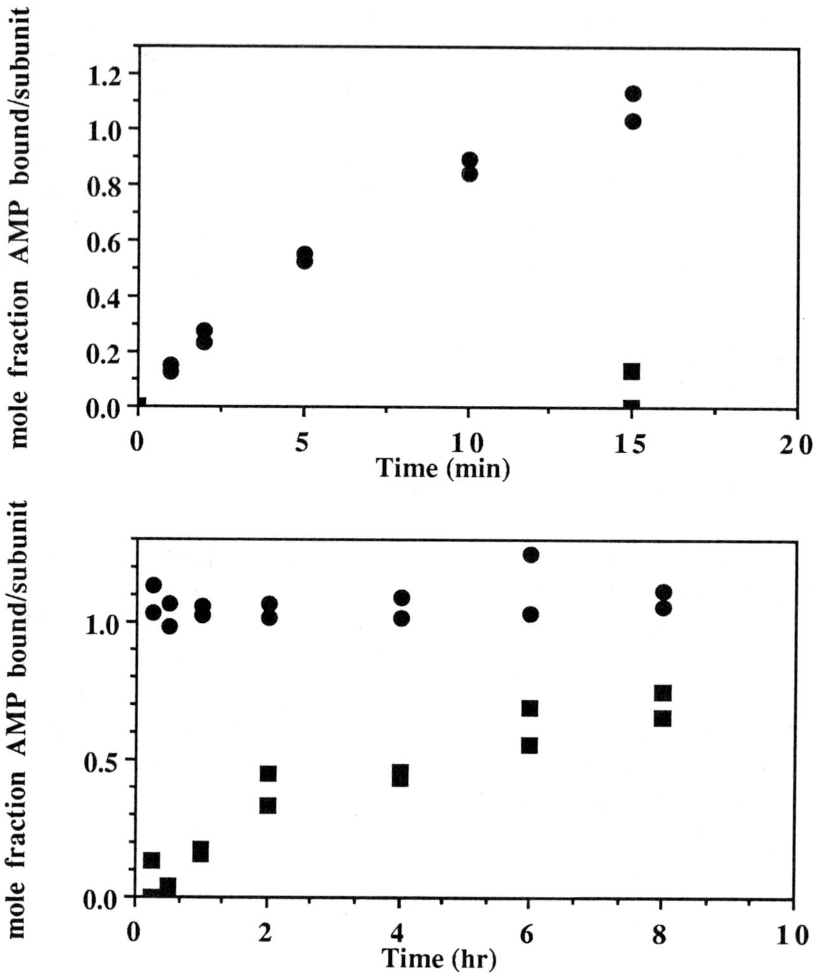

Figure 1. Time dependent incorporation of ^{32}P-AMP into GS. Samples of *E.coli* (●) and *Anabaena 7120* (■) GS were subjected to adenylylation reaction conditions. Aliquots were removed at various times and protein bound radioactivity was determined (See Methods). Upper graph shows reaction progress in first 15 minutes and lower graph shows progress in eight hours. At 24 hours mole fractions AMP bound/ subunit were 1.0 and 0.8 for *E.coli* and *Anabaena 7120* GS respectively.

TABLE I. AMINO ACID SEQUENCES AROUND TYROSINE-397 OF VARIOUS GLUTAMINE SYNTHETASES AND *IN VIVO* ADENYLYLATION INFORMATION

GS SOURCE	RESIDUES 391 TO 411*	ADENYLYLATED IN...	
		E.coli?	OTHER?
Ec	A-M-D-K-N-L-**Y**-D-L-**P-P**-E-E-A-K-E-I-P-Q-V-A	yes	yes(host)
St	P-M-D-K-N-L-**Y**-D-L-**P-P**-E-E-A-K-E-I-P-Q-V-A	yes	yes(host)
Ab	A-M-D-K-N-L-**Y**-D-L-**P-A**-E-E-L-A-K-V-P-T-V-C	yes	n/a
Tf	A-M-D-K-N-L-**Y**-D-L-**P-A**-E-E-Q-A-N-I-P-G-V-A	yes	n/a
Sc	P-I-D-K-D-L-**Y**-E-L-**A-P**-E-E-H-A-N-V-A-Q-V-P	no	yes(host)
Rl	A-M-D-K-D-L-**Y**-D-L-**P-P**-K-E-L-K-K-I-P-T-V-C	n/a	yes(Kp)
An	P-L-D-K-N-I-**Y**-E-L-**S-P**-E-E-L-A-K-V-P-S-T-P	no	no
Bs	P-I-D-R-N-I-**Y**-V-M-**S-K**-E-E-R-M-E-N-G-I-V-D	no	no

Abbreviations used in this table: Ec=*Escherichia coli*; St=*Salmonella typhimurium*; Ab=*Azospirillum brasilense*; Tf=*Thiobacillus ferrooxidans*; Sc=*Streptomyces coelicolor*; Rl=*Rhizobium leguminosarum*; An=*Anabaena 7120*; Bs=*Bacillus subtilis*; Kp=*Klebsiella pneumoniae*; n/a=no available information. Sequence data obtained from reference 3.
*Based on the *E. coli* numbering system.

minimization of the Tyr-397 area might provide reasons for the kinetic differences observed in Figure 1 and the data reported by Fisher et al (5).

The initial step in this computational analysis involved determining the portion of glutamine synthetase to be incorporated into the potential energy minimizations of the wild type and computer-generated mutant enzymes, the focus being on the covalently-modified Tyr-397. The decision on how much of the protein to include was based on a two factors. Firstly, the aforementioned differences in *Anabaena 7120* and *E. coli* GS in close proximity to Tyr-397; viz., Ile-396, Glu-398, and Ser-400 (Table I). Secondly, the crystallographic data show the presence of a highly solvated loop protruding from the periphery of dodecameric GS. In fact, two of these loops (each from diagonally-oriented subunits) appear to cap a solvent-containing cavity formed by internal portions of the protein. Tyr-397 is centrally located along this loop and points out into solution. With this information at hand, a residue range was selected that included the previously discussed positions and any nearby amino acids of possible influence; the 21 residues making up the most exposed portion of the loop (aa 391-411) met these criteria.

There are a number of prominent features of this loop, the most striking of which is its isolation from the rest of the protein. Thus, the loop and the Tyr-397 residue have potentially high accessibility to the active site of ATase. In addition to Tyr-397 pointing out into solution, Lys-394 and Asp-398, Glu-402,403 are also exposed, providing a positively-charged and negatively-charged 'patch', respectively, on either side of the covalently-modified tyrosine. These patches may be involved in the proper anchoring of ATase prior to adenyl transfer. Pro-400 is also outwardly directed and located adjacent to another proline (401; inwardly directed) which initiates a seven-residue α-helix (aa 401-407). Pro-400 is of particular interest, not only due to its substitution by serine in *Anabaena 7120* GS, but as a result of its absence from other enzymes which likewise are not adenylylated in *E. coli* (Table I). Worthy of mention is that the presence of a proline at position 401 is not a requirement for adenylylation. Figure 2 is a ribbon diagram illustrating the protruding loop of *E. coli* GS with Lys-394, Tyr-397, and Pro-400 highlighted; relevant inter-residue distances are listed in Table II.

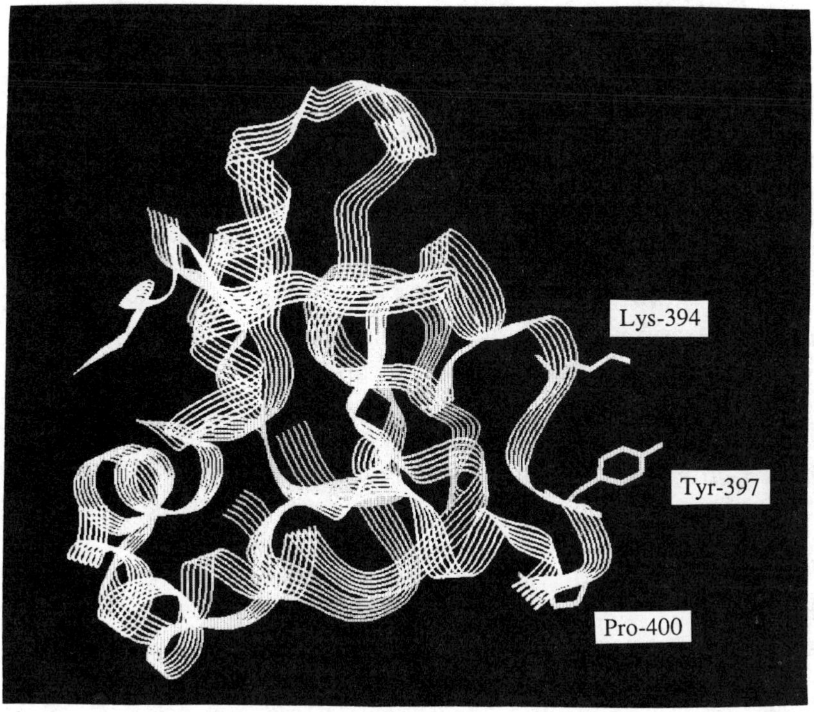

Figure 2. Ribbon diagram of a partial subunit of *E. coli* (*S. typhimurium*) glutamine synthetase with Lys-394, Tyr-397 and Pro-400 displayed on the protruding loop.

TABLE II. INTER-RESIDUE DISTANCES IN THE TYR-397 REGION

TYR-397 (CZ)	to	PRO-400 (CG)	10.0 Å
TYR-397 (CZ)	to	PRO-401 (CG)	14.3 Å
TYR-397 (OH)	to	LYS-394 (NZ)	9.3 Å
TYR-397 (OH)	to	ASP-398 (CG)	7.0 Å
TYR-397 (OH)	to	GLU-402 (CD)	11.1 Å

Computer-generated mutations of the *E. coli* GS loop were prepared in accordance with the unadenylylated enzyme substitutions in that region, in an attempt to gain insight into the location of recognition domains for ATase, as well as to aid in the design of an efficient protocol for laboratory mutagenesis. In each case prior to minimization, the appropriate hydrogens were added to the loop fragment and it was immersed in enough water to sufficiently hydrate the residues; mutations were made before hydrogen and water additions. Varied constraint forces were applied to the last three residues at each end of the loop, the end residues being constrained to the greatest extent. This was to reflect the loop's initial penetration of the interior of the protein.

The first set of mutations to *E. coli* GS were centered around the Pro-400,401 sites. In order to represent the three enzymes known to resist adenylylation in *E. coli* (*S. coelicolor*, *Anabaena 7120*, and *B. subtilis*), Pro-400 was replaced by serine and alanine, respectively. An additional alteration, P401A, was made to reflect *T. ferrooxidans* GS which is adenylylated in *E. coli.*

Comparisons of the peptide backbone in the minimized structures of P400S, P400A, and P401A mutants with minimized wild type enzyme revealed similar orientations throughout; perturbations of less than 1 Å were observed. In addition, no appreciable changes in the Lys-394, Tyr-397, and wild type proline positions were evident. Indeed, the most notable feature did not involve atomic shifts, but rather the mere substitution of Pro-400 by serine or alanine (Figure 3). One could argue that the introduction of the hydroxymethyl moiety of serine disrupts the ATase binding pocket; however, its absence in the Ala-400 mutant provides the same result: no adenyl transfer. Therefore, it is highly likely that the outer proline itself (Pro-400) is an important recognition site for ATase, perhaps necessary for proper alignment of Tyr-397 for covalent adenylylation.

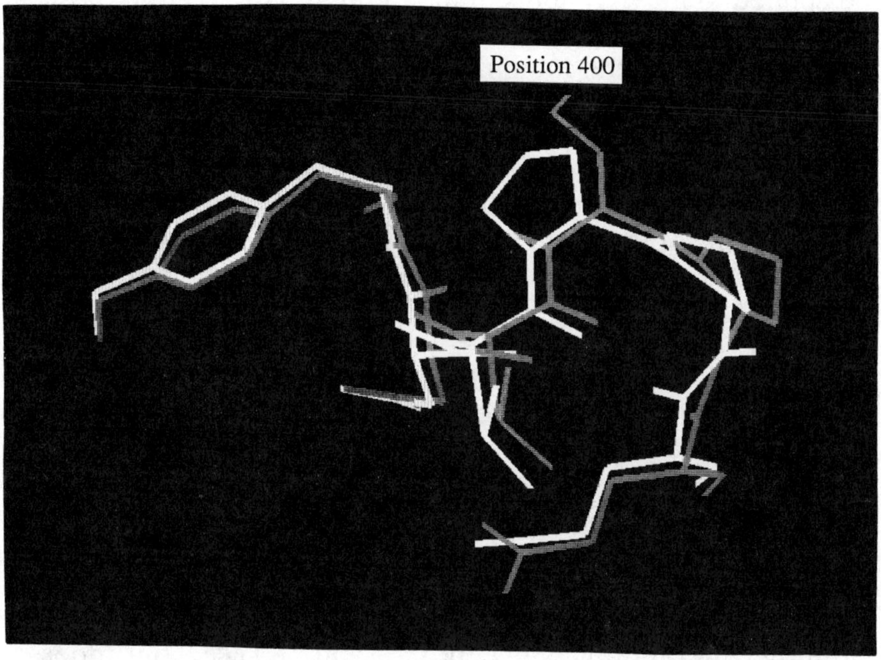

Figure 3. Overlay of residues 397 to 402 in wild type *E. coli* GS (white) and the computer-generated P400S mutant (gray).

Prior to reaching any absolute conclusions, consideration must be given to the possible limitations of the minimization protocol used in this study. In each case, the amino acid replacement was made following the dihedral angles of the wild type residue and then analyzed. An improved method, which will be applied to these systems in the future, involves an initial search for the lowest energy conformation(s) at the mutation site, followed by a general minimization of the entire structure (9). It is not clear whether this procedure is necessary for a system of this size (21 residues) or location. Gross changes may not be anticipated due to the apparent conformational freedom of this loop, a result of a lack of salt-bridging and hydrogen bonding with other regions of the protein.

IV. CONCLUSIONS

The data in this paper show that ATase from *E. coli* derives a substantial amount of its specificity for GS from the amino acids in the immediate region around Tyr-397, the amino acid that is adenylylated. Pro-400 seems to be a crucial residue for proper 'active site' recognition of the Tyr-397 by ATase. GS from *Anabaena 7120* was shown to be a 'slow' substrate for *E. coli* ATase and position 400 has a serine in place of proline. Molecular modelling studies suggest that Pro-400 provides a recognition and binding interaction between GS and ATase, but does not substantially alter the spatial disposition of the group undergoing modification; viz., Tyr-397.

REFERENCES

1) Reitzer, C.J. and Magasanik, B. (1987) in *Escherichia coli and Salmonella typhimurium* (Neidhart, F.C., ed.) Vol. 1, 302, American Society for Microbiology, Washington, D.C.

2) Stadtman, E.R. and Ginsburg, A. (1974) in *The Enzymes* (Boyer, P., ed.) 10, 755-807.

3) Wray, L.V., Jr., and Fisher, S.H. (1988) Gene 71, 247-256.

4) Colombo, G. and Villafranca, J.J. (1986) J. Biol. Chem. 261, 10587-10591.

5) Fisher, R., Tuli, R. and Haselkorn, R. (1981) Proc. Natl. Acad. Sci. USA 78, 3393-3397.

6) Yamishita, M.M., Almassy, R.J., Janson, C.A., Cascio, D. and Eisenberg, D. (1989) J. Biol. Chem. 264, 17681-17690.

7) Brooks, B., Bruccoleri, R. E., Olafson, B. D., States, D. J., Swaminathan, S. and Karplus, M. (1983) J. Comp. Chem. 4, 187-217.

8) Reimann, E. M., Walsh, D. A. and Krebs, E. G. (1971) J. Biol. Chem. 246, 1986-1995.

9) Shih, H. H.-L., Brady, J. and Karplus, M. (1985) Proc. Natl. Acad. Sci., USA 82, 1697-1700.

31

Structural Analysis of RNase T1 Native and Site-Directed Mutants by Capillary Electrophoresis

John E. Wiktorowicz
Kenneth J. Wilson
Applied Biosystems Inc.
San Jose, CA 95134 and

Bret A. Shirley
Biochemistry Department
Texas A & M University
College Station, TX 77843

I. Introduction

Proteins exhibit a hierarchy of complex structures which derive from their amino acid sequences. Comprehension of the interplay among these higher order structures permits an understanding of the relationship between protein structure and function. An appreciation of this relationship is the first step in controlling function by creating or manipulating structure, and predicting structure from function.

In addition to the many tools available for the analysis of protein structure, capillary electrophoresis (CE) contributes speed, accuracy, and high resolving power. Current technology allows a variety of separation motifs which may be tailored to the separation and analysis of specific proteins. The simplest format in principle, free solution capillary electrophoresis (FSCE), accomplishes separations under high field strengths without the use of anticonvective media. Because analyses are performed in aqueous media, separation conditions which maintain native structure can be utilized. This provides the opportunity to relate electrokinetic mobility with native protein structure using carefully selected model

Techniques in Protein Chemistry II

systems. Thus, beyond the ability to simply separate closely migrating proteins, CE is capable of generating data from which structural information can be extracted.

Aspergillus oryzae ribonuclease T1 (E.C. 3.1.27.3) is an acidic protein (pI=2.95) of about 11,000 daltons (1). Its three dimensional structure has been deduced (2, 3) and expression of the cloned gene has been obtained in *Escherichia coli* (4). In addition, several mutants have been created via site-directed mutagenesis (5). Kinetic analyses of the mutants as well as thermal denaturation data have been published (6), making the RNase T1's an excellent model system with which to study the relationship between electrokinetic behavior and protein higher order structure.

II. Materials and Methods

All chemicals were reagent grade. Ribonuclease T1 and site-directed mutants were purified as described earlier (7).

CE was performed on an Applied Biosystems Model 270A Capillary Electrophoresis System in fused silica capillaries. Specific separation conditions are enumerated in the figure legends.

III. Results and Discussion

The substitutions in each of the different isoenzymes are shown in Figure 1.

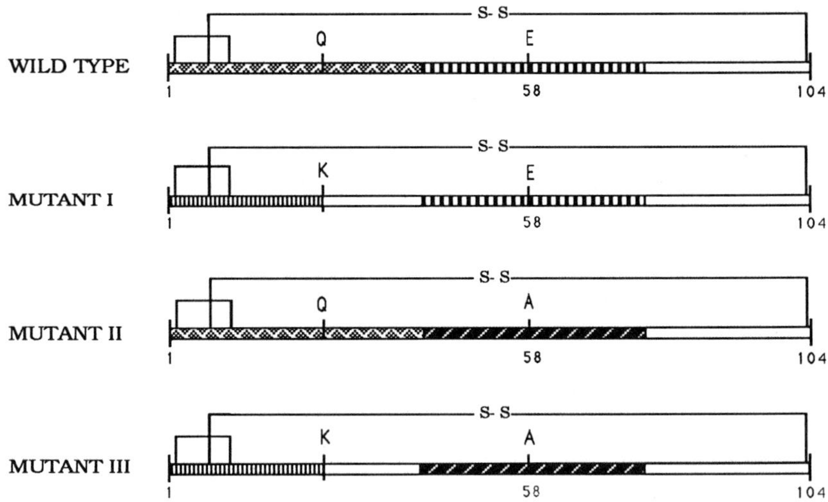

Figure 1. Wild-type and mutant forms of RNase T1. Substitutions at positions 25 and 58, as well as disulfide bridges are indicated. Similarly shaded fragments reflect identical tryptic peptides.

Each isoform contains two disulfide bridges which play critical roles in maintaining the native structure of the proteins (8, 9). In all cases, analyses were performed in the absence of disulfide reducing agents.

Since separations in FSCE are exquisitely sensitive to the effective charge of the proteins to be analyzed, it is illustrative to predict the relative electrokinetic behavior of the isoforms from their primary structure. Accordingly, the substitution of lysine for glutamine at position 25 in mutant I alters the charge:mass ratio of the wild-type (WT) enzyme by +1. Likewise, the alanine substitution in mutant II changes the charge:mass ratio by +1 through the elimination of the glutamic acid residue at position 58. Mutant III, containing both substitutions, exhibits a net charge:mass ratio difference from WT of +2. The behavior of the tryptic peptides of each isoform upon CE analysis was consistent with the nature of the respective substitution (data not shown).

Based on the amino acid sequences of the isoforms, their separation should follow the pattern shown schematically in Figure 2. Since all of the proteins are negatively charged at pH 7, their migration towards the anode (positive electrode) is countered by the overwhelming endosmotic flow. Thus separations exhibit a hydrodynamic component unique to FSCE. The WT enzyme should

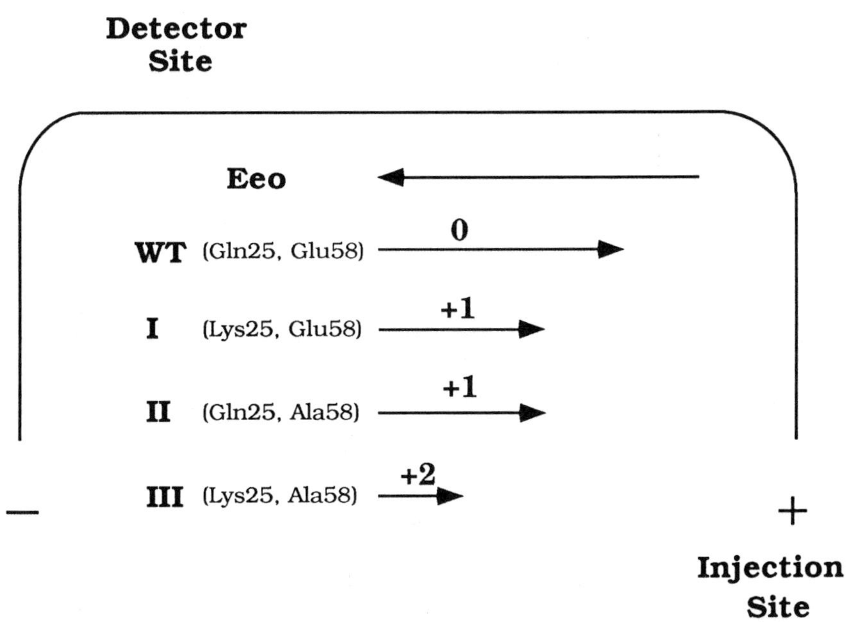

Figure 2. Prediction of RNase T1 wild-type and mutant isoforms separation.

be the most successful in countering the endosmotic flow due to its charge, and therefore appear later in the electropherogram than any of the isoforms. This peak should be followed by the single mutants (I and II), which should migrate with identical mobilities. Mutant III, containing a charge difference from WT of +2 is least successful in migrating against the endosmotic flow and appears first in the electropherogram. The actual migration pattern is shown in Figure 3.

Despite equal charge:mass ratios predicted from their amino acid sequences (Figure 1), mutants I and II are completely resolved in their native conformations by CE. This behavior suggests that the substituted residues in the isoforms exist in different ionic environments, provided that the overall tertiary structure of the isoforms is not altered by the mutations. That the gross tertiary structure remains unchanged in the mutants is supported by enzyme kinetic data which show that the K_m's of mutants I, II, and III are undiminished from the wild-type enzyme (6). The Ala^{58} substitutions in II and III affect only catalytic efficiency, resulting in greatly decreased k_{cat}'s due to the substitution of Glu^{58}, one of the catalytic residues.

Structural studies by X-ray crystallography (3) indicate that Gln^{25} in the wild-type enzyme exhibits near 70% solvent accessibility, whereas Glu^{58} exhibits less than 5%. Since these mutations induce little if any changes in the tertiary structure of the isoenzymes, the solvation of Lys^{25} in mutant I is greater than the solvation of Ala^{58} in mutant II (70% vs 5%, respectively). Thus, while the charge difference between mutant I and WT at residue 25 is closer to +1, the Ala^{58} substitution in mutant II results in a charge difference from WT significantly less than +1. This analysis predicts that mutant I (Lys^{25}) will be less acidic than mutant II (Ala^{58}) which will have a mobility closer to WT. Mutant III exhibits all of these effects, consistent with its structure.

As summarized in Table 1, the differences between the mobilities at 30° of WT and mutant II (0.01) are nearly the same as between I and III (0.02); the only difference between WT and II, and I and III is the Ala^{58} substitution. Likewise the mobility differences of WT and mutant I (0.21), and II and III (0.22) are similar, since the only difference in sequence between the pairs is the Lys^{25} substitution.

In order to further investigate the dependence of mobility on tertiary structure of the isoforms, electrophoresis was performed at denaturing temperatures (Figure 4). Protein species migrate as sharp peaks when: (1) all molecules exhibit homogeneous structure - and therefore identical mobilities; (2) when diffusion is minimal, and (3) when migration through the capillary is unimpeded by capillary surface interactions (10). During the 7 - 10 minute separation time (Figure 3), diffusion of the molecules is minimal, even at higher temperatures. At pH 7.0, the acidic side chains on the surface of the molecule are largely unprotonated, imparting a net negative charge to the molecular surface. Molecules are

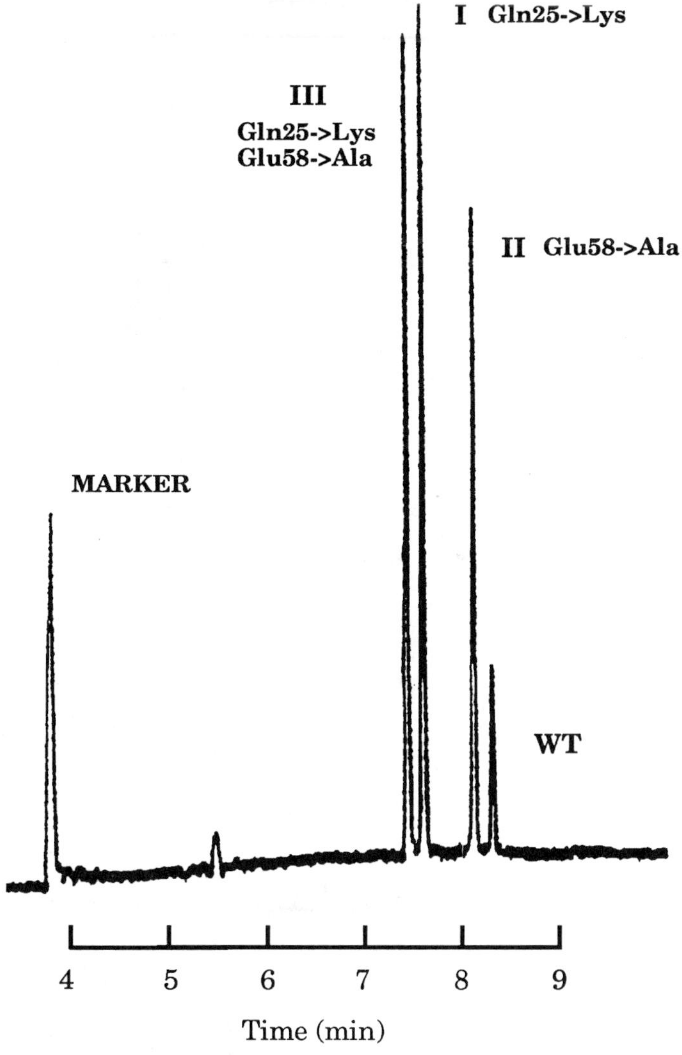

Figure 3. Separation of RNase T1 isoforms by CE. The marker peak represents a 1 sec. vacuum injection of a 1:1000 dilution of mesityl oxide. In addition to comparison of mobilities from individual runs, each RNase T1 isoenzyme was added to a mixture of the others at different concentrations (20 to 60 μg/ml) to aid in identification. Field, 430 V/cm.; current, 32 μA; buffer, 25 mM sodium phosphate (pH 7.0); capillary length, 70 cm. (68 cm. to detector); i.d., 50 μm; injection, vacuum, 0.5 sec.; temp., 30°C; detection 200 nm.

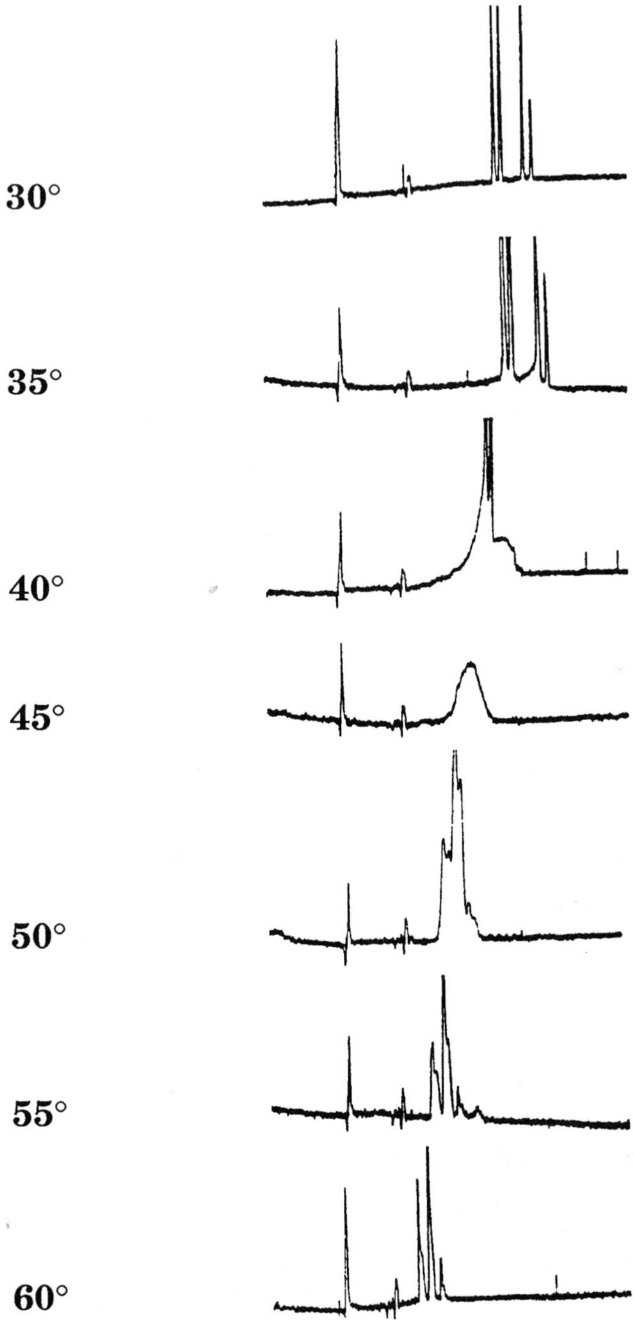

Figure 4. Temperature analysis of RNase T1 isoforms. Separations were performed in 20 mM TES / 5 mM NaCl (pH 7.0). Other run conditions were as described in Figure 3.

Table 1. RNase T1 Mobilities[a]

Temperature	*W T*	*I*	*II*	*III*
30°	-2.57	-2.36	-2.56	-2.34
35°	-2.65	-2.47	-2.71	-2.49
40°	-2.81	-2.60	-2.84	-2.62
45°	-2.92	-2.70	-2.76	-2.81
50°	-2.94	-2.75	-2.66	-2.46
55°	-2.88	-2.69	-2.66	-2.46
60°	-2.97	-2.75	-2.75	-2.48

[a]Mobilities (x 10^4 cm^2 / V - sec) were measured using mesityl oxide as an endosmotic flow marker. The analyses were performed in 20 mM TES (N-Tris(hydroxymethyl)methyl-2-aminoethanesulfonic acid) / 5 mM NaCl (pH 7.0).

therefore repelled from the negatively charged silica surface, precluding any coulombic interaction. Thus, peak broadening at higher temperatures is not primarily due to artifactual causes relating to surface interactions and diffusion, but, most likely, to localized disruption of hydrogen bonds in any given molecule, resulting in altered mobilities. Even though these changes may be transitory, their effect on mobility will be cumulative, with a more diffuse peak as the net result. This effect will be greatest when the protein is at its melting temperature, T_m. Once this transition temperature is exceeded, the molecules acquire increasingly homogeneous character and mobility; i.e., as temperature increases past the melting transition, electrophoretic peak sharpness increases until it reaches its maximum value when the protein is fully denatured.

If our contention that the isoform mobility differences are due to the solvation of the altered residues is correct, then disruption of the tertiary structure of the proteins should cause them to migrate strictly according to their charge:mass ratios. More specifically, mutants I and II should exhibit identical mobilities in their fully denatured state (since their mass and charge are the same). Moreover, the mobility difference between WT and I should be the same as the mobility difference between I and III. Both predictions are borne out by Table 1 and illustrated graphically in Figure 5.

IV. Conclusions

The analyses presented are examples of how protein structural information may be extracted from free solution electrokinetic behavior. In this example, without prior knowledge of the three-dimensional structure, but knowing primary structure, amino acid substitutions, and kinetic effects, it would be possible to predict the

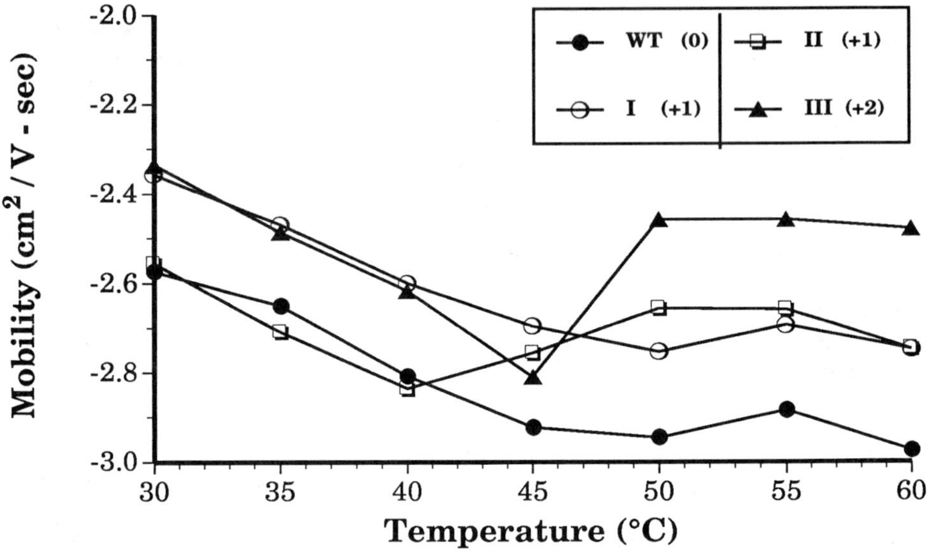

Figure 5. RNase T1 mobilities vs. temperature. From Table 1.

solvation differences of altered residues of mutant proteins. With further study, quantification of the solvation differences may also be possible. In general, continued study of well-characterized protein systems will expand our understanding of the general rules which relate protein electrokinetic behavior in free solution (at non-denaturing pH's) with higher order structure.

ACKNOWLEDGEMENTS

The authors would like to thank C. Nick Pace of the Biochemistry Department at Texas A & M University for helpful discussions.

REFERENCES

1. Takahashi, K. (1965) J. Biol. Chem. **240**, 4117-4119.
2. Heinemann, U., and Saenger, W. (1982) Nature **299**, 47-31.
3. Sugio, S., Amisaki, T., Ohishi, H., and Tomita, K. (1988) J. Biochem. (Tokyo) **103**, 554-366.
4. Quass, R., McKeown, Y., Stanssens, P., Frank, R., Blocker, H., and Hahn, U. (1988) Eur. J. Biochem. **173**, 617-622.
5. Stanssens, P., Opsomer, C., McKeown, Y., Kramer, W., Zabeau, M., and Fritz, H. (1989) Nucl. Acids Res. **17**, 4441-4454.

6. Shirley, B. A., Stanssens, P., Steyaert, J., and Pace, C. N. (1989) J. Biol. Chem. **264**, 11621-11625.
7. Shirley, B. A., and Laurents, D. V. (1990) J. Biochem. Biophys. Meth. **20**, 181-188.
8. Oobatake, M., Takahashi, S., and Ooi, T. (1979) J. Biochem. (Tokyo) **86**, 65-70.
9. Pace, C. N., Grimsley, G.R., Thomson, J.A., and Barnett, B. J. (1988) J. Biol. Chem **263**, 11820-11825.
10. McManigill, D., Swedberg, S. A. (1989) In "Techniques in Protein Chemistry" (T. Hugli, ed.), Chap. 45. Academic Press, New York.

SECTION V

NMR/X-Ray Workshop

32

NMR/X-RAY WORKSHOP: AN OVERVIEW

Juliette T. J. Lecomte

Chemistry Department
The Pennsylvania State University
University Park, Pennsylvania 16802

Single crystal X-ray diffraction has, until the past few years, played a unique role in providing detailed structural information on proteins. Recently, nuclear magnetic resonance (NMR)[1] spectroscopy has emerged as an efficient method for the determination of high-resolution structures of proteins in solution (Clore and Gronenborn, 1989; Wüthrich, 1989). This new method continues to develop at a rapid pace and, thanks to remarkable advances in spectrometer and computer technologies, it is being applied in an increasing number of laboratories. The NMR/X-ray Workshop, organized and chaired by R. Klevit,[2] covered current topics in structural determination of proteins with an emphasis on the NMR procedure.

---

[1]Abbreviations used: BPF4, bovine platelet factor 4; d-g, distance geometry; HMQC, heteronuclear multiple quantum coherence; INEPT, insensitive nucleus enhancement by polarization transfer; IL-8, Interleukin-8; NMR, nuclear magnetic resonance; NOE, nuclear Overhauser effect; NOESY, two-dimensional nuclear Overhauser spectroscopy; REVINEPT, reverse INEPT; RMSD, root mean square deviation; TOCSY, totally correlated spectroscopy; 2-D, two-dimensional; 3-D, three-dimensional.

[2]R. E. Klevit, Biochemistry Department, University of Washington, Seattle, WA 98195.

Techniques in Protein Chemistry II

I. X-RAY METHOD

The determination of a structure by the X-ray method depends on the formation of crystal, diffracting to adequate resolution, and on the correct phasing of the reflections. For large molecules, two approaches are commonly considered to solve the phase problem: isomorphous replacement and molecular replacement. When suitable crystals of heavy atom derivatives cannot be obtained, one must resort to the second option using a known structure as closely related as possible to that under investigation.

Molecular Replacement Structure Solution of Interleukin-8 Using the NMR Model[3]

E. Baldwin bridged the fields of protein crystallography and protein NMR with his work on Interleukin-8 (IL-8). IL-8 is a 8,000-dalton cytokine which has been recently crystallized (Baldwin *et al.*, 1990). The IL-8 crystals diffract to 1.6 Å but the initial search for suitable heavy atom derivatives has not met with success. Thus, molecular replacement was applied using as model the mean restrained minimized NMR structure of IL-8 (Clore *et al.*, 1990). Precedence for this approach has been reported on Tendamistat, a protein of the same size (Braun *et al.*, 1989).

The IL-8 dimer was first oriented with the molecular two-fold axis along the crystallographic two-fold axis. The cross-rotation function provided Euler angles of $\alpha = 26.5$, $\beta = 106.5$, and $\gamma = 121.0$ after refinement. The correctly oriented dimer was subjected to a stepwise one-dimensional translation along the cell diagonal and the structure was refined at each step. The origin was identified by the lowest R factor and the largest correlation factor. Parameters of $x = y = 0.63$ and a $P3_121$ space group were obtained. The same convergence was arrived at using an alternative starting model, *i.e.* the X-ray structure of the homologous protein bovine platelet factor 4 (St. Charles *et al.*, 1989). The structure was then rebuilt and refined to an R factor of 0.187 with data to 1.6 Å. Forty-three water molecules were included. The details of the procedure with both IL-8 and BPF4 models are reported by Baldwin and coworkers in their contribution to this volume.

---

[3]Paper presented by E. T. Baldwin, R. St. Charles, and A. Wlodawer. National Cancer Institute-FCRDC, Frederick, MD 21701.

The symmetric dimer of IL-8 forms a 6-stranded antiparallel β-sheet and two long antiparallel α helices which are located on one side of the sheet; this structure is reminiscent of that of the α_1/α_2 domain of the human class I histocompatibility antigen HLA-A2. Interestingly, at this stage of refinement the structure obtained from the diffraction data differs from the NMR structure used for molecular replacement. In solution, the two helices are 14 Å apart whereas in the solid state, they are closer and the β-sheet presents less of a twist. The origin of this distortion is unclear. It is noted that within a narrow window of residues, the local structure defined by the NMR model agrees well with that of the X-ray data and the possibility that small errors, not noticeable at the local level, are propagated and amplified was raised. Calculation of the two-dimensional nuclear Overhauser spectroscopy maps from the X-ray coordinates ("back-calculation", see below) confirms that there are discrepancies between the two structures.

II. NMR METHOD

The process of structural determination by NMR spectroscopy involves three steps: identification and assignment of signals in the spectra, inter-proton distance measurements, and computation of the structure. Most of the current efforts are devoted to the improvement of the first and last aspects; both were addressed in the remainder of the workshop.

A. *Heteronuclear Three-Dimensional NMR: Towards the Structure Determination of Medium Sized Proteins*[4]

The assignment of individual protein resonances to specific protons is a difficult and lengthy task. The commercial availability of spectrometers operating at high magnetic fields and the advent of two-dimensional NMR spectroscopy has rendered the process feasible for small rigid systems.

[4]Paper presented by E. R. P. Zuiderweg, K. W. Mollison, W. Mandecki, G. W. Carter, L. P. McIntosch[†], F. W. Dahlquist[†], H. L. Eaton, A. M. Petros, E. T. Olejniczak, and S. W. Fesik. Pharmaceutical Discovery Division, Abbott Laboratories, Abbott Park IL 60064. [†]Institute of Molecular Biology, University of Oregon, Eugene, Oregon 97403.

The two-dimensional experiments used for assignment fall in two groups: correlated experiments which rely on *J*-coupling and provide through-bond information, and nuclear Overhauser effect (NOE) experiments which provide through-space information. Correlated experiments are crucial in that they allow for the identification of the signals belonging to a same residue.

A molecular weight of approximately 10,000 daltons is often quoted as the limit for structural determination involving a reasonable amount of work. Two-dimensional homonuclear proton techniques start to fail when applied to larger proteins because the number of resonances and their linewidth become excessively large. Both of these factors contribute to intractable spectral overlap. Furthermore, the accelerated relaxation rate due to the slow tumbling time not only broadens the lines but also lowers the sensitivity of the experiments and renders the traditional homonuclear coherence transfer experiments useless.

E. Zuiderweg presented a summary of the recent advances in data collection which increase the range of accessible molecular weight and the ease of data interpretation. The new methods alleviate the problem of overlap by expanding the experiments into a third frequency domain. Three-dimensional experiments are a logical extension of the two-dimensional ones (Griesinger *et al.*, 1987). They can be viewed as a combination of two 2-D experiments such that the detection period of the first experiment serves as the evolution period of the second. The general scheme is summarized as Preparation – Evolution (t_1) – Mixing 1 – Evolution (t_2) – Mixing 2 – Detection (t_3). In this fashion, it is possible to correlate three spins, for example $H_1 \rightarrow H_2 \rightarrow H_3$. The introduction of the third dimension improves the resolution but does not circumvent the second obstacle, namely the loss of homonuclear coherence transfer. As a consequence the usefulness of the homonuclear 3-D experiment is limited. However, 3-D NMR spectroscopy becomes an extremely powerful tool for spectral assignments when a heteronuclear step is included.

Two heteronuclei of general interest for protein studies are ^{15}N and ^{13}C, whose levels can readily be enriched in most proteins which are cloned and expressed in bacterial systems. Since most *J*-coupling values involving ^{15}N and ^{13}C are considerably larger than $J_{H\text{-}H}$ values, the evolution times in the heteronuclear step can be made short and the proton signals are not lost during the experiment. Various combinations have been developed: HMQC–COSY, HMQC–NOESY, NOESY–HMQC, TOCSY–HMQC... They have been reviewed recently as well as their application to peptides, C_5a (8.5 kDa) and T_4 Lysozyme (18.7 kDa) (Fesik and Zuiderweg, 1990, and the contribution of Zuiderweg and coworkers to this volume). For the sake of illustration, a typical heteronuclear 3-D experiment could be a combination of HMQC and NOESY experiments. In this case, the heteronuclear shift is indirectly detected in t_1, the protons coupled to the heteronuclei in t_2, and the protons in dipolar contact with those selected in t_2, during t_3. This experiment is particularly useful with ^{15}N labels as it provides a series of two-dimensional NOESY subspectra each at the frequency of a ^{15}N nucleus. The wide dispersion of the ^{15}N chemical shift helps in resolving overlapping backbone proton resonances. Although the HMQC–NOESY experiment can be performed in two dimensions, the sensitivity of the 3-D version is superior.

Correlation can also be achieved through ^{13}C nuclei. As with ^{15}N, several experiment variations are possible; in addition, the ^{13}C–^{13}C one-bond coupling can be exploited to relay the coherence. This is illustrated by the C → C → H TOCSY–REVINEPT experiment (Fesik *et al.*, 1990). The first step is a ^{13}C isotropic mixing correlating all covalently attached ^{13}C, the information is then transferred to protons with a reverse INEPT experiment. This procedure avoids coherence transfer through small *J* values. Another possibility involves H → C → C → H transfer (Kay *et al.*, 1990).

The decomposition of the data into subspectra facilitates greatly the assignment process and reduces the risks of errors. With these improvements, the size limit has been advanced to an estimated 25 kdaltons. Given the sensitivity of modern spectrometers, the useful concentration of labeled protein can be reduced below 1 mM. Since many proteins of larger molecular weight present independent domains of size within 10 to 25 kdaltons, the heteronuclear 3-D methods constitute a most important advance in the field.

B. *Structure Determination of a Protein in Conformational Equilibrium*[5]

The internuclear distances needed in the third phase of the structural determination are extracted from NOESY data. In slowly tumbling rigid systems, the rate at which the NOE intensity builds up is simply proportional to r^{-6}. This model is clearly a simplification of the physical reality as proteins are not rigid objects. It is therefore necessary to evaluate how the various internal motions exhibited by proteins in solution affect the intensity of the NOEs. J. Prestegard focused on these complications.

Three types of regimes are generally distinguished according to the rate of the motion with respect to the overall tumbling rate of the protein: fast, intermediate, and slow. For rapid, isotropic and uncorrelated motions

---

[5]Paper presented by J. H. Prestegard. Chemistry Department, Yale University, New Haven, CT 06511.

(for example unhindered methyl rotation), it has been shown that the distance separating the mean nuclear positions (as in the rigid model) is an acceptable approximation of the effective distance calculated when taking into account angular and distance fluctuations (LeMaster *et al.*, 1988). Fast internal motions have negligible influence on the final structure.

The other extreme is represented by slow exchange between two or more conformers yielding distinct sets of resonances. Slow conformational equilibrium in the native state has been demonstrated for a number of proteins, including among others the acyl carrier protein from spinach (Kim and Prestegard, 1990) and staphylococcal nuclease (Loh *et al.*, article in this volume). The existence of more than one conformer increases the complexity of the spectrum. Multiple exchanging forms can be confirmed by comparing the number of observed resonances to that of expected resonances, by applying experiments designed to demonstrate exchange (NOESY and rotating frame NOESY), or by perturbing the position of the equilibrium with varying solution conditions. It is important to realize that the observation of one single set of peaks does not guaranty the presence of one single conformer. Indeed, exchange may take place between distinct species at a rate which averages the chemical shifts. This intermediate situation is the most difficult to confront, in part because it is not immediately recognizable.

In the presence of exchange, it would be erroneous to confine the molecule into one conformation reflecting the average NOE. In general, distance and correlation time fluctuations (and occasional conformational diversity) are taken into account by determining upper bounds to rather than well-defined values for the distances derived from NOE data. Since tightening the range of distance should lead to more precise structures, a dynamic model is devised to consider conformational averaging explicitly. In the first attempt at structure building, a single static model is used. The systematic appearance of NOE constraints violation in a defined region of the structure is taken to indicate conformational equilibrium. To investigate this possibility, structures from the first set which exhibit distinct extremes in violations are considered pairwise and in varying proportions. The refinement process is pursued further and the different linear combinations are compared. This method leads to improvements of the root mean square deviations (RMSDs) and generally lowers the penalty associated with constraint violations. Application of this method to the acyl carrier protein from *Escherichia coli* identified regions of the protein which are particularly mobile (Kim and Prestegard, 1989). Caution has to be exercised, for this method can be viewed as increasing the number of parameters to be adjusted and as such, it is likely to produce a structure with fewer violations.

C. *Toward the Structure Determination of the HIV-1 Nucleocapsid protein by NMR*[6]

A growing number of procedures and algorithms are available for the calculation of a protein structure based on a set of distance constraints derived from NOE data (Clore and Gronenborn, 1989; Wüthrich, 1989). A common protocol is to generate distance-geometry (d-g) structures which satisfy all the experimental constraints, then to refine the structures with the application of molecular dynamic methods. The final result is presented as a superposition of independent solutions with minimum RMSDs compared to a single structure. The size of the RMSDs provides a measure of the precision of the structure determination. How can the exactness of the NMR structure be evaluated? M. Summers discussed the topic of precision versus accuracy and emphasized the use of back-calculation with application to the HIV-1 nucleocapsid protein.

Back-calculation is the simulation of the NOESY spectra corresponding to the calculated structure; a back-calculated spectrum resembling closely the experimental spectrum is taken as evidence for an exact structure. Back-calculations can also be used to guide the process of structural determination. As developed by Summers and coworkers (Summers *et al.*, 1990), a judicious adjustment of minimal internuclear distances helps in excluding non-observed effects from the theoretical spectrum obtained from d-g structures with minimized penalty. Modified bounds files are thus generated which can be improved in an iterative fashion. As stressed in the contribution of Summers and coworkers to this volume, back-calculation is not crucial for convergence to low RMSDs; however, convergence does not imply exactness. It was also observed that in the HIV capsid peptide, an alanine residue presents incompatible observed contacts indicative of conformational exchange. The design of improved back-calculation algorithms could include explicit consideration of conformational equilibria.

---

[6]Paper presented by M. F. Summers, T. L. South, P. R. Blake, D. R. Hare[†], R. Sowder[‡], L. O. Arthur[‡], and L. E. Henderson[‡]. Department of Chemistry and Biochemistry, University of Maryland Baltimore County, Baltimore, MD 21228. [†]Hare Research Inc., Woodinville, WA 98072. [‡]Retroviral protein Biochemistry Laboratory, AIDS-Vaccine Program, NCI-FCRDC, Frederick, MD 21701.

III. CONCLUDING REMARKS

The X-ray/NMR Workshop discussed new directions in high-resolution structure determination of proteins. It illustrated that sophisticated improvements can now be brought to the NMR method. Thus, for the data collection aspect, the use of third and higher dimensionality with isotopically labeled material gives access to large molecular weight proteins, facilitates the assignment procedure for small ones, and raises the possibility of efficient automatized interpretation. For the data analysis aspect, simulation of the spectra and incorporation of internal motions aim towards rapid convergence to an exact structure and realistic representation of molecules in solution. Continued progress in these areas is expected in the near future from the participants to the Workshop and other workers in the field.

In its early days, the NMR method relied heavily on the information provided by X-ray crystallography to grow and become established. As the technique matures, it is now seen that NMR information can be directly used by X-ray crystallographers to help in their structural determinations. This new trend promises to be most productive. The close combination of both methods will expand our knowledge and understanding of proteins by providing complementary perpectives on their structural and dynamic properties.

REFERENCES

Baldwin, E. T., Franklin, K. A., Appella, E., Yamada, M., Matsushima, K., Wlodawer, A., and Weber, I. T. (1990) *J. Biol. Chem. 265*, 6851-6853.

Braun, W., Epp, O., Wüthrich, K., and Huber, R. (1989) *J. Mol. Biol. 206*, 669-676.

Clore, G. M., and Gronenborn, A. M. (1989) *Crit. Rev. Biochem. Mol. Biol. 24*, 479-564.

Clore, G. M., Appella, E., Yamada, M., Matsushima, K., and Gronenborn, A. M. (1990) *Biochemistry 29*, 1689-1696.

Fesik, S. W., and Zuiderweg, E. R. P. (1990) *Q. Rev. Biophys. 23*, 97-131.

Fesik, S. W., Eaton, H. L., Olejnicsak, E. T., and Zuiderweg, E. R. P. (1990) *J. Am. Chem. Soc. 112*, 886-888.

Griesinger, C., Sørensen, O. W., and Ernst, R. R. (1987) *J. Magn. Reson. 73*, 574-579.

Kay, L. E., Ikura, M., and Bax, A. (1990) *J. Am. Chem. Soc. 112*, 888-889.

Kim, Y., and Prestegard, J. H. (1989) *Biochemistry 28*, 8792-8797.

Kim, Y., and Prestegard, J. H. (1990) *J. Am. Chem. Soc. 112*, 3707-3709.
LeMaster, D. M., Kay, L. E., Brunger, A. T., and Prestegard, J. H. (1988) *FEBS Lett. 236*, 71-76.
St. Charles, R., Walz, D. A., and Edwards, B. F. P. (1989) *J. Biol. Chem. 264*, 2092-2099.
Summers, M. F., South, T. L., Kim, B., and Hare, D. R. (1990) *Biochemistry 29*, 329-340.
Wüthrich, K. (1989) *Science 243*, 45-50.

33

MOLECULAR REPLACEMENT STRUCTURE SOLUTION OF INTERLEUKIN-8 USING THE NMR MODEL

Eric T. Baldwin, Joel Sussman, Robert St. Charles and Alexander Wlodawer

Crystallography Laboratory, National Cancer Institute-FCRDC, ABL-Basic Research Program, Frederick, MD.

I. INTRODUCTION

The initial phases for a protein crystal structure determination are routinely established by one of two methods. Multiple isomorphous replacement (MIR) phases are obtained by making heavy atom derivatives. Alternatively, a model of a homologous structure could be correctly rotated and translated in the unknown cell and the phases calculated. If heavy atom derivatives cannot be prepared and a model of sufficient accuracy is unavailable, then the X-ray structure will remain unsolved. The advent of protein NMR technology opened a new avenue for crystal structure determination. Instead of relying on homologous models derived from other structure solutions, a model of the protein obtained by NMR could be used in the molecular replacement calculation. In principle an NMR model could be vastly superior to a model based upon a structural homologue and might greatly enhance the chance for successful calculations. However, because the information content of the NMR experiment is limited to 5 Å nearest neighbor distance constraints, some doubts were expressed whether an NMR structure could accurately described longer range organization of the protein molecule. These concerns were supported by the large rms deviations of early NMR models from high resolution X-ray structures. Recently, NMR structure determination has been significantly improved when stereospecific restraints at β-methylene hydrogens became incorporated into the structure calculation (Nilges *et al.*, 1990). The qualitative analysis of NOE and coupling constant data give estimates of the torsion angles which constrain the three-dimensional structure and decrease the rms deviation among the calculated structures to about half of that observed previously. Thus, current NMR structures may be even better models for molecular replacement structure solutions.

Techniques in Protein Chemistry II

In test calculations performed using crambin (Brünger *et al.*, 1987a) and tendamistat (Braun *et al.*, 1989), the NMR-derived models were sufficient to phase the X-ray data. These experiments also demonstrated the importance of using for the molecular replacement calculations the average structure, or a structure closest to the average. Finally, their results emphasized the well-known influence of small structural changes on the success or failure of the molecular replacement. In well-ordered crystals like crambin deletion of side chains can decrease the rotation function signal. On the other hand, including side chains in incorrect conformations can seriously hamper the investigation, as was observed in the tendamistat study.

We report here the first successful combination of NMR and X-ray methods for the molecular replacement determination of an unknown crystal structure. We solved Interleukin-8 (IL-8), which is a small chemotactic factor released by monocytes and a number of other cell types in response to inflammatory stimuli (Matsushima and Oppenheim, 1989). The factor promotes the directed migration of neutrophils, basophils and T-cells along concentration gradient to the inflammatory focus. IL-8 is related to a number of immune system proteins, all of which show a conserved region containing four cysteines forming two disulfide bridges.

II. DATA COLLECTION AND ROTATION FUNCTION SOLUTION

Recombinant human IL-8 used to grow crystals was purified from *E. coli* (Furuta *et al.*, 1989). Crystals were grown from ammonium sulfate (Baldwin *et al.*, 1990a) and belonged to the trigonal space group $P3_121$ (or $P3_221$) with unit cell parameters a = b = 40.3 and c = 90.1 Å. Data have been collected using the Siemens area detector to 1.6 Å resolution.

The structure of IL-8 has been solved with the use of molecular replacement. Two independent sets of atomic coordinates were used as starting points. The coordinates of bovine platelet factor 4 (BPF4) refined at 3 Å resolution were provided by B. Edwards (St. Charles *et al.*, 1989), while IL-8 coordinates resulting from the NMR structure determination (Clore *et al.*, 1989,1990) were contributed by M. Clore and A. Gronenborn. The BPF4 molecule forms a crystallographically independent tetramer consisting of two almost-perfect dimers, which are not related to themselves by a proper two-fold rotation. The IL-8 model was a dimer with perfect 2-fold symmetry.

As an initial step, a perfect dimer of BPF4 was created by duplication of molecule 1 and was oriented with its 2-fold axis parallel to z. IL-8 model was oriented in an identical manner, to help in the interpretation of the molecular rotation (Fehlhammer *et al.*, 1975). The C_α coordinates of these two models differed by 1.65 Å rms after the superposition of 121 pairs of atoms.

We have noticed that in order to accommodate a dimer in the lattice of space group $P3_121$ ($P3_221$) it was necessary to align the molecular 2-fold of IL-8 with the crystallographic 2-fold axis, found in the (x,y) plane of the unit cell. For the conventions employed in the molecular replacement package MERLOT (Fitzgerald, 1987), the model two-fold along z is correctly superimposed upon the crystallographic two-fold by a 30° α rotation and a 90° β rotation. The γ rotation defines the rotation of the model about the two-fold axis. None of the cross rotation results described below corresponded to this ideal rotation. This was the first indication that there were quaternary structural differences between BPF4, the IL8-NMR model, and IL-8 in the crystal (Baldwin *et al.*, 1990b).

Four models were used in the molecular replacement calculations. Two of these were derived from BPF4 model and two from the IL-8 NMR model. BPF4 model 1 was the full monomer which had been oriented with the two-fold axis along z. BPF4 model 2 was prepared from model 1 by trimming selected sidechains back to the β carbon. Based upon our sequence comparison those residues whose sidechains were larger in BPF4 than those of IL-8 were removed (Q-24, N-35, R-37, H-38, P-52, P-54, K-60, K-61, K-65, Q-70, Q-71, R-72, P-73, Y-75, K-76, K-77, K-80, K-81, and Q-85). In both cases the BPF4 model begins at residue 24 and extends through residue 85. The IL-8 model 1 is the full NMR monomer without hydrogens and includes residues 2 through 72. Following the example of the test calculations above the mean restrained minimized NMR structure consisting of 582 non-hydrogen atoms was chosen. Since the average NMR structure was derived from 30 dimer coordinates resulting from the NMR structure determinations, an rms deviation ($\overline{u}^2$) for each atom was computed. For consistency these values can be converted to temperature factors, $B = 8\pi^2\overline{u}^2$. Hence, an atomic deviation of 1 Å corresponds to a B-factor of 78.9 $Å^2$. IL-8 NMR model 2 was derived from model 1 by removing all of the atoms with B-factors higher than 75. This model included residues 6 through 72 and consisted of 460 non-hydrogen atoms. All of the deleted side chains were on the surface. The only main chain atoms deleted from the model were at the amino terminus. All four models were oriented identically and the model structure factors calculated for an 80 Å cubic cell in space group P1.

The Crowther fast cross rotation function was calculated for each model over six resolution ranges with various radii of integration (ARAD). The results of selected experiments are recorded in Tables I through III. For each experiment the peaks greater than 75% of the maximum signal in the function were examined. Since we expected the α and β rotations to be near 30 and 90° respectively, the peak list for each experiment was culled and only those rotations within the acceptance window were recorded in Tables I-III. For the α rotation values from 15 to 45° were accepted, while for the β rotation values from 65 to 115° were accepted. All values of γ were accepted.

Resol.	ARAD	rank	α	β	γ	percent	rms
40 to 3	17	1	35.0	70.0	300.0	100.0	3.83
		2	25.0	110.0	120.0	99.9	3.82
10 to 3	17	1	35.0	75.0	300.0	100.0	5.05
		2	25.0	105.0	120.0	99.8	5.04
8 to 3	17	1	35.0	75.0	300.0	100.0	5.94
		2	25.0	105.0	120.0	99.8	5.93
40 to 4	23	1	25.0	105.0	120.0	100.0	4.88
		2	35.0	75.0	300.0	99.7	4.86
10 to 4	23	1	25.0	105.0	120.0	100.0	5.50
		2	35.0	75.0	300.0	99.5	5.47
8 to 4	23	1	25.0	105.0	120.0	100.0	6.11
		2	35.0	75.0	300.0	98.4	6.08

Table 1: Cross rotation results using the IL-8 model with all atoms (model 1). The resolution and radius of integration are in Å.

The result of the cross rotation function with IL-8 NMR models 1 and 2 were essentially identical (Model 2 not shown). For the experiments listed in Tables I the correct rotation and its two-fold related symmetry mate are the top peaks. These peaks were found in all calculations but were not always the highest. The rotation functions generally became noisier as the radius of integration was decreased (data not shown). Although α and β varied considerably in all of the calculations (sometimes producing solutions just outside the acceptance window), the γ solutions varied by less than 10°. Subsequently, the Lattman rotation function (Lattman, 1972) as implemented in MERLOT was calculated. The refined rotation angles were α = 26.5, β = 106.5 and γ = 121.0°.

The results of the cross rotation function with BPF4 model 1 and model 2 are shown in Tables II and III. With BPF4 model 1 the correct solution was never the highest peak in the list and was often down on the list of peaks. For those experiments which excluded data below 8 Å the correct solution consistently ranked higher. The distribution of the γ rotation was broader than that observed for the NMR models. In addition a spurious γ rotation near 0/180° was frequently observed (40 to 3 Å resolution in Table II). BPF4 model 2 performed better in the cross rotation function than the full model. As is indicated in Table III the correct peaks were the highest for data from 40 to 4 Å resolution with a radius of integration of 23 Å. Other experiments (not shown) also had the correct solution as the top peak. In addition, experiments utilizing model 2 generally ranked the correct solution higher in the peak list than the full model. Occasionally, an incorrect γ rotation was observed but not as frequently as that observed with model 1. Model 1 and model 2 were used in Lattman rotation analysis to fine-tune the orientation prior to translation search. The full model gave an α rotation equal to 30.5, β = 108.5 and γ = 119.5°. Model 2 gave a refined rotation with α = 27.5, β = 100.5 and γ = 123.0°.

Resol.	ARAD	rank	α	β	γ	percent	rms
40 to 3	17	18	28.3	75.0	355.0	81.9	2.42
		19	31.7	105.0	175.0	81.8	2.41
		22	26.7	75.0	0.0	80.1	2.36
10 to 3	17	13	43.3	115.0	310.0	86.2	3.60
		14	16.7	65.0	130.0	86.0	3.59
		19	28.3	75.0	305.0	82.9	3.45
		20	31.7	105.0	125.0	82.7	3.45
8 to 3	17	3	23.3	75.0	125.0	98.5	4.61
		4	36.7	105.0	305.0	98.3	4.60
		5	41.7	110.0	310.0	96.6	4.52
		6	18.3	70.0	130.0	96.5	4.52
40 to 4	23	9	30.0	75.0	305.0	92.0	3.27
		11	31.7	105.0	125.0	90.9	3.24
		13	15.0	100.0	125.0	89.4	3.18
		14	45.0	80.0	305.0	88.6	3.16
		19	18.3	65.0	130.0	83.8	2.99
		20	41.7	115.0	310.0	83.8	2.99
10 to 4	23	5	28.3	75.0	305.0	98.3	3.75
		6	31.7	110.0	120.0	98.2	3.74
		20	18.3	65.0	130.0	87.7	3.33
		21	41.7	115.0	310.0	87.6	3.32
		48	38.3	115.0	125.0	81.3	3.07
8 to 4	23	3	25.0	80.0	125.0	96.7	4.31
		4	35.0	100.0	305.0	96.2	4.28
		7	18.3	65.0	130.0	93.9	4.18
		8	41.7	115.0	310.0	93.8	4.18
		14	41.7	115.0	130.0	89.1	3.97
		16	20.0	65.0	305.0	88.9	3.96
		19	31.7	110.0	120.0	86.8	3.87
		22	28.3	70.0	300.0	85.1	3.79

Table 2: Cross rotation function results using the full BPF4 model.

The signal for the correct solution for the rotation function using BPF4 model was poor. In a blind analysis the correct rotation would not have likely been the first or even second choice. The assumption of the coincidence of the crystallographic two-fold and the molecular dimer axis narrowed the choices of rotation solutions enough to permit a correct interpretation of the rotation function. When the BPF4 models were in general orientations (data

Resol.	ARAD	rank	α	β	γ	percent	rms
40 to 3	17	37	30.0	105.0	125.0	78.4	2.05
10 to 3	17	3	30.0	75.0	305.0	97.1	3.94
		4	30.0	105.0	125.0	96.7	3.92
8 to 3	17	4	28.3	75.0	305.0	96.4	4.31
		5	31.7	105.0	125.0	95.9	4.29
		6	43.3	115.0	310.0	94.9	4.24
		12	23.3	75.0	125.0	82.2	3.68
		15	36.7	105.0	305.0	82.0	3.67
40 to 4	23	1	15.0	100.0	125.0	100.0	3.70
		2	43.3	80.0	305.0	99.5	3.68
		3	30.0	100.0	125.0	98.9	3.66
		4	30.0	80.0	305.0	98.7	3.65
		12	18.3	70.0	130.0	77.9	2.89
		13	41.7	110.0	310.0	77.9	2.89
10 to 4	23	5	30.0	75.0	305.0	99.0	3.73
		6	31.7	105.0	125.0	98.3	3.70
		13	16.7	100.0	125.0	91.8	3.45
		15	18.3	70.0	130.0	91.3	3.43
		17	41.7	110.0	310.0	91.2	3.42
		19	43.3	80.0	305.0	91.2	3.42
8 to 4	23	3	33.3	105.0	125.0	94.8	4.06
		4	26.7	75.0	305.0	94.1	4.03
		7	18.3	70.0	130.0	91.1	3.90
		9	41.7	110.0	310.0	91.1	3.90

Table 3: Cross rotation results using BPF4 model with selected side chains deleted.

not shown), the rotation calculations never produced consistent or convincing solutions. Translation solution based upon these incorrect orientations could not be refined. On the other hand, the IL-8 NMR models were better in the rotation analysis than either of the BPF4 models. The correct solution was often the only solution in the peak list and would have been easily chosen without assuming that the asymmetric unit contained half of a dimer.

The rotation results underscore the dramatic importance of this experiment to X-ray structure solution methodology. In the absence of the NMR data this structure could only have been solved because of a fortunate symmetry constraint. The BPF4 monomer in all likelihood could not have been closely enough related to the IL-8 to allow us to solve it by molecular replacement. Hence, the NMR model was superior to the structural homologue and insured the success of the molecular replacement calculation.

III. TRANSLATION FUNCTION CALCULATIONS

The correct translation for the monomer for all four models was determined in a non-standard way using the rigid-body refinement option in the program CORELS (Sussman *et al.*, 1977). Because the asymmetric unit contains a half of a dimer, the possible translations are greatly restricted. The cross rotation function applied the necessary rotations about the dimer center of mass to correctly orient the dimer axis along the crystallographic two-fold direction. This rotation axis lies in the $z = 0$ plane and is coincident with the $x = y$ diagonal line. Thus, the correct translation can be determined by merely calculating an R-factor for various grid points along the $x = y$ line. In this case the monomer was placed at successive position 2.5 Å apart and the rotations and translations of the monomer refined to convergence, with data for 15 to 9 Å (63 reflections). An important new modification of CORELS allowed this translation search to be carried out efficiently.

Initial x=y	R_f	CC_f	x	y	z
0.00	54.7	0.421	-0.070	0.123	0.040
0.05	52.4	0.476	-0.035	0.174	0.035
0.10	60.2	0.247	0.070	0.324	0.047
0.15	60.0	0.246	0.077	0.325	0.046
0.20	61.0	0.297	0.143	0.332	0.030
0.25	60.9	0.229	0.249	0.446	0.029
0.30	58.8	0.359	0.250	0.450	0.029
0.35	53.7	0.456	0.320	0.520	0.038
0.40	53.4	0.463	0.332	0.496	0.038
0.45	47.0	0.619	0.496	0.691	0.052
0.50	48.5	0.584	0.485	0.704	0.044
0.55	49.1	0.556	0.499	0.704	0.048
0.60	42.9	0.625	0.550	0.732	0.044
0.65	45.2	0.628	0.549	0.731	0.045
0.70	64.2	0.095	0.638	0.863	0.030
0.75	53.6	0.307	0.753	0.987	0.042
0.80	55.0	0.322	0.754	0.987	0.040
0.85	54.9	0.322	0.754	0.987	0.043
0.90	59.0	0.307	0.877	1.022	0.050
0.95	46.8	0.594	0.926	1.116	0.035

Table 4: Translations for the IL-8 NMR (model 1) in space group $P3_121$. R_f is the refined R-factor, while CC_f is the correlation coefficient. Column one indicates the dimer center of mass while x,y,z are the refined center of mass coordinate for the monomer.

The translation search was carried out in both enantiomorphic space groups for all four models (Tables IV and V). The results clearly indicated that $P3_121$ was the correct space group and all four models converged to the same solution. Surprisingly, the IL-8 NMR model (Table IV) had the lowest correlation coefficient (0.62) among all four models. The correlation coefficient for the full BPF4 model and NMR model 2 were the same (0.665 and 0.670 respectively, not shown). BPF4 model 2 had the best correlation coefficient 0.713 (Table V).

Initial x=y	R_f	CC_f	x	y	z
0.00	55.7	0.398	-0.106	0.082	0.025
0.05	59.5	0.312	-0.040	0.199	0.021
0.10	53.3	0.405	-0.002	0.223	0.033
0.15	52.8	0.408	0.032	0.225	0.036
0.20	60.9	0.092	0.048	0.295	0.034
0.25	61.0	0.260	0.128	0.316	0.025
0.30	59.2	0.301	0.248	0.413	0.023
0.35	61.4	0.236	0.231	0.497	0.033
0.40	52.4	0.471	0.339	0.539	0.029
0.45	53.3	0.466	0.341	0.539	0.029
0.50	46.1	0.723	0.501	0.701	0.034
0.55	41.5	0.713	0.502	0.686	0.040
0.60	40.6	0.713	0.499	0.697	0.038
0.65	40.9	0.707	0.536	0.697	0.040
0.70	50.6	0.526	0.499	0.820	0.051
0.75	57.5	0.301	0.699	0.893	0.029
0.80	52.9	0.503	0.659	0.878	0.043
0.85	58.1	0.182	0.728	0.967	0.033
0.90	59.5	0.338	0.814	0.998	0.043
0.95	55.6	0.379	0.898	1.021	0.040

Table 5: Translation of BPF4 model with selected side chains deleted in space group 152.

Thus far, only the NMR model 1 has been refined to higher resolution. Work on the other three models is in progress. Rigid body refinement of the model with data to 4 Å reduced the R-factor to 50.1%. This was followed by positional refinement with 10-3 Å data with X-PLOR (Brünger *et al.* 1987b) which reduced the R-factor to 40.7%. Next, a cycle of crystallographic refinement by simulated annealing with heating to 2000 K reduced the R- factor to 27.9%. Refinement was continued with the restrained least squares program PROLSQ (Hendrickson, 1985). Several cycles of refinement, interspersed with manual rebuilding using computer graphics and the addition of solvent molecules, led to the current model, characterized by an

R-factor of 18.7% for data to 2.0 Å with the deviation of bond lengths from standard values of 0.019. All main chain torsion angles are found in the allowed regions of the Ramachandran plot. The description of this structure is presented elsewhere (Baldwin *et al.*, 1990b).

ACKNOWLEDGEMENTS

Research sponsored by the National Cancer Institute, DHHS, under contract NO. NO1-CO-74101 with ABL and a NRSA post-doctoral award to E.T. Baldwin. The contents of this publication do not necessarily reflect the views or policies of the Department of Health and Human Services, nor does mention of trade names, commercial products, or organizations imply endorsement by the U.S. Government.

REFERENCES

Baldwin, E. T., Franklin, K.A., Appella, E., M. Yamada, Matsushima, K., Wlodawer, A. and Weber, I.T. (1990a) *J. Biol. Chem.* **265**, 6851-6853.

Baldwin, E. T., Weber, I.T., St.Charles, R., Xuan, J-C., Appella, E., Yamada, M., Matsushima, K., Edwards, B.P.F., Clore, G.M., Gronenborn, A.M. and Wlodawer, A. (1990b) *Proc. Nat. Acad. Sci. USA*, submitted for publication.

Brünger, A.T, Campbell, R.L., Clore, G.M., Gronenborn, A.M., Karplus, M., Petsko, G.A. and Teeter, M.M. (1987a) *Science* **235**, 1049-1053.

Brünger, A., Kuriyan, J. and Karplus, M. (1987b) *Science* **235**, 458-460.

Braun, W., Epp, O., Wüthrich, K., and Huber, R. (1989) *J. Mol. Biol.* **206**, 669-676.

Clore, G.M., Appella, E., Yamada, M., Matsushima, K., and Gronenborn, A.M. (1989) *J. Biol. Chem.* **264**, 18907-18911.

Clore, G. M., Appella, E., Yamada, M., Matsushima, K., and Gronenborn, A. (1990) *Biochemistry* **29**, 1689-1696.

Fehlhammer, Schiffer, M., Epp, O., Colman, P.M., Lattman, E.E., Schwager, P., Steigemann, W. and Schramm, H.J. (1975) *Biophys. Struct. Mechanism* **1**, 139-146.

Fitzgerald, P.M.D. (1987) *J. Appl. Crystallogr. 21*, 273-278.

Furuta, R., Yamagishi, J., Kotani, H., Sakamoto, F., Fukui, T., Matsui, Y., Sohmura, Y., Yamada, M., Yoshimura, T., Larsen, C.G., Oppenheim, J.J. and Matsushima, K. (1989) *J. Biochem. (Tokyo)* **106**, 436-441.

Hendrickson, W.A. (1985) *Methods Enzymol.* **115**, 252-270.

Lattman, E.E. (1972) *Acta Crystallogr.* **B28**, 1065-1068.

Matsushima, K. and Oppenheim, J.J. (1989) *Cytokine* **1**, 2-13.
Nilges, M., Clore, G.M., and Gronenborn, A.M. (1990) *Biopolymers* **29**, 813- 822.
St. Charles, R., Walz, D. A. and Edwards, B. F. P. (1989) *J. Biol. Chem.* **264**, 2092-2099.
Sussman, J., Holbrook, S.R., Church, G.M. and Kim, S-H. (1977) *Acta Crystallogr.* **A33**, 800-804)

34

Is 2D-NOESY Back-Calculation Necessary for NMR-Based Structure Determination? A Systematic Study Involving a Retroviral-Type Zinc Finger Peptide

Paul R. Blake[1]
Dennis R. Hare[2]
Michael F. Summers[1,*]

Department of Chemistry and Biochemistry, University of Maryland Baltimore County, Baltimore, Maryland 21228,
and
Hare Research Inc., 14810 216th Avenue N.E., Woodenville, Washington 98072

I. Introduction

NMR spectroscopy is becoming an increasingly popular tool for structural studies of biomacromolecules (1-9). Although the utility of this new technique is now well established, the ultimate capabilities and limitations of the method, including the level of precision attainable and limiting protein molecular weights, have yet to be determined. Recently, we described a new approach for protein structure determination that involves back-calculation of 2D NOESY spectra for structures generated by distance geometry (DG) calculations (3). This approach enabled the application of relatively tight interproton distance constraints in the refinement process. Although the objective of obtaining simulated spectra for comparison with experimental data is clearly worthwhile (indeed, several other

1UMBC
2Hare Research, Inc.
*To whom correspondence should be addressed.
Supported by NIH Grant GM42561-01 (M.F.S.)

Techniques in Protein Chemistry II

laboratories are developing similar approaches (9-12)), the consequences of our approach on the "resolution" (defined historically as the convergence (pairwise RMSDs) obtained for several superpositioned DG structures generated from different starting coordinates) relative to approaches that do not utilize back-calculations has not been studied.

To determine the consequences of back-calculations and tight distance constraints on structure refinement and degree of convergence, we have carried out a systematic study involving the progressive generation of structures (a) in the absence of NOE-derived distance constraints, (b) using only loose distance constraints for observed cross peaks (2.0-2.5 Å, 2.0-3.5 Å, and 2.0-4.5 Å for qualitatively observed strong, medium, and weak cross peak intensities, respectively), (c) with addition of lower distance constraints (4.0 Å) to limit the approach of protons that do not give rise to cross peaks in the experimental spectrum but do give cross peaks in the back-calculated spectra generated for models in step b (above), and (d) using tight interproton distance constraints to generate models with NOE build-up curves that match the experimental NOE build-up data. The degree of convergence and "goodness" (defined as the ability to reproduce the experimental NOESY data) obtained at each step is described.

II. Materials and Methods

A. NMR DATA

All NMR data were collected with a GE GN-500 (500 MHz, ^{1}H) spectrometer, transferred to Silicon Graphics 220-GTXB and Personal Iris computers via ethernet, converted to readable files with an in-house program (GENET), and processed with FTNMR™ (Hare Research, Inc.). 2D rotating frame Overhauser effect (ROESY) data were used to discriminate between direct NOEs and spin diffusion cross peaks. Details regarding the collection and processing of NMR data for Zn(HIV1-F1) have been described (3).

B. DISTANCE GEOMETRY AND BACK-CALCULATIONS

Distance geometry and 2D NOESY back-calculations were performed with DSPACE™ and BKCALC™ (Hare Research, Inc.). All calculations were performed with the Silicon Graphics workstation computers. An overview of the general approach and details regarding determination of parameters for DSPACE™ and BKCALC™ have been described (3). In this work, structures were generated with several different sets of experimental constraints as indicated in the Introduction. Atomic radii and bonded distance constraints used to prepare a bounds matrix are listed in Table 1. Constraints were included to enforce planarity for the backbone and side chain amide, carboxyl, and aromatic groups, and for the Zn–His bond (Table 1). Constraints were also included to enforce Zn-S(Cys)-Cβ(Cys) bond angles of 109.5°. To allow potential hydrogen-bonding atoms to approach to distances less than the sum of their van der Waals distances, backbone and side chain NH protons were allowed to approach carbonyl and sulfur atoms to within 75% of the sum of the van der Waals radii (Table 1). However, no specific hydrogen bonding or zinc coordination geometry constraints were included in the bounds matrix.

Macro programs were utilized within DSPACE™ to generate initial atom positions and to perform automated refinement (see Scheme 1). Initial atom positions were embedded randomly into 3-space and subjected to simulated annealing and conjugate gradient minimization (13-15). Additional structures were generated by performing two sequential 10 Å randomizations of atom positions, which provide for improved sampling of conformational space compared to the difference-embed method (16).

Penalty minimization was achieved via application of several steps of simulated annealing and conjugate gradient minimization (Scheme 1). These algorithms were employed to reduce the penalty for each structure (penalty = sum of the squared differences between interatomic distances in the structure and the bounds matrix). Penalties were evaluated with the expressions,

$$P_{total} = P_{covalent} + P_{experimental} + P_{nonbonded}$$

$$P_{covalent,\ experimental\ \text{or}\ nonbonded} = \sum K_{type}\,(d\text{-}d_0)^2$$

where K_{type}, the constant used for evaluation of penalty functions for different types of constraints, is set to 1.0 (i.e., covalent, experimental, and nonbonded constraints are treated equally).

TABLE 1. Parameters Used for Distance Geometry Calculations

Bonded Constraints / Planarity Constraints

Atom	Distance in Å	Atoms	Distance in Å
C—N	1.335	Cα[n]---Cα[n-1]	3.802
N—Ca	1.449	HN[n]---O[n-1]	3.149
C—O	1.229	HN[n]---Cα[n-1]	2.564
Ca—C	1.552	Cα[n]---O[n-1]	2.775
Zn—S	2.300	Zn---His(Nδ)	4.130
Zn—N	2.000	Zn---His(Cγ)	4.130

Atomic Radii

Atom	Radius
H	0.950
C	1.500
N	1.300
O	1.300
S	1.850
Zn	0.74

Adjusted Radii

Atoms pairs	% of radius used as a lower limit
amide---carbonyl	75
amide---sulphur	75
Gln(amide)---carbonyl	75
Asn(amide)---carbonyl	75

III. RESULTS

A. STRUCTURES GENERATED WITHOUT NOESY CONSTRAINTS.

Structures were generated with only the primary structure and Zn-ligand bonds as constraints. Six low-penalty structures (ca. 0.02 Å^2) are shown in Fig. 1A. Clearly, the primary constraints are insufficient to define the folding of Zn(HIV1-F1). The structures in Fig. 1A superimpose with RMSDs for backbone atoms (C,Cα,N,HN and O

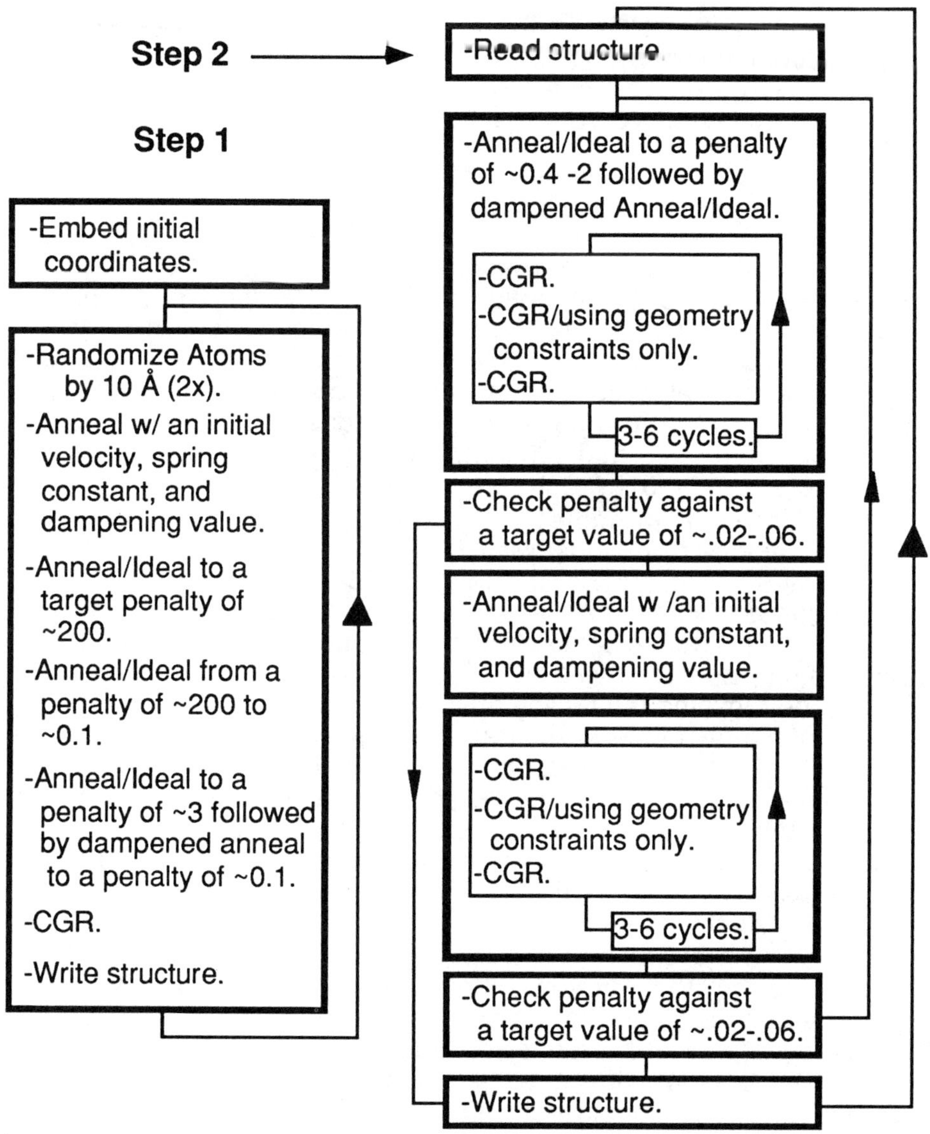

Scheme 1: Flow diagrams representing the procedure used for DG-structure generation and refinement. Loop counters were employed to eliminate infinite loops. Penalty target values were not necessarily achieved for all structures. For step 1, values for the initial velocity, spring, and dampening constants of 0.08, 0.005, and 0-0.2 were employed. Other definitions: CGR = conjugate gradient minimization, anneal/ideal = simulated annealing with idealized bond lengths (±0.01 - ±0.5 Å) and bond angles (±0.1- ±5.0°).

of residues C(3)-C(16)) in the range 2-3 Å. A 2D NOESY back-calculated spectrum generated for a representative structure is shown in Fig. 2A. This spectrum clearly lacks the features observed in the experimental NOESY spectrum given in Fig. 2E.

B. STRUCTURES GENERATED WITH LOOSE NOESY CONSTRAINTS

The most widely employed approach for generation of protein structures involves the utilization of loose constraints that represent cross peaks due to direct NOE pathways. In this study, constraints of 2.0-2.5, 2.0-3.5, and 2.0-4.5 Å were used to represent strong, medium and weak cross peaks, respectively. (Note: upper distance constraints of 3.5 and 4.5 Å were used due to the high sensitivity obtained with our 15 mM sample). Six structures generated with the procedure outlined in Scheme 1 were minimized to give final structures with penalties of 0.048 ± 0.001 Å^2, and were superpositioned to give pairwise RMSD values (for backbone atoms of residues C(3)-C(16)) in the range 0.06-0.18 Å. This remarkably high degree of convergence was not expected, and is probably due to several factors including (a) the large number of long-range NOE constraints, and (b) the significant contributions of van der Waals radii in determining the folding of such a compact structure.

Although these structures were generated to apparent high "resolution", *or more accurately, to high convergence,* the back-calculated spectra (Fig. 2B) indicate that several structural elements are not consistent with the experimental data. In particular, an NOE cross peak of moderate intensity is observed for the HN(8)-HN(6) proton pair (see arrow, Fig 2B.), reflecting the interproton distance of ca. 3.6 Å in the model structures. A similar cross peak was not observed in the experimental spectrum, Fig. 2E. Since the HN(8) and HN(6) protons are well behaved (i.e., they exhibit normal, expected cross peaks to several nearby, non-exchanging protons and are not in rapid exchange with the water protons), the back-calculated data provide good evidence that aspects of the model conformations are incorrect. Other obvious differences between experimental and simulated spectra include the relative intensities of the HN(5)-HN(4) and HN(3)-HN(6) cross peaks (see Figs. 2B and 2E), as well as relative intensities of several cross peaks for non-exchangable protons (data not shown).

C. STRUCTURES GENERATED WITH LOOSE NOESY CONSTRAINTS AND 2D NOESY BACK-CALCULATIONS

In order to generate structures that were more consistent with experimental NOESY data, constraints were added to limit the approach of proton pairs that gave rise to cross peaks in the back-calculated spectra which were not observed in the experimental spectra. In general, distances were limited to a minimum of 4.0 Å, except for cross peaks associated with moderately exchanging amide protons for which lower values were used. It is important to point out that, at this stage of the refinement process, only minimum distance constraints are added and no narrowing of the observable interproton distances is applied. Six superpositioned structures are shown in Fig. 1C, and the downfield region of the back-calculated NOESY spectrum for a representative structure is shown in Fig. 2C. Although the convergence of the C structures (pairwise RMSDs of 0.22-0.26 Å) equals that of the B structures, the C structures clearly provide a better fit to the experimental data. For example, in the downfield region of the back-calculated NOESY spectra, cross peak intensities for the HN(8)-HN(6) and HN(6)-HN(3) proton pairs have been reduced.

1A.

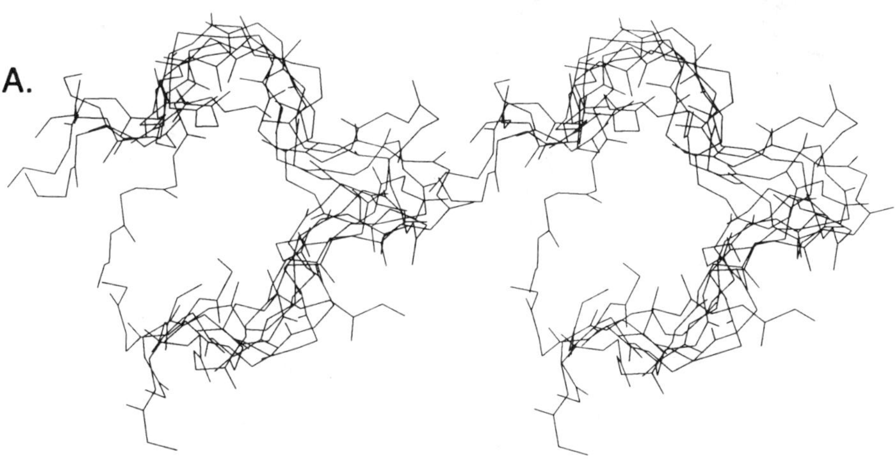

Figure 1. Six superpositioned (C, Cα, & N atoms of residues C(3)-C(16)) DG structures obtained (A) with only primary and Zn-ligand constraints, (B) with loose (strong, medium, weak) distance constraints, (C) with loose distance constraints and additional constraints, based on back-calculations of B structures, to limit the approach of specific proton pairs, and (D) with tight constraints based on quantitative fitting of NOESY build-up data. Only the C, Cα, N and O atoms are displayed.

1B.

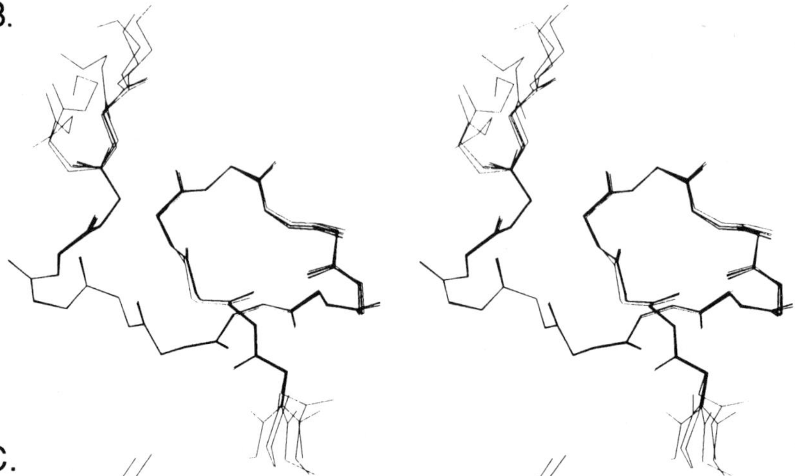

1C.

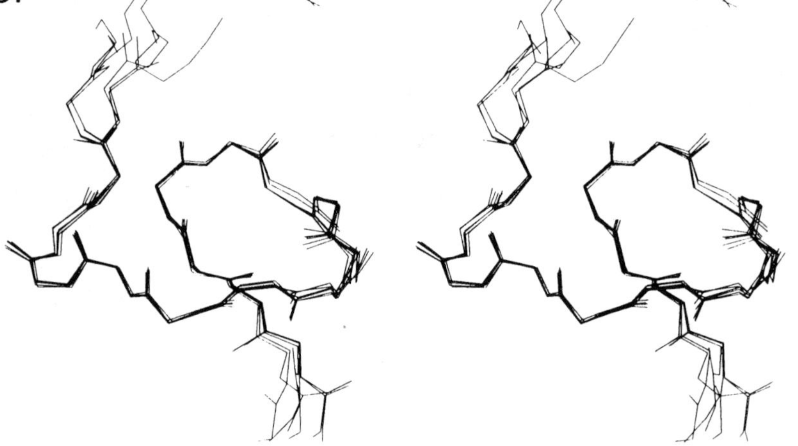

1D.

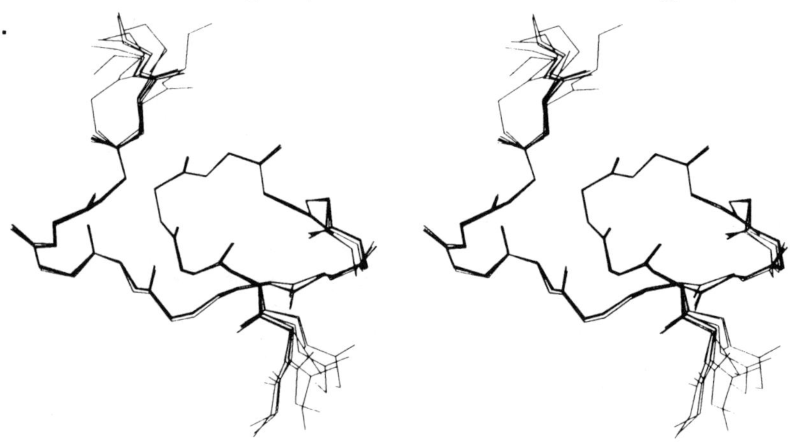

Figure 1 (continued).

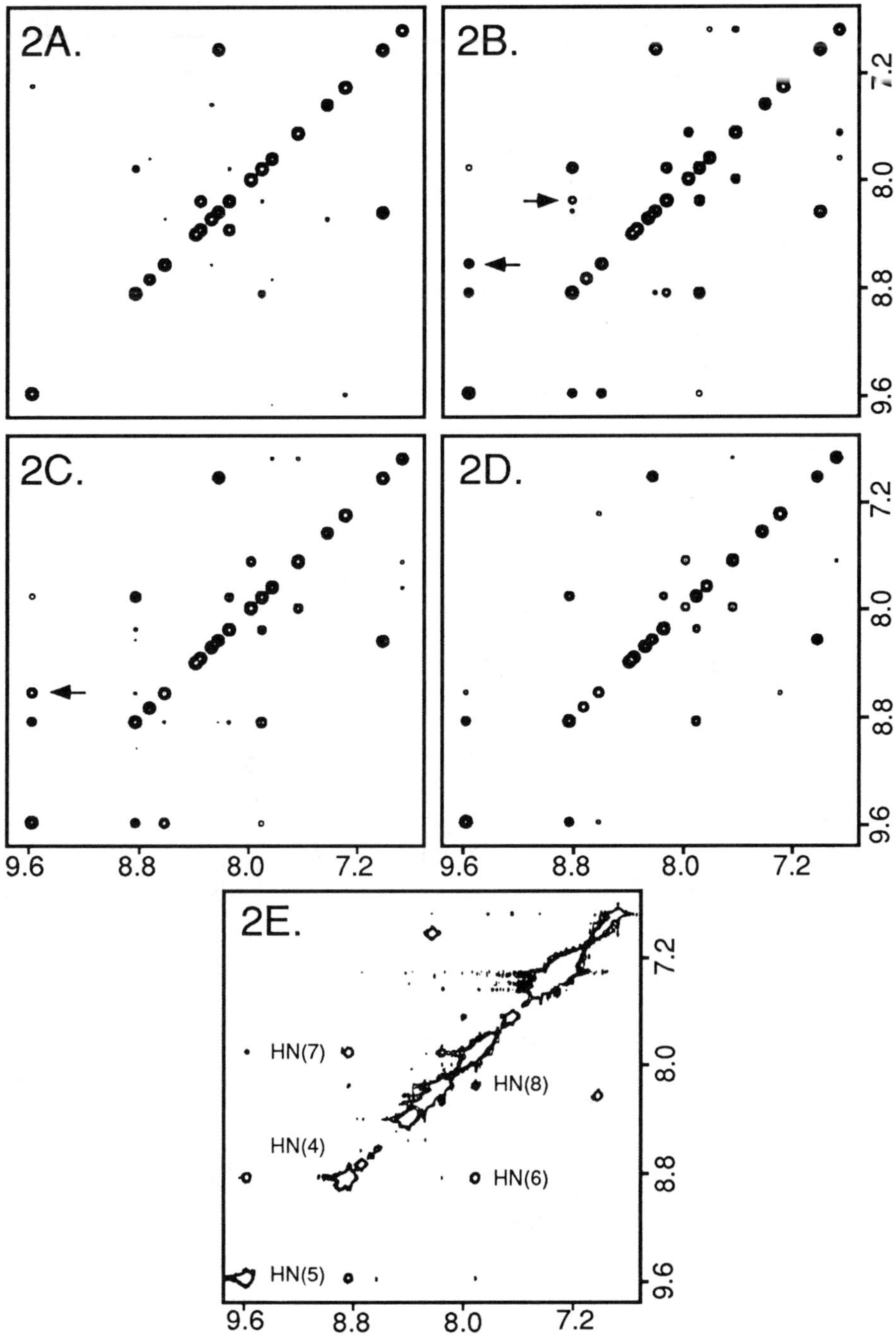

Figure 2. A-D: 2D-NOESY back-calculated spectra representative DG structures A-D, respectively, shown in Fig. 1. E: experimental NOESY spectrum.

Although the "goodness" of the C structures has been improved, differences between the theoretical and experimental NOESY spectra still exist. For example, in Fig. 2C, the strong intensity of the back-calculated HN(4)-HN(5) cross peak relative to the experimental cross peak (Fig. 2E) is clearly visible. Several other differences exist when the NOE build-up profiles are compared (data not shown). (Complete experimental NOE build-up and diagonal peak decay curves have been published (3)). Thus, at this stage of the refinement process, the back-calculated and experimental 2D NOESY spectra are qualitatively similar, but cross peak intensities do not necessarily match. The most significant structural changes resulting from the addition of back-calculation constraints involve reorientation of the amide linkages between residues 7-8 and 9-10.

D. STRUCTURES GENERATED VIA COMPLETE SIMULATION OF THE NMR RELAXATION MATRIX

Although the above C structures gave back-calculated spectra that qualitatively match the experimental data, differences in cross peak intensity are clearly visible. The most notable differences involve the intensities of the NH(4)-HN(5) cross peaks (Fig. 2C). To account for intensity differences, a full simulation of the relaxation matrix was performed. NOESY data were collected with mixing times of 5, 50, 100, 300, and 500 msec, and constraints for directly observed NOE cross peaks were adjusted and narrowed in order to quantitatively match the experimental NOE buildup and diagonal peak decay data. This step of the refinement process has been described at length (3). Eight final structures and a portion of the theoretical NOESY spectrum for a representative final structure are shown in Figs. 1D and 2D, respectively. The back-calculations for the final structures included individual z-leakage rate parameters to account for variations in amide-solvent exchange rates. The final structures superimpose with RMSDs for backbone atoms in the range 0.06 – 0.19 Å. The structural differences between the C and D models are small, with the most significant changes involving slight reorientations of the 6–7 and 9–10 amide linkages, Fig. 1. Unlike the B and C structures, the backbone carbonyl of residue Cys(3) is oriented in a manner consistent with hydrogen bonding to the Gly(7) backbone amide proton. NH····O distances of 2.99, 3.29 and 2.24 Å were obtained in the B, C, and D structures, respectively.

IV. DISCUSSION

Initial inspection of the structures shown in Fig. 2 might lead one to conclude that back-calculations are not required for NMR based structure determination, since the structural variations between the B-D models is relatively small. It appears that, for a system such as Zn(HIV1-F1) which contains a large number of long-range and sequential NOE interactions within a small stretch of amino acids, 2D NOESY back-calculations are not necessary to define the overall folding, at least when distance geometry algorithms are properly employed for minimizing penalties. However, as described in our earlier work (3), back-calculations can be very useful for "directing" folding to a low-penalty conformation. This is especially important when performing simulated annealing and minimization with slower computers or without the aid of several new algorithms (e.g. "idealized" annealing and conjugate gradient minimization; see Scheme 1). Thus, in our earlier work, initial structures with modest penalty values (ca. 1.0 Å^2) were modified to give low penalty structures by addition of constraints dictated by the back-calculated spectra. The earlier approach utilizing back-calculations to direct folding to low penalty structures appears to be more time consuming now that faster computers and new DG algorithms are available.

The low penalty B structures, generated with loose NOESY distance constraints, exhibit pairwise RMSD values similar to those of the D structures which were generated with very tight distance constraints. We believe that this is due to the influence of van der Waals radii in determining the conformation of this tightly folded structure. Despite the high convergence obtained for the B structures, it is clear from the back-calculated spectra that aspects of the 3D structure are not correct. For example, the backbone NH proton of Gly(7) is hydrogen bonded to the Cys(3) carbonyl in the D structures but not in any of the A-C structures. The slow solvent exchange observed for the Gly(7) amide proton at elevated temperatures provides good evidence that this proton is indeed involved in hydrogen bonding.

It is significant to note that none of the B structures generated via Scheme 1 gave penalties as low as those observed for the C structures, despite the fact that the C structures were generated with more than twice as many experimental constraints. We attribute this to deficiencies inherent in the refinement process, where initial structures with moderate penalties (< ca. 1.2 Å^2, after the First Step indicated in Scheme 1) are refined with the use of small initial velocities (low heating). With this approach, low penalty barriers can be surmonted, but high penalty barriers are circumvented as shown in Fig. 4. The penalty barriers that separate the B and C structures appear to be associated with the Gly(7) and Gly(10) residues which are involved in tight turns. It should be pointed out that the C structures, when refined using the B-structure bounds matrix, exhibited penalties of ca. 0.02 Å^2, which are lower by a factor of 2 than the penalties obtained with the B-structure constraints and are slightly lower than the penalties obtained with the C structure constraints. Thus, on addition of constraints (from B to C), we have modified the phi(ϕ)-psi(ψ)-penalty surface function, eliminating or

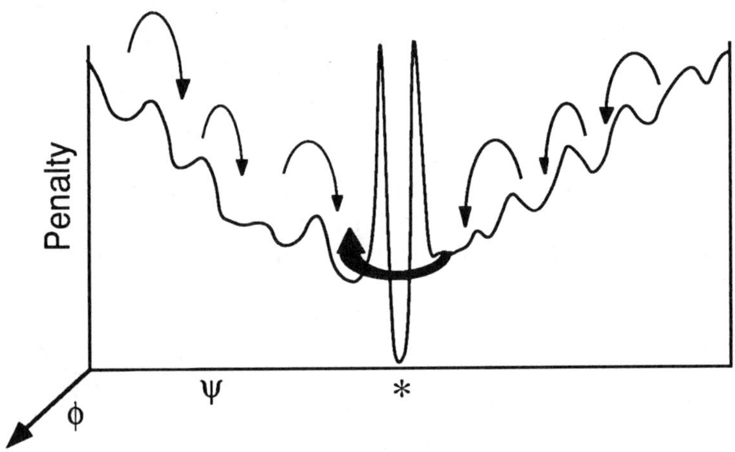

Figure 4. Diagram showing a slice of a hypothetical psi-phi-penalty function surface. With the DG algorithms we employed, small penalty barriers are surmounted but large penalty barriers are circumvented. The net result is convergence on a low penalty minimum that is not necessarily the "global minimum" (*).

raising the penalty of several local minima but not altering significantly the penalty value for what appears to be the "global minimum" (indicated by an asterisk).

Is 2D NOESY back-calculation necessary? To obtain highly converged structures, back-calculations are not necessary, at least for structures as compact as Zn(HIV1-F1). We believe that this is due to the nature of our DG structure generation approach, where high-velocity annealing (severe heating) is initially employed to rapidly search conformational space for the correct general folding (resulting in moderately low penalties), and then low-velocity annealing with idealized primary geometry constraints is employed to surmount low-penalty barriers in search for minimum penalty conformations. Thus, at the second stage of refinement, annealing velocities are not great enough to overcome high-penalty conformational barriers. This results in the generation of a highly converged set of structures. However, as shown above, highly converged structures are not necessarily "good" structures in terms of their consistence with experimental data. The high degree of convergence ("resolution") is in fact misleading, and the common use of the term "resolution" to describe the degree of convergence should probably be avoided.

2D NOESY back-calculations appear to be necessary for establishing orientations of some amide linkages, at least when additional constraints, such as user-defined hydrogen bonding, are not included in the bounds matrix. For Zn(HIV1-F1), lack of regular secondary structure common to most proteins (ie. α-helices & β-sheets) precluded the implementation of user defined hydrogen bond constraints. Thus, for this molecule, several hydrogen bonds (NH···O and HN···S) could only be identified subsequent to DG structure generation, and one hydrogen bond (Cys(3)O-Gly(7)NH) was only identified after complete simulation of the NOESY build-up profiles.

REFERENCES

1. Hard, T., Kellenbach, E., Boelens, R., Maler, B. A., Dahlman, K., Freedman, L. P., Carlstedt-Duke, J., Yamamoto, K. R., Gustafsson, J., Kaptein, R. (1990). Science **249**, 157.
2. Driscoll, P. C., Gronenborn, A. M., Wingfield, P. T., Clore, G. M. (1990). Biochemistry **19**, 4668.
3. Summers, M. F., South, T. L., Kim, B., Hare, D. R. (1990). Biochemistry **29**, 329.
4. Lee, M. S., Gippert, G. P., Soman, K. V., Case, D. A., Wright, P. E. (1989). Science **245**, 635.
5. Wüthrich, K. (1989). Science **243**, 45.
6. Wüthrich, K. (1986). "NMR of Proteins and Nucleic Acids", Wiley, N.Y.
7. Billeter, M., Schaumann, T., Braun, W., Wüthrich, K (1990). Biopolymers **29**, 695.

8. Gronenborn, A. M., Clore, G. M. (1989). Biochemistry **28**, 5978.
9. Boelens, R., Koning, T. M. G., van der Marel, G. A., van Boom, J. H., Kaptein, R. (1989). J. Magn. Reson. **82**, 290.
10. Banks, K. M., Hare, D. R., Reid, B. R. (1989). Biochemistry **28**, 6996.
11. Nerdal, W., Hare, D. R., Reid, B. R. (1989). Biochemistry **28**, 10008.
12. Borgias, B. A., James, T. L. (1990) J. Magn. Reson. **87**, 475.
13, Havel, T. F., Wüthrich, K. (1985). J. Mol. Biol. 21, 5129.
14. Havel, T. F., Crippen, G. M., Kuntz, I. D. (1979). Biopolymers **18**, 73.
15. Crippen, G. M. (1981). "Distance Geometry and Conformational Calculations", Research Studies Press/Wiley, Chichester, U.K.
16. Metzler, W. J., Hare, D. R., Pardi, A. (1989). Biochemistry **28**, 7045.

35

HETERONUCLEAR THREE-DIMENSIONAL NMR: TOWARDS THE STRUCTURE DETERMINATION OF MEDIUM SIZED PROTEINS IN SOLUTION

Erik R.P. Zuiderweg, Edward T. Olejniczak, Robert T. Gampe, Jr., Andrew M. Petros, Hugh L. Eaton and Stephen W. Fesik
NMR Research, Pharmaceutical Discovery Division, Abbott Laboratories, Abbott Park, Illinois 60064

INTRODUCTION

The determination of the three-dimensional structures of small proteins (MW < 10,000) in solution by NMR spectroscopy has become a well established technique (for a review see ref 1). The protocol consists of the assignment of the resonances of the proton NMR spectrum by two dimensional (2D) techniques such as scalar-correlated spectroscopy, (COSY, DQF-COSY, TOCSY/HOHAHA) in which the amino acid spin systems are catalogued by transfer of magnetic coherence through (series of) three-bond proton-proton couplings, followed by the identification of nuclear Overhauser effect cross-peaks between the resonances of sequentially adjacent amino acids from 2D NOESY spectra. This assignment forms the basis of the subsequent search for NOE distance constraints between sequentially remote protons using NOESY spectra. The obtained NOE distance constraints, which represent proton-proton distances in the range 2.2 - 5.0 A, are used as input for a distance geometry algorithm of the metric matrix or dihedral angle space format. Such a program derives a series of three-dimensional structures compatible with the distance constraints (and scalar coupling constraints) and with protein stereochemistry. Typically, the obtained structures are further refined by use of restrained molecular dynamics programs.

This approach breakes down for medium-sized proteins (10,000 < MW < 25,000), because of two reasons: i, for medium sized proteins the number of resonances becomes so large that 2D spectra become unresolved and hence cannot be unambiguously analysed; and ii, the natural linewidth of the resonances increases to over 10 Hz (because of the slower molecular tumbling) which causes coherence transfer processes over proton-proton scalar couplings (typically < 10 Hz) to become ineffective. Thus, for medium sized proteins it is hard, if not impossible, to obtain the classical scalar correlated spectra, and therefore the assignments. Even if backbone resonance assignments were obtained with selective N-15 isotopic labeling (2), overlap in the aliphatic region of the NOESY spectrum would preclude analysis of distance constraints for structure determination. As an example of an overlapping NOESY data set, Fig. 1 shows the amide proton region of a 2D NOESY spectrum of T4 lysozyme, a protein of MW 18,700. It is clear that complete analysis of this

Techniques in Protein Chemistry II

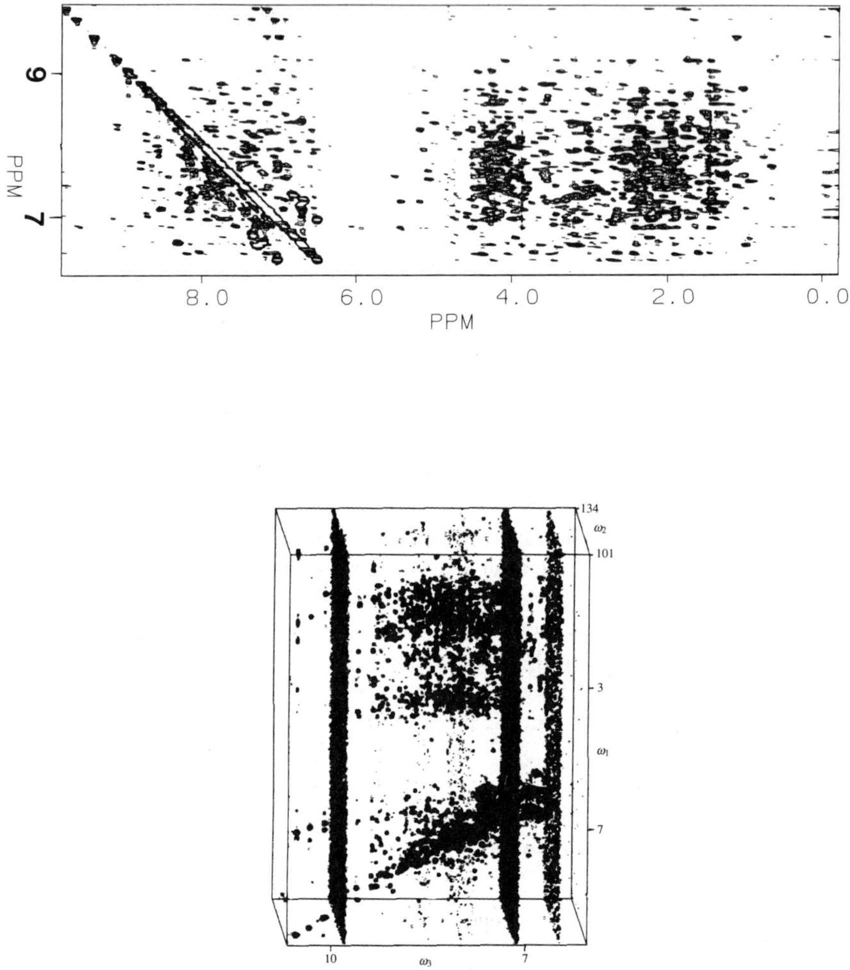

Figure 1: The amide proton region of a two-dimensional NOE spectrum and of a three- dimensional NOE spectrum (3D [H,N-15,H] NOESY-HMQC) of a 4 mM sample of N-15 labeled T4-lysozyme (from ref. 15, permission by Cambridge University Press)

data is not possible because of the multitude of overlapping cross peaks.

At the time that two-dimensional NMR was developed, it was already realized that extension of the methods into more dimensions would lead to additional resolution (3); as two-dimensional spectra are better resolved than one- dimensional spectra, so are three-dimensional spectra better resolved than two-dimensional datasets. Thus, the above mentioned problem of overlapping resonances for larger proteins can in principle be resolved by three dimensional (3D) spectroscopy. 3D NMR experiments are constructed by recording a series of 2D experiments as a function of another 2D experiment as shown in Fig. 2a. The first reports on 3D NMR combined two homonuclear (proton) experiments such as NOESY with COSY or NOESY with HOHAHA/TOCSY (4-6). Although these experiments show that 3D spectroscopy is possible, they also show a very low sensitivity even when applied to relatively small proteins. This is caused by the low efficiency of the homonuclear scalar transfer (COSY or HOHAHA/TOCSY) in the 3D experiments. For medium sized proteins this transfer step becomes even more restrictive as described above. Thus, homonuclear 3D experiments for medium sized proteins appear to be of limited use (with the possible exception of NOESY-NOESY (7)).

In heteronuclear 3D experiments, heteronuclear scalar transfers replace the homonuclear transfers. Heteronuclear transfers are governed by large one-bond coupling constants (90 - 150 Hz) and are therefore very efficient even for large proteins. Consequently, heteronuclear 3D NMR experiments provide the sensitivity and the resolution (because of the higher dimensionality) necessary to study medium sized proteins. Typically, the experiments consist of a combination of a proton-proton NOESY experiment and a one-bond heteronuclear scalar correlated experiment such as HMQC and variations thereof (8). The first heteronuclear 3D experiment was performed on a tripeptide (N-15 HMQC-NOESY) (8) but applications to small (9) and medium (10) sized proteins were demonstrated quickly after. The general appearance of such a 3D spectrum is shown in Fig.1 where it is seen that the 2D spectrum is spread in another dimension (N-15). A major requirement for these experiments is that the proteins need to be labeled (uniformly for most cases) with N-15 and/or C-13 isotopes. The labeling is generally carried out by expressing the proteins in microorganisms that grow on minimal media with N-15 ammonium salts and C-13 glucose as a sole nitrogen and carbon source, respectively. The quantity of protein needed is the same as for 2D NMR studies, i.e. approximately a 0.5 ml 1 mM solution. Here we will discuss a series of heteronuclear 3D experiments in an order corresponding to the order of application of these experiments in a novel NMR protocol for the determination medium sized protein structures in solution, rather than in the order of historical development.

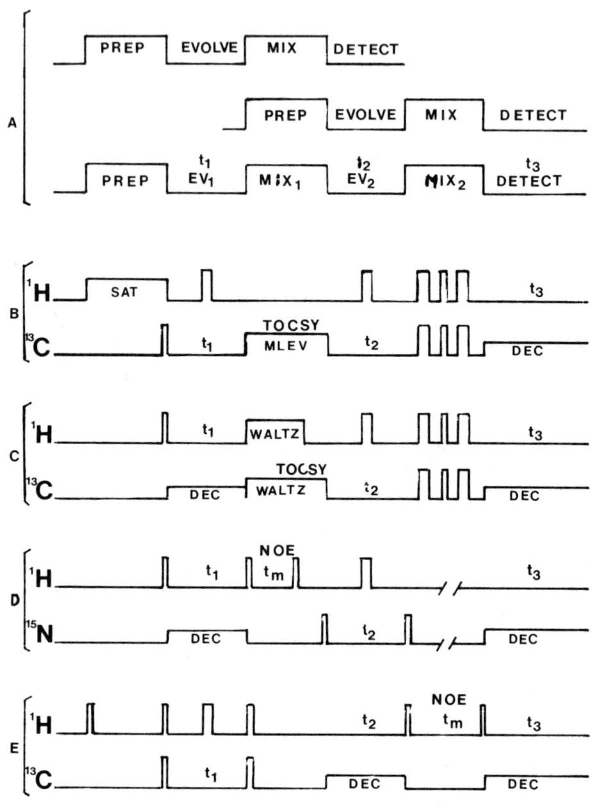

Figure 2: Pulse sequences for generating heteronuclear three-dimensional NMR spectra. Lane a, the general construction principle of 3D sequences; Lane b, 3D [C-13,C-13,H] TOCSY-REVINEPT ;Lane c, 3D [H,C-13,H] HEHOTOCSY; Lane d, 3D [H,N-15,H] NOESY-HMQC ; Lane e, 3D [C-13,H,H] HSMQC-NOESY.

SPIN SYSTEM ASSIGNMENT: 3D C-13 C-13 TOCSY TRANSFER

In this method amino acid spin systems are assigned by coherence transfer through single chemical bonds. The correlation patterns are therefore independent of the spatial conformation of the residues (11). The original pulse sequence for this experiment (12) is shown in Fig. 2b. Here we will consider a closely related sequence, derived from experiments reported in refs 12, 13 and 14, (Fig 2c) which results in (f1,f3) spectra that are reminiscent of the classical proton TOCSY/HOHAHA data. As an example, we describe the events occurring in a threonine spin system, by considering the fate of the magnetization of the proton attached to the CB carbon. HB proton coherence is generated by the proton excitation pulse and its frequency is encoded by the variable evolution time t1. During the next step, a heteronuclear cross-polarization sequence is applied, by which the HB coherence is transferred to the attached C-13, CB. The following C-13 - C-13 isotropic mixing sequence (TOCSY) distributes the C-13 magnetization over all attached C-13 nuclei (CA, CB and CG). In the following evolution period (t2) the C-13 frequencies of these atoms are encoded. The next step is to bring the coherence back to the protons by a a reverse INEPT sequence, after which the HA, HB and HG resonances are acquired (t3). Upon three-dimensional Fourier transformation, one obtains the following 3D peaks (f1,f2,f3) (HB,CB,HB); (HB,CG,HG) and (HB,CA,HA). The latter two 3D resonances establish the important connectivities which delineate the Thr spin system. Fig. 3 shows several C-13 planes of the described 3D experiment performed with a 4 mM sample of C-13 labeled T4-lysozyme (MW 18,700). The vertical axis is the f1 dimension, the horizontal the f3 dimension while the f2 dimension (C-13) is given by the different planes. In a plane corresponding to a Thr CB frequency, we observe peaks with the frequency coordinates (HA, CB, HB), (HB, CB, HB) and (HG,CB,HB). Together with planes at the CG and CA frequencies which show the same set of f1 resonances, all six resonance frequencies of the Thr spin system are obtained and their association established. The delineation of (parts) of a Ile spin system is shown in Fig. 3 as well. Upon full analysis, this type of experiment (and also the experiment in its original form, in which magnetization starts on one of the carbons (12)) yields assignments for the proton and C-13 resonances of the amino acids. The key to the success of this experiment is the transfer of coherence via a carbon coupling network. C-13 - C-13 one-bond scalar couplings are between 35 and 55 Hz and provide therefore quite efficient transfers for proteins smaller than 25,000 MW (linewidth smaller than 30 Hz).

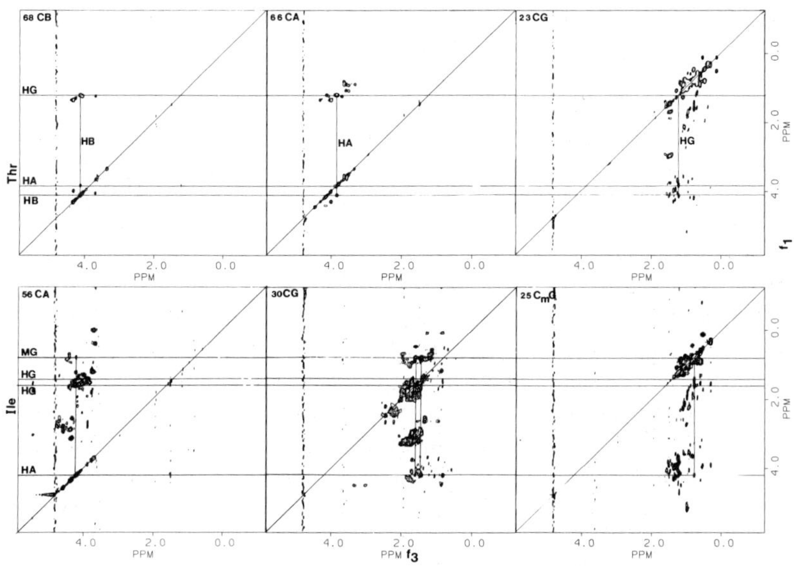

Figure 3: Cross-sections (f1,f3) through a 3D [H,C-13,H] HEHOTOCSY spectrum of C-13 labeled T4-Lysozyme. The top three panels show the spin system identification of a threonine residue. The proton resonances are indicated in f1 and f3; the carbon frequencies are identified by the C-13 resonance position of the f2 planes. The lower three panels show a similar delineation of parts of a Ile spin system.

SEQUENTIAL ASSIGNMENT AND SECONDARY STRUCTURE: 3D N-15 RESOLVED NOESY

Once the spin system identifications are obtained with the above methods, it is necessary to identify their position in the amino acid sequence. In principle, we make use of the sequential assignment procedure (1) by searching for NOE's between amide proton and other backbone and HB resonances. As is apparent from Fig. 1, such a search cannot be unambiguously made in a 2D spectrum of a larger protein. By performing the NOESY experiment as a function of a HMQC experiment in the sequence (8-10) shown in Fig. 2d, t1 frequency encoded proton magnetization (for example, the amide proton of residue i+1) is transferred by NOE to e.g. amide proton i. This proton magnetization is transferred by the HMQC experiment to the attached N-15 labeled amide nitrogen i, the frequency of which is encoded during t2. In the final part of the sequence, the signal is reverted back to the amide proton i. As a result, a 3D signal is obtained at frequencies (f1,f2,f3) (HN(i+1), N(i), HN(i)). Similarly, proton magnetization starting on HA or sidechain protons (e.g. HA(i), HA(i+1), HB(i+1)) can end on the same (f2,f3) "skewer" : (HA(i), N(i),HN(i)), (HA(i+1), N(i),HN(i)) and (HB(i+1), N(i),HN(i)). Thus the NOESY spectrum is resolved with respect to the N-15 frequency of the amide nitrogen attached to the amide proton resonating in f3. Fig. 4 shows some planes out of a N-15 resolved NOESY spectrum of T4 lysozyme. It is apparent that a dramatic improvement in spectral resolution is obtained by this method; especially the search for NOEs between amide protons is greatly facilitated. The spectrum is analyzed by searching for skewers (on different N-15 planes) which have NOE's in common (15); the horizontal lines indicate the NOE matches, the vertical lines the skewers. The distribution of NOE intensities in these skewers is dependent on the secondary structure of the protein; as has been extensively described for 2D NMR signals (1), intense cross peaks between amide protons indicate alpha helical secondary structure, while strong sequential amide proton - alpha proton NOEs indicate extended structure.

SEQUENTIAL ASSIGNMENT AND SECONDARY STRUCTURE: 3D C-13 RESOLVED NOESY

The N-15 resolved NOESY experiment edits the amide protons with respect to the chemical shift of the attached N-15 nucleus. The majority of sequential connectivities, however, involves aliphatic protons and the overlap between the resonance frequencies of these protons interferes with the proper identification of the involved NOEs. Therefore, it is necessary to further characterize the shifts of these aliphatic protons by editing the NOESY spectrum with respect to the chemical shifts of the attached C-13 labeled carbon atoms. A pulse sequence which achieves this (16) is shown in Fig. 2e; it is basically the inversion of the sequence in Fig. 2d. Fig. 5 shows some of the C-13

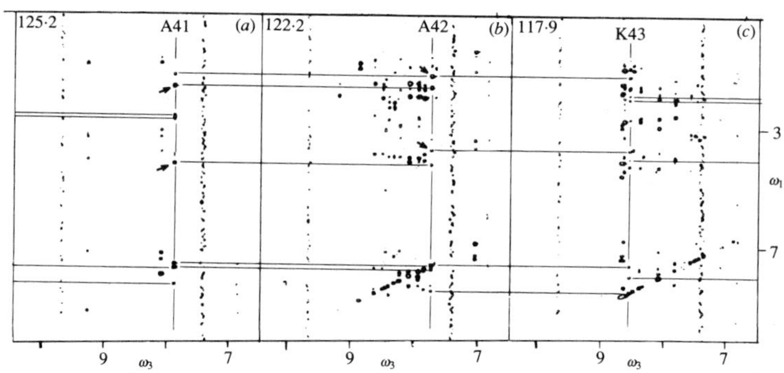

Figure 4: Cross sections (f1,f3) through a 3D [H,N-15,H] NOESY-HMQC spectrum of N-15 labeled T4-Lysozyme (a 3D contour plot of this spectrum is shown in Fig.1). Sequence specific assignments are shown for 3 residues, obtained by matching (horizontal lines) the skewers (vertical lines) in the three planes. The N-15 chemical shifts of the planes are indicated at the top (From Ref. 15, permission by Cambridge University Press)

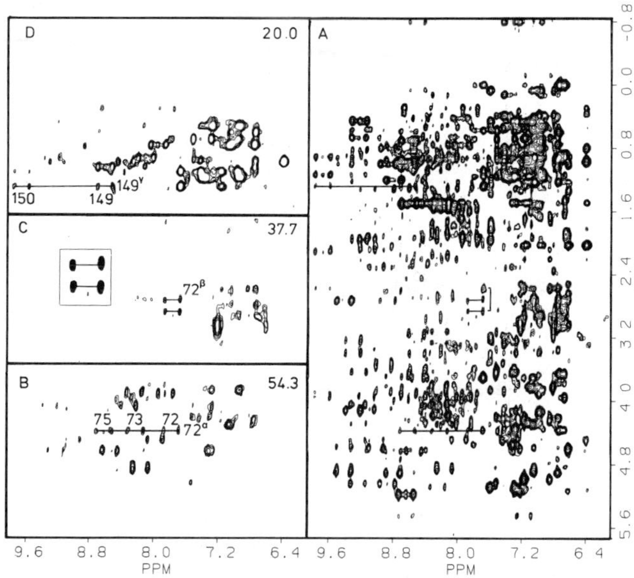

Figure 5: Projection (A) and cross sections (f2,f3) (B-D) through the amide-aliphatic region (horizontal-vertical) of a 3D [C-13,H,H] spectrum of N-15, C-13 labeled T4 lysozyme, in which approximately 50 amide protons have not been exchanged for deuterons. The spectrum is N-15 coupled. In panel B (C-13 shifts are indicated in the corner) the connectivities Val 75 NH, Ala 73 NH - Asp 72 HA and the intra residue connectivity Asp 72 NH - HA are indicated; in panel B the intra residue connectivity Asp 72 NH HB1, HB2 is indicated and enlarged; in panel D the connectivities Ile 150 NH - Val 149 HG and Val 149 NH - Val 149 HG are shown. The locations of these resonances are also indicated in the projection (From Ref. 16, permission by Academic Press).

planes of a 3D C-13 resolved NOE spectrum of T4 lysozyme. The spectrum contains resonances caused by e.g. the transfer process HA(i) -> CA(i)(t1) -> HA(i)(t2) -> HN(i+1)(t3). A skewer (f1,f2) through such a resonance is oriented horizontally and typically contains the resonances in f3 of HN(i), HN(i+1), and for alpha helices, HN(i+3) and HN(i+4) as illustrated in Fig. 5. Thus, well resolved skewers align the NOEs originating from the same aliphatic proton and overlap and interpretation difficulties are largely alleviated. It needs to be pointed out, though, that the C-13 - resolved NOESY spectrum lacks NOEs between amide protons since neither of the pair is attached to C-13. Therefore, it is necessary to have access to N-15 resolved data (see above) as well, especially for alpha helical proteins. The evaluation of the NOE intensities for secondary structure interpretation in both spectra is complicated to some extent by the efficiency of the heteronuclear coherence

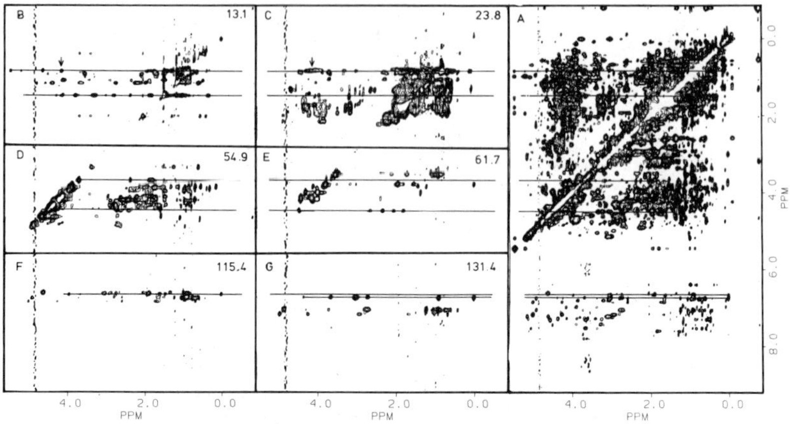

Figure 6: Projection (A) and cross sections (f2,f3) (B-G) through the aliphatic - aliphatic / aromatic region (horizontal-vertical) of a 3D [C-13,H,H] spectrum of N-15, C-13 labeled T4 lysozyme. NOE skewers which delineate NOEs originating from a single proton or proton group are indicated in the different panels which were taken at the C-13 chemical shift indicated in the corners. The locations of these resonances are also indicated in the projection. Arrows indicate NOE cross peaks which have two chemical shifts in common (f2 and f3) and could be resolved by the 3D experiment only. (From Ref. 16, permission by Academic Press).

transfers. However a semi-quantitative evaluation of the NOEs can easily be made and should be sufficient to delineate the elements of regular secondary structure.

LONG-RANGE NOE IDENTIFICATION AND TERTIARY STRUCTURE: 3D C-13 RESOLVED NOESY

NOEs between protons of residues which are remote in the amino acid sequence (long-range NOEs) are the most restrictive distance constraints for the determination of tertiary structure. With the exception of inter-beta sheet NOEs, these NOEs will for the majority of cases occur between side-chain protons. The right-hand panel of Fig. 6 shows the search region for these NOEs in a 2D NOE spectrum of T4 lysozyme. It is evident that, even when all resonance assignments would be available, the spectrum cannot be interpreted because of severe overlap. The other panels of this figure show some planes of a 3D C-13 resolved NOESY spectrum (16) of this molecule. The lines indicate skewers which delineate those NOEs which originate from a single proton. Such a proton can thus be identified on the basis of two frequencies: the proton frequency itself and the C-13 frequency of the attached carbon (the plane on which the skewer occurs). It is only in the case that simultaneous degeneracy of both these frequencies occurs,that the proton cannot be uniquely identified. It is for this interpretation of the 3D NOE spectrum in terms of both H-1 and C-13 shifts, that the availability of C-13 assignments (see C-13 C-13 TOCSY, above) is important. If the C-13 resonance frequencies were only to be used for spreading the NOESY data, only resolution, but not interpretation would improve. It is anticipated that a large number of distance constraints can be identified from this type of spectrum. Even for smaller proteins it appears worthwile to obtain a 3D C-13 resolved NOE spectrum since some overlap in the aliphatic spectrum is common for such systems as well.

STRUCTURE DETERMINATION OUTLOOK

In addition to NOE identification, heteronuclear 3D NMR spectra yield the resolution to allow the determination of the values of heteronuclear three-bond J-coupling constants. A simple example is given in the inset of Figure 5, where two NOEs between the amide proton and beta protons of residue 72 are enlarged; since no N-15 decoupling was employed, the one bond HN - N J-coupling causes a splitting of the NH resonance. The offset in vertical dimension between the connected resonances is caused by the three-bond coupling N-15 - HB (17). The value of this coupling constant is determining for the sidechain dihedral angle X. Similar coupling constants can be determined throughout the 3D spectrum, and the associated angles can be used as input in structure determination programs. In summary, a novel protocol for the determination of protein structures in solution by NMR spectroscopy can be outlined as: i) obtain spin system assignments from 3D C-13 C-13 TOCSY spectra ii) obtain sequential assignment and secondary structure from N-15 and C-13 resolved 3D NOE spectra iii) identify long-range NOE distance (and dihedral angle) constraints from C-13 resolved 3D NOE spectra. This protocol is applicable to proteins of molecular weights up to at least 25,000; for larger systems the C-13 - C-13 TOCSY transfer, which relies on coherence transfers using interactions of 35 Hz, will become inefficient because of signal decay (linewidth). It is anticipated that the structure determination will further proceed along well established lines of distance geometry and restrained molecular dynamics programs. It can also be anticipated that the 3D spectra provide the resolution to allow the evaluation of theoretical NOESY spectra back-calculated on the basis of the computed three-dimensional structure(s).

ACKNOWLEDGEMENT

We thank Drs. L.P. McIntosh and F.W. Dahlquist for the preparation of a sample of C-13, N-15 labeled T4 Lysozyme.

REFERENCES

1) Wuthrich, K. "NMR of Proteins and Nucleic Acids", (1986) Wiley, New York.
2) McIntosh, L.P., Dahlquist, F.W. and Redfield, A.G. (1987) J. Biomol. Struct. Dyn. **5**, 21.
3) Ernst, R.R., Bodenhausen, G. and Wokaun, A. "Principles of Nuclear Magnetic Resonance in One and Two Dimensions" (1987) Clarendon Press, Oxford
4) Griesinger, C., Sorensen, O.W. and Ernst, R.R. (1987) J.Am. Chem. Soc. **109**, 7227
5) Oschkinat, H., Griesinger, C., Kraulis, P.J., Sorensen, O.W., Ernst, R.R., Gronenborn, A.M., and Clore, G.M. (1988) Nature **332**, 374
6) Vuister, G.W., Boelens, R. and Kaptein, R. (1988), J. Magn. Reson. **80**, 176
7) Boelens, R., Vuister, G.W., Koning, T.M.G., and Kaptein, R. (1989) J. Am. Chem. Soc. **111**, 8525
8) Fesik, S.W. and Zuiderweg, E.R.P. (1988), J. Magn. Reson. **78**, 588
9) Zuiderweg, E.R.P. and Fesik, S.W. (1989), Biochemistry **28**, 2387
10) Marion, D., Kay. L.E., Sparks, S.W., Torchia, D.A. and Bax, A., (1989) J. Am. Chem. Soc. **111**, 1515
11) Oh, B.H., Westler, W.M., Darba, P., and Markley, J.L., (1988) Science **240**, 875
12) Fesik, S.W., Eaton, H.L., Olejniczak, E.T., Zuiderweg, E.R.P., McIntosh, L.P., and Dahlquist, F.W. (1990), J.Am. Chem. Soc. **112**, 886
13) Kay, L.E., Ikura, M. and Bax, A. (1990) J.Am. Chem. Soc. **112**, 888
14) Zuiderweg, E.R.P., J. Magn. Reson. (1990) in press
15) Fesik, S.W. and Zuiderweg, E.R.P., (1990) Q. Rev. Biophys. **23**, 97
16) Zuiderweg, E.R.P., McIntosh, L.P., Dahlquist, F.W. and Fesik, S.W. (1990) J. Magn. Reson. **86**, 210
17) Montelione, G.T., Winkler, M.E., Rauenbuehler, P. and Wagner, G. (1989) J. Magn. Reson. **82**, 198

36

STRUCTURAL ANALYSIS OF THE AMINO-TERMINAL DOMAIN OF APO-CALMODULIN BY ^{1}H-HOMONUCLEAR 2D-NMR

Ross C. Hoffman and Rachel E. Klevit

Department of Biochemistry
University of Washington
Seattle, WA 98195

I. INTRODUCTION

Calmodulin (CaM) is a small, thermostable, acidic protein that functions as a calcium signal transducer in eukaryotic cells. In response to elevated levels of intracellular calcium, CaM binds 4 Ca^{2+} ions and undergoes a conformational change that allows it to interact with and activate its targets [for comprehensive reviews see refs. 1-4]. Although the crystal structure of $4Ca^{2+}$-CaM has been solved and refined at 2.2Å (5), there is no direct structural information regarding the Ca^{2+}-free form of the protein. Thus, the precise nature of the conformational change that imparts CaM's activity is not well understood.

Two-dimensional ^{1}H-homonuclear NMR has been used in this study to characterize apo-CaM. Due to the size of CaM and the degree of internal sequence homology, spectral resolution of the approximately 800 resonances is poor and therefore complete assignment of CaM by 2D ^{1}H-homonuclear methods has proved impossible. A general approach to the problems presented by a large protein is to separate the protein into its component domains and study them individually. Previous studies on tryptic fragments of CaM have shown that each of the two globular domains (TR1=1-77, TR2=78-148) retain their

---

Abbreviations used: EGTA, Ethyleneglycol-bis-(β-aminoethyl ether)N,N,N',N'-tetraacetic acid; TSP, Sodium 3-Trimethylsilylpropionate-2,2,3,3,-d4; NOE, nuclear Overhauser effect.

Techniques in Protein Chemistry II

folded structures and their affinities for Ca^{2+} independently (6-8). However, even using the single 77-residue domain, TR1, complete assignments are not feasible using ^{1}H-homonuclear NMR, due to the internal sequence homology of the EF hand pair and to a preponderance of one type of secondary structure (α-helix).

In this report we will demonstrate that even under these conditions some useful structural information is obtainable. The assignments made in apo-TR1 allow identification of secondary structural elements and the identification of a number of tertiary contacts between these secondary structural elements, defining the tertiary fold.

II. MATERIALS AND METHODS

Sample preparation: The generation, purification, and preparation of the apo-TR1 samples was performed essentially as previously described (6). Apo-TR1 was dissolved in 400μL of 50mM imidazole-d4 (98.4 atom% MSD Isotopes) pH6.1, 50μM EGTA, 100mM KCl, and 100μM TSP (MSD Isotopes, as a chemical shift reference).

NMR Experiments: All spectra were recorded at 35°C on a Bruker AM-500 NMR Spectrometer. Spin system assignments were made using magnitude COSY and RELAY spectra collected with H_2O presaturation and confirmed by TOCSY spectra collected with SCUBA H_2O suppression. Sequential assignments were accomplished using NOESY spectra collected in the TPPI mode.

Distance Measurements: NOESY spectra were collected for a 2.7mM apo-TR1 D_2O sample with τ_{mix}=50, 75, and 100msec. Spectra were transformed using a skewed sinebell window function in t_2 and t_1 and a 5th-order polynomial baseline correction in t_2. Crosspeaks were scaled to the internal TSP reference. The crosspeak build-up rate calculated by linear regression was compared to the build-up rate calculated for several geminal $\beta\beta'$ pairs with a fixed distance of 1.54Å to obtain the experimental distance. The experimental distances were not corrected for methyl contribution or the possibility of coherence effects in the ruler build-up rate and may therefore slightly underestimate the actual distances.

Table I. ^{1}H NMR Chemical Shifts of Bovine Apo-TR1[a]

Residue	NH	C^{α}	C^{β}	C^{γ}	C^{δ}	C^{ϵ}	C^{ζ}
	chemical shift (ppm)[b]						
A 10	7.98	4.16	1.52				
E 11	7.80	4.14	2.05,2.13				
F 12	8.76	4.93	3.51,3.64		7.11	7.28	
K 13	9.29	3.79	1.82,1.95				
E 14	8.07	4.06	2.21				
A 15	7.68	4.31	1.72				
F 16	8.54	3.73	2.93,3.18		6.52[c]	6.91[c]	7.40[c]
S 17	8.40	4.26					
L 18	7.39	3.93	1.73	1.65	0.77,0.90		
F 19		4.34	2.40,3.18		7.32	7.20	
D 24	8.73[c]	4.49	2.57,2.81				
G 25	10.09[c]	3.92					
		4.26					
T 26	7.60[c]	5.52[c]	4.08	1.08			
I 27	8.34[c]	4.79[c]	1.77	0.84,1.11	0.36[c]	-0.05($C^{\gamma}H_3$)[c]	
T 28	8.33[c]	5.08[c]	4.67	1.32			
V 35	7.98	3.35	1.91	0.54[c],0.74			
S 38	8.15	4.45	4.20,4.25				
V 55	7.53	4.33	2.24	0.96,1.00			
N 60	8.91[c]						
G 61	9.45[c]	3.94					
		4.18					
T 62	7.64[c]	5.15[c]	4.10	1.11			
I 63	8.95[c]	4.60[c]					
D 64	8.21[c]	5.59[c]	2.78,3.26				
F 65	8.71	3.84	2.41,2.96		6.51[c]	7.16[c]	7.40[c]
E 67	8.03	4.08	1.83				
F 68	8.51	4.12	3.11,3.20		6.94[c]	7.01[c]	7.11[c]
L 69	8.30	3.30	1.58	1.07	0.66,0.91		

[a]T=35°C, pH 6.1
[b]chemical shifts are given relative to TSP
[c]previously assigned resonances, see refs. 6,7,13,14

Fig. 1. Connectivity diagram for apo-TR1 showing sequential short-range NOEs and i,i+3 medium-range NOEs.

III. RESULTS AND DISCUSSION

A comparison of the 2DNMR spectra of the two tryptic fragments of CaM and whole CaM confirms the earlier 1D-NMR and Ca^{2+}-binding studies which indicated that the domains retain their native conformation after proteolysis and purification by HPLC. Even in two dimensions, the crosspeaks from TR1 and TR2 are coincident with the respective crosspeaks in whole CaM (data not shown.) Sequence-specific assignments were obtained for a number of resonances using conventional methods (9,10) and are presented in Table I. Although sequence-specific assignments could be made for several unique residues throughout the sequence, it was not always possible to "walk" along the chain due to poor resolution of the NH and CαH resonances.

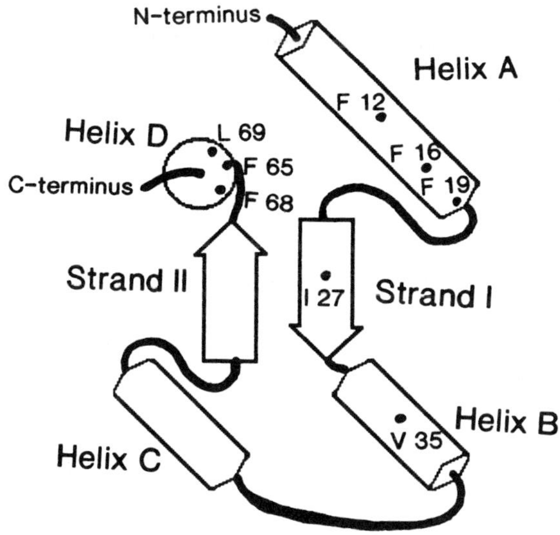

Fig. 2. Schematic of TR1 showing the labelling of the secondary structural elements and the location of assigned aromatics and aliphatics.

The short and medium range NOEs that have been unambiguously assigned are shown in the connectivity diagram (Fig. 1). Patterns of connectivity indicative of specific types of secondary structure are observed: α-helical patterns are observed for A10 through L18, T29 through E31, V35 through S38, and F65 through L69 while an anti-parallel β-sheet connectivity pattern is evident for G25 to T28 and G61 to F65.

The main secondary structural elements present in the N-terminal domain of Ca^{2+}-bound CaM as observed in the crystal structure are 4 α-helices from residues T5 to F19, T29 to L39, A46 to V55, and F65 to K77 and the short anti-parallel β-sheet between residues T26 to T28 and T62 to D64. Although only limited resonance assignments have been obtained for apo-TR1, NOE patterns indicative of regular secondary structure are observed and suggest that TR1 has the same secondary structural elements in the Ca^{2+}-free and Ca^{2+}-bound states. The implication is that the conformational transition that accompanies metal-binding is therefore a redisposition of pre-existing secondary structural elements. This hypothesis was suggested by Hennessey *et al.* (11) whose vacuum UV CD results show only a 4% increase in the amount of α-helix present in CaM when Ca^{2+} binds. A rearrangement of

Table II: Proton-proton Distances in the N_t Domain of CaM

Proton-proton Pair	Distance, Å Measured	Apo-CaM (model)	Ca-CaM (X-ray)
I) Intrahelical			
i) Helix D			
F65 δ - L69 δMe	2.9	3.6	5.1
F65 ϵ - L69 δMe	1.9	2.6	3.5
F68 δ - L69 δMe	*	2.2	2.6
II) Interhelical			
i) Helices A & B			
F12 δ - V35 γ'Me	*	6.5	9.3
F16 δ - V35 γ'Me	*	4.0	10.1
F16 δ - V35 γMe	*	3.1	9.7
F16 ζ - V35 γ'Me	*	6.9	12.1
F19 w - V35 γ'Me	2.2	3.8	3.4
F19 w - V35 γMe	1.8	3.3	3.1
ii) Helices A & D			
F12 w - L69 δMe	*	2.6	2.6
F16 ϵ - F65 α	2.1	3.0	3.0
F16 ϵ - F68 β	2.5	4.2	4.2
F16 ϵ - F68 β'	2.2	3.6	2.8
F65 ϵ - K13 α	1.8	2.5	2.4
F65 ζ - F12 β	2.2	3.4	3.5
F65 ζ - F12 β'	2.2	2.8	2.2
F65 ζ - K13 α	2.5	3.6	3.2
F68 w - F12 β	2.5	4.2	3.6
F68 w - F16 β	3.0	4.5	3.4
III) Helix-Sheet			
i) Helix A - Sheet			
F16 δ - I27 γMe	3.4	3.3	7.1
F16 δ - I27 δMe	2.1	4.1	3.2
F16 ϵ - T26 α	3.4	3.0	3.9
F16 ϵ - I27 γMe	2.2	3.7	6.6
F16 ϵ - I27 δMe	2.2	2.9	3.2
F16 ϵ - D64 α	3.8	4.2	3.1
F16 ζ - T26 α	1.8	2.3	2.4
F16 ζ - I27 γMe	*	4.0	6.9
F16 ζ - I27 δMe	*	4.4	4.9
F16 ζ - D64 α	2.1	2.9	2.7
F19 δ - I27 γMe	2.2	5.1	3.5
F19 δ - I27 δMe	2.1	4.8	3.6
ii) Helix B - Sheet			
V35 γMe - I27 δMe	*	2.2	6.1

* indicates a crosspeak is only observed at longer mixing times($\geq$150msec).

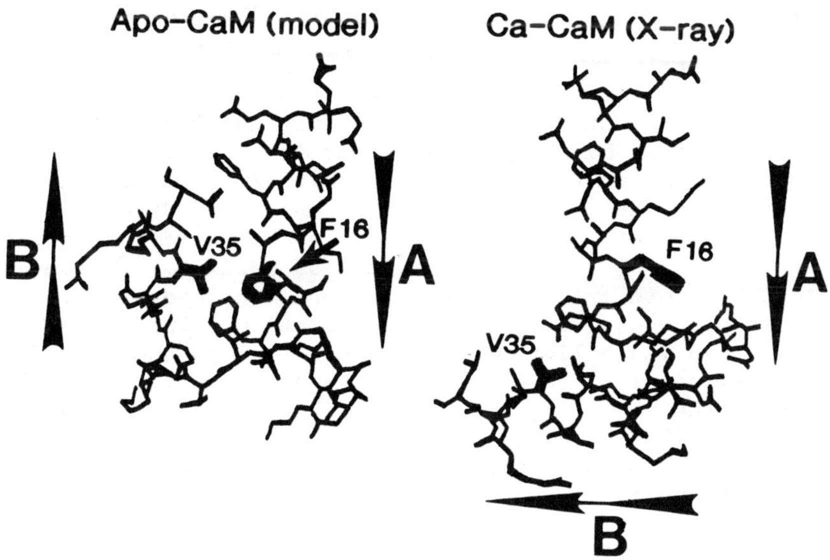

Fig. 3. Line drawing of the first EF hand of TR1 from the apo-CaM model (left) and the 4Ca2+-CaM crystal structure (right) showing the orientation of the two helices and the relative disposition of F16 and V35.

existing α-helices is also suggested by the apo-CaM model (12). In TR1, the transition predicted by the model is a pivoting of the helix B-linker-helix C region in towards helices A and D when Ca^{2+} dissociates resulting in a more closed conformation. The labelling of secondary structural elements is shown in fig. 2.

Long-range NOEs between residues distant in the sequence can be used to identify tertiary interactions among defined secondary structural elements. In the case of apo-TR1, where spectral resolution is poor and resonance assignments are incomplete, few unambiguous interactions can be identified. Thus, it is not possible to use a distance geometry algorithm to determine the structure. However, it is possible to calculate distances for resolved NOEs with assigned parents and compare these distances to the crystal structure of $4Ca^{2+}$-CaM and the apo-CaM model. Table II compares the experimentally obtained distances between aromatic ring protons and other aliphatic protons to the distances measured in the crystal structure and model. The location of the aromatic and aliphatic residues can be seen in Figure 2.

Most of the aromatic-aliphatic interactions between helices A and D and the sheet are very similar in both the $4Ca^{2+}$-CaM crystal structure and the apo-CaM model. There are however interactions between helices A and B and between helix B and the sheet that would give rise to NOEs in the apo conformer proposed by the model and not in the calcium-bound conformer. In the crystal structure, helix B lies with its helical axis in the plane of the sheet and perpendicular to helix A (5). In this conformation V35 in the center of helix B interacts with F19 at the C-terminus of helix A. The apo-CaM model has more extensive interaction among hydrophobic sidechains because of its closed conformation. Since helix B would now have its helical axis perpendicular to the sheet and run anti-parallel to helix A, V35 would have interactions with F16 in the center of helix A and I27 in the sheet, as well as with F19 (12). This can be seen in figure 3.

NOEs are observed at longer mixing times (150msec) in apo-TR1 between F16 and V35. Even if these are secondary NOEs, they indicate that these residues are within approximately 6Å of one another. These NOEs would not be observed if apo-TR1 were in a conformation similar to $4Ca^{2+}$-CaM where the distance between F16 and V35 is roughly 11Å. Although it is not possible to conclusively determine the structure of apo-TR1 from this data, it is possible to conclude that apo-TR1 retains a unique, folded structure in the absence of metal and that this structure contains regions of secondary structure similar to the calcium-bound conformer. Furthermore, apo-TR1 has different tertiary interactions than the calcium-bound conformer and these tertiary interactions support the validity of the apo-CaM model proposed by Strynadka and James.

IV. ACKNOWLEDGEMENTS

This work was supported by NIH NIDDK R01 DK35187, by an American Heart Association Established Investigatorship (R.E.K.) and by NIH Molecular Biophysics Training Grant #T32-GM08268-02 (R.C.H.).

Special thanks to Natalie Strynadka and Mike James for supplying us with the model coordinates of apo-CaM.

V. REFERENCES

1. Scharff, O. (1981).Cell Calcium **2**, 1
2. Vanaman, T.C. (1980).*In* "Calcium and Cell Function"(W.Y.Cheung, ed.), Vol.I, p. 41. Academic Press, NY
3. Klee, C.B. and Vanaman, T.C. (1982). Adv. Protein.Chem. **35**, 213
4. Cohen, P. and Klee, C.B. (1988)."Calmodulin", Elsevier Science Publ., Amsterdam
5. Babu, Y.S., Bugg, C.E., and Cook, W.J. (1988).J.Mol.Biol. **204**, 191
6. Dalgarno, D.C., Klevit, R.E., Levine, B.A., Williams, R.J.P., Dobrowolski, Z., and Drabikowski, W. (1984). Eur.J.Biochem. **138**, 281
7. Ikura, M., Hiraoki, T., Hikichi, K., Minowa, O., Yamaguchi, H., Yazawa, M. and Yagi, K. (1984).Biochem. **23**, 3124
8. Minowa, O. and Yagi, K. (1984)J.Biochem. **96**, 1175
9. Wuethrich, K., Wider, G., Wagner, G. and Braun, W. (1982).J.Mol.Biol. **115**, 311
10. Englander, S.W. and Wand, A.J. (1987).Biochem. **26**, 5953
11. Hennessey, J.D., Parthasavathy, M.W., Johnson, C.W., Malencik, D.A., Anderson, S.A., Schimerlik, M.I., and Shalifin, Y. (1987). Biopolymers **26**, 561
12. Strynadka, N.C.J. and James, M.N.G. (1988).Proteins **3**, 1
13. Krebs, J. and Carafoli, E. (1982).Eur.J. Biochem. **124**, 619
14. Ikura, M., Minowa, O., Yazawa, M., Yagi, K., and Hikichi, K. (1987).FEBS **219**, 17

37

The Three-Dimensional Solution Structure of Human Anaphylatoxin C3a[1]

Matthew W. Kalnik, Walter J. Chazin, & Peter E. Wright
Department of Molecular Biology, Research Institute of Scripps Clinic
La Jolla, CA 92037, USA

I. Introduction

Human C3a anaphylatoxin (M_r 8900; 77 amino acids) is a biologically active fragment of the C3 component within the complement pathway involved in the inflammatory response. C3a induces smooth muscle contraction and increases vascular permeability simultaneous with complement activation, as well as, histamine release from mast cells. Synthetic peptides comprising the carboxy-terminal fragment (LGLAR) of human C3a define the basic effector portion of C3a and presumably this segment is the site of interaction with the cellular receptor. Native C3a is ~50-fold more potent than synthetic peptides in competitive spasmogen assays (1).

NMR[2] structural studies of human C3a were conducted to determine a detailed three-dimensional model in the hope of elucidating the mechanism by which the amino-terminus and core of the C3a molecule enhance the biological activity of the LGLAR effector segment by 50 to 100-fold.

Advances in multi-dimensional NMR have provided a methodology by which the complete set of ^1H spins in small proteins (<15 kD) can be uniquely identified and thus the local interactions between specific spins can be determined. Analysis of the NOE, which entails exchange of magnetization between spins located ~<5.0 Å apart in space, provides a set of semi-quantitative distance constraints. These constraints along with the covalent geometries are used to reconstruct a 3D molecule in cartesian space via the implementation of distance geometry (2,3). A solid-state 3D structure for C3a has been determined by X-ray crystallography to 3.2 Å resolution (4). The first 12 residues are disordered in the crystalline state and are missing in the X-ray model.

---

1. This work was supported by two grants from the National Institutes of Health: PO1-HL16411-17 & RO1-GM36643.

2. *Abbreviations*: NMR, nuclear magnetic resonance; NOE, nuclear Overhauser effect; 3D, three-dimensional; TPPI, time-proportional phase incrementation; FID, free induction decay; NOESY, nuclear Overhauser effect spectroscopy; COSY, correlated spectroscopy; E. COSY, exclusive correlated spectroscopy.

The complete set of ^1H resonance assignments of human C3a have been previously determined in this laboratory (5). Taking these assignments as the starting point, the experimental methodology used in constructing the distance and torsion angle constraints and the initial structural refinements will be the focus of this report.

II. Material and Methods

Human C3a was isolated from human sera (6,7) and NMR samples were prepared as described previously (5). ^1H NMR spectra were acquired from a single C3a sample (2.0 mM protein, 10 mM NaCl, 0.5 mM NaN_3, pH 5.5, 300 K) with either a Bruker AM-500 or AM-600 spectrometer. The data were processed offline on a Convex C-240 or Silicon Graphics Personal Iris 4D/20 with FTNMR software (Hare Research, Inc., Woodinville, WA). All spectra were collected in a phase-sensitive sequential acquisition mode. Quadrature detection in ω_1 was achieved with TPPI[2] (8). All data that were cosine modulated in ω_1 were treated for t_1 ridge suppression by dividing the first FID by 2 (9) and baseline corrected with a spline function in both dimensions.

A NOESY[2] (10) build-up series was collected with mixing periods of 15, 30, 45, 60 & 90 ms in 95% ^1H_2O/ 5% ^2H_2O (for convenience referred to as ^1H_2O solution). A Hahn-echo observation pulse scheme was used for detection to eliminate baseline artifacts (11). These NOESY spectra were acquired with 32 scans per t_1 value and 450 t_1 values ($t_{1\,max}$= 40 ms).

Four other NOESY experiments, from which the distance constraints were derived, were collected to optimize sensitivity. Two experiments in ^1H_2O and two in ^2H_2O were acquired with mixing periods of 75 and 150 ms, 96 scans per t_1 increment, and 550 t_1 values ($t_{1\,max}$ = 51 ms). All NOESY data were zero-filled to a final 2K x 2K matrix giving a digital resolution of 3.1 Hz/point in ω_2 and 2.6 Hz/point in ω_1. The NOESY data were also treated with a lorentz-to-gauss transformation to optimize sensitivity and reduce baseline roll.

2D COSY (12) in ^1H_2O were folded in the ω_1 dimension to maximize the digital resolution (2.9 Hz/point in ω_2 and 4.0 Hz/point in ω_1 prior to zero-filling) and acquired with 48 scans per t_1 value and 750 t_1 values ($t_{1\,max}$ = 125 ms). Strong sine-bell window functions were applied to suppress baseline distortions. The final data matrix consisted of a submatrix containing the HN $\rightarrow$ H^α spectral region (6.0 –10.0 ppm in ω_2 & 3.0 – 6.0 ppm in ω_1) which was zero-filled to yield a final digital resolution of 0.18 Hz/point in ω_2 and ω_1. Analysis of the amide/alpha crosspeaks proved to be somewhat problematical since C3a is mostly α-helical and has $^3J_{HN\alpha}$ values < 6 Hz, while the typical resonance linewidth in C3a is on the order of 8 Hz. The method of Kim & Prestegard (13) was used to determine the $^3J_{HN\alpha}$ values reducing the uncertainty of the measurement due to cancellation within the resonance multiplet.

E. COSY (14) experiments were recorded in ^2H_2O to obtain accurate values of the $^3J_{\alpha\beta}$ coupling constants to aid in the stereo-assignment of the prochiral β-methylene protons (15,16). The $^3J_{\alpha\beta}$ values were determined by measurement of peak-to-peak separation within the crosspeak multiplet. E. COSY spectra were folded in the ω_1 dimension to maximize the digital resolution (3.3 Hz/point in ω_2 and 5.9 Hz/point in ω_1 prior to zero filling) and acquired with 144 scans per t_1 value and 580 t_1 values ($t_{1\,max}$ = 85 ms). The final data matrix consisted of a submatrix containing the H^α $\rightarrow$ H^β spectral region (2.0 – 6.0 ppm in ω_2 & 1.0 – 5.0 ppm in ω_1) which was zero-filled to yield a final digital resolution of 0.41 Hz/

Table I. Summary of Distance Constraints for C3a at 300 K

Type of constraint	Set 1	Set 2	Set 3
Intraresidue	335	321	363
± 1 residue	179	194	218
± 2 residues	10	18	18
± 3 residues	74	92	114
± 4 residues	8	23	33
≥ 5 residues	62	91	118
H-bond	15	15	15
β – methylenes[a]	0	7	7
γ, δ – methyl groups[b]	0	0	3[c]

[a] Unambiguous stereo-assignments taken directly from the output of HABAS (17).
[b] Stereo-assignments of prochiral methyl groups in leucine and valine.
[c] Assigned after 'floating-chirality' analysis of Set 2 structures (20).

point in ω_2 and ω_1. The program HABAS (17) was used to automate the stereo-assignment of the prochiral β-methylene protons.

NOE distance constraints were determined by integration of crosspeak intensities using an FTNMR utility which incorporates ellipsoid footprinting of each crosspeak. Constraint generation and assignment database management was semi-automated with in-house software[3] including intelligent error-checking, internal consistency of crosspeak assignment and data format conversion.

Distance geometry embedding, conjugate gradient minimization and simulated annealing using distance-driven dynamics were conducted with the program DSPACE (Hare Research, Inc., Woodinville, WA).

III. Results and Discussion

The generation of constraints for the distance geometry embed and the structural refinement consisted of an iterative procedure, where initially well-resolved crosspeaks were assigned based solely on chemical shifts at 300 K. Further assignments are based on the structures generated from this initial data set. The first round of assignments yielded 668 NOE-derived distance constraints (Table I). Ambiguities due to chemical shift degeneracy were resolved by using the initial structures and only considering proton pairs separated by ≤ 7 Å. This second set contained 739 NOE-derived distance constraints. These second round structures were again used to filter crosspeak assignment ambiguities and a third set of 864 NOE-derived distance constraints was generated.

The intensities of the NOE crosspeaks were separated into four categories (Table II) based on characteristic distance markers such as $d_{\alpha N}(i,i+1)$ connectivities in helical segments which are approximately 3.5 Å apart, strongest $d_{\beta N}(i,i+1)$ connectivities which are ~2.5 Å apart, $d_{\alpha N}(i,i+3)$ which are ~3.3 – 3.4 Å apart in 3_{10}- and α-helices, and d_{NN} connectivities in helical regions which are between ~2.6 – 2.8 Å apart (18). The NOE crosspeak intensities derived from spectra acquired in ^1H_2O and ^2H_2O were scaled based on the averaged ratio of 20 resolved diagonal peaks and strong isolated crosspeaks detected between nonexchangeable protons (data not shown). The upper bounds of each category were extended

3. GENBOUND, Dr. M. W. Kalnik, unpublished.

Table II. NOE Distance Constraint Calibration for C3a at 300 K

	NOEs involving	
Scaled Volumes[a]	Backbone Protons only	Side-chain Protons
> 240	1.8 – 2.6 Å	1.8 – 3.0 Å
120 – 240	1.8 – 3.0 Å	1.8 – 4.0 Å
0 – 120[b]	2.5 – 4.0 Å	2.5 – 5.0 Å
observed @ 150 ms	2.5 – 5.0 Å	2.5 – 5.5 Å

[a] Arbitrary units measured from a NOESY collected with a 75-ms mixing period.
[b] A lower bound of 2.5 Å was included since no $d_{\alpha N}(i,i+4)$ NOEs were observed in α-helical regions at mixing time of 75 ms.

for NOEs involving side-chain protons to account for the differential mobility of the side-chains and backbone according to the uniform averaging model (19). Since hydrogen-bonding can only be strictly determined in α-helices (rather than 3_{10}- helices), the following criteria were used to identify regions of α-helix: slowed amide proton exchange, $^3J_{HN\alpha}$ values < 6 Hz, and detection of $d_{\alpha N}(i,i+2)$, $d_{\alpha N}(i,i+3)$, and $d_{\alpha N}(i,i+4)$ NOEs. This analysis allowed 4 hydrogen-bonds in Helix II (Leu19 to Gln29) and 11 hydrogen-bonds in Helix IV (Glu47 to Arg65) to be included as distance constraints. The Van der Waals radii of the backbone amide protons and carbonyl oxygens were reduced to 0.75 Å accounting for the polarizability of these atoms.

The general protocol for structure determination begins with the definition of a bounds matrix which contains the full set of distance constraints. The upper and lower bounds are then smoothed by application of the triangle inequality

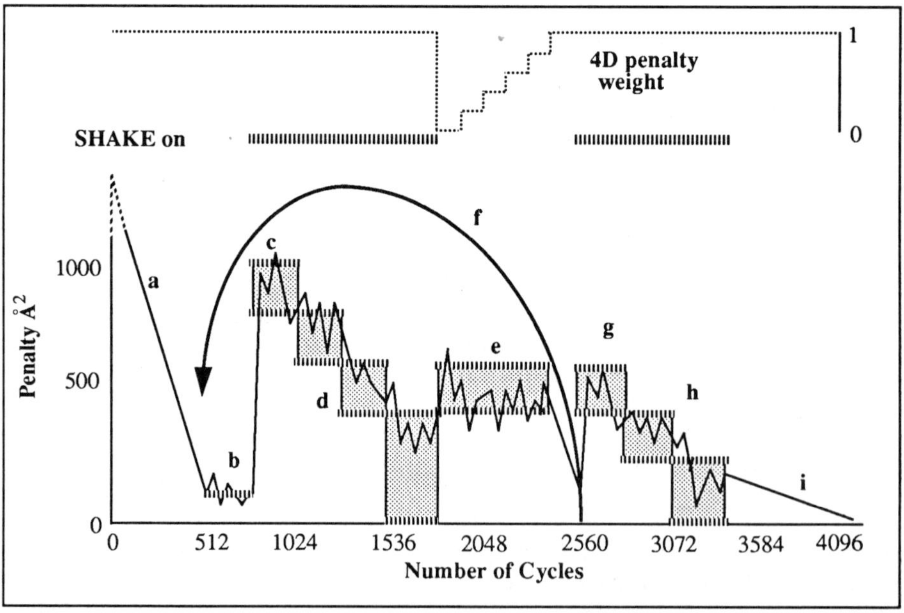

Figure 1. C3a Refinement Protocol using Dspace

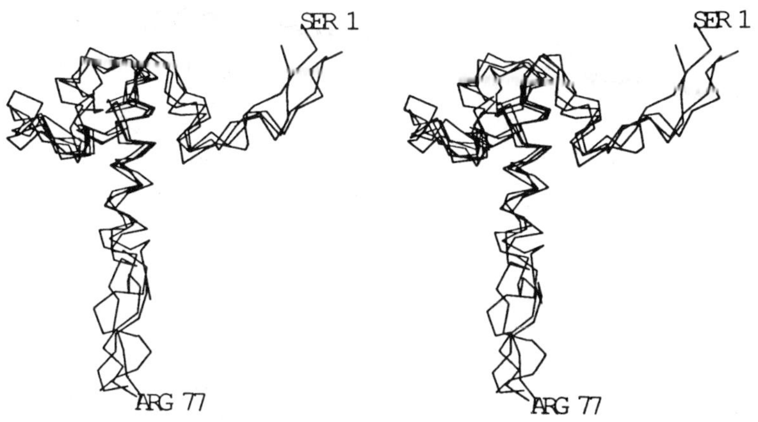

Figure 2. Superposition of four converged C3a solution structures after refinement using the protocol outlined in Figure 1. Pairwise RMSDs for the backbone atoms of residues 6–65 were all less than 1.0 Å, all atom pairwise RMSDs for residues 6–65 were all less than 3.5 Å.

(21). The metric matrix (3) was generated by choosing random trial distances for each pair of atoms which lie between the upper and lower bounds and calculating the metric which is defined as the scalar product of the two vectors from the center of mass of the molecule to each of the two atoms of interest. The eigenvalues of this metric matrix were then calculated and the 4 largest eigenvalues were used to generate a structure in 4-dimensional space.

The refinement of the embedded structures is outlined in Figure 1: ***a***) 512 steps[4] of conjugate gradient minimization. ***b***) 256 equilibration steps of undamped dynamics without any added velocity in three dimensions. ***c***) 128 steps of 'high-penalty' dynamics for conformational sampling with SHAKE (22) constraints[5] enabled. ***d***) 384 steps of damped dynamics to allow molecule to 'anneal' slowly into a low-penalty conformation. ***e***) 128 steps of dynamics with 4-dimensional sampling followed by 512 steps of 'cooling' into 3-dimensional space (no SHAKE constraints). ***f***) test of residual penalty at this stage of refinement and if above 20 Å^2 total penalty, repeat stages ***b–f*** up to 3 more times. ***g***) 128 steps of intermediate level dynamics with SHAKE enabled. ***h***) 256 annealing steps of damped dynamics. ***i***) 1024 steps of CG minimization. The shaded regions is Figure 1 represent the range of penalty values allowed during that particular stage of refinement.

Figure 2 depicts the superposition of the α-carbon traces of 4 converged structures after being subjected to the above refinement protocol. The criteria for convergence were: no single constraint violations greater than 0.5 Å, each pairwise root-mean square deviation for the backbone atoms[6] less than 1.0 Å, and a residual total penalty of less than 10 Å^2. The structures reveal a similar global

4. To reduce the amount of computational time the penalty surface was recalculated every 16 steps of refinement and the first 256 steps of minimization considered only covalent geometries and VDW interactions.
5. 0.005 Å allowable bond length deviation and 0.010 Å allowable geminal distance deviation.
6. Residues 6–65 were used for the superpositioning since very few inter-residue constraints were measured in the amino- and carboxy-terminal regions of C3a (unpublished observation and see (5)).

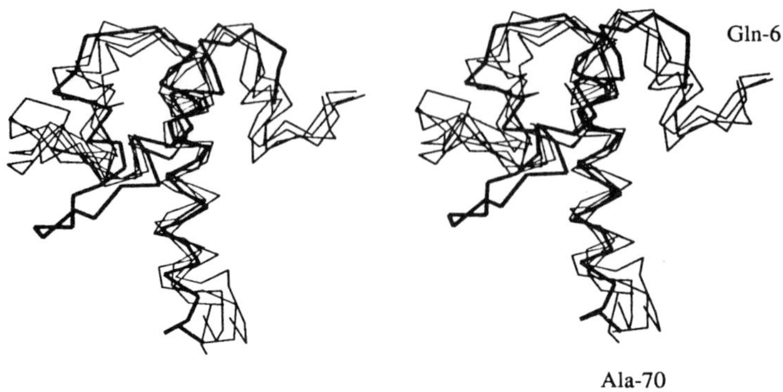

Figure 3. Superposition of the four converged C3a solution structures (thin; residues 6–70) with the X-ray structure determined at 3.2 Å resolution (thick; residues 13–70).

fold of C3a as determined previously by a qualitative interpretation of the NMR data and by single crystal X-ray crystallography (4). The solution state structure has well-defined α-helix extending from Leu19 to Gln29 (Helix II) and from Glu47 to Arg65 (Helix IV), whereas in the solid state structure, Helix II extends from residue Tyr15 to Met27 and Helix IV extends from residue Gly46 to Ser71. The superposition of the X-ray crystallographic structure (thick) and of the 4 converged solution state structures (thin) is given in Figure 3.

In the solution structures, Tyr15 is packed between Val12, Leu19, Cys23 and Val52, while in the solid state Tyr15 is directed away from Helix II & IV (Figure 4). This favorable packing of Tyr15 in the solution state may be the driving force behind the shift in the register of Helix II from residues 15–27 in the solid state to residues 19–29 in the solution state. Another significant difference between the solution and solid-state structures is the extent of the C-terminal helix (IV). In the solution state this helix begins fraying at Arg65, while in the solid state the helix extends through to Ser71 (Figure 5). This difference is very significant

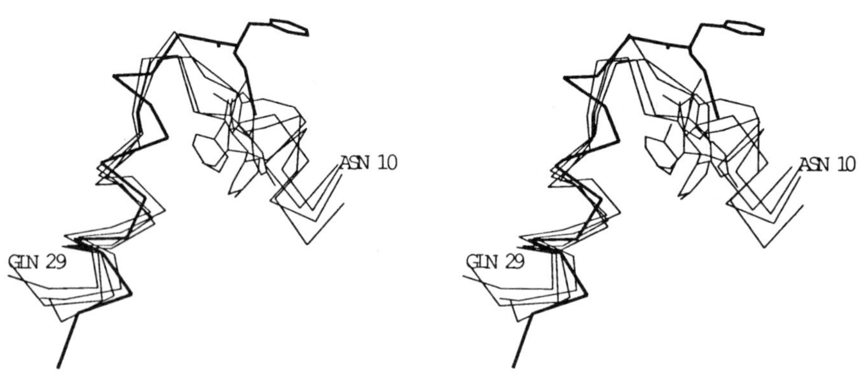

Figure 4. Superposition of the four converged C3a solution structures (thin; residues 10–29) with the X-ray structure (thick; residues 13–29). Tyr15 side-chain is shown, highlighting the differences in the packing of this side-chain in solution versus the solid state.

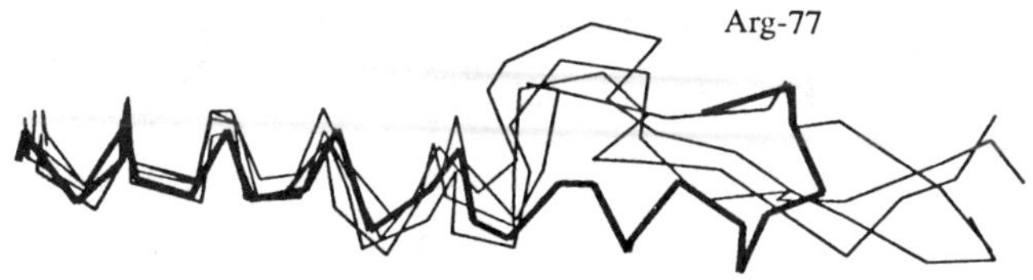

Figure 5. Superposition of the four converged C3a solution structures (thin) with the X-ray structure (thick) highlighting the difference in the C-terminal Helix IV (residues 47–77). This helix in the solution state becomes increasingly disordered from residue Arg65 to the C-terminus, whereas in the solid state Helix IV is well ordered to Ser71.

since the effector LGLAR residues of C3a are found at the end of the of this C-terminal helix in the solid state, whereas all of the solution data indicate that the last 10 C-terminal residues are able to sample a large part of the available conformational space.

The short helical region from Cys36 to Arg38 in the crystal structure is not well determined in the solution state. Significant differences are in fact observed between the solution and crystalline conformations of the entire loop (Asn30 to Gly46). These differences are still somewhat ambiguous at the present stage of analysis owing to the lack of multiple long range constraints anchoring this loop to the core of C3a. Finally, we note that the amino terminal residues Glu6 to Val12 in the solution state contain 'helical'-like turns consistent with the qualitative interpretation of a transient helix from residues 6 to 12 (5). Residues 1 to 12 are disordered in the crystal structure (4).

Another iteration of assignment of new constraints and generation of a much larger set of converged structures will enable a much more quantitative comparison of the solution and solid state structures and will hopefully resolve any ambiguities in the structure of the loop region and the packing of the amino terminal 'helix' (residues 6–12).

The binding of C3a to its receptor has to be critically re-examined in light of the data presented here in that the LGLAR effector segment of the molecule is free to sample many conformations and is not conformed to a region at the end of a C-terminal helical tether.

Acknowledgments

We wish to thank Dr. Tony Hugli for providing the C3a sample, Dr. Paul Weber, Mikael Akke & Garry Gippert for many helpful discussions on distance geometry and simulated annealing.

References

1. Hugli, T. E. (1981) CRC Crit. Rev. Immunol. **1**(*4*), 321–366.
2. Havel, T. F., Crippen, G. M., & Kuntz, I. D. (1979) Biopolymers, **18**, 73–81.
3. Crippen, G. M. (1981) Distance Geometry & Conformational Calculations, Research Studies Press, John Wiley & Sons, New York, NY.
4. Huber, R., Scholze, H., & Deisenhoffer, J. (1980) HoppeSeyler's Z. Physiol. Chem., **361**, 1389–1399.
5. Chazin, W. J., Hugli, T. E., & Wright, P. E. (1988) Biochemistry, **27**, 9139–9148.
6. Hugli, T. E., Vallota, E. H., & Muller-Eberhard, H. J. (1975) J. Biol. Chem., **250**, 1472–1478.

7. Hugli, T. E., Gerard, C., Kawahara, M., Scheetz, M. E., Barton, R., Briggs, S., Koppel, G., & Russell, S. (1982) Mol. Cell. Biochem., **41**, 59–66.
8. Marion, D. & Wüthrich, K. (1986) Biochem. Biophys. Res. Commun., **113**, 967–974.
9. Otting, G. Widmer, H., Wagner, G., & Wüthrich, K. (1986) J. Magn. Reson., **66**, 187–193.
10. Jeener, J., Meier, B. H., Bachmann, P. & Ernst, R. R. (1979) J. Chem. Phys., **71**, 4546–4553.
11. Rance, M. & Byrd, R. A. (1983) J. Magn. Reson., **54**, 221–240.
12. Aue, W. P., Bartholdi, E., & Ernst, R. R. (1976) J. Chem. Phys., **64**, 2229–2246.
13. Kim, Y. & Prestegard, J. H. (1989) J. Magn. Reson., **84**, 9–13.
14. Greisinger, C., Sørensen, O. W., & Ernst, R. R. (1987) J. Magn. Reson., **75**, 474–492.
15. Hyberts, S. V., Märki, W., & Wagner, G. (1987) Eur. J. Biochem., **164**, 625–635.
16. Zuiderweg, E. R. P., Boelens, R., & Kaptein, R. (1985) Biopolymers, **24**, 601–611.
17. Güntert, P., Braun, W., Billeter, M., & Wüthrich, K. (1989) J. Am. Chem. Soc., **111**, 3997–4004.
18. Wüthrich, K. (1986) NMR of Proteins & Nucleic Acids, John Wiley & Sons, New York.
19. Braun, W., Bösch, C., Brown, L. R., Go, N., & Wüthrich, K. (1981) Biochim. Biophys. Acta, **667**, 377–396.
20. Weber, P. L., Morrison, R., & Hare, D. (1988) J. Mol. Biol., **204**, 483–487.
21. Havel, T. F., Kuntz, I. D., & Wüthrich, K. (1983) Bull. Math. Biol., **45**, 665–720.
22. Rycaert, J. P., Ciccotti, G., & Berendsen, H. J. C. (1977) J. Comp. Phys., **23**, 327–341.

38

THREE-DIMENSIONAL SOLUTION STRUCTURE OF CALBINDIN D_{9k}[1]

Mikael Akke[2], Nicholas J. Skelton, Johan Kördel[2] and Walter J. Chazin[3]

Department of Molecular Biology, Research Institute of Scripps Clinic,
La Jolla, CA 92037.
and
Department of Physical Chemistry 2, Lund University,
S-221 00 Lund, Sweden.

I. INTRODUCTION

In the last few years three-dimensional structure determination of proteins in solution by nuclear magnetic resonance (NMR) spectroscopy has become an established technique (reviewed in Wüthrich, 1986; Clore & Gronenborn, 1989). This technique complements the traditional crystal X-ray diffraction method in several ways. One important advantage of structure determination by NMR in solution, that pertains specifically to our interest in calcium binding proteins, is the possibility to study proteins that do not crystallize.

The long-term objective of our research on the calmodulin superfamily of calcium regulatory proteins is to understand the fundamental role of conformation and dynamics in determining specificity and function. These proteins are characterized by a common helix-loop-helix motif for the Ca^{2+} binding sites, termed EF-hands. A pair of EF-hands constitutes the functional unit of the proteins in the calmodulin superfamily. It has not been possible to crystallize any of these proteins with more than one level of Ca^{2+} occupancy. The only available model for the conformational changes that accompany the binding of Ca^{2+} in these proteins is based on a comparison of the Ca^{2+}-free (apo) N-terminal domain and the Ca^{2+}-loaded C-terminal domain of troponin C in the crystalline state (Herzberg et al., 1986). Recent experimental evidence from Ca^{2+}-binding studies of mutant forms of troponin C support this model (Fujimori et al., 1990; Grabarek et al., 1990). However, in order to understand fully the conformational and dynamical consequences of Ca^{2+} binding, a specific binding domain must be examined at different levels of Ca^{2+} occupation. To this end, we have applied high-resolution 2D ^1H NMR spectroscopy to study the solution structure and dynamics of calbindin D_{9k} in the apo and Ca^{2+}-bound states. Calbindin D_{9k} (M_r 8600) is the smallest member of the calmodulin superfamily, and contains a functional pair of EF-hands. Its structure in the Ca^{2+}-loaded crystalline state (Szebenyi & Moffat, 1986) resembles the Ca^{2+} binding domains of the larger

[1] We gratefully acknowledge financial support from the National Institutes of Health (grant GM 40120 to W. J. C.), the Swedish Natural Sciences Research Council (graduate fellowship to J. K.) and the Swedish Royal Academy of Sciences (scholarship to M. A.).

[2] Lund University.

[3] Address correspondence to this author at the Research Institute of Scripps Clinic.

proteins calmodulin and troponin C. The protein used in our studies is obtained from a synthetic bovine calbindin gene expressed in *Escherichia coli* (Brodin et al., 1986).

At an early stage of the 2D ^1H NMR studies, it was discovered that calbindin D_{9k} exists in two forms due to *cis-trans* isomerization of the G42-P43 peptide bond (Chazin et al., 1989), that give rise to two sets of resonances in the NMR spectrum. The P43→G mutant (P43G) exhibits a single set of resonances and has very limited structural differences from the wild-type (Kördel et al., 1990). This justifies the use of this mutant for three-dimensional structure determination, greatly facilitating the analysis of NOESY spectra and making possible more accurate measurements of the NOE volumes.

Here, we present our current procedure for producing solution structures from NMR data by distance geometry (DG) and restrained molecular dynamics (rMD) calculations, exemplified by our work on the apo and Ca^{2+}-loaded P43G bovine calbindin D_{9k} and on the Ca^{2+}-loaded native porcine calbindin D_{9k}.

II. METHODS

Experimental procedures. The quality of the solution structures generated from experimental NMR data ultimately depend upon the number of NOE cross-peaks and the precision with which their volumes can be extracted from 2D NOESY spectra. The following discussion highlights some of the data acquisition and processing procedures used in our laboratory to obtain high quality NOESY spectra.

In the process of setting up the NOESY experiment, various measures are utilized to minimize baseline artefacts that produce systematic errors in the subsequent quantitation of cross-peak volumes. ω_2 ridges emanating from imperfect suppression of the solvent (H_2O) resonance can be minimized by using low power irradiation that is phase coherent with the hard pulses of the NOESY experiment (Zuiderweg et al., 1986). Owing to the sequential acquisition of data on Bruker spectrometers, further ω_2 baseline "tilts" arising from zero-order phase corrections can be avoided by adjusting the receiver reference phase to acquire the data with pure sine or cosine modulation in ω_2 (Marion & Bax, 1988). Baseline distortions in ω_2 arising from imperfect low-pass filters can be reduced by increasing the filter widths. This can be conveniently achieved by digitally over-sampling by a factor of two (Delsuc & Lallemand, 1986). The introduction of a short Hahn-echo period after the final observe pulse further reduces any baseline artefacts (Davis, 1989). Correct choice of the refocussing delay after the echo removes the need for any first-order phase correction. The use of a Hahn-echo period also allows the ADC to be gated open before the top of the echo is reached (by an amount corresponding to 2 or 4 dwell times). Processing then requires a left shift of each row by 2 or 4 points (1 or 2 quadrature pairs) prior to Fourier transformation. In this way artefacts in the first few data points arising from turning on the ADC can be removed, and the associated baseline roll avoided.

After collection of the data a simple spline baseline correction in ω_2 is applied to remove ω_2 ridges (arising from different DC offsets from one FID to the next), and the intensity of the first point in each ω_1 interferogram is halved so as to reduce ω_1 ridges (Otting et al., 1987). When a non-linear phase correction is applied in ω_1 (to correct for precession during the finite pulses and phase-shifting delays), a baseline distortion is produced, this being most severe near the diagonal at the edges of the spectrum (Figure 1a). This baseline distortion arises because of the TPPI method of data acquisition in t_1. Linear prediction of the first point in each interferogram has been utilized to obviate the need for phase correction, and hence remove the baseline distortion (Marion & Bax, 1989). In principle, a baseline correction scheme could also be applied to this problem.

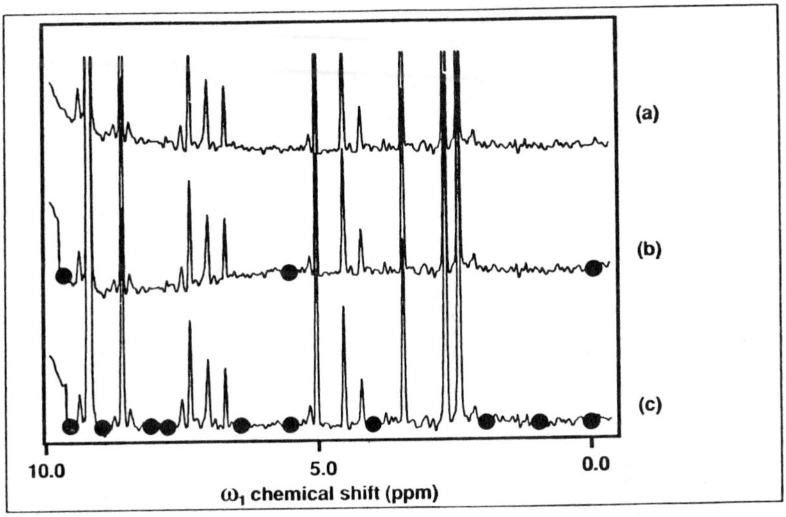

Figure 1. Cross-sections at ω_2 = 9.15 ppm (Val61 NH) through a 60 ms NOESY spectrum of apo P43G calbindin D_{9k} in H_2O. The filled circles indicate the points used for fitting the baseline correction function. The uncorrected spectrum (a), and the effect of performing a baseline correction using resonance-free points common to all columns (b), and points common to columns in a localised strip of the spectrum (c), are shown.

However, baseline corrections are complicated by the lack of uniformity of the baseline shape in different columns of the spectrum, and by the inability to pick baseline points suitable for fitting a baseline correction function anywhere but at the edges of the spectrum, due to the high density of peaks in the spectrum. We have overcome these problems by splitting the spectrum into 30 vertical strips, and correcting the baseline in each of these strips separately. Because of the lower peak density within any one strip, a greater number of baseline points can be used in the correction process. This produces a less distorted baseline (Figure 1c) than in the original spectrum (Figure 1a), or in a spectrum where only points at the edges of the spectrum were used to define the baseline (Figure 1b).

For P43G calbindin D_{9k}, a series of NOESY spectra with mixing times ranging from 20 to 200 ms were acquired in H_2O and D_2O solution. The build-up of NOE intensity with mixing time was monitored for various sequential and medium range NOEs in these spectra (Figure 2). It was found that the cross-peaks in the 60 ms spectra had a reasonable signal to noise ratio, whilst still being in the linear build-up regime (i.e. the cross-peaks are not dominated by spin diffusion). Regions of the proteins defined previously as being helical (Kördel et al., 1989; Skelton et al., 1990) provided the basis for a volume-to-distance calibration within the 60 ms spectrum, based on NOE intensities for sequential and medium range proton pairs of known minimal separation in helices. Efforts were then concentrated on identifying *all* cross-peak positions (and measuring the corresponding cross-peak volume) within this spectrum. The NOE contacts were then assigned one of five upper distance bounds on the basis of the cross-peak volume and the determined volume-to-distance calibration curve.

The longer mixing time NOESY spectra were then inspected for cross-peaks not present in the shorter mixing time data. These cross-peaks are subjected to larger spin diffusion contributions which prohibit a direct volume-to-distance conversion. A few cross-peaks clearly subject to spin diffusion [e.g.

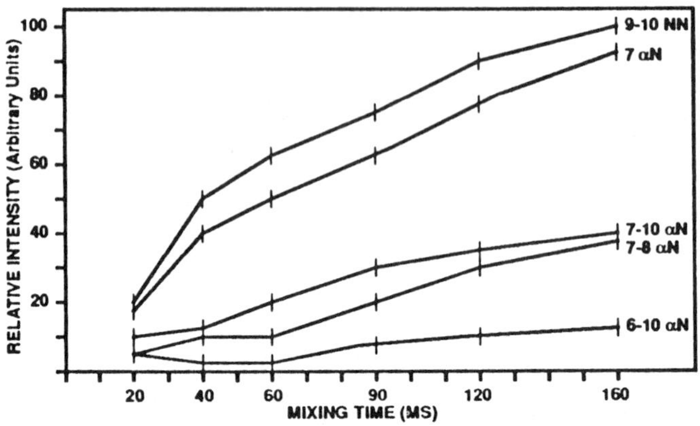

Figure 2. Build-up curves of the NOE for a selection of representative short and medium range connectivities in a helical region of apo P43G calbindin D_{9k}.

$d_{N\alpha}$(i,i+1)] can, however, be used to estimate a maximum upper bound to be used for all cross-peaks present in *only* the longer mixing time NOESY spectra. According to the triangle inequality, the upper limit for the bound can be taken as the sum of the upper bounds for the separate steps in a possible spin diffusion path [e.g. $d_{N\alpha}(i,i+1) = d_{NN}(i,i+1) + d_{N\alpha}(i,i)$]. By identifying cases where the latter two bounds had been determined in the 60 ms spectrum, it was possible to estimate an upper bound for the new peaks observed in the 200 ms spectrum.

The task of assignment of the observed cross-peaks to particular proton pairs was greatly simplified by the use of the program GENBOUND (kindly provided by Dr. Matthew Kalnik). At this early stage in the analysis, only those NOEs which can be unambiguously assigned are used in the structure calculation. Proton-proton pairs whose NOE cross-peaks are not unambiguously assignable on the basis of the spectra alone can often be identified once three-dimensional structures are available. These NOE constraints can then be used in subsequent rounds of DG calculations.

Structure calculations. The initial local NOE distance constraints and J-couplings provide a basis for making stereospecific assignments of the C^{β}-methylene protons and identifying allowed ranges of the dihedral angles ϕ, ψ and χ_1. This is performed using the grid search program HABAS (Güntert et al., 1989), which also provides a check for internal consistency of the input. The HABAS output of dihedral-angle constraints is used in conjunction with experimental hydrogen bond and NOE-derived distance constraints as input for the DG program DISGEO (Havel & Wüthrich, 1984). For cases where no stereospecific assignments of prochiral protons can be obtained, the standard pseudoatom representation (Wüthrich et al., 1983) is used. However, a modified pseudoatom correction to the experimental upper bounds has been implemented (Akke et al., 1990). For cases where NOE connectivities are observed from a single proton to both protons or methyl groups attached to a prochiral carbon atom, the corresponding upper bound to the pseudoatom point is calculated by the law of cosines, and the resulting value is used in the DG calculations if it is lower than the value obtained from the standard pseudoatom correction (Wüthrich et al., 1983). All experimental distance constraints that are redundant with basic steric constraints (Wüthrich et al., 1983) are filtered out prior to the DG calculations. Experimentally assigned hydrogen bonds are enforced during the DG calculations by constraining the NH-O distance to 1.8-2.0 Å and the N-O distance to 2.7-3.0 Å. The lower bound for the NH-O distance of all other possible hydrogen bond pairs is set to 1.8 Å, since otherwise the close approach of the NH and O atoms will be inconsistent with the lower bound given by the sum of the van der Waals radii. This modification, which was pointed out to us by Dr. Paul Weber of Hare Research Inc., significantly improves the quality of the secondary structure elements at the DG level. The ranges between the upper and lower bounds of the local distance matrix were further reduced prior to the bounds smoothing procedure. These bounds were obtained by examining the extreme distances allowed between atoms directly bonded to sequential C^{α} atoms, as determined by pairwise rotation of ϕ_{+} and ψ_{-} in dipeptide units (G.P. Gippert, unpublished). The DG calculations make full use of the refinement procedures available in DISGEO for chirality, planarity and dihedral angle constraints. Several cycles of DG calculations are performed; after each cycle the structures are used to screen the ambiguous NOEs for possible assignments. A statistical procedure for obtaining additional stereospecific assignments is also applied (Kline et al., 1988), whereby the structures are inspected to identify cases where only one of the two possible stereospecific assignments is consistent with the NOE constraints. The DG structures are subsequently subjected to restrained molecular dynamics and restrained energy minimization (rEM), in order to sample further the available conformational space as well as refine the structures.

The rMD and rEM calculations are performed with AMBER (Singh et al., 1986), modified by Dr. David A. Case to include a potential function to penalize interproton distances that deviate from the experimental constraints: $E_d = K_d \cdot (d - d_0)^2$ for $d > d_0$, and $E_d = 0$ otherwise. d_0 is the upper limit of the distance constraint. Deviations from the dihedral angle constraints are penalized by the function: $K_{\phi} \cdot [1 - \cos(\phi - \phi_0)]$ for ϕ outside of the allowed range, and 0 otherwise. ϕ_0 is the nearest endpoint of the allowed range and $K_{\phi} = 32$ kcal·mol^{-1}.

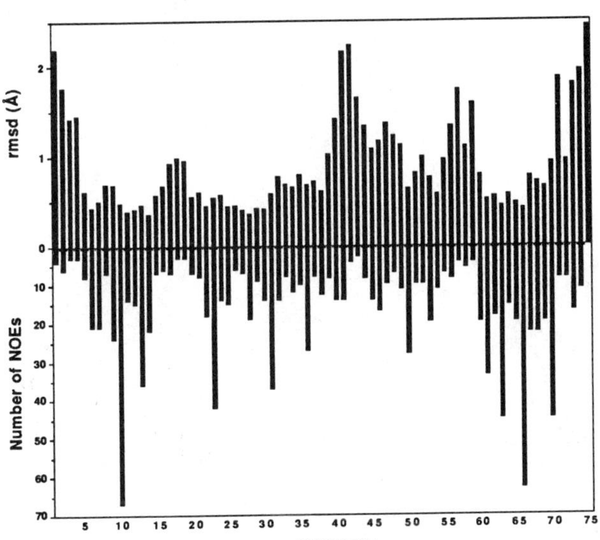

Figure 3. Average local rms deviations from the average coordinates for the backbone atoms of a set of 5 unrefined structures from the first round of DG calculations (upper panel) and number of NOE constraints per residue (lower panel) for apo P43G calbindin D_{9k}.

In addition, the torsional force constant for the ω dihedral angle is increased relative to the standard AMBER potential (Weiner et al., 1986), in order to keep the peptide units close to planar (Lee et al., 1989). Distance constraints to non-stereospecifically assigned protons are enforced with r^{-6} averaging of the upper bounds (Clore et al., 1986). Hydrogen bonds are not constrained during the rMD calculations. The rMD refinement protocol involves an initial rEM followed by an "annealing" cycle, which employs a temperature regulated MD algorithm (Berendsen et al., 1984). The first phase consists of 1 ps of heating the system towards a target temperature of 1200 K with a temperature relaxation time, τ, of 0.6 ps. During the second phase, the system is equilibrated at 1200 K for 4 ps with $\tau = 0.2$ ps. The last phase involves 6 ps of cooling towards 0 K with $\tau = 2.0$ ps. Finally, rEM produces the refined structures. K_d is incremented by a factor of 2 at regular intervals during the initial rEM, heating and equilibration phases, to a final value of 32 $kcal \cdot mol^{-1} \cdot Å^{-2}$, and is thereafter kept constant during the cooling phase and final rEM. Several aspects concerning the choices of parameters during the "annealing" cycle have previously been discussed (Gippert et al., 1990).

III. RESULTS AND DISCUSSION

In this section we present some selected results from the on-going structure calculations of the apo P43G and the Ca^{2+}-loaded porcine calbindin D_{9k} that point at the relationship between the input data and the quality of the resulting structures. The correlation between the number of NOE constraints per residue and the definition of the resulting DG structures as described by the average local rms deviation from the average coordinates is shown in Figure 3. The apo P43G data presented here are preliminary, and we expect to increase significantly the number of stereospecific assignments, improve the dihedral angle constraints and resolve a large number of ambiguous NOEs on the basis of these initial structures. Figure 4 shows a set of 5 unrefined apo P43G calbindin D_{9k} structures, where the lack of constraints in the loop regions (residues 15-21, 37-43 and 55-59) and at the ends of the protein is clearly manifested as poorer definition of these residues (cf figure 3). Several interior side-chains (not shown) are well defined, allowing for stereospecific assignments of methylene and methyl protons by the statistical procedure described above.

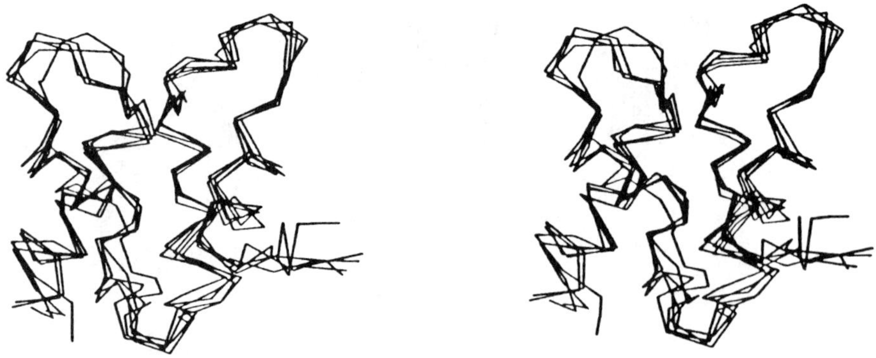

Figure 4. Stereoview of 5 unrefined structures from the first round of DG calculations of apo P43G calbindin D_{9k}.

Table I. Structural statistics for porcine calbindin D_{9k}.

	DG	rMD
Average rms deviation from average coordinates		
Backbone, helices	0.73	0.64
Backbone, all residues	1.07	1.58
All atoms, helices	1.29	1.19
All atoms, all residues	1.71	2.03
Average residual NOE constraint violations		
Maximum violation (Å)	0.66	0.44
number of violations≥ 0.5 Å	2.60	0.00

By comparing the DG and rMD structural statistics and residual violations for the porcine calbindin D_{9k} (Table I), it is readily apparent that the rMD refinement leads to structures that fulfill more closely the experimental constraints. Correspondingly, well-constrained regions of the protein become better defined, whereas the opposite is true for regions with few constraints. These latter effects reflect the more efficient sampling properties of rMD. The limited sampling on the DG level is inherent to our present implementation of the embed algorithm. Recent advances have been presented that allow more extensive sampling of conformational space with both the DG (Havel, 1990) and rMD (Torda et al., 1990) methods.

Future directions include refinement of the structures based on comparisons of the back-calculated and experimental NOESY spectra and build-up curves, using the program REMARC (Yip & Case, 1990), and the determination of heteronuclear coupling constants for generating additional torsion angle constraints.

ACKNOWLEDGMENTS

We are very grateful to Garry P. Gippert and David A. Case for help with the DISGEO and AMBER calculations, respectively.

REFERENCES

Akke, M., Drakenberg, T., & Chazin, W.J. (1990) *In preparation.*

Berendsen, H.J.C., Postma, J.P.M., van Gunsteren, W.F., DiNola, A., & Haak, J.R. (1984) *J. Chem. Phys.* **81,** 3684-3690.

Brodin, P., Grundström, T., Hofmann, T., Drakenberg, T., Thulin, E., & Forsén, S. (1986) *Biochemistry* **25,** 5371-5377.

Chazin, W.J., Kördel, J., Drakenberg, T., Thulin, E., Brodin, P., Grundström, T., & Forsén, S. (1989b) *Proc. Natl. Acad. Sci. U.S.A.* **86,** 2195-2198.

Clore, G.M., Brünger, A.T., Karplus, M., & Gronenborn, A.M. (1986) *J. Mol. Biol.* **191,** 523-551.

Clore, G.M., & Gronenborn, A.M. (1989) *CRC Crit. Rev. Biochem.* **24,** 479-564.

Davis, D.G. (1989) *J. Magn. Reson.* **81,** 603-607.

Delsuc, M.A., & Lallemand, J.Y. (1986) *J. Magn. Reson.* **69,** 504-507.

Fujimori, K., Sorenson, M., Herzberg, O., Moult, J., & Reinach, F.C. (1990) *Nature* **345,** 132-135.

Gippert, G.P., Yip, P.F., Wright, P.E., & Case, D.A. (1990) *Biochem. Pharm.* **40,** 15-22.

Grabarek, Z., Tan, R., Wang, J., Tao, T., & Gergely, J. (1990) *Nature* **345,** 132-135.

Güntert, P., Braun, W., Billeter, M., & Wüthrich, K. (1989) *J. Am. Chem. Soc.* **111,** 3997-4004.

Havel, T., & Wüthrich, K. (1984) *Bull. Math. Biol.* **46,** 673-698.

Havel, T.F. (1990) *Biopolymers* **29,** 1565-1585.

Herzberg, O., Moult, J., & James, M.N.G. (1986) *J. Biol. Chem.* **261,** 2638-2644.

Kline, A.D., Braun, W., & Wüthrich, K. (1988) *J. Mol. Biol.* **204,** 675-724.

Kördel, J., Forsén, S., & Chazin, W.J. (1989) *Biochemistry* **28,** 7065-7074.

Kördel, J., Forsén, S., Drakenberg, T., & Chazin, W.J. (1990) *Biochemistry* **29,** 4400-4409.

Lee, M.S., Gippert, G.P., Soman, K.V., Case, D.A., & Wright, P.E. (1989) *Science* **245,** 565-680.

Marion, D., & Bax, A. (1988) *J. Magn. Reson.* **79,** 352-356.

Marion, D., & Bax, A. (1989) *J. Magn. Reson.* **83,** 205.

Marion, D., & Wüthrich, K. (1983) *Biochem. Biophys. Res. Commun.* **113,** 967-974.

Otting, G., Widmer, H., Wagner, G., & Wüthrich, K. (1987) *J. Magn. Reson.* **66,** 187-193.

Torda, A.E., Scheek, R.M., & van Gunsteren, W.F. (1990) *J. Mol. Biol.* **214,** 223-235.

Singh, U.C., Weiner, P.K., Caldwell, J.W., & Kollman, P.A. (1986) AMBER 3.0, University of California, San Francisco, CA.

Skelton, N.J., Forsén, S., & Chazin, W.J. (1990) *Biochemistry* **29,** 5752-5761.

Szebenyi, D.M.E., & Moffat, K. (1986) *J. Biol. Chem.* **261,** 8761-8777.

Weiner, S.J., Kollman, P.A., Nguyen, D.T., & Case, D.A. (1986) *J. Comp. Chem.* **7,** 230-252.

Wüthrich, K., Billeter, M., & Braun, W. (1983) *J. Mol. Biol.* **169,** 949-961.

Wüthrich, K. (1986) *NMR of Proteins and Nucleic Acids*. Wiley, New York.

Yip, P., & Case, D.A. (1989) *J. Magn. Reson.* **83,** 643.

Zuiderweg, E.R.P., Hallenga, K., & Olejniczak, E.T. (1986) *J. Magn. Reson.* **70,** 336-343.

39

ISOTOPIC LABELING OF SPECIFIC AMINO ACID TYPES AS AN AID TO NMR SPECTRUM ASSIGNMENT OF THE METHIONINE REPRESSOR PROTEIN

David W. Hoffman and Leonard D. Spicer
Departments of Biochemistry and Radiology,
Duke University Medical Center, Durham, North Carolina 27710

I. INTRODUCTION

NMR spectroscopy has evolved over the last several years into a powerful method for the study of protein structure. Before NMR data can be effectively used in a structural study however, it is first necessary to assign the resonances to the specific protons from which they arise. After a significant fraction of the NMR spectrum has been assigned a protein structure may be determined that is consistent with the nuclear Overhauser effect (NOE) data and coupling constants, which provide inter-proton distance and bond angle restraints. The spectrum assignment process can be simplified by incorporating isotopic labels into the protein structure. Here we describe methods for labeling specific amino acid types with N-15 and deuterium.

The complete and accurate assignment of a protein spectrum is a difficult task, even in molecules with less than 100 amino acids. Incorrectly assigned protons lead to erroneous NOE constraints and can result in an incorrectly determined protein structure. One particular problem in NMR spectrum assignment comes in associating amide protons with specific amino acid types. This is usually done using various types of homonuclear "through-bond" correlation experiments, such as DQF-COSY (1) and Hartmann-Hahn spectroscopy (2), which identify sets of coupled protons belonging to the same spin system. The coupling pattern observed within a spin system is indicative of the amino acid type, however 14 amino acid types have a CH_2 unit in the beta position and therefore exhibit similar coupling patterns. Lys, Arg, Glu and Gln are usually difficult to distinguish in homonuclear correlated spectra of moderate sized proteins. Even the amino acid types that in principle have unique coupling patterns, such as Ala, Val, Thr, Gly and Ile, can be easily misidentified if there is a coincidental chemical shift degeneracy or weak coupling constant.

In the case of proteins with molecular weights over 15 kDaltons, homonuclear through-bond correlation experiments become much less useful for identifying spin systems and assigning them to amino acid types. There are two primary reasons for the difficulty. First, the two dimensional spectra are crowded with many overlapping peaks, so complete spin systems often cannot be identified. Second, the T_2 relaxation time of a large protein is significantly shorter than the time required for magnetization transfer by through-bond

Techniques in Protein Chemistry II

coupling, resulting in many weak or undetectable cross-peaks. As a means for overcoming these difficulties, it is clearly desirable to have alternatives to homonuclear proton correlated spectroscopy for assigning resonances to specific amino acid types.

When a protein is expressed in bacteria, selected types of deuterium (3) or N-15 labeled amino acids (4) may be incorporated into the protein by growing the appropriate over-producing strain in a defined medium, providing an alternative method for identifying the amino acid type associated with specific resonances. In one possible approach, selective deuteration can be used to remove the resonances of a specific amino acid type from a proton-detected NMR spectrum. Alternatively, resonances of an amino acid type can be identified in a proton-nitrogen heteronuclear multiple quantum coherence (HMQC) spectrum (5,6) of a protein which has been selectively enriched in the N-15 labeled amino acid. Recently the spectra of several proteins which are quite large by NMR standards such as T4 lysozyme (7), E. coli thioredoxin (8), tryptophan repressor (9) and staphylococcal nuclease (10) have been substantially assigned using strategies which have included deuterium, N-15 and C-13 labeling of amino acid types. Isotopic labeling is also a requirement for the newest 3-D and 4-D methods (11,12) which show great promise for extending the molecular weight range of protein NMR.

Although selective amino acid type labeling has proven to be a valuable tool in solving the assignment problem in proteins, labeling methods frequently used in the past have several important disadvantages. First, the preparation of isotopically labeled samples can be time consuming and expensive. Second, the natural metabolic pathways of the bacteria can scramble an isotope to other amino acid types, yielding a non-specifically labeled product. The bacteria may also synthesize amino acids from other precursors in the growth medium, thus diluting the isotopic label. The latter problems may be reduced by transforming the expression system into an auxotrophic bacterial strain (8), however different auxotrophs may be required for each amino acid type being labeled.

In our laboratory the methionine repressor protein of E. coli (13-15) is being studied using NMR techniques and assignment of the spectrum is in progress. This 24 kDalton symmetric dimer, with 104 amino acids in each subunit, is a large molecule by NMR standards. In the course of our investigation, we have identified general methods for specifically incorporating a variety of nitrogen-15 and deuterium labeled amino acid types into the protein molecule, using relatively inexpensive commercially available isotope sources. We have found that by appropriate design of the bacterial growth medium we are able to label selectively several different amino acid types without excessive dilution or scrambling of the isotopic label and without resorting to auxotrophic bacterial strains. This avoids most of the difficulties usually associated with type-specific amino acid labeling. Once a spin system is assigned to a specific type of amino acid, it becomes a starting point for making sequence-specific assignments. The spectra of these selectively labeled protein samples have been used to obtain many sequence specific assignments in the methionine repressor spectrum. The methods for isotopic labeling of specific amino acid types described in this paper should be applicable to other proteins which are expressed in prototrophic strains of E. coli.

II. METHODS

Methionine repressor protein was obtained from the over-producing strain of E. coli BL21(DE3) using the T7 RNA polymerase expression system (16,17). Typically, one liter cell cultures were grown, yielding 50 to 100 mg of pure methionine repressor protein. Cells were induced with IPTG when the culture reached an absorbance of 1.0 measured at 550 nm. The growth medium contained (per liter) 6 g Na_2PO_4, 3 g KH_2PO_4, 1 g NH_4Cl, 0.5 g NaCl, 2 mM $MgSO_4$, 0.1 mM $CaCl_2$, 1 mM $FeCl_3$, 100 mg ampicillin, 40 mg of each nucleotide, 5 mg thiamine, 1 mg biotin, 1 mg choline chloride, 1 mg folic acid, 1 mg niacinamide, 1 mg pantothenate, 1 mg pyridoxal, 0.1 mg riboflavin. The nucleotides and vitamins are included to increase the protein yield and growth rate of the bacteria, without significant dilution of isotope. For uniformly N-15 labeled protein 99 % N15 enriched NH_4Cl was used. For proteins with specific isotopically labeled amino acids this medium was supplemented as follows. *N-15 glycine and serine*: 3.6 g dextrose, 200 mg of each amino acid except glycine, 25 mg N-15 glycine at innoculation, and an additional 225 mg N-15 glycine at induction. *N-15 glycine, serine, cysteine and tryptophan*: 3.6 g dextrose, 200 mg of each amino acid except glycine, serine, cysteine and tryptophan, 25 mg N-15 glycine at innoculation, and an additional 225 mg N-15 glycine at induction. *Deuterated alanine methyl groups*: dextrose was reduced to 500 mg, 2 g DL-alanine (d_3-methyl, 90 % deuterated) at innoculation and 500 mg at induction. Isotopically enriched amino acids were obtained from Cambridge Isotopes. Typically, the protein concentration in the NMR samples was 2 mM and NMR data was collected at 42 °C. All NMR experiments were carried out at 11.9 Tesla on a General Electric GN500 spectrometer equipped with an inverse detection probe. NMR data were processed using FTNMR software obtained from Hare Research.

III. RESULTS AND DISCUSSION

N-15 labeling. The nitrogen-proton HMQC spectrum of uniformly N-15 enriched methionine repressor protein, shown in Figure 1, correlates the amide

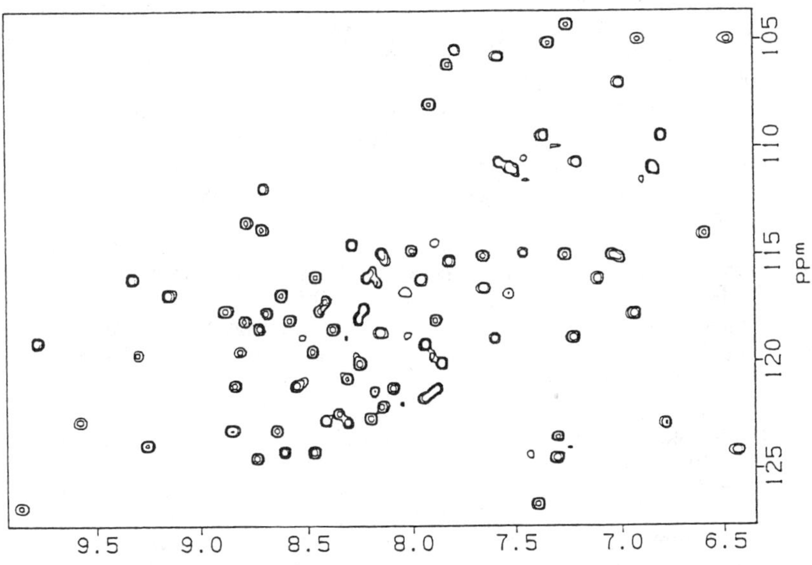

Figure 1. Proton-nitrogen HMQC spectrum of uniformly N-15 labeled methionine repressor protein. Peaks correlate the chemical shifts of amide protons with the amide nitrogens of the same amino acid.

proton chemical shifts with the amide nitrogen of the same amino acid. This uniform label HMQC spectrum served as a template against which selectively N-15 labeled samples were compared. The HMQC experiment has a significant sensitivity advantage over proton-proton correlation experiments when applied to large proteins: single bond N-H correlations are less affected by short T_2 relaxation times than three bond H-H correlations because of the large 90 Hz N-H coupling constant compared with 3 to 9 Hz for three bond H-H couplings. A strong amide N-H correlation peak is observed for nearly every amino acid except proline in the methionine repressor, while in the homonuclear COSY spectrum only 65 percent of the amide protons have detectable three-bond couplings.

Methionine repressor protein with N-15 labeled glycine and serine was prepared by growing the over-producing strain of bacteria in a medium containing all 20 amino acids, with only glycine being N-15 enriched. N-15 serine was synthesized from glycine by the bacteria. While the labeled serine was diluted by serine from the medium, the enrichment obtained was still sufficient for heteronuclear NMR experiments. Increased enrichment can, however, be obtained by deleting serine from the growth medium. The chemical shifts of the amide protons and nitrogens of glycine and serine are easily identified in the HMQC spectrum, shown in Figure 2. There was a slight labeling of the amine protons of glutamine or asparagine, but these peaks are easily distinguished as doublets in the HMQC spectrum. Once the amide protons of the glycines and serines were identified, some sequence-specific assignments were made by identifying NOEs to sequential amino acids of known types in NOE and HMQC-NOE spectra.

We extended this straightforward approach to type-specfic labeling by preparing methionine repressor protein with N-15 labeled glycine, serine, cysteine and tryptophan. The bacteria were grown in a medium containing the other 16 unlabeled amino acids plus N-15 glycine. Under these growth conditions serine, cysteine and tryptophan all receive their alpha nitrogen from glycine. The proton-nitrogen HMQC spectrum of this preparation, shown in Figure 3, is similar to the spectrum of Figure 2 except for the appearance of three new peaks which were assigned to the cysteine and two tryptophans of the methionine repressor. Sequence specific assignments were again made by identifying NOEs to sequential amino acids.

Additional protein samples with a single labeled amino acid type were prepared with N-15 lysine, valine, leucine, threonine and methionine using a bacterial growth medium containing an excess of all twenty amino acid types, only one of which was N-15 enriched. In each case the N-15 label was incorporated without scrambling, and the amide protons of each amino acid type were identified in the HMQC spectrum. With valine and leucine the biosynthetic pathways were not entirely repressed, resulting in a significant dilution of the isotopic label. Using a similar growth medium, attempts to prepare protein specifically labeled with tyrosine and phenylalanine failed. With these two metabolically more active amino acid types the isotope was badly scrambled by non-repressible transaminases.

Deuterium labeling. Methionine repressor protein containing deuterated alanine methyl groups was prepared to solve a specific assignment problem. In principle, alanine and threonine belong to separate spin classes, and should be distinguishable in NMR experiments that detect H-H through-bond couplings. However, if the threonine alpha and beta protons either have degenerate chemical shifts or are weakly coupled, as is the case in several threonines in the methionine repressor, the alanines and threonines cannot be distinguished by spin system class.

Alanine and threonine spin systems can be distinguished if the alanine methyl groups can be selectively deuterated. A strategy for preparing protein with deuterated alanine, however, is not obvious. Alanine is extremely active

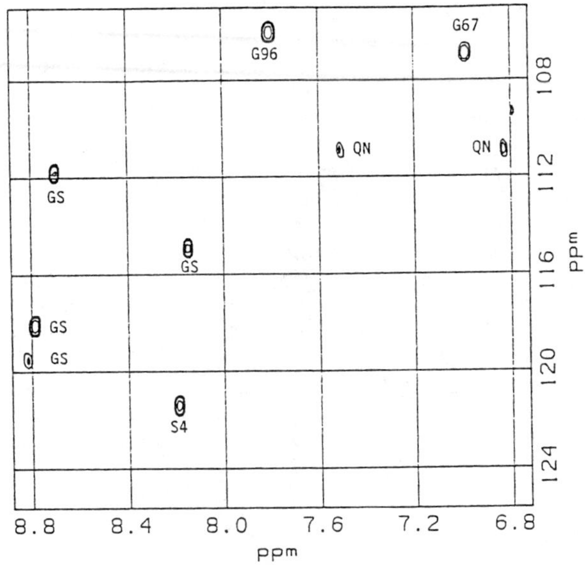

Figure 2. HMQC spectrum of the methionine repressor protein containing N-15 labeled glycine and serine, indicating the chemical shifts of the amide protons and nitrogens of the glycines and serines. Peaks which have been assigned to specific amino acids are numbered. Peaks associated with glycines and serines which are not yet specifically assigned are labeled "GS". Peaks associated with glutamine or asparagine are labeled "QN".

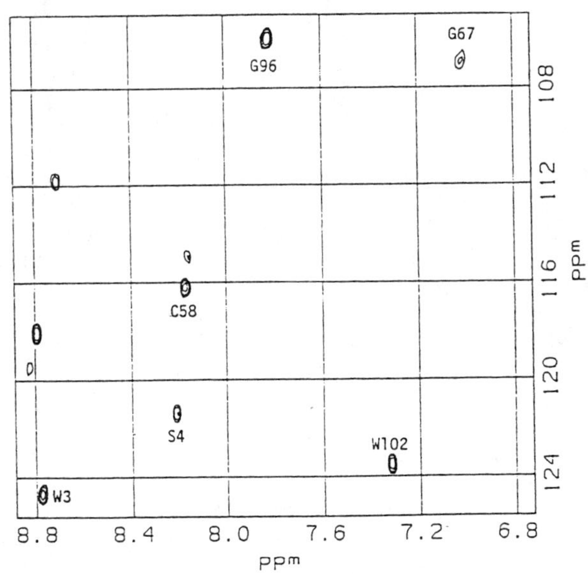

Figure 3. HMQC spectrum of the methionine repressor protein containing N-15 labeled glycine, serine, cysteine and tryptophan. Peaks indicate the chemical shifts of the amide protons and nitrogens of these four amino acid types.

metabolically, continuously exchanging its methyl group with pyruvate via a non-repressible pathway, and pyruvate is abundant within the cells, being a product of glycolysis. In addition, at present there is no alanine auxotroph of E. coli. A growth medium containing 19 unlabeled amino acids plus an excess of deuterated alanine yielded a protein sample without significant deuteration of the alanine methyl groups. We therefore decided to use a medium containing deuterated alanine, no other amino acids, and a growth-limiting amount of dextrose. These conditions were designed to use the alanine to deuterate the pyruvate pool within the cells, so that exchange with pyruvate will not dilute the labeled alanine. This medium resulted in deuteration of several amino acid types in addition to alanine, however since the threonine methyl group does not arise from either alanine or pyruvate our goal of distinguishing the spin systems of these two amino acids was accomplished. Comparison of the proton DQF-COSY spectra of Figure 4 shows that the alanine methyl couplings are significantly reduced in the deuterated sample, while the threonine methyl couplings are relatively unaffected. After the alanine and threonine methyl groups were distinguished we were able to make specific assignments by identifying NOEs to sequential amino acids using proton NOE and HMQC-NOE spectra.

IV. CONCLUSIONS

A reasonable spectrum assignment strategy for medium and large size proteins may include the preparation of samples containing specifically labeled amino acid types. By performing relatively straightforward NMR experiments, the resonances associated with the labeled amino acids can be identified, thus simplifying the spectrum assignment problem. The choice of labeled samples can be dictated by the complexity and amino acid composition of the particular protein being studied. Isotopic labeling permits resonances to be assigned to amino acid types with a higher degree of confidence than is likely to be obtainable using homonuclear correlation methods alone, reducing the chance of errors in assigning NMR spectra and in protein structure determinations.

For proteins that are expressed in E. coli, samples specifically labeled with one or more different amino acid types can often be prepared simply by an appropriate choice of growth medium. In the cases cited here, the cost of the isotope label is also very reasonable. For example, four amino acid types (gly, ser, cys, and trp) were N-15 labeled in the amide nitrogen position using glycine as the sole N-15 source. When a choice exists, labeling with N-15 is usually preferable to deuterium for two reasons: (i) with N-15 a signal is introduced into an empty background, rather than removed from a crowded background, and (ii) amide proton chemical shifts can be easily identified in nitrogen-proton HMQC experiments, which are inherently more sensitive than proton-proton correlation experiments because of the large N-H coupling constant.

V. ACKNOWLEDGEMENTS

We would like to thank Drs. Thomas W. Kirby and Ronald C. Greene for supplying the methionine repressor over-producing strain of E. coli. Support for this work from the National Institutes of Health under grant GM 41829 is gratefully acknowledged. The Duke NMR Center was established with grants from the NIH, NSF, and North Carolina Biotechnology Center.

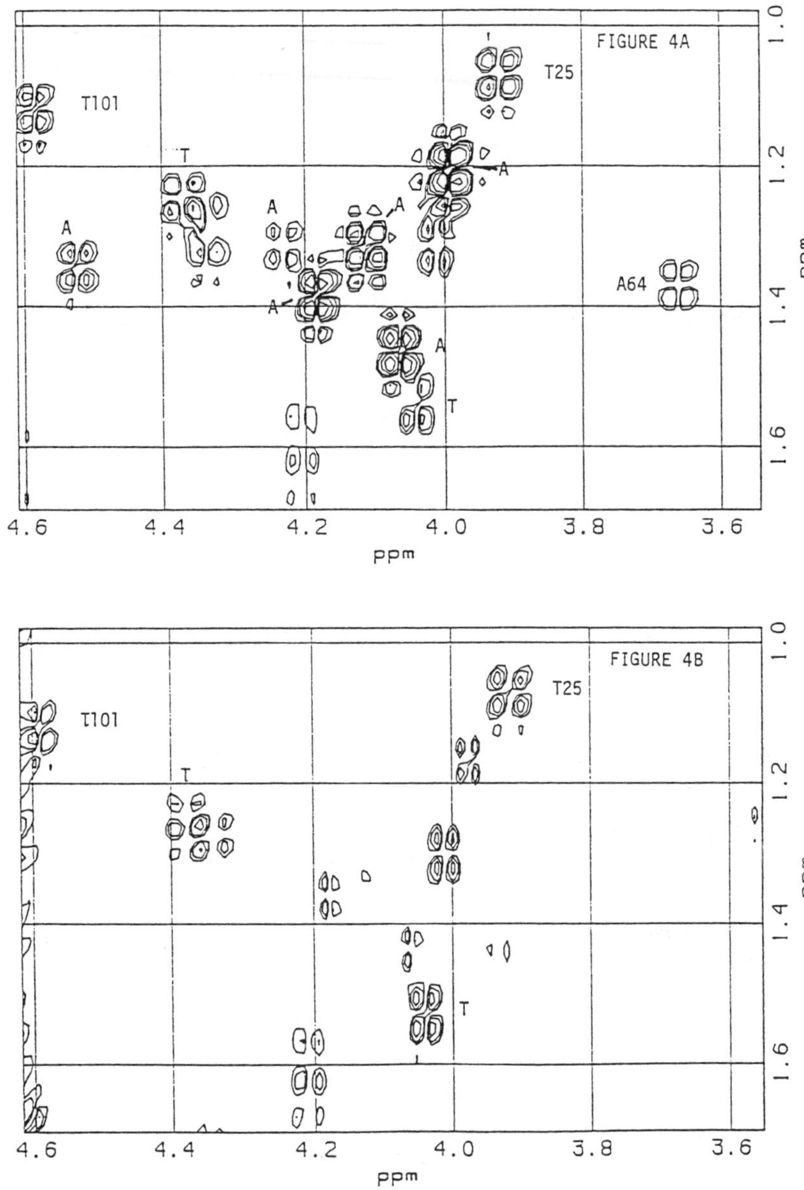

Figure 4. A section of the DQF-COSY spectrum of the methionine repressor protein containing the alanine methyl to alpha proton couplings and threonine methyl to beta proton couplings, as well as a few peaks corresponding to other amino acid types. A) Methionine repressor containing protonated alanine methyl groups. B) 90% deuterated alanine methyl groups.

VI. REFERENCES

1. M. Rance, O. W. Sørensen, G. Bodenhausen, G. Wagner, R. R. Ernst and K. Wüthrich (1983) *Biochem. Biophys. Res. Commun.*, **117**, 479-485.

2. D. G. Davis and A. Bax (1985) *J. Amer. Chem. Soc.*, **107**, 2821-2822.

3. D. L. LeMaster (1990) *Quart. Rev. Biophys.*, **23**, 133-174.

4. D. C. Muchmore, L. P. McIntosh, C. B. Russell, D. E. Anderson and F. W. Dahlquist (1989) *Meth. Enzym.*, **177**, 44-73.

5. A. Bax, R. H. Griffey and B. L. Hawkins (1983) *J. Mag. Reson.*, **55**, 301-315.

6. A. M. Gronenborn, A. Bax, P. T. Wingfield and G. M. Clore (1989) *FEBS Lett.*, **243**, 93-98.

7. L. P. McIntosh, A. J. Wand, D. F. Lowry, A. G. Redfield and F. W. Dahlquist (1990) *Biochemistry*, **29**, 6341-6362.

8. D. M. LeMaster and F.M. Richards, *Biochemistry*, **27**, 142-150 (1988).

9. C. H. Arrowsmith, R. Pachter, R. B. Altman, S. B. Iyer and O. Jardetzky (1990) *Biochemistry*, **29**, 6332-6341 (1990).

10. J. Wang, A. P. Hinck, S. N. Loh and J. L. Markley (1990) *Biochemistry*, **29**, 102-113.

11. S. W. Fesik and E. R. P. Zuiderweg (1990) *Quart. Rev. Biophys.*, **23**, 97-131.

12. L. E. Kay, G. M. Clore, A. Bax and A. M. Gronenborn (1990) *Science*, **249**, 411-414.

13. R. Shoeman, B. Redfield, T. Coleman, N. Brot, H. Weissbach, R. C. Greene, A. A. Smith, I. Saint-Girons, M. M. Zakin and G. N. Cohen (1985) *BioEssays*, **5**, 210-213.

14. J. B. Rafferty, W. S. Somers, I. Saint-Girons and S. E. V. Phillips (1989) *Nature*, **341**, 705-710.

15. I. Saint-Girons, J. Belfaiza, Y. Guillou, D. Perrin, N. Guiso, O. Barzu, and G. N. Cohen (1986) *J. Biol. Chem.*, **261**, 10936-10940.

16. F. W. Studier, B. A. Moffet (1986) *J. Mol. Biol.*, **189**, 113-130.

17. A. H. Rosenberg, B. N. Lade, D.-S. Chui, S.-W. Lin, J. J. Dunn and F. W. Studier (1987) *Gene*, **56**, 125-135.

SECTION VI

Mass Spectrometry Workshop

40

A Brief Overview of Mass Spectrometric Methods For the Analysis of Peptides and Proteins

Bradford W. Gibson[1]

Department of Pharmaceutical Chemistry and Mass Spectrometry Facility
University of California, San Francisco, CA 94143-0446

I. Introduction

Over the last decade, mass spectrometry has evolved from a primarily peripheral technique in protein chemistry to a central and now indispensable component of the overall methodology of protein structure determination. The reasons behind this change are complex, but one crucial factor can be quickly identified. Although some progress had been made in the late 1970's on ionization processes that allowed underivatized peptides to be ionized without thermal degradation (such as field desorption and plasma desorption mass spectrometry), the introduction of fast atom bombardment (FAB) (Barber et al., 1981) or liquid secondary ion mass spectrometry (LSIMS) (Aberth et al., 1982) was clearly a major turning point. By bombarding the peptide sample dissolved in a liquid matrix (e.g. glycerol) with an atom or ion beam of ≈ 5-20 keV, both protonated or deprotonated molecular ions could be easily generated. The ease of use and compatibility of FAB and LSIMS with magnetic sector and quadrupole mass spectrometers rapidly advanced and expanded the LSIMS methodologies for peptide and protein analysis (for review, see Martin and Biemann, 1987), and these techniques quickly dominated the field of protein mass spectrometry.

Some of the first applications of LSIMS and FAB to peptide and protein analysis are still widely used. Molecular weight determinations of peptides or peptide mixtures can be used to identify posttranslational modifications, map a protein structure, or simply confirm a peptide's composition (Gibson and Biemann, 1984). Alternatively, an examination of the relatively weak peptide backbone fragments in these spectra can generate a partial sequence. Or, if one is lucky, a complete sequence can sometimes be obtained if a nanomole or so of peptide is used. However, a complete sequence is the exception and, if one has

[1]Financial support was provided by NIH Division of Research Resources and the NSF Biological Instrumentation Program.

Techniques in Protein Chemistry II

less than a nanomole, the extent of fragmentation is generally too low even for a partial sequence. Nevertheless, with the concurrent advances in instrumentation over the last decade, particularly high field magnets and higher mass range quadrupole analyzers arranged in tandem configurations, LSIMS has truly matured as a technique for peptide and protein sequence analysis. This fact was clearly evident in this year's mass spectrometry workshop, where LSIMS-based tandem sequencing strategies clearly dominated the sequencing methodologies, although this is likely to change with the recent introduction of electrospray and matrix assisted laser desorption techniques.

II. Tandem Peptide Sequencing

When combined with tandem or MS/MS instruments, collisional induced dissociation (CID) of LSIMS (or FAB) sputtered peptide molecular ions can readily produce fragment ion spectra of peptides of up to m/z 2500 in the low picomole levels (Walls *et al.*, 1990; Hill *et al.*, 1989). During this years workshop, A.L. Burlingame from U.C.S.F. and Don Hunt from the University of Virginia presented strategies for applying tandem mass spectrometry to problems in protein and glycoprotein sequence analysis. In A.L. Burlingame's presentation, methodologies were described for peptide sequencing using tandem mass spectrometry on a high field, four sector magnetic instrument (Kratos Concept IIHH) at high collision energies (4-6 keV) (Walls *et al.*, 1990). It was emphasized that this sequencing methodology is very fast, versatile and capable of handling various posttranslational modifications, and is particularly well suited for dealing with picomole amounts of proteins, levels that are typically obtained from 2-D gels. To emphasis this latter point, a brief description was presented of an experiment that involved the loading of 50 picomoles of fatty acid binding protein (≈ 15 kDa) onto a SDS-PAGE gel, its separation, elution from the gel, proteolytic digestion and partial sequencing. It is quite clear from the work being carried out in this and other laboratories that a combination of rapid low picomole sequencing by tandem mass spectrometry of proteins eluted from 1- and 2-D gels is a technique that will a major focus in the development of mass spectrometric strategies over the next decade.

Using the U.C.S.F. tandem mass spectrometer, Falick and coworkers have described a procedure to identify variants of hemoglobin that differ by only one amino acid. The tandem CID spectra of peptides generated after tryptic digestion of intact hemoglobins that either did not match the expected mass or HPLC elution position were subjected to MS/MS analysis. Interpretation of these spectra was relatively straightforward, and the identities of the substituted amino acids quickly identified. Buechter *et al.* used a similar strategy to isolated and sequence a peptide from creatine kinase containing the affinity label epoxycreatine. This moiety had been introduced as a covalent label to identify the specific amino acids involved in catalysis. Again, the final MS/MS spectrum provided unequivocal proof of a specific cysteine residue (Cys282) as the major site of epoxycreatine modification.

In the workshop presentation, D.F. Hunt described a somewhat different sequencing strategy using low energy CID tandem spectra obtained on a triple quadrupole mass spectrometer. In this scheme, unit mass resolution of the resulting peptide fragment ions is not obtained since a the entire molecular ion cluster, and not just the ^{12}C-containing monoisotopic component, is transmitted through the first quadrupole. However, unit mass resolution and accuracy is apparently not necessary as methyl ester and N-acetylated derivatives of the same peptide are also analyzed, which then allows one to identify fragments as N- or C-terminal and to determine the number of carboxyl and/or amino functional groups present on these fragment ions. Taken together, the data sets offer a means to define the sequence, with exception of Leu and Ile assignments which at present can only be distinguished under high energy CID conditions (Johnson *et al.*, 1987). A more complete description of this sequencing strategy is presented in chapter 43 of this volume by Hunt *et al.*

It was also readily apparent that the time required to obtain these tandem spectra is also quite impressive, typically taking no more than a few minutes. The potential sequencing throughput of such instruments then becomes as high as 100-200 residues per hour (or even greater). However, this dramatic increase in spectral information has placed new demands on spectral interpretation and sequencing strategies. Therefore, it is not surprising to see various computer-based sequencing algorithms starting to appear (see, for example, Johnson and Biemann, 1989). At least one of these computer-based strategies was used to help sequence the test peptides as described in the workshop by A. L. Burlingame (Hines *et al.*, unpublished results). That strategy essentially relies on the assignment of possible ion types to various fragments, and then seeks out related ions for both N- and C-terminal fragments. Another strategy is presented in more detail by Yates *et al.* as a chapter in this section, and was designed to interpret lower energy type fragmentation (approx. 50-100 eV) that occurs under the CID conditions employed in triple quadrupole instruments.

III. Protein Mass Determination

One clear limitation of LSIMS and FAB techniques has been the inability to sputter high mass peptides or proteins with any reasonable degree of efficiency. Although a few examples of peptides or small proteins in the 10 - 25 kDa mass range have appeared in the LSIMS literature (Loo and Green, 1987; Gibson *et al.*, 1987; Martin and Biemann, 1987), the sensitivity has been generally quite low and these analyses far from routine. However, two recent innovations in high mass analysis have virtually eliminated this limitation overnight.

The first of these two techniques is called matrix-assisted laser desorption mass spectrometry (LDMS) (Karas and Hillenkamp, 1988, Chait and Beavis, 1990a,b). This technique takes advantage of various UV-absorbing matrices, primarily nicotinic acid and cinnamic acid derivatives (Beavis and Chait, 1989), to absorb energy from a laser pulse in order to put into the gas phase and ionize proteins that have been previously mixed with this solid matrix. The ionized proteins are mass-analyzed by precise measurement of the time period between the original laser pulse and the arrival times of ions using a time-of-flight mass spectrometer, where there is in principle no upper limit to the mass range and virtually all ions may be detected. This relatively efficient process has now made routine mass determination possible of proteins as large as 200 kDa using as little as 1 picomole (or less). A fairly typical spectrum is shown below (see Fig. 1), presented by Brian Chait during the symposium workshop, of immunoglobulin G1(average M_r 148140). With their current prototype LDMS instrument, Chait and Beavis are able to obtain molecular weight measurements of proteins with an accuracy of 0.01%. Of course, this is a measurement of the unresolved peaks that constitute the molecular ion cluster obtained at a resolution of ≈300-500 FWHM, a value that will need to be improved if more accurate and precise molecular weights are to be determined for even larger or more heterogeneous proteins. Indeed, this point was raised after Chait's talk i.e., what ultimate resolution is necessary or even desirable for protein molecular weight determination? For the analysis of different states of protein phosphorylation, it was argued that a resolution of perhaps 3000-4000 FWHM is required if the protein has a mass of around 50 KDa. Otherwise, one would not be able to resolve the mass differences of 80 Da that would correspond to the addition of phosphate *eg*. M_r 50,000 vs. 50,080. While in general agreement with this statement, Chait countered that going much above a resolution of 3,000 for additional improvements in accuracy would produce questionable gains and one would only encounter further problems of microhetreogeneity (such as partial oxidations of Met residues) and the natural isotopic distribution of the molecular ion of the protein itself.

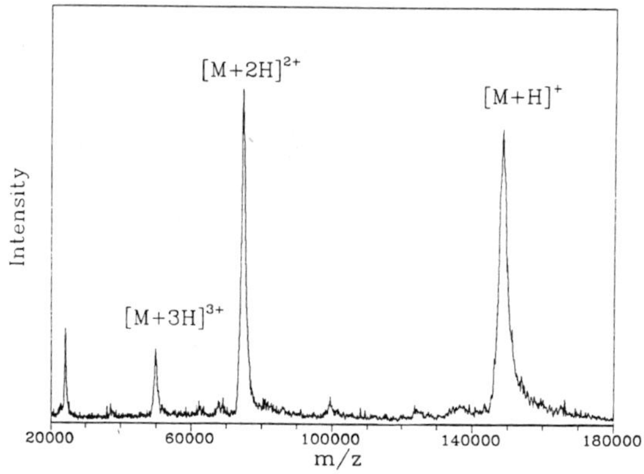

Figure 1. LDMS spectrum taken on a time-of-flight mass spectrometer of ≈1 picomole of monoclonal immunoglobulin G1 (*mus musculus*) (M_r 148,140). Time scale is displayed in 50-ns time bins. Heterogeneity of observed peaks is due to heterogeneity of N-linked glycosylations and that caused by photochemically generated adduct ions. Reprinted from Beavis and Chait, 1990a with permission from the American Chemical Society.

A second major innovation of the last two or three years that was clearly evident at this years meeting was electrospray mass spectrometry (Fenn *et al.*, 1989; Loo *et al.*, 1990). Briefly, this technique requires that the peptide or protein sample be delivered as a high velocity liquid spray at atmospheric pressure, but at high voltages with respect to the mass analyzer (≈3 kV). This 'electrospray' or 'ionspray' then undergoes differential pumping to remove the solvent molecules (usually methanol/H_2O with acetic acid) and the remaining highly charged protein molecules are then focused into the mass analyzer. For proteins, multiply charged ions are created from the intact molecule, with charge states ranging from $(M+H_7)^{10+}$ to $(M+H_{12})^{20+}$ for a protein such as ecotin (McGrath *et al.*, 1991) as shown in Fig. 2. These electrospray devices have been traditionally linked to quadrupole mass analyzers, which separate ions according to their mass-to-charge ratios, m/z. The multiply charged protein ions typically have m/z values below 2000, well within the range of this type of mass analyzer. Simple algorithms can then convert these series of multiple charged ions to a single mass (or molecular weight) value (Mann *et al.*, 1989). There are several papers in this section that have used this technique for an array of protein molecular weight determinations. For example, Geisow and his cowokers have presented a strategy to characterize recombinant proteins such as human serum albumin for quality assessment purposes. By careful and accurate mass assignments and close attention to smaller components, they were able to offer a rapid and surprisingly thorough assessment of several recombinant proteins.

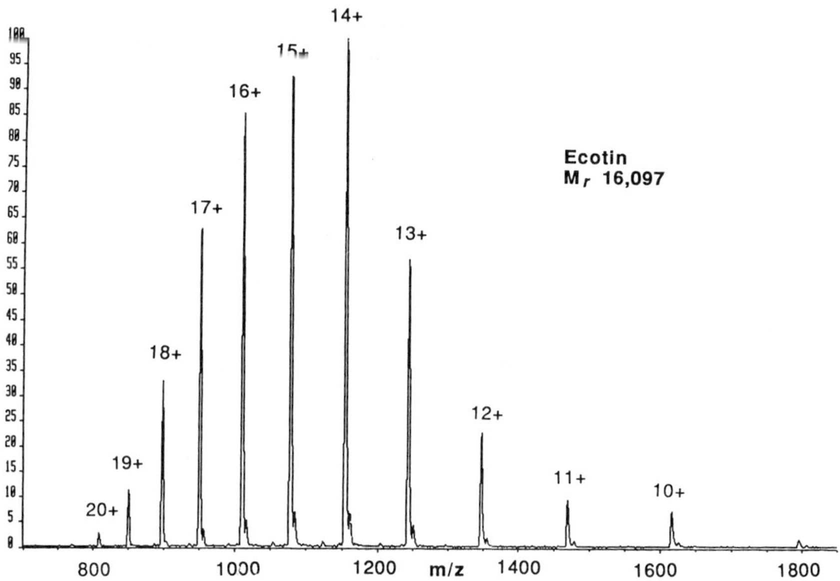

Figure 2. The electrospray mass spectrum of ecotin (M_r 16,097) taken on a prototype VG Bio-Q mass spectrometer. The charge states of the protein are listed above each peak.

IV. Towards Higher Efficiency Peptide Analysis

One additional benefit of the recent success of matrix-assisted LDMS and electrospray mass spectrometry for high-mass protein analysis, has been that these techniques are also well suited for efficient lower mass peptide analysis. In his workshop presentation. Dr. B. Chait described a strategy for the molecular weight mapping of proteins after CNBr treatment. In the matrix-assisted LDMS spectrum shown below, apolipoprotein AI was digested with CNBr for 2 h in 1M HCl producing the expected four major fragments with (M+H)+ peaks ranging in mass from ≈3,000 to 11,000 Da. Such an experiment is generally not possible for LSIMS and FAB techniques due to the suppression effects one sees of one peptide over another and the much lower ion abundances at these high masses. And when this same protein was treated with CNBr only briefly, Chait and Beavis were able to see the intermediate fragments that correspond to incomplete cleavages and, therefore, they could assign the proper order. In this volume, Caprioli *et al.* have also presented a similar LDMS-based strategy for the anaysis of peptide mixtures generated by proteolytic digests of proteins.

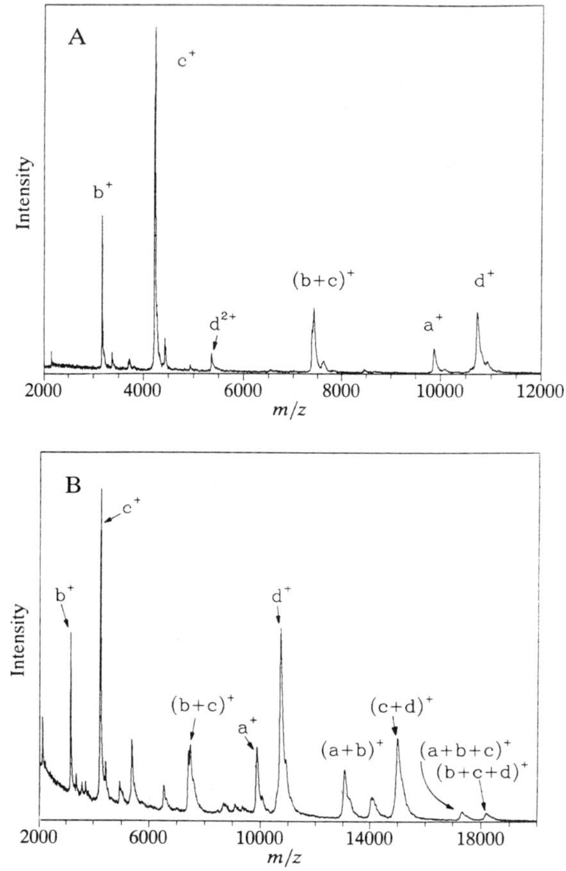

Figure 3. LDMS-TOF mass spectrum of two different CNBr digests of human apolipoprotein AI (243 residues) recorded on a TOF mass spectrometer (A) after 2 h (nearly complete digest) and (B) after one or two minutes (incomplete). For clarity, only the singly charged molecular ions are labeled in spectrum B. All peaks except those containing the C-terminal peptide (peptide d) correspond to the homoserine lactone product. Peptides a-d correspond to residues 1-85 (peptide a), 86-112 (peptide b), 113-148 (peptide c), and 149-243 (peptide d). Reprinted with permission from Beavis and Chait, 1990b.

In electrospray MS, similar peptide molecular weight maps can also be readily obtained, either by analyzing the complete mixture or separated on-line by HPLC/MS. During the workshop, Hunt described his setup for analyzing peptide mixtures using microcapillary HPLC columns directly coupled to an electrospray quadrupole system. In this way, molecular weights can be analyzed using as little as 1 picomole of sample (total injection). Partial and perhaps complete sequencing can also be accomplished by fitting the HPLC to a triple quadruple MS, with low energy CID spectra obtained using 5-10 picomoles per component peptide. However, the interpretation of tandem CID peptide doubly or triply charged molecular ions (i.e., (M+2H)++ and (M+3H)+++) is not altogether as straightforward as singly charged peptides since the charge states of fragments are not necessarily known. This is, however, a very powerful and sensitive approach and we can expect to see some major advances and innovations in this area during the next few years.

With all these new advances, the future for peptide sequencing analysis by mass spectrometry appears to be set for another big jump in sensitivity. The large improvements in ionization efficiency using electrospray and laser desorption mass spectrometry have clearly not yet been fully exploited. The coupling of electrospray and matrix assisted laser desorption ionization methodologies with more efficient analyzers such as the recently introduced ion trap mass spectrometer and magnetic sector instruments with extended array detectors should yield even more promising results in the very near future. For example, an ion trap mass spectrometer is quite capable of analyzing low picomole and possibly femtomole signals, and can carry out MS/MS and even $(MS)^n$ type experiments on a peptide molecular ion (Jardine *et al.*, 1990). Alternatively, a laser desorption source on a tandem magnetic sector instrument should be able to supply much larger efficiencies of molecular ions from much lower sample amounts for CID fragmentation without the discrimination effects seen in LSIMS. However, the pulsed nature of the laser system requires simultaneous detection capabilities as provided by an array type detector. Since several four-sector magnetic mass spectrometers are currently fitted with array detectors, it would appear to be only a matter of time before this capability is demonstrated.

References

Aberth, W., Straub, K.M. and Burlingame, A.L. (1982) *Anal. Chem.* **54**, 2029.

Barber, M., Bordoli, R.S., Sedgwick, D.R., and Tyler, A.N. (1981) *J. C. S. Chem Commun.*, 325.

Beavis, R. and Chait, B.T. (1989) *Rapid Commun. Mass Spectrom.* **3**, 432.

Beavis, R. and Chait, B.T. (1990a) *Anal. Chem.* **62**, 1836.

Beavis, R. and Chait, B.T. (1990b) *Proc. Natl. Acad. Sci. USA* **87**, 6873.

Bell, D. and Green, B.N. (1987) *Abs. 35th Conf. Mass Spectrom. Allied Topics*, Denver, Colorado, pp 540-541.

Fenn, J.B., Mann, M., Meng, C.K., Wong, S.F., Whitehouse, C.M. (1989) *Science* **246**, 64.

Gibson, B.W. and Biemann, K. (1984) *Proc. Natl. Acad. Sci. USA* **81**, 1956.

Gibson, B.W., Walls, F.W., Falick, A.M., Lipka, J.J., Waskell, L.A. and Cottrell, J. (1987) *Abs. 35th Conf. Mass Spectrom. Allied Topics*, Denver, Colorado, pp. 66-67.

Hill, J.A., Biller, J.E., Martin, S.A. and Biemann, K. (1989) *Int. J. Mass Spectrom. Ion Proc.* **92**, 211.

Jardine, I., Hail, M., Lewis, S., Zhou, J. and Whitehouse, C. (1990) *Abs. 38th Conf. Mass Spectrom. Allied Topics*, Tucson, Arizona, pp. 16-17.

Johnson R. , Martin, S.A., Biemann K., Stults, J.T. and Watson, J.T. (1987) *Anal. Chem.* **59**, 2621.

Johnson, R.S. and Biemann, K. (1989) *Biomed. Environ. Mass Spectrom.* **18**, 945.

Karas, M. and Hillenkamp, F. (1988) *Anal. Chem.* **60**, 2299.

Loo, J.A., Edmonds, C.G., and Smith R.D. (1990) Science 248, 201.

Mann, M., Meng C.K. and Fenn, J.B. (1989) *Anal. Chem.* **61**, 1702.

Martin, S.A., and Biemann, K. (1987) *Mass Spectrom. Revs.* **6**, 1.

McGrath, M.E., Hines, W.M., Sakanan, J.A., Fletterick, R.J. and Craik, C.S. (1991) *J. Biol. Chem.*, submitted

Walls, F.C., Baldwin, M.A., Falick, A.M., Gibson, B.W., Gillece-Castro, B.L., Kaur, S., Maltby, D.A., Medzihradszky, K.F., Evans S. and Burlingame A.L., In "Biological Mass Spectrometry" (A. L. Burlingame and J. A. McCloskey, eds.) Elsevier, Amsterdam (in press).

41

MASS SPECTROMETRY WORKSHOP: AN OVERVIEW

John E. Shively
Beckman Research Institute of the City of Hope, Duarte, CA 91010.

I. INTRODUCTION

This is a brief overview of the Mass Spectrometry Workshop held at the Fourth Symposium of the Protein Society in San Diego on August 12-15, 1990. The purpose of the workshop was to bring protein chemists up to date views of the types of problems which can be solved by mass spectrometry with an emphasis on real samples. In order to stress this aspect of the workshop, a test sample, designated symposium test peptide-4 (STP-4), was distributed among a variety of mass spectrometry groups for analysis. A detailed analysis of these results is presented in companion papers in this book. Presentations were given by Brian Chait at Rockefeller University, Al Burlingame at the University of California at San Francisco, Don Hunt at the University of Virginia at Charlottesville, and Terry Lee at the City of Hope at Duarte, CA. The presentations covered a variety of topics and generated a good deal of discussion. The following overview presents some of the points raised during the workshop, and includes a liberal commentary by the author. No attempt has been made to give a verbatim transcript of the proceedings.

II. ANALYSIS OF INTACT PROTEINS

Brian Chait discussed some of the virtues of the ideal mass spectrometer for the analysis of proteins. They include high sensitivity, high resolution, the ability to analyze mixtures and high molecular weight proteins, and the means to fragment samples and to analyze the fragments. In many respects the

Techniques in Protein Chemistry II

laser desorption/time of flight mass spectrometer has these virtues. The instrument described was built by Brian Chait and is quite simple in operation. It consists of a UV laser (ND-YAG, 355 nm output) focused on a sample stage, and a linear time of flight mass spectrometer (1). Samples are dissolved in a solution containing a matrix such as sinapinic acid (3,5-dimethoxy-4-hydroxycinnamic acid), the solvent is removed and the resulting solid sample is pulsed with the laser beam. The matrix absorbs light in the UV range and transfers energy to the protein sample which ionizes principally as the singly charge species. The sensitivity for proteins such as cytochrome C, trypsinogen, and myoglobin is in the femtomole range, with as little as 0.5 pmoles applied to the sample stage resulting in signal noise ratios >10. For these three examples the mass accuracy is within 1 mass unit (2). High molecular weight examples were shown such as immunoglobulin (Mr about 150,000). The overall mass accuracy appears to be about 0.01%, equivalent to the measurements made on small proteins such as myoglobin by electrospray/triple quadrupole instruments (3).

One of the amazing features of these mass measurements is that they can be made on samples containing micromolar to millimolar amounts of salts, and in the presence of complex mixtures containing other proteins, lipids, and salts. For example, one can obtain peaks for major proteins present in whole serum or milk. The sample may also be treated with enzymes to generate peptides, remove carbohydrate, etc. The mass changes can be compared to the intact sample and allow one to deduce additional information about the sample. Some interfering substances have been identified such as SDS and glycerol. On the other hand, spectra have been obtained in the presence of urea and guanadinium HCl. Given the newness of matrix assisted laser desorption in mass spectrometry, it will take considerable experimentation to determine the appropriate range of matrices, allowable contaminants, and nature of interfering substances. One can already predict that this technique will present new opportunities for the molecular weight analysis of proteins, especially large proteins.

Several points can be made here. The first is that protein chemists have longed dreamed of a truly accurate, high resolution method for the mass measurement of proteins. In many cases, such a measurement can be valuable as a proof of structure. If the calculated mass and observed mass don't agree, one can suspect either errors in the original structural analysis (either protein or cDNA data) or the presence of posttranslational modifications (proteolysis, glycosylation, phosphorylation, etc.). The demonstrated accuracy of 0.01% means that one can potentially observe most single amino acid changes in a protein of Mr of 10-20,000, and can have molecular weights of proteins at the 100,000 range accurate to within 10 mass units. It should be noted that currently, the resolution of LD spectra are quite low (<1000) and whereas the mass accuracy may be within 10 mass units, it is not possible to separate the molecular ion signal for two 100,000 MW proteins that differ by ten mass units and some average value for the mass would be obtained. The next point concerns instrumentation. In theory, a time of flight mass spectrometer is one of the simplest to construct, and thus could be commercialized for prices as low as $100-200,000. If instruments do become available in this price range, most protein facilities would be able to afford one. Plasma desorption time of flight mass spectrometers have already become popular for the analysis of peptides because of their relative low prices and ease of operation. One can predict that the laser desorption instruments would open the world of proteins to accurate mass measurements. The other technique that presently competes with LDMS in the mass analysis of proteins is electrospray in its various forms on quadrupole and magnetic sector instruments (see below). For the moment it would appear that LD has the advantage for very large proteins (>50,000 MW) and samples that are mixtures. It is hard to predict the future, but one can anticipate further developments for these instruments. Perhaps the most exciting is further increases in sensitivity. If LD/TOF/MS can currently analyze low femtomole sample amounts, will it soon be possible to analyze attamole amounts? In this regard we can be optimistic, since these instruments are capable of very high ion transmission. Thus, we can look forward to further breakthroughs in this arena.

III. FAB/MS ANALYSIS OF PEPTIDES ON MAGNECTIC SECTOR INTRUMENTS

Fast atom bombardment (FAB) or more correctly liquid secondary ion mass spectrometry (liquid SIMS) was first used for the analysis of peptides some nine years ago. Since then, it has played a dominant role in the structural analysis of peptides and proteins, particularly with respect to sequence analysis. Most protein chemists are very familiar with this technique and have come to rely on it as a routine method for obtaining highly accurate mass measurements of peptides usually in the mass range of 500-5000. Al Burlingame presented some of the basics in fast atom bombardment/mass spectrometry, especially as used on tandem magnetic sector instruments. Peptides up to mass 10,000 can be analyzed by the instruments, but sensitivity is generally less than for PD, LD and ES for the larger structures. Samples are dissolved in dilute acetic acid, HPLC solvents such as trifluoroacetic acid/water/acetonitrile, or DMSO (for more hydrophobic peptides), mixed with a matrix such as glycerol, thioglycerol, a mixture of DTE/DTT, or nitrobenzyl alcohol, and bombarded with a beam of neutral atoms (xenon) (FAB) or metal ions (cesium) (liquid SIMS). The secondary ion mass spectrum is not dependent on whether the primary beam is composed of ions or atoms. For the purposes of this discussion the two terms can be used interchangeably.

The matrix in FAB/MS plays an all important role. In mixtures, not all of the samples will be sampled with the same efficiency due to competition for the liquid surface. Thus, ion intnesities are a function of the distribution of the sample between the interior and surface of the liquid drop. Although either positive or negative ions can be analyzed, more sensitivity is usually obtained for positive ions. For scanning magnetic sector instruments, sensitivities for molecular wieght determinations can be as low as 10 picomoles of sample analyzed. A factor of 10 to 100 more sample is needed for sequence analysis. However, in instruments equipped with array detectors where large portions of the spectrum can be recorded at a time, the sensitivities may be improved by a factor of 10-100. Peptides in the low mass range (100-500) may be obscured by matrix ion peaks. Thus, this technique is less sensitive for di-tetrapeptides. Since very hydrophillic (especially small hydrophillic) peptides often do poorly in FAB/MS, it may be necessary to convert them to more hydrophobic derivatives such as their octyl esters. Burlingame gave several examples where this approach increased the sensitivity of FAB/MS by a factor of 10-50.

Magnetic sector instruments have good resolution and isotope clusters can be fully resolved. In tandem four sector systems, monoisotopic peaks may be selected for further analysis by collision activation decomposition (CAD) and mass spectrometry. In recent years, four sector MS/MS instruments have been used in a number of labs for biomolecular structure analysis. (4-6). These instruments are capable of obtaining sequence information on peptides ranging from 2-20 amino acids. Often 1-2 nmoles of sample are required to obtain complete sequence information, but with array detectors, the amount of sample levels can be reduced to 10-100 pmole. The CAD experiments are performed at high energy (1-10 KeV), and usually allow one to distinguish between leucine/isoleucine which have identical masses. MS/MS experiments have the further advantage of being able to select one ion from a mixture, thus obviating one of the banes of protein chemists: the analysis of mixtures.

The chief drawback of the magnetic instruments is their size and cost. Two sector instruments cost about $500,000 and four sector instruments over $1,000,000. The high cost has been offset by the high throughput, allowing dozens of samples to be analyzed in a single day. While the average protein chemistry lab may not be able to afford to have its own system, there is an increasing number of facilities which researchers can collaborate with.

Interpetation of mass spectral sequence data is another problem. The analysis may take several hours to days even for a skilled mass spectrometrist. Obviously, the solution is to computerize the analysis to ensure that all possible solutions are considered. In this regard, both Biemann and Burlingame have developed programs to interpret peptide sequence data from tandem four sector systems. One may ask if these instruments can always determine complete sequences. The philosophy underlying the tandem four sector approach is that all of the information needed to solve the structure in contained in the mass spectrum. While true for many peptides and perhaps a majority of peptides, often the sequence information is not complete particularly for larger structures and limited sample amounts. In these cases ancillary information is necessary, such as Edman degradation, acetylation of amino groups, esterification of carboxyl groups, cleavage to smaller, more easily analyzed pieces, etc. These approaches are well described in accompanying reports on STP-4.

What about the future of these instruments? New ion sources such as electrospray (which produce multiply charged species) are being adapted to the instruments, thus increasing their capabilities into the analysis of proteins. These instruments have been the workhorses in the protein field to date, and can be expected to continue to play a central role in the years to come.

IV. FAB/MS AND ELECTROSPRAY TRIPLE QUADRUPOLE INSTRUMENTS

Don Hunt described his experience with two types of ion sources on triple quadruple instruments. The first is a FAB/MS source which has been the mainstay of the lab (7). By selecting a mass window of 2-6 units (as opposed to the monoisotopic ion selection for MS/MS in the magnetic sector work), he has increased sensitivity and adapted the ms/ms approach to the triple quadrupole instrument. Low energy CAD (10-300 eV) is performed in the region of the second quad, and the third quad analyzed the resulting fragments. Hunt led the workshop group through the process of a complete analysis of a typical peptide. In general, experiments must be performed to help distinguish the N- from the C-terminal series ion fragments. His approach includes running the sample before and after derivatization with acetic anhydride to count free amino groups (the N-terminus if unblocked, and lysines), and with MeOH/HCl to count carboxyl groups (the C-terminus if unblocked, and aspartic and glutamic acids). The approach has been successful for a large number of peptides in the range of 4-20 amino acids. Occasionally, he must also resort to one or more cycles of subtractive Edman chemistry (ie. mass spectral analysis of the remaining peptide). The sensitivity of the method is 10-20 picomoles per analysis. Interpetation of the MS/MS spectra is possible by computer programs developed by John Yates at Cal Tech and Joe Zhou at Finnigan.

The second ion source described is electrospray which allows the introduction of samples at atmospheric pressures directly from an HPLC (flow rates in the range of 1-50 uL/min). The nature of the source promotes the formation of multiply charged ions. Small tryptic peptides will be observed as doubly charged ions due to the presence of the amino group and one basic residue (in acidic HPLC solvents such as TFA/acetonitrile). The fragmentation of these peptides in the triple quad instruments usually leads to a distinctive N-terminal and C-terminal fragment ion series with enough sequence information to deduce complete structures for peptides in the size range of 2-15 amino acids. Complete sequence information is often lacking, and further work is required (fragmentation with proteases, Edman chemistry, etc.). Those peptides that carry more than two charges yield ms/ms spectra that are much more difficult to interpret. It appears that the two techniques, FAB/MS and electrospray may be complementary, each supplying some unique information. Although the technique is too new to fully judge its capabilities, there is considerable excitement because of its obvious adaptation to HPLC and capillary electrophoresis, and its high sensitivity. As little as 1 picomole of sample is sufficient to give molecular weight information, and 10 picomoles for MS/MS information.

Another area of excitement for electrospray sources is its utility in analyzing protein samples (8). The electrospray source yields molecular ions with a distribution of charges. The number of charges for most proteins is such that the m/z values fall well within the normal mass range of a quadrupole analyzer (m/z 400-2000). Each multiply charged species is an independent measure of the mass of the protein and taken together yield a very accurate measure of the mass (0.01%). Computer algorithms are available to transform the multiple peak signal of a protein into a single peak which facilitates the analysis of spectra of mixtures. As sample molecular weights exceed 100,000, the resolution of the analysis is insufficient to seaparate the various charge states. Thus, very large protein structures cannot be analyzed by electrospray at this time. It remains to be seen what the molecular weight limit will be for routine analyses.

Finally, the electrospray source gives very low backgrounds compare to FAB/MS, making it an excellent choice for the analysis of small peptides. One can easily predict that electrospray will become as important as FAB/MS for the analysis of proteins in the next few years.

Perhaps of equal importance is that electrospray has rekindled interest in quadrupole instruments (heretofore shunned for peptide analysis due to their limited mass ranges). These instruments cost about $300-400,000 for triple quads, and significantly less for single quads. The single quad instruments may be most useful for obtaining molecular weight analysis of peptides and proteins eluted from HPLC or capillary electrophoresis in the low picomole to femtomole range.

Hunt spent the last part of his presentation describing preliminary results obtained on ion traps in labs at Purdue and Oakridge. These devices can perform ms^n type of experiments and utilize ion sources such as FAB or electrospray. Peptides and proteins have been analyzed with sensitivities in the femtomole range. Mass accuracies are much less than can presently be obtained using quadrupole analyzers and a commercial product is probably at least three years away.

The audience was left with the impression that in the future mass spectrometry and protein chemistry will intimately connected. We can expect faster, more sensitive analyses of whole proteins, and their peptide fragments. We can look forward to a new generation of mass spectrometrists who are cross-trained in protein chemistry, and protein chemists who are cross-trained in mass spectrometry.

V. REFERENCES

1. Beavis, R.C., and Chait, B.T., (1989) Rapid Comm. Mass Spec., **3**, 432-435
2. Beavis, R.C., and Chait, B.T., (1990) Anal. Chem., **62**, 1836-1840.
3. Mann, M., Meng, C.K., and Fenn, J.B., (1989) Anal. Chem. , **61**, 1702-1708.
4. Gibson, B.W., Webb, J.W., Yamasaki, R., Fisher, S.N. Burlingame, A.L., Mandrell, R.E., Schneider, H., and Griffiss, J.M., (1989) Proc. Natl. Acad. Sci. USA, **86**, 17-21.
5. Biemann, K., and Scoble, H.A., (1987) Science, **237**, 992-998.
6. Carr, S.A., Roberts, G.D., and Hemling, M.E., Mass Spectrometry of Biological Materials (C.N. McEwen and B.S. Larsen, eds.) Marcel Dekker, New York, 1990, pp. 87-136.
7. Hunt, D.F., Shabanowitz, J., Yates, J.R., Griffin, P.R., and Zhu, N.Z., Mass Spectrometry of Biological Materials (C.N. McEwen and B.S. Larsen, eds.) Marcel Dekker, New York, 1990, pp. 169-195.
8. Smith, R.D., Loo, J.A., Edmonds, C.G., Barinaga, C.J., and Udseth, H.R., (1990) Anal. Chem. ,**62**, 882-889.

42

Sequencing of the Society Test Peptide IV by High Performance Tandem Mass Spectrometry

Katalin F. Medzihradszky, Steven C. Hall, David A. Maltby, Wade M. Hines and Alma L. Burlingame

Department of Pharmaceutical Chemistry and Mass Spectrometry Facility, University of California, San Francisco, CA 94143-0446

I. INTRODUCTION

In the last decade, new methods in mass spectrometry have become increasingly powerful and versatile tools for structural characterization of proteins [1]. For example, tandem mass spectrometry has proven its ability to determine peptide and protein sequences [2-6]. Using a high performance four sector instrument, protonated molecular ions of peptides in a given sample are formed with low internal energy in the first mass spectrometer using liquid secondary ion mass spectrometry (LSIMS) [7]. The ^{12}C isobar of any given components' precursor ion is then selected by MS-I and allowed to undergo high energy collision with an inert gas, usually helium. This collision process enhances the selected ions' internal energy and induces a suite of unimolecular dissociation reactions characteristic of the sequence and structure of the molecule. These fragment ions are recorded in MS-II with an accuracy of ±0.2 Da and taken together comprise the resulting collision-induced dissociation (CID) spectrum.

High kinetic energy collisions are able to induce both dissociation reactions of the peptide backbone [8] and amino acid residue side chain fragmentation processes up to relatively high mass (≤ 2,500). These side chain fragmentation processes are essential in permitting differentiation between, and thus identifying, the isobars leucine and isoleucine [9]. In CID spectra, several series of sequence related ions [10] may be formed depending upon the location of basic residues within a given amino acid sequence which tend to protonate and "fix" charge preferentially [11]. One additional isobaric pair of amino acids, namely lysine and glutamine, may be readily distinguished by acetylation and location of the 42 Da shift associated with each acetylated lysine residue.

In this paper we present the primary structure elucidation, using a combination of microbore HPLC, proteolytic digestion and high performance tandem mass spectrometry with array detection, of the peptide sample distributed to participants in the Mass Spectrometry Workshop held at the Fourth Symposium of the Protein Society, San Diego, California, August 11-15, 1990.

Techniques in Protein Chemistry II

II. Experimental

Mass Spectrometry — Liquid secondary ion mass spectrometry experiments were performed on a VG 70SE mass spectrometer, operating in the positive ion mode, scanning over the mass range m/z 2700-300, at a scan rate of 10 sec/decade. Thioglycerol/glycerol (1:1) containing 1% TFA was used as liquid matrix. Collision-induced dissociation analysis was performed using a Kratos Concept IIHH four sector tandem mass spectrometer equipped with a cesium ion source [7, 12] and a multichannel array detector that allows simultaneous detection of ions over a 4% mass window [13]. CID spectra were acquired while operating at a mass resolution of 1000 for MS-II . The collision cell was floated at 4 keV and the helium pressure was adjusted until the parent ion intensity was decreased to 30% of its initial value [13]. The resulting CID spectra were interpreted both manually and with an automated sequencing algorithm [14].

Tryptic digestion — The entire content of the polypropylene vial containing the peptide sample was incubated in 70 mM NH_4HCO_3 (pH 8.2) at 37°C for 4h, at 1:25 (w/w) enzyme:substrate ratio (It was assumed that 100 picomoles of peptide were present). The digest was injected onto an ABI Aquapore OD-300 C_{18} reverse phase column (1.0 x 100 mm) equilibrated with 100% A. The percentage of solvent B was increased using a linear gradient from 0% to 100% in 100 minutes at a flow rate of 50 µl/min, beginning ten minutes after the injection. Solvent A was water containing 0.1% trifluoroacetic acid (TFA). Solvent B consisted of 70% acetonitrile, 30% water and 0.1% TFA. The chromatographic separation was performed using an ABI 140A Dual Syringe Solvent Delivery System. Ultraviolet detection of the separated components was accomplished by monitoring the column effluent absorbance at 215nm with an ABI 783A Programmable Absorbance Detector.

III. Results and Discussion

LSIMS analysis was employed to determine the molecular weight of the test peptide, yielding a protonated molecular ion (MH^+) at m/z 1819.1 (data not shown). The spectrum showed another ion at m/z 1835.2, 16 mass units higher than the major component of the mixture, indicating the presence of a partially oxidized methionine in the peptide. Acetylation on the probe was performed using an acetic anhydride/methanol (1:3) mixture. After the acetylation, the protonated molecular ion shifted to m/z 1861.2. The observed 42 mass units increase of the initial MH^+ value corresponded to incorporation of one acetyl group, suggesting that the peptide contained only one free amino group, which might be the N-terminus or an ε-amino group of a lysine residue, if the N-terminus were blocked. Residual TFA from the peptide diluent (0.5%) may have interfered with complete acetylation. After determining the peptide's molecular weight, CID analysis was carried out choosing the ^{12}C isobar of protonated molecular ion as the parent ion (see Fig. 1). The immonium ions [15] indicated the presence of proline (m/z 70), leucine/isoleucine (m/z 86), methionine (m/z 104), histidine (m/z 110), phenylalanine (m/z 120), and tryptophan (m/z 130 and 159). The presence of arginine and lysine/glutamine was questionable, since not all of the characteristic low mass ions (m/z 59, 70, 87, 100, 112 and 129 for arginine; 84, 101 and 130 for lysine) were observed. The CID spectrum exhibited abundant ion pairs 28 mass units apart in the low mass range. These ion pairs at m/z 157/185, 167/195, 184/212, 246/274, 257/285 corresponding to dipeptide

	Val	Ser	Trp	Ala	Met	Phe	Pro	Asn	Gly	Lys
v			819							
w		1004			616			258		
x					719				230	
y		1037	950	764	693	562	415	318	204	147
z		1021	934			546				
a	72	159	345			694	791	905		
b		187	373	444	575	722	819			

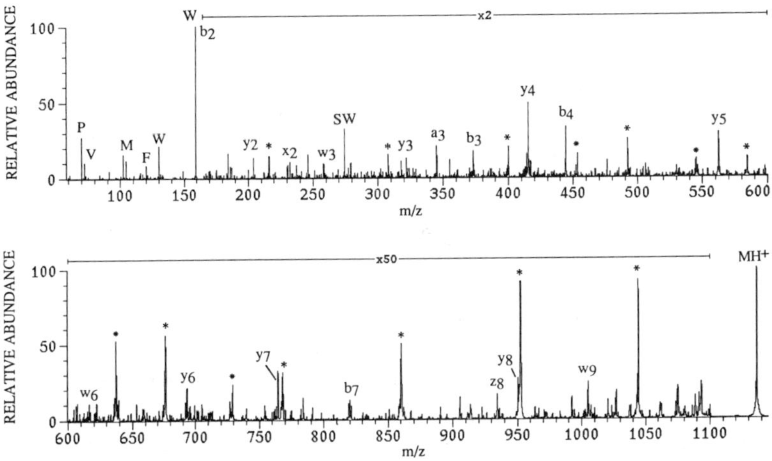

Figure 1. High energy CID spectrum of the intact test peptide (^{12}C MH^+ at m/z 1819.1) Fragment ions belonging to the same ion series are listed in rows. Underlined numbers indicate that both the **a** and **a+1** fragments were observed. Numbers in italics represent **a+1** ions. The table contains all the observed fragment ions; on the spectrum only the most abundant ones are labelled.

moieties which could be Ala-Leu (**a_2/b_2**), His-Gly, Pro-Asn, Ser-Trp, Phe-His or vice versa, respectively, appeared due to extensive occurrences of multiple backbone cleavages yielding these internal sequences. These internal sequence ions prevented the identification of the N-terminal sequence from the CID spectrum of the intact peptide. The comparison of the CID and the LSIMS spectra of the intact test peptide permitted the identification of some **c**-ions, which are commonly observed in the LSIMS spectra, but usually absent in CID data. Ions at m/z 798 (**c_7**), 885 (**c_8**), 1071 (**c_9**), 1142 (**c_{10}**), 1273 (**c_{11}**) and 1517 (**c_{13}**) were identified in this manner. Consideration of all information gathered thus far, resulted in the following partial working sequence: His-Phe-{Gln-Leu}-Arg-Val-Ser-Trp-Ala-Met-Xxx-Xxx-Asp-Xxx-Xxx. The presence and position of the arginine was suggested by a **d** ion at m/z 569.

The putative presence of an arginine in the sequence suggested carrying out a tryptic digestion. Molecular ions were expected to be seen at m/z 700 and 1136 in the resulting digest mixture. The tryptic digestion was performed on the entire content of a second vial of sample and the resulting components were separated using reverse phase microbore HPLC (see Fig. 2). Aliquots of the tryptic components were subjected to LSIMS analysis. Two fractions gave the expected protonated molecular ions at m/z 700.4 and 1136.5, respectively, and were subsequently analyzed by tandem mass spectrometry.

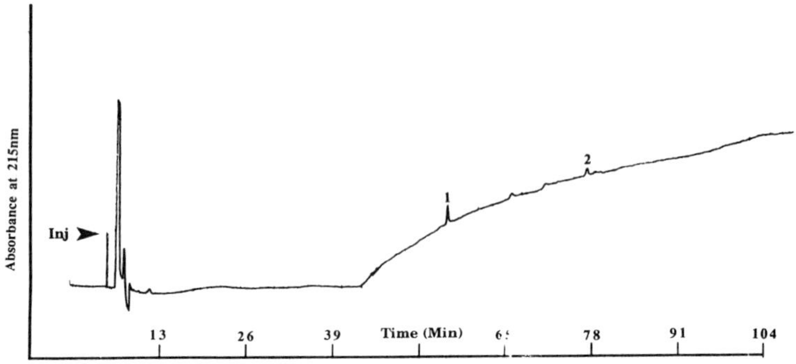

Figure 2. Reverse phase microbore HPLC chromatogram of the test peptide tryptic digest. LSIMS analysis on Fraction 1 gave ^{12}C MH^+ at m/z 700.4, and Fraction 2 showed ^{12}C MH^+ at m/z 1136.6.

The smaller tryptic peptide (MH^+ m/z 700.4) featured an abundant protonated molecular ion and its high energy CID spectrum readily provided the necessary information to obtain an unambiguous amino acid sequence (see Fig. 3). Both manual and computer interpretation of the CID spectrum suggested the same sequence: Ala-Leu-Phe-His-Gly-Arg. The automated sequencing algorithm designed in our laboratory [14] utilizes a pattern recognition based method to identify sets of sequence ions (i. e., $\mathbf{a_4, b_4, y_{n-4}}$) whose relationships are independent of amino acid sequence. It further records these sets noting the mass of the **y** ion (**y** centers) that would belong to that set, links sets whose theoretical **y** ions differ by an amino acid residue mass and recursively traces through the linked "**y** centered" families to establish candidate sequences. Thus, completed sequences are evaluated based on similarity of the patterns and abundances of the individual **y** families. Scores based on summed abundances of identified sequence ions, the number of sequence ions found and similarity of successive **y** families all favored the sequence Ala-Leu-Phe-His-Gly-Arg as best choice with dipeptide transpositions and substitutions following with lower rank.

The MH^+ ion for the C-terminal tryptic peptide was only moderately abundant. After manual interpretation of its CID spectrum (see Fig. 4) the Val-Ser-Trp-Ala-Met-Phe-Pro-Asn-Gly-Gln sequence was deduced. Its CID spectrum exhibited very abundant fragment ions 92 mass units apart corresponding to glycerol loss from a "peak-at-every-mass" background ion with the same nominal mass as the protonated molecular ion of the peptide [16]. Initially, the presence of these abundant artifact peaks made the interpretation more difficult. The absence or low intensity of the expected **v** and **w** ions (if there were a lysine at the C-terminus) taken together with the earlier result of acetylation on the probe led us to misidentify the C-terminal residue as its isobar glutamine. Summarizing our CID analyses, the sequence Ala-Leu-Phe-His-Gly-Arg-Val-Ser-Trp-Ala-Met-Phe-Pro-Asn-Gly-Gln was proposed for the intact peptide.

Acknowledgements

This work was supported by grants from the National Institutes of Health National Center for Research Resources Grants No. P41RR01614 (to A. L. Burlingame) and

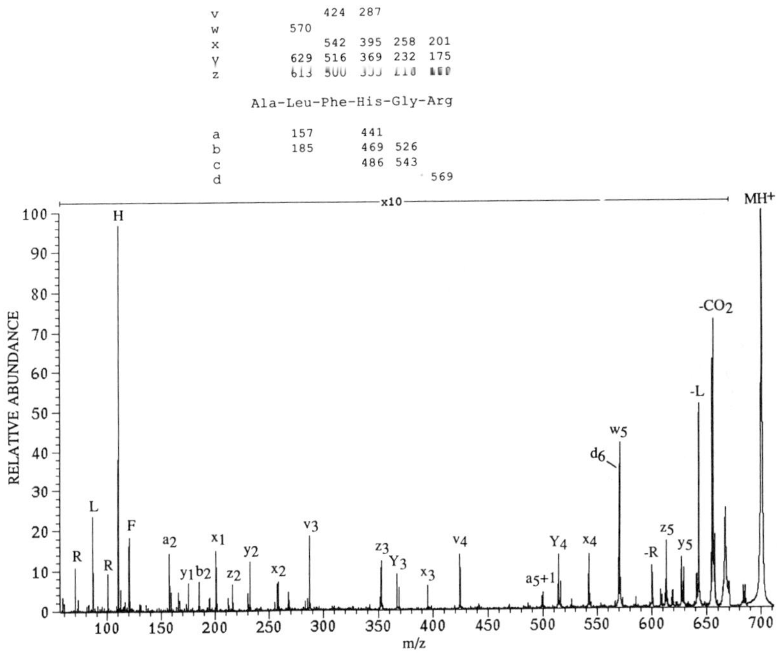

Figure 3. High energy CID spectrum of the smaller tryptic peptide (^{12}C MH^+ at m/z 700.4). Fragment ions belonging to the same ion series are listed in rows.

RR04112 (to G. L. Kenyon), and the National Science Foundation Grant No. DIR8700766 (to A. L. Burlingame) and National Institutes of Health Training Grant No. NS07219 (to S. C. H.).

References

1. Burlingame, A. L., Millington, D. S., Norwood, D. L., and Russell, D. H. (1990) *Anal. Chem.*, **62,** 268R-303R.
2. Johnson, R. S.and Biemann, K. (1987) *Biochemistry,* **26,** 1209-1214.
3. Mathews, W. R., Johnson, R. S., Cornwell, K. L., Johnson, T. C., Buchanan, B. B., and Bieman, K. (1987) *J. Biol. Chem.,* **262,** 7537-7545.
4. Griffin, P. R., Kumar, S., Shabanowitz, J., Charbonneau, H., Namkung , P. C., Walsh, K., Hunt, D. F., and Petra, P. H. (1989) *J. Biol. Chem.,* **264,** 19066-19075.
5. Gibson, B. W., Yu, Z., Aberth, W., Burlingame, A. L., and Bass, N. M. (1988) *J. Biol. Chem.*, **263**, 4182-4185.
6. Kaur, S., Medzihradszky, K. F., Yu, Z., Baldwin, M. A., Gillece-Castro, B. L., Walls, F. C., Gibson, B. W., and Burlingame, A. L. (1990) in *Biol. Mass Spectrom.*, Burlingame, A. L., and McCloskey, J. A., Eds., Elsevier, Amsterdam, 285-314.
7. Aberth, W., Straub, K. M., and Burlingame, A. L. (1982) *Anal. Chem.,* **54,** 2029-2034.
8. Alexander, A. J., Thibault, P., Boyd, R. K., Curtis, J. M. and Rinehart, K. L. (1990) *Int. J. Mass Spectrom. Ion Proc.,* **98,** 107-134.

	Ala	Leu	Phe	His	Gly	Arg	Val	Ser	Trp	Ala	Met	Phe	Pro	Asn	Gly	Lys
v						1191										
w						1190		1004						258		
x			1659					1063						344	230	
y				1486	1349	1292		1037					415	318	204	
z″			1618			1277			935	749	678					
a		157			498	654	753	840	1026	1097	*1230*	*1376*	*1473*	*1587*	1643	
b		185	332	469	526	682	781	868	1054	1125	1256			1614	1671	
c							798	885	1071		1273		1517	1631		
d						569	739	824			1168			1543		

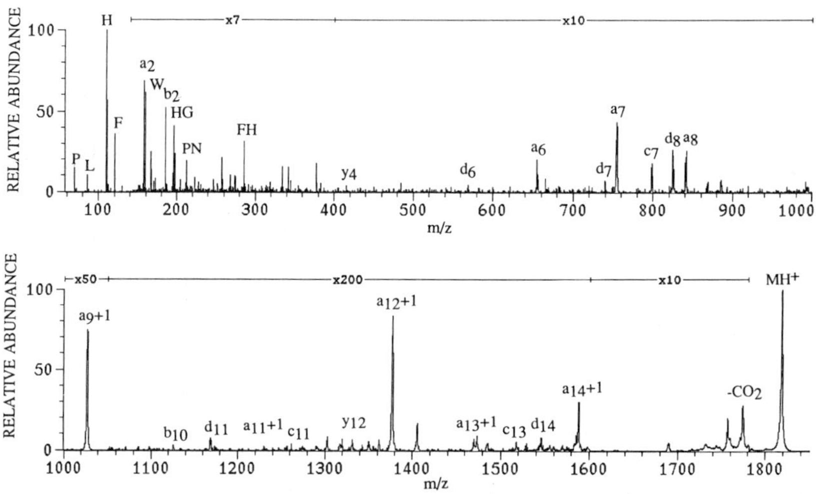

Figure 4. High energy CID spectrum of the C-terminal tryptic peptide (^{12}C MH^+ at m/z 1136.6). Fragment ions belonging to the same ion series are listed in rows. Ions labelled with an asterisk are artifact peaks related to a glycerol cluster series occurring at the same nominal mass.

9. Stults, J. T., and Watson, J. T. (1987) *Biomed. Environ. Mass Spectrom.* **14,** 583-686.
10. Throughout the text the nomenclature first proposed by Roepstorff, P., and Fohlman, J., (1984) *Biomed. Mass Spectrom.* **11,** 601-603, and modified by Biemann, K., (1988) *Biomed. Environ. Mass Spectrom.,* **16,** 99-111, is used for the peptide fragments.
11. Johnson, R. S., Martin, S. A., and Biemann, K. (1988) *Int. J. Mass Spectrom. Ion Proc.* **86,** 137-154.
12. Falick, A. M., Wang, G. H., and Walls, F. C. (1986) *Anal. Chem.* **58,** 1308-1311.
13. Walls, F. C., Baldwin, M. A., Falick, A. M., Gibson, B. W., Kaur, S., Maltby, D. A., Gillece-Castro, B. L., Medzihradszky, K. F., Evans, S., and Burlingame, A. L. (1990) in *Biol. Mass Spectrom.*, Burlingame, A. L., and McCloskey, J. A., Eds., Elsevier, Amsterdam, 197-215.
14. Hines, W. M., Gibson, B. W., Falick, A. M., and Burlingame, A. L. in prep.
15. Johnson, R. S., and Biemann, K. (1989) *Biomed. Environ. Mass Spectrom.* **18,** 945-957.
16. Falick, A. M., Medzihradszky, K. F. and Walls, F. C. (1990) *Rapid Commun. Mass Spectrom.* **4,** 318-322.

43

MASS SPECTROMETRIC METHODS FOR PROTEIN AND PEPTIDE SEQUENCE ANALYSIS

Donald F. Hunt, Janice E. Alexander, Ashley L. McCormack, Paul A. Martino, Hanspeter Michel, Jeffrey Shabanowitz, and Nicholas Sherman

Department of Chemistry, University of Virginia, Charlottesville, Virginia 22901

M. Arthur Moseley, James W. Jorgenson, and Kenneth B. Tomer

Laboratory of Molecular Biophysics, National Institute of Environmental Health Sciences, Research Triangle Park, NC 27709

I. INTRODUCTION

The role of mass spectrometry in the structural characterization of proteins continues to grow at a rapid rate. Molecular weight measurements on proteins at the femtomole/picomole level in the mass range up to 200 kDa by laser desorption on a time-of-flight instrument have been described recently (1-4). Similar results have also been reported with electrospray ionization on a quadrupole instrument (5,6). Sequence analysis of peptides in mixtures by the combination of liquid secondary ion mass spectrometry and collision activated dissociation on triple quadrupole and tandem four sector instruments is now routine at the 40-300 pmol level (7). From an initial 0.1 - 1.0 nmol sample of a 50 kDa protein, it is usually possible to obtain sequence information covering 30-70% of the molecule with only 4-5 days of effort. Total instrument time for the above analysis is only 3-4 hours. Recently developed computer algorithms that facilitate interpretation of peptide CAD spectra recorded on the triple quadrupole instrument promise to reduce the

Techniques in Protein Chemistry II

analysis time even further.

Here we discuss the utility of collision activated dissociation experiments on the triple quadrupole mass spectrometer for locating sites of phosphorylation in proteins. Also presented are results from a recent effort to explore the possibility of combining both capillary electrophoresis and microcapillary HPLC with tandem mass spectrometry via the electrospray ionization technique. Sequence analysis of peptides at the 1-10 pmol level is reported. Use of the technique to deduce the structure of developmentally-regulated, posttranslational modifications of the protein, β-tubulin, is also described.

II. EXPERIMENTAL

Mass spectra were recorded on either a TSQ-70 triple qudrupole instrument (Finnigan-MAT, San Jose, CA) or a quadrupole Fourier transform instrument constructed at the University of Virginia. Operation of the latter instrument has been described previously (8). Sample ionization and volatilization by particle bombardment on the TSQ-70 instrument were accomplished with a cesium ion gun (Antek, Palo Alto, CA) operated at 6 keV (7). For ion detection, the conversion dynode of this instrument was operated at 15 keV. Samples for analysis on either instrument were prepared by adding 0.5 ul of 0.1% trifluoroacetic acid solution containing peptide(s) (10-100 pmol) to 0.5 ul of a monothioglycerol matrix on a gold-plated stainless-steel probe. Electrospray mass spectra were recorded on the TSQ-70 instrument equipped with the newly developed Finnigan electrospray ion source. The electrospray needle was operated with a voltage differential of 3-5 kV and a sheath flow of 2-6 ul/min of 3/1 mixture of methanol/0.5% acetic acid. Collision activated dissociation experiments were conducted at energies of 20-25 eV for doubly charged ions and 15-18 eV for triply charged ions. Argon at a pressure of 3-5 mtorr was employed as the collision gas.

CZE experiments were conducted with chemically modified fused silica columns having a inside diameter of 75 microns and a length of 1 meter. Columns were treated with aminopropyl trimethoxysilane to place positively charged groups on the wall surfaces. Peptide separations were performed with 0.01 M acetic acid (pH = 3.4) as the buffer at a flow rate of 0.5 ul/min and a voltage and current of -30 kV and 3 uA, respectively. The injection volume was 0.5 to 10 nl.

Miro-capillary HPLC experiments were conducted with fused silica columns having and inside diameter of 75 microns and a length of 70 cm. The last 10 cm of the column was filled with C-18 packing material. Peptides were eluted with a gradient of 0-80% acetic acid (0.5%)/acetonitile over a 15 min period at a flow rate of 1-2 ul/min.

Protein (1-10 pmol) was digested with trypsin for 12 h at room temperature in a 13 cm piece of 75 u i.d. fused silica having an approximate volume of 1 ul. The short piece of fused silica was then butt connected to the micro capillary HPLC column and the tryptic peptides were washed onto and eluted from the the C-18 packing material with the solvent gradient specified above.

Methyl esters of peptides were generated as previously described (9).

III. RESULTS AND DISCUSSION

A. Sequence Analysis of Phosphorylated Peptides.

Bovine myelin basic protein (MBP) has been found to be an excellent *in vivo* substrate for MAP kinase (Mitogen-Activated Protein kinase) (10). The latter protein has been implicated in the cell proliferative process that results from exposure to a variety of mitogens, including EGF, platelet-derived growth factor, insulin-like growth factor I, thrombin, and phorbol esters (11). To probe the substrate recognition sequence of this enzyme, MBP was phosphorylated by MAP kinase, and then digested with the enzyme, endo-Lys-C. To search for phosphopeptides generated in the proteolytic digest, the mixture of peptides was first fractionated by reverse phase HPLC. Mass spectra were then recorded on aliquots of all the resulting fractions. Only one HPLC fraction afforded a mass spectrum containing an (M+H)+ ion that failed to match a predicted endo-Lys-C fragment. This ion occurred at m/z 1572.5, a value slightly higher than that predicted (m/z 1571.8) for the (M+H)+ ion of the phosphorylated fragment containing residues 91-104 of MBP. A spectrum recorded on the methyl ester of the peptide sample (Fig. 1, inset) showed (M+H)+ ions at both 1586.2 and 1601.2 (ave. mass) and confirmed that above HPLC fraction contained a mixture of amidated and deamidated forms of the phosphopeptide containing residues 91-104.

To deduce the site of phosphorylation, the (M+H)+ ions for the above peptides were subjected to collision activated dissociation on the triple quadrupole instrument (9). In this experiment the first mass filter of the TSQ-70, quadrupole 1, is set to pass all ions within a 6 mass unit window centered around the (M+H)+ ions for the amidated and deamidated forms of the phosphopeptide, m/z 1571.8 and 1572.8, respectively. On transmission to the collision chamber, the (M+H)+ ions of both species suffer 10-100 low energy (15-40 eV) collisions with argon present at a pressure of 3-5 mtorr. During this process, kinetic energy is converted to vibrational energy and the peptide (M+H)+ ions undergo fragmentation, more or less randomly, at the various amide bonds along the backbone of the linear peptide chain. Elimination of small molecules such as water, ammonia, and phosphoric acid, also occurs from the side

115.0	228.1	327.2	428.3	525.3	681.4	862.5	959.5	1056.6	1153.6	1240.6	1368.7	1425.7	1571.8	b_n
Asn	Lxx	Val	Thr	Pro	Arg	Thr-P	Pro	Pro	Pro	Ser	Gln	Gly	Lys	
	1457.8	1344.7	1245.6	1144.6	1047.5	891.4	710.4	613.3	516.3	419.2	332.2	204.1	147.1	y_n

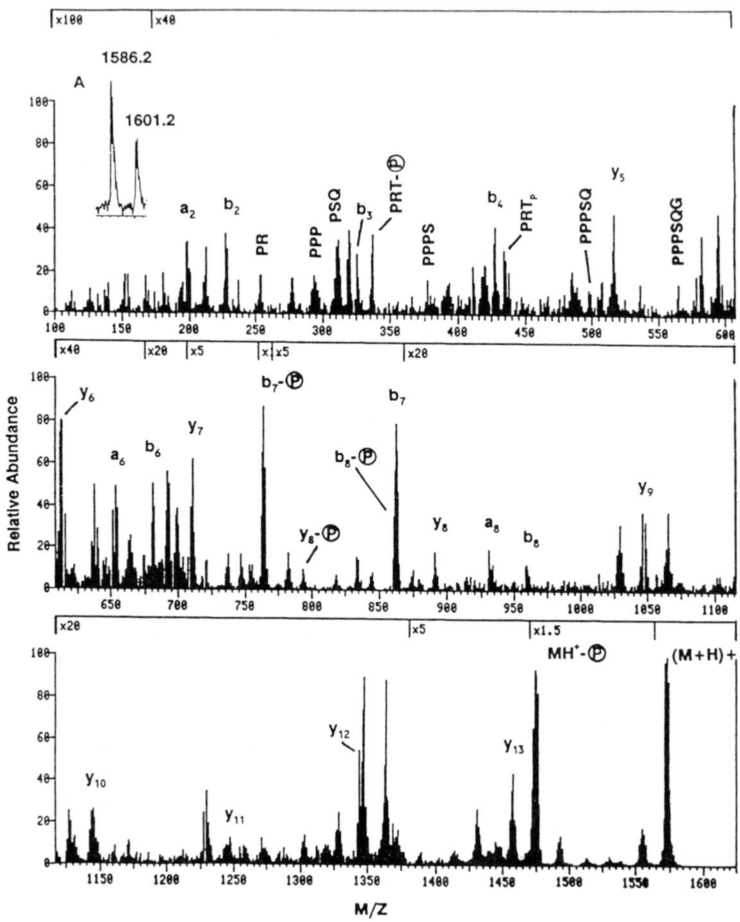

Figure 1. Collision activated dissociation mass spectrum recorded on the (M+H)+ ions (m/z 1571.8 and 1572.8) from a mixture of amidated and deamidated forms (inset) of the underivatized, phosphorylated, endo-Lys-C fragment of myelin basic protein (MBP). The derived sequence deduced from analysis of fragment ions of type b and y is shown; –Ⓟ indicates an ion resulting from loss of phosphoric acid (98 Da). Inset: Fourier transform mass spectrum recorded on the methyl ester derivatives showing the amidated and deamidated forms of the phosphorylated endo-Lys-C fragment.

chains of some amino acids. A collection of neutrals and ions is produced by the above processes. Neutrals are pumped away by the vacuum system and the charged fragments are transmitted to a third mass filter, quadrupole 3, that separates them according to mass. To maximize ion transmision (sensitivity), quadrupole 3 is also usually operated at less than unit resolution.

To generate the spectrum in Fig. 1, the mass range from 100-1600 Da was scanned 16 times, data was acquired in the centroid mode, and the resulting spectra were then summed. Total instrument time to acquire the data was 64 s. Signals for each of the fragment ions in the spectrum appear in clusters because the data system is attempting to resolve signals that are more then 1 mass unit wide, the clusters contain contributions from ^{13}C and other isotopes in each fragment, and the amidated and deamidated forms of the peptide are both present in the ion cluster that has been subjected to collision activated dissociations.

The sequence of amino acids in the sample is deduced by subtracting m/z values for peaks that occupy the same relative positions within two cluster patterns and are separated by at least 57 and no more than 186 mass units, the residue masses of the lightest and heaviest amino acids (9) (phosphotyrosine has a residue mass of 243 Da). Presence of additional signal multiplets corresponding to the loss of ammonia (17 Da), water (18 Da), phosphoric acid (98 Da), and carbon monoxide (28 Da) from the principal fragment ion often facilitates recognition of the next highest member of a particular fragment ion series (9).

Predicted masses (m/z values) for ions of type y for the amidated form of the phosphopeptide containing residues 91-104 of MBP are shown above the structure in Fig. 1. These ion types all contain the C-terminal residue plus one, two, three, four, etc. additional residues. Those observed in the spectrum are underlined. Subtraction of m/z values for any two fragments that differ by a single amino acid, NHCH(R)CO, generates a value that specifies the mass and thus the identity of the extra residue in the larger fragment. In the present example, ions of type y facilitate identification of the first nine residues in the peptide chain. The mass difference observed between the y_8 and y_7 ions (m/z 891.4 and 710.4, respectively) is 181 Da. This is the expected value for the residue, phosphothreonine. Elimination of phosphoric acid (98 Da) occurs from y_8 but not from y_7 as expected.

Predicted fragment ions of type b for the amidated form of the phosphopeptide are shown above the structure in Fig. 1. Those observed in the spectrum are underlined. These ions all contain the amino terminal residue plus one or more additional residues. Subtraction of m/z values for any two fragments of type b that differ by a single amino acid, NHCH(R)CO, generates a value that identifies the extra residue present in the larger fragment. In the present example, ions of type b confirm the identity of the first eight amino acids in the peptide chain. The mass difference observed between the b_4

and b_3 ions (m/z 428.3 and 327.3, respectively) is 101 Da, the expected value for threonine. The mass difference between the b_7 and b_6 ions (m/z 862.5 and 681.4, respectively) is 181 Da, the expected value for a phophothreonine residue. The abundant ion at m/z 764.5 results from the loss of phosphoric acid from the side chain of the phosphothreonine residue in the b_7 ion and provides additional confirmation of the assignment of Thr-97 as the site of phosphorylation in MBP. Most of the unlabeled ions in Fig. 1 can be explained by loss of either ammonia or water from the side chains of amino acids in ions of either type b or type y. Several ions of type a also appear in the spectrum and are labeled as such. These ions differ from type b ions by the loss of carbon monoxide. Fragments resulting from multistep cleavage of the peptide chain at proline residues (9) are labeled with the single letter code for the amino acid residues contained within the fragment.

B. Capillary Zone Electrophoresis (CZE)/ Electrospray Ionization Mass Spectrometry.

To further extend the utility of tandem mass spectrometry for protein sequence analysis, we have recently explored the possibility of interfacing capillary zone electrophoresis (CZE) to the mass spectrometer via the electrospray ionization source (6). To date, most of the problems encountered with the above configuration stem from the fact that CZE is normally performed at high positive potential with buffer systems at high pH to minimize absorption of peptide samples on the negatively charged walls of fused silica capillaries. Since collision activated dissociation experiments to generate sequence information from peptides are best performed on positively charged ions, it is necessary to protonate the negatively charged peptides emerging from the CZE column prior to the ion generation step in the electrospray source. Unfortunately, introduction of strong acids at this point often promotes electrical breakdown of the electrospray and causes the resulting sample current to fluctuate badly on samples at the fmol level. When the CZE column is operated with low pH buffers and negative potential to separate positvely charge molecules, peptides containing basic residues exhibit band broadening due to absorption on the walls of the fused silica. Derivatization of the fused silica with aminopropyl-trimethoxysilane overcomes this problem.

Shown in Fig. 2a is a mass spectrum recorded on β-endorphin (MW = 3461) as it elutes from a CZE column at the 160 fmol level. Ions formed hy the addition of four, five, and six protons to the molecule dominate the spectrum. Displayed in Fig. 2b is a collision activated dissociation spectrum recorded on the $(M+H)^+$ ion of β-casomorphin (1 pmol) as it elutes from the CZE column. Signals corresponding to fragment ions of type b and y are labeled as such on the spectrum. The observed fragmentation is sufficient to define

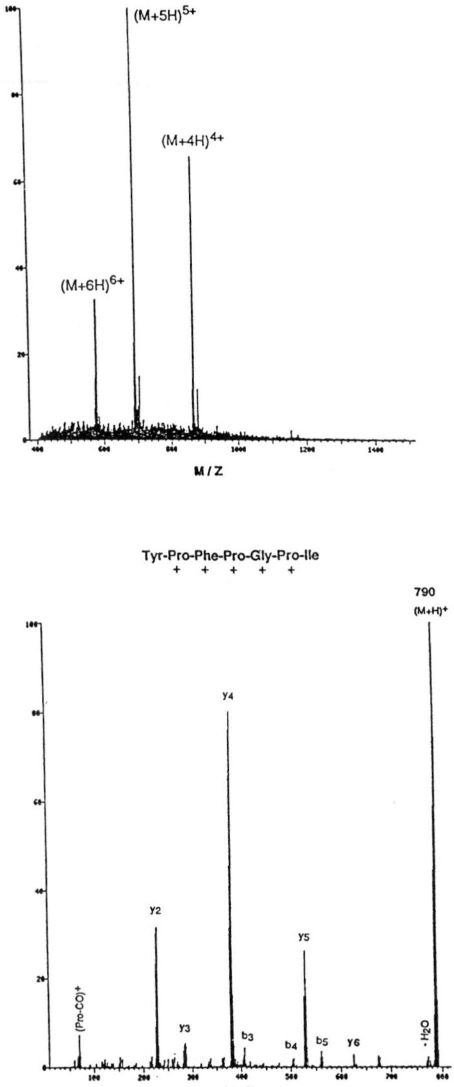

Figure 2. (A) CZE/electrospray ionization mass spectrum recorded on 160 fmol of β-endorphin (MW = 3461); (b) Collision activated dissociation spectrum recorded on the (M+H)+ ion of β-casomorphin (1 pmol) as it elutes from the CZE column.

the complete amino acid sequence of β-casomorphin at the 1 pmol level.

At present, an inability to preconcentrate sample prior to CZE chromatography limits the analytical utility of the above methodology. Fused silica columns must be dipped into a miminum volume of 1-2 ul although only 10-20 nl of this are actually loaded onto the column. Injection of a peptide into the mass spectrometer at the 1-2 pmol level requires that a total sample of 50-100 pmol be available for analysis.

C. Microcapillary HPLC/ Electrospray Ionization Mass Spectrometry.

Use of microcapillary HPLC in conjunction with electrospray mass spectrometry allows one to circumvent the problem discussed in the above paragraph. Fused silica columns (75 cm x 50 u i.d.) are also used in this experiment but the last 10 cm of the column is packed with 10 u particles coated with C_{18}-adsorbent. Mixtures of peptides (1-10 pmol) in dilute acetic acid solution (30-50 ul of 0.5% acetic acid) are passed through the fused silica and absorbed at the head of the micro C_{18}-column. Proteolytic digestions can also be carried out at the 1-10 pmol level inside small sections (15 cm) of unpacked fused silica columns. These are then butt-connected to the micro-capillary HPLC column. Products of the proteolytic digests are washed onto the head of the micro-capillary column with dilute acetic acid. Peptides are then eluted from the column into the mass spectrometer with a 15 min gradient of 0% to 80% acetonitrile.

Shown in Fig. 3a is a reconstructed ion chromatogram recorded on the peptides produced from a tryptic digest (fused silica tube) of β-lactoglobulin at the 10 pmol level. Peptide sequences corresponding to the signals observed in Fig. 3a are listed in Fig. 3b. Shown in Fig. 4a is the mass spectrum recorded on the peptide corresponding to residues 92-101 in β-lactoglobulin. The major ion observed in the spectrum corresponds to the doubly protonated species, $(M+2H)^{++}$, at m/z 596.6. The collision activated dissociation mass spectrum recorded on this doubly charged ion is shown in Fig. 4b. Signals of type y dominate the spectrum and provide the necessary information to determine the complete sequence of amino acids present in this peptide. Total quantity of protein required for the above analysis was only 5-10 pmol.

D. Developmentally Regulated Posttranslational Modifications of Neuronal β-Tubulin.

Microtubules, assembled from two similar 50 kDa proteins designated α- and β-tubulin, are involved in a number of important biological processes including segregation of chromosomes during cell division, cell motility, organelle transport, and maintenance of cell shape (12). Class III β-

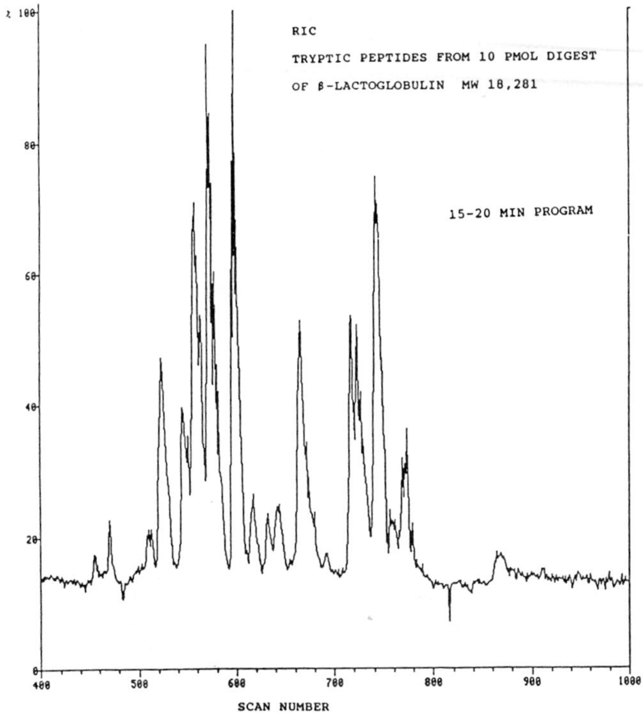

β-LACTOGLOBULIN TRYPTIC PEPTIDES

	M/Z	RESIDUE	SEQUENCE
+2	337	9- 14	GLDIQK
+2	460	84- 91	IDALNENK
+2	420	142-148	ALPMHIR
+2	468	1- 8	LIVTQTMK
+2	452	76- 83	TKIPAVFK
+2	596	92-101	VLVLDTDYKK
+3	546	125-138	TPEVDDEALEKFDK
+2	533	92-100	VLVLDTDYK
+3	601	61- 75	WENGECAQKKIIAEK
+2	857	149-162	LSFNPQTLEEQCHI
+3	772	41- 60	VYVEELKPTDEGDLEILLQK
+3	903	15- 40	VAGTWYSLAMAASDISLLDAQSAPLR

Figure 3. (A) Reconstructed ion chromatogram from a microcapillary HPLC run on a tryptic digest of β-lactogloglobulin (10 pmol); (B) Tryptic peptides from β-lactoglobulin listed in order of their elution from the microcapillary HPLC column.

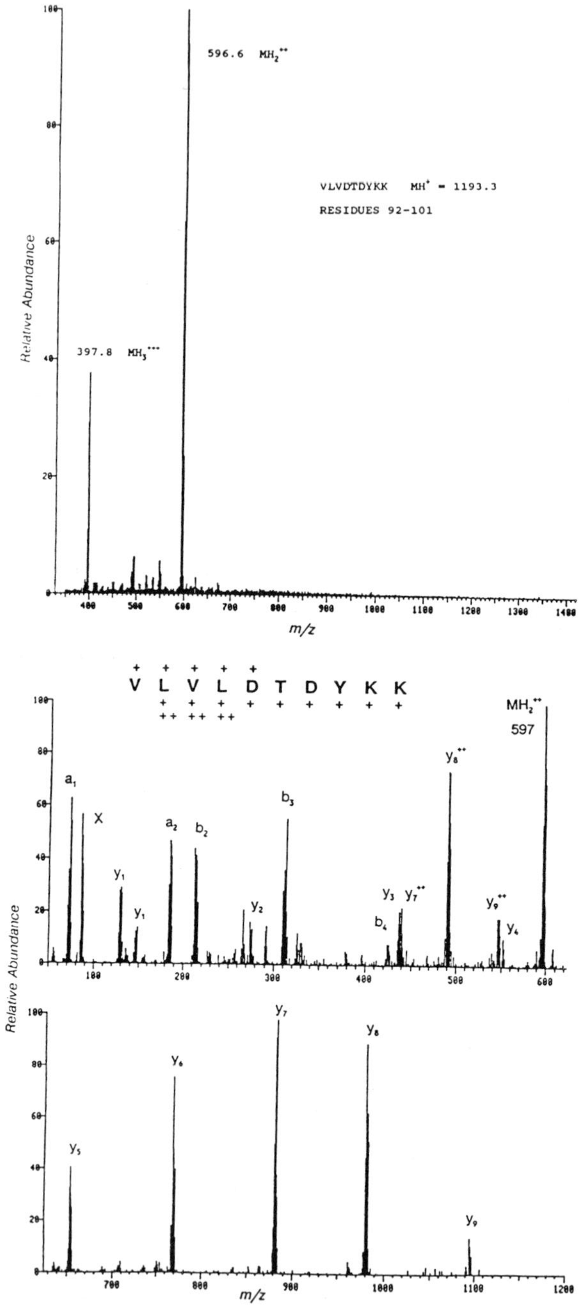

Figure 4. (A) Mass spectrum recorded on 5–10 pmol of a tryptic peptide, residues 92–101 of β-lactoglobulin, as it elutes from the microcapillary HPLC column; (B) Collision activated dissociation mass spectrum recorded on the (M+2H)++ ion of the peptide corresponding to residues 92–101 of β-lactoglobulin.

tubulin, a mamallian isotype found only in neurons and cells in the testis, undergoes a developmentally regulated increase in heterogeneity in the former but not the latter tissue (13,14). At least six additional isoforms of neuron specific Class III β-tubulin are detected in the adult brain of both rat and cattle. Previous work localized the site of heterogeneity in the protein to the C-terminal 20 amino acids, a highly acidic region thought to be involved in the interaction of microtubules with both calcium ions and microtubule associated proteins (MAPs).

To further characterize both the nature and location of structural modifications responsible for the observed heterogeneity, Class III β-tubulin from bovine brain was subjected to amino acid sequence analysis by mass tandem spectrometry. Posttranslationally modified peptides from the Class III β-tubulin were obtained by treating the complete mixture of tubulin isotypes and isoforms from bovine brain with cyanogen bromide. The resulting peptides were then incubated with the a monoclonal antibody, TuJ1, that binds to an epitope on the C-terminus of the Class III β-tubulin isotype selectively. Unbound peptides were removed from the solution by filtration. Bound peptides were liberated on treatment of the antigen-antibody complex with dilute acid and fractionated further by reverse phase HPLC. All fractions were treated with methanolic HCl to convert the peptides to the corresponding methyl esters. Mass spectra recorded on an aliquot from one of the HPLC fractions showed (M+H)+ ions at m/z 1738.6, 1961.8, 2105.1, 2248.4, and 2390.2. Each of the latter four signals is separated by 143 Da, the incremental value expected if the peptides corresponding to a pair of adjacent ions differ in composition by a single glutamic acid residue in the form of its methyl ester. The second ion in the series differs from the first by the addition of both phosphate (80 Da) and a glutamic acid methyl ester residue (143 Da).

To deduce the amino acid sequences of the above peptides, an additional aliquot of the particular HPLC fraction was subjected to analysis by the microcapillary HPLC/electrospray mass spectrometry procedure described above. Collision activated dissociation spectra recorded on the doubly charged ion, (M+2H)++, at m/z 870 (M+H+ = 1738.6) confirmed the structure of the unmodified C-terminal peptide of β-tubulin as YEDDEEESEAQGPK. Predicted ions of types b and y for this sequence (nominal masses) are displayed on the structure below.

164	307	436	565	708	851	994	1081	1224	1295	1423	1480	1577	1737
Tyr	Glu	Asp	Asp	Glu	Glu	Glu	Ser	Glu	Ala	Gln	Gly	Pro	Lys
1737	1574	1431	1302	1173	1030	887	744	657	514	444	315	258	161

Shown in Fig. 5 is the collision activated dissociation spectrum recorded on the (M+2H)++ ion, m/z 981 (M+H+ = 1960.8) that was assumed to contain an additional Glu residue plus phosphate. Note that the first 12 ions of type y found for the unmodified C-terminal peptide are also present in this mass spectrum. We conclude that it is the first two residues at the amino terminus of the chain that bear the posttranslational modifications. Ions of type b and a for the amino terminal tyrosine are found at m/z 244 and 216 rather than at m/z 164 and 136. The observed values are those expected for phosphotyrosine. Note that the expected y_{13} ion (m/z 1575) containing an unmodified Glu residue at its N-terminus is absent from the spectrum in Fig. 5. Found in its place is a signal at m/z 1543 that results from the elimination of an intact glutamic acid methyl ester (175 Da) from an ion of type y (y_{13}) containing an extra glutamic acid residue attached through the side chain of the residue at position 2 in the sequence. We conclude that the structure of this posttranslationally modified β-tubulin isoform contains a phosphorylated tyrosine and glutamylated Glu residue at positions 437 and 438, respectively. Other isoforms have been shown to be phosphorylated at positions 437 and/or 444 and to contain up to six glutamic acid residues attached to position 438. Efforts to understand the role that these modifications play in the life cycle of the cell are in progress.

E. Developments for the Future.

Recent improvements in the design and operation of the ion trap mass spectrometer (15) promise to extend the utility of the tandem mass spectrometric approach to sequence analysis even further. Initially the ion trap mass spectrometer was designed as an inexpensive detector for small organics eluting from a gas chromatograph. Because of its limited mass range (650 Da), the ion trap attracted little attention within the biochemical community. Now, however, Cooks has shown that the mass range of the ion trap can be extended to accomodate molecules up to mass 30,000 (16). Cooks has also demonstrated that the instrument can be operated as a tandem mass spectrometer for sequence analysis of peptides at the femtomole and sub-femtomole level (17). Recently, Glish and McGluckey have interfaced an electrospray ion source to the ion trap and shown that a number of peptides and proteins can be detected at the low femtomole level (18). Direct sequence analysis of peptides in mixtures at the subpicomole level by the combination of either CZE or microcapillary HPLC and the ion trap mass spectrometer appears to be an exciting possibility. Dissociation of intact proteins to afford amino acid sequence information at both the N-terminus and C-terminus of the molecule for the purpose of constructing oligonucleotide probes also seems doable on this relatively inexpensive instrument.

244.1	530.2	659.2	788.3	931.3	1074.4	1217.4	1304.5	1447.5	1518.6	1646.6	1703.7	1800.7	1960.8	b_n
Tyr-P	Glu	Asp	Asp	Glu	Glu	Glu	Ser	Glu	Ala	Gln	Gly	Pro	Lys	
	$(Glu)_1$ 1717.8	1431.7	1302.6	1173.6	1030.5	887.5	744.4	657.4	514.3	444.3	315.2	258.2	161.1	y_n

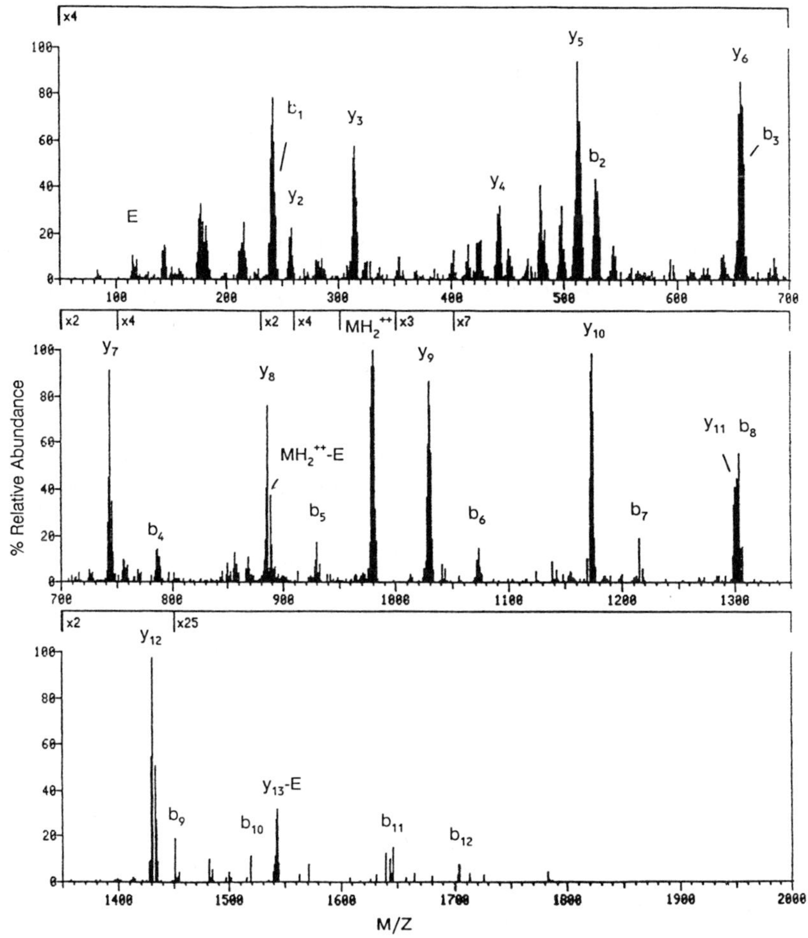

Figure 5. Collision activated dissociation mass spectrum recorded on the (M+2H)++ ion, m/z 981 (M+H+ = 1960.8) assumed to result from the C-terminal cyanogen bromide peptide, YEDDEEEASEGPK, that has both been phosphorylated and glutamylated.

IV. REFERENCES

1. Karas, M. and Hillenkamp, F. (1988) Anal. Chem. 60, 2299.
2. Karas, M., Bahr, U., Ingendoh, A. and Hillenkamp, F. (1989) Angew. Chem. 101, 805.
3. Beavis, R.C. and Chait, B.T. (1989) Rapid Commun. Mass Spectrom., 3, 233.
4. Beavis, R.C. and Chait, B.T. (1989) Rapid Commun. Mass Spectrom., 3, 432.
5. Fenn, J.B., Mann, M., Meng, C.K., Wong, S.F. and Whitehouse, C.M. (1989) Science, 246, 64.
6. Smith, R.D., Loo, J.A., Edmonds, C.G., Baringa, C.J. and Udseth, H.R. (1990) Anal. Chem., 62, 882-899.
7. Hunt, D.F., Shabanowitz, J., Griffin, P.R., Yates, J.R., III, Martino, P.A., and McCormack, A.L. (1990) In: Proceedings of the Second International Symposium on Mass Spectrometry in the Health and Life Sciences, A.L. Burlingame and J.A. McCloskey, Eds. Elsevier, Amsterdam), In Press.
8. Hunt, D.F., Shabanowitz, J., Yates, J.R., III, Zhu, N.Z., Russell, D.H., and Castro, M.E. (1987) Proc. Natl. Acad. Sci., U.S.A., 84, 620-623.
9. Hunt, D.F., Yates, J.R., III, Shabanowitz, J., Winston, S. and Hauer, C.R. (1986) Proc. Natl. Acad. Sci., U.S.A., 83, 6233-6237.
10. Erickson, A.K., Payne, D.M., Martino, P.A., Rossomando, A.J., Shabanowitz, J., Weber, M.J., Hunt, D.F., and Sturgill, T.W. (1990) J. Biol. Chem., In Press.
11. Rossomando, A.J., Payne, D.M., Weber, M.J., and Sturgill, T.W. (1989) Proc. Natl. Acad. Sci., U.S.A., 86, 6940-6943.
12. Dustin, P. (1984) Microtubules (Springer Verlag, N.Y.)
15. Stafford, G.C., Kelley, P.E., Syka, J.E.P., Reynolds, W.E. and Todd, J.F.J. (1984) Int. J. Mass Spectrom. Ion. Proc., 60, 85.
16. Kaiser, R.E., Louris, J.N., Amy, J.W. and Cooks, R.G. (1989) Rapid Commun Mass Spectrom., 3, 225.
17. Louris, J.N., Amy, J.W., Ridley, T.Y. and Cooks, R.G. (1989) Int. J. Mass Spectrom. Ion Proc., 88, 97.
18. Van Berkel, G.J., Glish, G.L. and McLuckey, S.A. (1990) Anal. Chem., 62, 1284-1295.

44

CHARACTERIZATION OF SYMPOSIUM TEST PEPTIDE-4

Donald F. Hunt, Janice E. Alexander,
Ashley L. McCormack, Paul A. Martino,
Hanspeter Michel, and Jeffrey Shabanowitz

Department of Chemistry
University of Virginia
Charlottesville, Virginia 22901

I. INTRODUCTION

Symposium Test Peptide-4 (STP-4) was provided as a lyophilized powder in the bottom of a plastic Eppendorf tube. The accompanying instructions suggested that the sample should be analyzed by the standard methods available in the participating laboratories. Quantity and purity of the sample were not specified.

At the meeting of the Protein Society it was revealed that STP-4 actually contained an N-terminally blocked tripeptide impurity at the 1.5 nmol level plus a sixteen residue peptide at the 200 pmol level. The impurity was not detected with our standard operating conditions for protein sequence analysis because the mass spectrometers in the laboratory are usually scanned from m/z 400-4,000. The molecular weight of the impurity was below this mass range.

Characterization of the high molecular weight component of Symposium Test Peptide-4 (STP-4) at the University of Virginia was accomplished on a Finnigan TSQ-70 triple quadrupole mass spectrometer. This sixteen residue peptide was found to have the sequence shown below.

NH_2-A X F H G R V S W A M F P N G K-COOH

Techniques in Protein Chemistry II

II. EXPERIMENTAL

Liquid secondary ion mass spectra were recorded on both Finnigan TSQ 70 triple quadrupole and tandem quadrupole Fourier transform instruments (1,2). Operation of these instruments for oligopeptide sequence analysis has been described previously (1-3). Collision activated dissociation mass spectra were recorded at the 30-50 pmol level. Molecular weight determinations were performed on the tandem quadrupole Fourier transform instrument at a sample level of 5-30 pmol.

Electrospray ionization/CAD spectra were recorded on the TSQ-70 instrument with the Finnigan-MAT electrospray ion source. Samples were introduced at the 5-10 pmol level via a microcapillary, fused silica HPLC column (75 mm x 50 u i.d.). The last 10 cm of this column was packed with 10 u particles coated with C-18 absorbent. STP-4 was washed onto the C-18 column with 30-50 ul of 0.5% acetic acid and then eluted into the electrospray ion source with a 15 min gradient of 0% to 80% acetonitrile at a flow rate of 5 ul/min.

III. RESULTS AND DISCUSSION

A. Summary of Chemical Experiments and Molecular Weight Determinations.

A liquid secondary ion mass spectrum of STP-4 recorded on the quadrupole Fourier transform instrument showed and abundant (M+H)+ ion at m/z 1817.6 (monoisotopic mass). A spectrum recorded under electrospray ionization conditions on the triple quadrupole instrument showed an abundant (M+3H)+++ ion at m/z 606-607 (average mass). Accordingly, a molecular mass of 1816.6 was assigned to STP-4.

Acetylation of STP-4 was performed by loading a 10 cm piece of 50 u i.d. fused silica with 5 pmol of sample in @ 1 ul of a 0.1% solution of acetic anhydride in 200 uM ammonium acetate. The mixture was allowed to stand at room temperature for several minutes and then butt connected to the microcapillary HPLC column. Acetylated STP-4 was then washed onto the C-18 absorbent and eluted into the electrospray ion source. The resulting mass spectrum showed an abundant (M+2H)++ ion at m/z 951. The calculated molecular mass of the acetylated product is thus 1900 Da.

Acetylation under the above conditions occurs only on amino groups in the peptide. Introduction of a single acetyl group increases the molecular weight by 42 Da. The observed shift of 84 Da suggested that the sample contained either (a) a free amino terminus plus a single Lys residue or (b) a blocked N-terminus and two Lys residues.

STP-4 (30 pmol) was converted to the corresponding methyl ester on treatment with 2N HCl/MeOH for 1 hr at room temperature. A liquid secondary ion mass spectrum recorded on the product of this transformation showed a new (M+H)+ ion at m/z 1831.5. Conversion of side chain or C-terminal -COOH groups to the corresponding methyl esters ($-COOCH_3$) increases the mass of the molecule by 14 units/COOH. The observed shift of 14 Da suggested that the peptide (a) contained a free C-terminus and was devoid of other acidic residues or (b) contained a blocked C-terminus and a single acidic residue.

B. Interpretation of Data Obtained from Collision Activated Dissociation (CAD) of (M+H)+ Ions.

To deduce the structure of STP-4, (M+H)+ ions from the sample were subjected to collision activated dissociation on the triple quadrupole mass spectrometer. Results are presented in Fig. 1. In this experiment the first mass filter of the TSQ-70 instrument, quadrupole 1, is set to pass all ions within a 6 mass unit window centered around the (M+H)+ ion at m/z 1817.6 (3). On transmission to the collision chamber, quadrupole 2, the (M+H)+ ions suffer 10-100 low energy (10-40 eV) collisions with argon present at a pressure of 3-5 mtorr. During this process, kinetic energy is converted to vibrational energy and the peptide (M+H)+ ions undergo fragmentation, more or less randomly, at the various amide bonds along the backbone of the linear peptide chain. Elimination of small molecules such as water and ammonia also occurs from the side chains of some amino acids. A collection of neutrals and ions is produced by the above processes. Neutrals are pumped away by the vacuum system and the charged fragments are transmitted to a third mass filter, quadrupole 3, that separates them according to mass. Quadrupole 3 is usually operated at less than unit resolution to maximize ion transmission (sensitivity).

To generate the spectrum shown in Fig. 1, the mass range from 50-2,000 Da was scanned seven times, data was acquired in the centroid mode, and the resulting spectra were then summed. Total instrument time to acquire the data was 45 s. Signals for each of the fragment ions in the spectrum appear as clusters both because the data system is attempting to resolve signals that are more than 1 mass unit wide and because the signal patterns contain contributions from ^{13}C and other isotopes present in each fragment.

Below are shown the predicted isotope patterns for three

72.0	185.1	332.2	469.3	526.3	682.4	781.4	868.5	1054.6	1125.6	1256.6	1403.7	1500.8	1614.8	1671.8	1817.9 b_n
Ala	Lxx	Phe	His	Gly	Arg	Val	Ser	Trp	Ala	Met	Phe	Pro	Asn	Gly	Lys
	1746.9	1633.8	1486.7	1349.7	1292.6	1136.5	1037.5	950.4	764.4	693.3	562.3	415.2	318.2	204.1	147.1 y_n

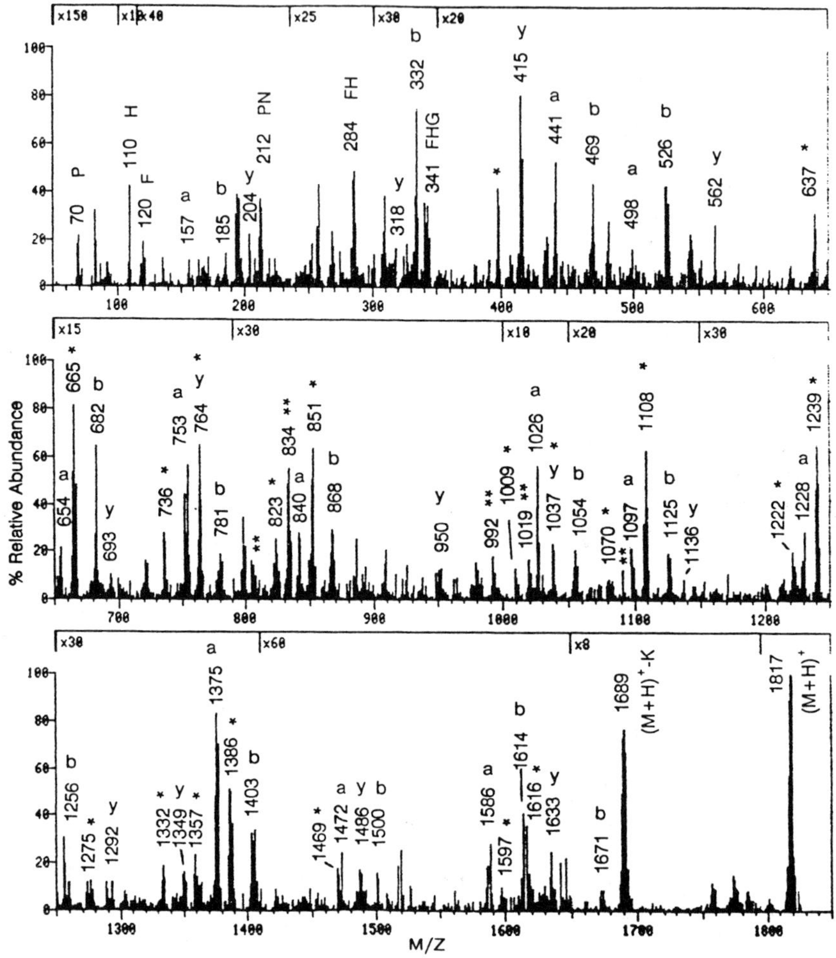

Fig. 1. Collision activated dissociation mass spectrum recorded on the (M+H)+ ions (m/z 1817.6) of STP-4. Predicted fragment ions of type b and type y are shown above and below the deduced structure of STP-4, respectively.

peptides containing 6, 10, and 14 amino acids, respectively. The actual peak shape recorded on the triple quadrupole instrument (profile mode) is shown on line 2 of the diagram. The third line of the diagram shows the experimentally observed pattern after the data system has summed the total signal intensity observed in each one mass unit window within the unresolved multiplet (centroid mode). Note that at m/z <800 the signal due to the monoisotopic mass of the ion is the most abundant in the cluster and has a value that is essentially identical to that of the rounded nominal mass. At m/z 1100 or so, the signal corresponding to the monoisopic mass remains the most abundant but is now about one mass unit higher than the rounded nominal mass. Near mass 1,600, the signal for the ^{13}C isotope becomes approximately equal to that of the ^{12}C isotope and the rounded nominal mass occurs at a value one mass unit below the first of the large peaks. In the discussion that follows we will use rounded nominal mass values to deduce the structure of STP-4. Accordingly, we will almost always select an ion on the low mass side of the most abundant signals in a particular cluster above m/z 1,000 and choose the most abundant ion in the cluster for signals below m/z 1,000.

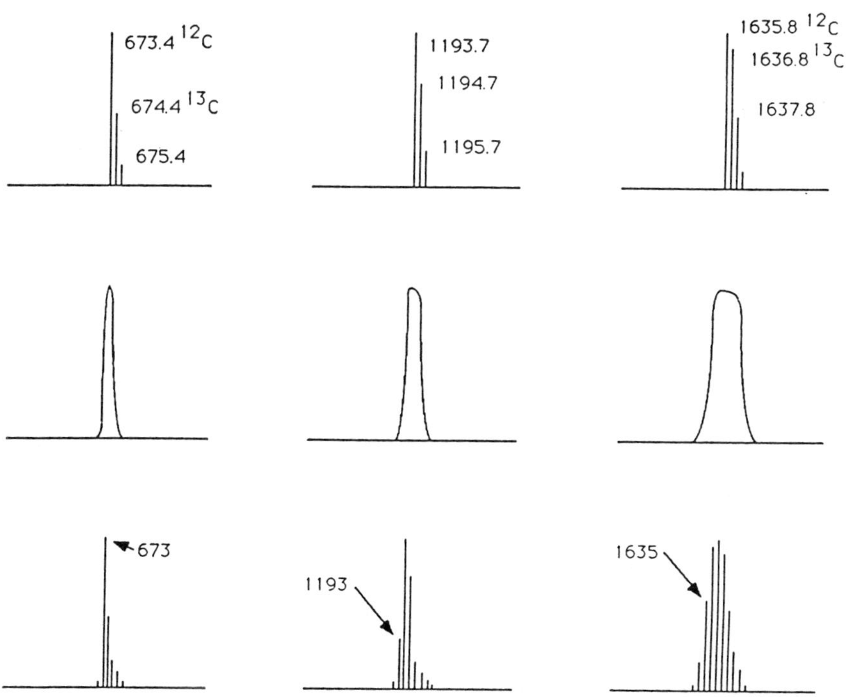

Two types of fragment ions are generally observed in the CAD spectra of peptide (M+H)+ ions recorded under the low energy collision conditions employed in triple quadrupole instruments. Ions of type b all contain the amino terminal residue plus one or more additional residues and have the general structure, $H-[NHCH(R)CO]_n^+$. Predicted nominal m/z values for fragments of this type derived from the deduced structure of STP-4 are shown above the sequence in Fig. 1. Note that subtraction of m/z values for any two fragments of type b that differ by a single amino acid, NHCH(R)CO, generates a value, the residue mass, that identifies the extra amino acid residue present in the larger fragment. Residue masses for the twenty common amino acids are shown in Table 1.

Table 1. Nominal mass values for peptide fragment ions.

Amino Acid			Residue Mass[a]	Mass of y_1 Ion[b]	Mass of a_1 Ion[c]
G	Glycine	Gly	57	76	30
A	Alanine	Ala	71	90	44
S	Serine	Ser	87	106	60
P	Proline	Pro	97	116	70
V	Valine	Val	99	118	72
T	Threonine	Thr	101	120	74
C	Cystine	Cys	103	122	76
L	Leucine	Leu	113	132	86
I	Isoleucine	Ile	113	132	86
N	Asparagine	Asn	114	133	87
D	Aspartic Acid	Asp	115	134	88
Q	Glutamine	Gln	128	147	101
K	Lysine	Lys	128	147	101
E	Glutamic Acid	Glu	129	148	102
M	Methionine	Met	131	150	104
H	Histidine	His	137	156	110
F	Phenylalanine	Phe	147	166	120
R	Arginine	Arg	156	175	129
C	Carboxymethy Cystine	Cmc	161	180	134
Y	Tyrosine	Tyr	163	182	136
W	Tryptophan	Trp	186	205	159

[a]Residue masses have the structure, -NH-CH(R)-CO-; [b]y_1 ions have the structure, $^+H_3NHCH(R)COOH$; [c]a_1 ions have the structure, $^+H_2N{=}CH{-}R$

Fragment ions that result from the loss of small neutral molecules such as water (18 Da), ammonia (17 Da) and carbon monoxide (28 Da) often accompany the formation of type b ions. The fragment ion produced by elimination of carbon monoxide from an ion of type b is designated as an ion of type a. Ions separated by 28 Da in a spectrum have a high probability of belonging to a type a-b pair.

Fragment ions of type y all contain the C-terminal residue plus one or more additional residues and have the general structure, $^{+}H-[NH_2CH(R)CO]nOH$. Predicted nominal m/z values for fragments of this type derived from the deduced structure of STP-4 are shown below the sequence in Fig. 1. Note that subtraction of m/z values for any two fragments of type y that differ by a single amino acid, NHCH(R)CO, generates a value, the residue mass (Table 1), that identifies the extra amino acid residue present in the larger fragment. Fragment ions that result from the loss of small neutral molecules such as water (18 Da), and ammonia (17 Da) but not carbon monoxide (28 Da) often accompany the formaion of type y ions.

One approach to the interpretation of the data in Fig. 1 is to search the high mass end of the spectrum for fragment ions of types a and type b. As indicated above, signals for these ions are easily recognized because they differ in mass by 28 Da. The following pairs of signals fit this criteria: 654-682, 753-781, 840-868, 1026-1054, 1097-1125, 1228-1256, 1375-1403, 1472-1500, 1586-1614. Note that each of these fragment ions suffers elimination of ammonia or water to generate additional signals (m/z values indicated by an *). Accordingly, the signals at m/z 682, 781, 868, 1054, 1125, 1256, 1403, and 1614 are assigned to ions of type b. The partial structure shown below is generated by subtracting the m/z values for adjacent pairs of the above ions. Subtraction of 1054 from 1125 generates the residue mass of Ala (71 Da), 1125 from 1256 = 131, the residue mass of Met, etc.

682	781	868	1054	1125	1256	1403	1500	1614	Type b Ions
	Val	Ser	Trp	Ala	Met	Phe	Pro	Asn	

Cleavage of any particular amide bond can give rise to either an ion of type b or y depending on which fragment retains the charge. If the mass of one type of ion is known, the other can be predicted by subtracting the observed value from the (M+H)+ ion and adding one Da, (y = (M+H)+ - b + 1, and b = (M+H)+ - y + 1). Predicted m/z values for ions of type y for the above partial sequence are shown below. Note that all of these type y ions are observed in the spectrum

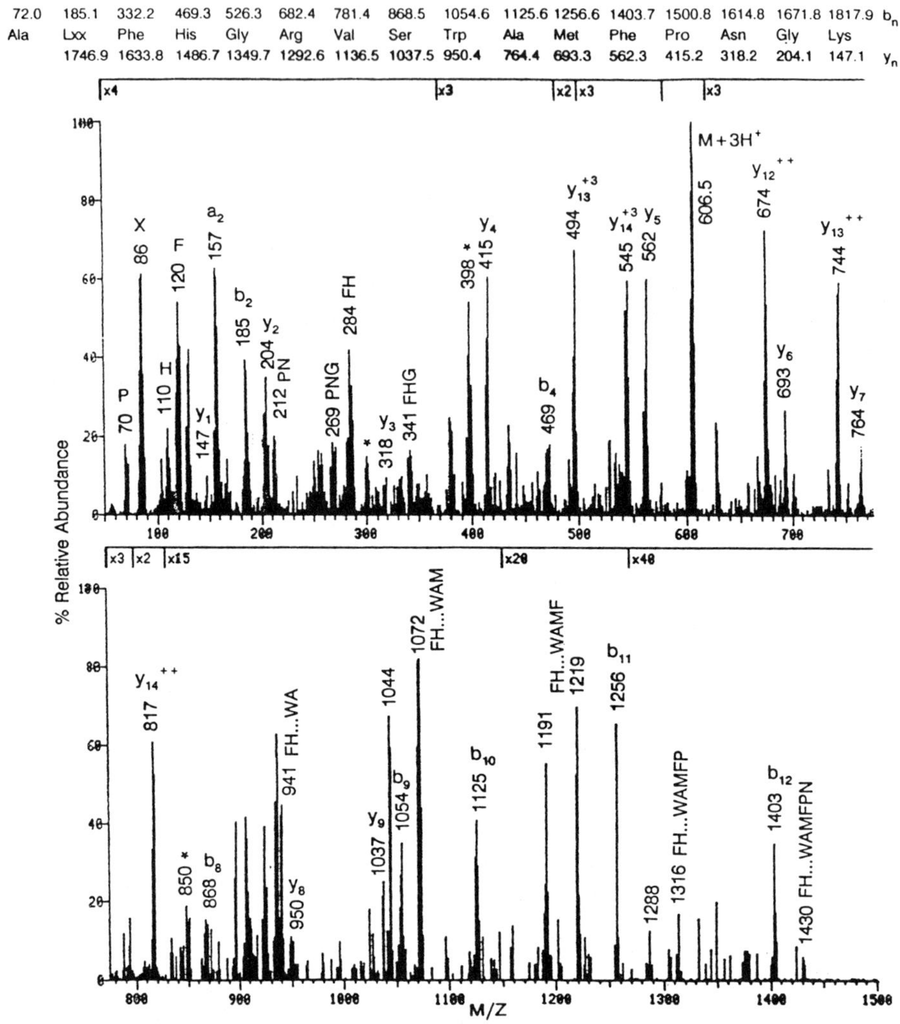

Fig. 2. Collision activated dissociation mass spectrum recorded on the (M+3H)+++ ions generated from STP-4 by the electrospray ionization process. Predicted fragment ions of type b and type y are shown above and below the deduced structure of STP-4, respectively.

although none of them are particularly abundant because they do not contain the strongly basic Arg residue.

682	781	868	1054	1125	1256	1403	1500	1614		Type b
	Val	Ser	Trp	Ala	Met	Phe	Pro	Asn		
	1136	1037	950	764	693	562	415	318	204	Type y

Additional type a-b doublets separated by 28 Da are observed at m/z 441-469, and 498-526. Note that these fragments do not lose ammonia or water and thus should not contain residues such as Ser, Thr, Asn, Gln, Lys or Arg. Subtraction of 526 from 682 and 469 from 526 generates the residue masses of 156 and 57, respectively. As a result, Arg and Gly are specified as the residues preceding Val in the sequence. Note that the ion of type b at m/z 682 contains Arg and does lose ammonia. Neither of the ions at m/z 469 or 626 contain Arg and neither ion loses ammonia.

Predicted values for ions of type y for the partial structure including these two additional residues are shown below. Note that both of the type y ions at m/z 1292 and 1349 are now observed and are accompanied by signals seventeen mass units to lower mass because both these type y ions now contain Arg, a residue that easily loses ammonia from its side chain.

469	526	682	781	868	1054	1125	1256	1403	1500	1614	
	Gly	Arg	Val	Ser	Trp	Ala	Met	Phe	Pro	Asn	
	1349	1292	1136	1037	950	764	693	562	415	318	204

To find additional ions of type y, we now look for unassigned pairs of signals separated by 17 Da that are located to the high mass side of m/z 1349 by at least 57 mass units (the residue mass of Gly, the lightest amino acid), but less than 186 mass units (the mass of Trp, the heaviest amino acid). Pairs of signals that satisfy this criteria are observed at m/z 1469-1486 and 1116-1633. Accordingly, m/z 1486 and 1633 are assigned as ions of type y and the corresponding ions of type b are calculated to occur at m/z 332 and 185. Since both of these ions occur in the spectrum, we conclude that the amino acids adjacent to Gly in the sequence are His and Phe, respectively.

185	332	469	526	682	781	868	1054	1125	1256	1403	1500	1614
	Phe	His	Gly	Arg	Val	Ser	Trp	Ala	Met	Phe	Pro	Asn
	1633	1486	1349	1292	1136	1037	950	764	693	562	415	318

The mass difference between the (M+H)+ ion and the type y ion at m/z 1633 is 184 Da, a value that corresponds to the combined residue masses for both Pro-Ser and Ala-Leu/Ile. Since a type y ion to differentiate these two possibilities is not present in the spectrum, the peptide was subjected to two cycles of manual Edman degradation. Mass spectra were recorded on a portion of the shortened peptide generated after each cycle. The (M+H)+ ion at m/z 1817.6 shifted to 1746.3 after the first cycle (-Ala) and to 1633.6 after the second cycle (-Leu/Ile). These results require that the first two residues be assigned as Ala-Lxx, where Lxx is either Leu or Ile, two residues of identical mass that can not be differentiated as yet on the triple quadrupole instrument.

Of the two remaining unassigned residues at the C-terminus of the sample, one has to be Lys because the acetylation experiment indicated the presence of two amine groups in the molecule. Only one, the N-terminus, has been located. The ion at m/z 1689 is formed by loss of a Lys from the C-terminus as a residue mass with retention of the C-terminal hydroxyl group on the charged fragment. Unaccounted for is a residue mass of 57 Da. Accordingly, the penultimate residue is assigned as Gly to complete the structure.

C. Interpretation of the CAD Mass Spectrum Recorded in the Electrospray Ionization Mode on the (M+3H)+++ ion from STP-4.

Shown in Fig. 2 is the CAD spectrum recorded on the (M+3H)+++ ion (m/z 607) of STP-4 generated under electrospray ionization conditions. In this technique sample in acidic solution is sprayed into the mass spectrometer as a mist from a needle maintained at high positive potential (4). Counter anions are stripped from the sample during this process and the peptide enters the mass analyzer with protons located on most if not all basic sites in the molecule. These include the N-terminal amino group, and the side chains of amino acids such as Lys, and Arg. Protonation of His also occurs in many cases. Since STP-4 is observed as an (M+3H)+++ ion (m/z 607), we expect to find at least three of the above groups in the sample.

The low mass end of the spectrum contains ions of type a_1. These have the composition, $+NH_2=CH(R)$, can originate from any position in the sequence, and are thus indicative of amino acid composition but not sequence. Signals at m/z 70, 86, 110, and 120 indicate the presence of Pro, Lxx, His, and Phe, respectively (Table 1). The relatively weak signal at m/z 147, corresponds to the y_1 ion for Lys at the C-terminus

(Table 1).

Note that the spectrum contains a series of ions of type y for the last nine residues in the sample. The y_{10} and y_{11} ions are absent whereas y_{12} through y_{14}, appear as doubly charged ions at m/z 674, 743, and 817, respectively. Note that all three of these fragment contain the two basic residues, Lys and Arg. Signals for triply charged ions of type y_{13} and y_{14} are observed at m/z 494 and 545.

Ions of type b also appear in the spectrum. Signals for b_8 through b_{12} are the most intense. Ions at m/z 941, 1071, and 1220, correspond to ions of type b containing FHGR- at the N-terminus and -WA, -WAM, and -WAMF at the C-terminus, respectively. All of the above data provide additional support for the deduced sequence.

D. Automated Interpretation of Triple Quadrupole CAD Spectra.

Computer algorithms for the automated interpretation of CAD spectra recorded on triple quadrupole instruments are now in the final stages of development (5). After the structure of STP-4 had been determined as described above, the data shown in Fig. 1 was analyzed with programs developed jointly by John Yates at the California Institute of Technology and Joseph Zhou at the Finnigan Corporation, San Jose, CA. The correct structure received the highest ranking by the data system.

IV. REFERENCES

1. Hunt, D. F., Yates, J. R., III, Shabanowitz, J, Winston, S., and Hauer, C. R. (1986) Proc. Natl. Acad. Sci., USA **83**, 6233-6237.
2. Hunt, D. F., Shabanowitz, J., Yates, J. R., III, Zhu, N. Z., Russell, D. H., and Castro, M. E. (1987) Proc. Natl. Acad. Sci., USA **84**, 620-623.
3. Hunt, D.F., Shabanowitz, J., Griffin, P.R., Yates, J.R., III, Martino, P.A., and McCormack, A.L. (1990) In: Proceedings of the Second International Symposium on Mass Spectrometry in the Health and Life Sciences, A.L. Burlingame and J.A. McCloskey, Eds., Elsevier, Amsterdam, In Press.
4. Smith, R.D., Loo, J.A., Edmonds, C.G., Baringa, C.J., and Udseth, H.R., (1990) Anal. Chem., 62, 882-899.
5. See the article in this volume by John R. Yates, III and Joseph Zhou.

45

American Protein Society Symposium Test Peptide-4 (STP-4): Composition and Analytical Results

Terry D. Lee and Michael Ronk
Beckman Research Institute of the City of Hope,
Division of Immunology, Duarte, California

Description of the Sample

STP-4 was designed as an unknown sample which could be analyzed by any protein chemistry lab and serve as a point of reference for the workshop discussions. Unlike previous symposium samples which were marvelous puzzels synthesized by Jean Rivier at the Salk Institute, the decision was made to provide a much simpler structure that would not require many man days to solve. It was expected that many labs would solve the structure. We decided to use a sixteen residue peptide which had recently been synthesized as a sequencing standard. The major feature of the peptide is that it contains all of the common amino acids that are missing from the N-terminal portion of beta-lactoglobulin which is used in many labs as a sequencing standard. Because it was unknown to the rest of the world, it seemed an ideal choice for the unknown peptide. Labs with sequencers could quickly determine most or all of the structure with one sequencing run. At the same time, the peptide was sufficiently small that it should be solvable by tandem mass spectrometry.

Techniques in Protein Chemistry II

At this point we became concerned that the problem might be too easy and decided to include a second peptide component. This was to be a small peptide with a blocked amino-terminus which could be obtained from a commercial source at modest cost. Having a blocked peptide would provide an additional challenge for mass spectrometrists. It was hoped that the mixture could be sequenced by tandem MS without prior separation of the components. Because it was blocked, the smaller peptide would go undetected by labs that use only Edman microsequencing alone. At the same time, it would not interfere with the Edman microsequence analysis of the larger peptide. The small peptide that was selected was an analog of Thyotropin Releasing Hormone. In addition to a pyroglutamic acid residue at the amino terminus, this tripeptide also had an unusual amino acid, namely methyl histidine. The additional challenge of a nonstandard amino acid was particularly appropriate for analysis by mass spectometry. Because we were concerned that such a small peptide might go unnoticed in FAB mass spectra that have an abundance of low mass ions due to matrix and fragments from larger structures, it was decided to use a much larger amount of the small peptide in the sample to decrease the chance that it might be overlooked. Accordingly, the composition of STP-4 was

Sequence	MW Monisotopic	MW Average	Amount
1. ALFHGRVSWAMFPNGK	1816.9	1818.1	0.2 nmole
2. pGlu-Me-His-Pro-NH2	376.2	376.4	2.0 nmole

Samples were distributed dry in 1.5 ml polypropylene centrifuge tubes to any investigator who requested one. Additionally, samples were mailed to various labs known to do mass spectrometry on peptides with the request that they analyze the sample and send in results even if they didn't plan to attend the meeting. It was expected that there would be some problem with the stability of the larger peptide because it contained methionine and tryptophane. In fact, many of the mass spectral results received showed varying degrees of oxidation. When stored dry at 5 deg. C. the sample is stable for months.

Analytical Results

HPLC Analysis

The two peptides in the STP-4 have very different chromatographic properties. The smaller peptide will elute with the solvent artifact on many reverse phase systems. On most reverse phase systems, the larger peptide requires 20-40% organic for elution. On a Phenomenex Ultracarb column, the tripeptide is retained sufficiently to separate it from the solvent front. In any event, unless investigators were careful to collect the large early eluting peak, the tripeptide part of the sample would be lost during HPLC purification.

Amino Acid Composition Analysis

Amino acid analysis data was received from several labs using four different methods on amounts ranging from 5 to 25% of the total sample. In each instance, there was a clear indication that the sample was a mixture, a large amount of two or three residues and a much smaller amount for the rest. Differences in the four methods were most apparent with respect to the tripeptide. By OPA, The 3-methyl His elutes in the same position as His and the Pro is not detected. By ion exchange chromatography with post-column ninhydrin derivatization, the Me-His elutes very close to His and could be easily mistaken for His. By precolumn derivatization with dansyl chloride, the Me-His is clearly separated from all of the common amino acids. By Pico Tag, the Me-His coelutes with Pro. Thus, conclusions concerning the composition of the tripeptide component would differ markedly depending on which method was used for the analysis (Table 1).

Table I. Amino acid composition analysis of the tripeptide component of STP-4

Method	Composition
Pre-column OPA - RP/HPLC	Glx-1, His-1
Post-column ninhydrin - Ion exchange/LC	Glx-1, Pro-1, His?-1
Pre-column dansyl chloride - RP/HPLC	Glx-1, Pro-1, ?-1
Pre-column Pico-Tag - RP/HPLC	Glx-1, Pro-2

Edman Microsequence Analysis

Although the large peptide was designed as a sequencing standard, it still poses a few problems for sequencers. Trp often yields a blank cycle when it occurs well into the sequence. This was the case for nearly all of the results reported for STP-4. The final few residues of a sequencing run are often lost due to washout of the sample; so when coupled to the problems of carryover, it is not always easy to determine the last cycle with certainty. These ambiguities are easily resolved if an accurate molecular weight has been determined by mass spectral analysis, and several groups were able to correctly determine the sequence of this peptide.

Mass Spectral Analysis for Molecular Weight

A variety of mass spectrometers with a number of different ion sources were used to analyze STP-4 (Table 2). In many instances, the mass range below m/z 400 was not examined and the presence of the small peptide was missed.

Generally, if the region below 400 was included, the presence of the tripeptide was readily apparent. One exception was the single laser desorption time-of-flight spectrum received where low mass ions obscurred the signal for the tripeptide. Many magnetic sector systems using FAB were unable to detect the molecular ion for the larger peptide if less than 10% of the sample was used. As would be expected, sector instruments with array detectors were more sensitive.

Spectra from plasma desorption time-of-flight (PD/MS) instruments showed the greatest variation in the species detected and the accuracy of the mass measurement. Carl March at Immunex was able to demonstrate that the signal for the larger peptide could be detected with as little as 1/500 of the sample and the response was linear with the amount of sample used.

Table II. Kinds of mass spectrometers used to analyze STP-4.

Instrument Type	Ionization Method
Magnetic Two Sector	FAB/Liquid SIMS Electrospray
Magnetic Four Sector	FAB/Liquid SIMS
Single Sector Quadrupole	FAB/Liquid SIMS Electrospray
Triple Sector Quadrupole	FAB/Liquid SIMS Electrospray
Tandem Quadrupole-Fourier Transform MS	FAB/Liquid SIMS
Time-of-Flight	Plasma Desorption Laser Desorption

Electrospray spectra were obtained on a variety of commercial and prototype sources on single and triple sector quadrupole instruments. One spectrum was obtained from an electrospray source on a magnetic sector instrument. On most systems, the tripeptide is observed as a intense singly charged ion at m/z 377. Surprisingly, in almost all spectra fragments of the tripeptide were usually observed at m/z 235 and 263. In normal operation, the electrospray source yields molecular ions with little fragmentation. In general, the most intense ion for the larger peptide component was the triply charged species at m/z 607. The doubly charged ion at m/z 910 was also always present. In one instance, the quadruply charged ion at m/z 456 was observed, and in a separate instance (Figure 1) the singly charged ion was observed.

Figure 1. Electrospray spectrum of STP-4 obtained using a triple sector quadrupole mass spectrometer.

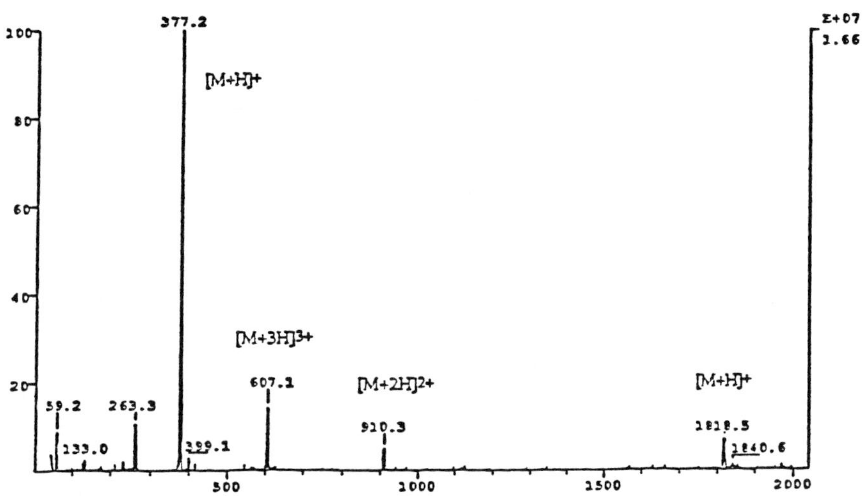

Mass Spectral Sequence Analysis

Mass spectral sequence analysis on the tripeptide component of STP-4 was attempted using a B/E linked scan on a two sector magnetic instrument using FAB, on a four sector magnetic instrument using FAB and on a triple quadrupole using electrospray. In every instance, the fragments at m/z 235 and 263 were evident corresponding to a and b series ions resulting from cleavage between the Me-His and Pro. On the tandem four sector system, ions were also observed at m/z 70, 84, and 124 corresponding to the imminium ions for each of the three amino acids in the tripeptide. From this information, John Stults and coworkers at Genentech correctly deduced the structure of the peptide.

The larger peptide component of STP-4 proved more difficult to sequence than expected. Results were presented by three groups using different methodologies. The analyses done at the University of Virginia by Don Hunt and coworkers are described in detail in this volume. Concisely stated, fragment ions in the collision activated decomposition (CAD) spectrum of the underivatized spectrum in a triple sector quadrupole instrument coupled with the molecular weight observed for the acetylated peptide was sufficient to unambiguously assign all but the first two amino acids at the amino terminus. By performing cycles of manual Edman degradation and measuring the mass of the remaining portion after each cycle, the first two residues were correctly assigned as Ala; and either Leu or Ile which have the same mass. The CAD spectrum was also obtained on the triply charged ion at m/z 606 in the electrospray spectrum. The various fragments were consistent with the previously assigned structure, but by themselves were insufficient to deduce the sequence.

John Yates and coworkers at the California Institute of Technology also used tandem mass spectrometry on a triple sector quadrupole to do sequence analysis on STP-4. In this instance, the data was correctly interpreted to give the right structure by a computer program developed by John Yates and Joe Zhou at Finnigan Corp.

Al Burlingame and coworkers at the University of California at San Francisco used a tandem, four sector, magnetic system for their sequence analysis. A detailed account of their work is also given in this volume. The high energy CAD spectrum on the intact peptide gave partial sequence information, but was not sufficient to solve the structure. The sample was then digested with trypsin and the products were separated by HPLC. The CAD spectra of the two tryptic peptides were readily interpreted except for the identity of the C-terminal residue which could be either Lys or Gln which have the same mass. An acetylation reaction was performed on the intact peptide. Unfortunately, the Lys residue at the C-terminus failed to acetylate and it was assigned as Gln. In the high energy collisions that occur in magnetic sector systems, it is possible to distinquish between Ile and Leu and the Leu residue was correctly assigned. Computer programs developed at the UCSF group where used to interpret the spectral data.

Summary

STP-4 proved to be more challenging then was originally intended and serves to illustrate the status of mass spectrometry in protein sequence analysis today. Many labs routinely use mass spectrometry to obtain molecular weight information. This nicely compliments data obtained by Edman microsequencing and amino acid analysis. New techniques such as electrospray and laser desorption have extended the mass range of accurate MW determinations beyond 100,000 Da. It is evident that mass spectrometry will be used more and more for protein sequence analysis. Past limitations of mass spectral methods have largely been overcome. Required sample amounts are well within the range currently used by those routinely doing protein structure analysis. The high cost of the instrumentation can be justified by the greater sample throughput. The time required for a trained mass spectrometrist to interpret the data has been greatly reduced by the use of computer interpretation programs. For any given peptide such as the larger component of STP-4, a single mass spectral analysis may not be sufficient to yield a complete sequence. Additional chemical and enzymatic manipulations may be necessary to complete the structure. However, a large fraction of the peptides generated from the enzyme digest of a protein can be sequenced without these complications. The complimentary nature of the various analytical methods should be emphasized and protein chemists should make increasing use of mass Spectrometry.

46

COMPUTER AIDED INTERPRETATION OF LOW ENERGY MS/MS MASS SPECTRA OF PEPTIDES

John R. Yates III
Patrick R. Griffin
Leroy E. Hood

Division of Biology
California Institute of Technology, Pasadena, California 91125

Joe X. Zhou

Finnigan MAT, San Jose, California 95134

I. INTRODUCTION

Since the introduction of the gas phase protein sequenator in 1981, N-terminal Edman degradation has been the most sensitive method for the determination of peptide and protein sequences (1). The development of direct protein sequencing strategies of electroblotted proteins has placed increased importance on separation and isolation procedures involving one and two dimensional gel electrophoresis (2). However, a full 30-50% of proteins isolated from SDS-PAGE gels are N-terminally blocked either biosynthetically prior to isolation or artifactually during isolation. *In situ* proteolytic digestion strategies on membranes or in polyacrylamide gels have simplified procedures for obtaining internal amino acid sequence from proteins (3, 4). However, the chemically or proteolytically fragmented proteins must be chromatographically separated in order to generate sequence information by sequential N-terminal degradation. Each peptide must be independently subjected to Edman degradation which requires 30-60 minutes for each residue. Consequently, the amino acid sequence determination of a 15 residue peptide can require 10-15 hours of instrument time.

Recent technological innovations to tandem mass spectrometers (MS/MS) have made the sensitivity of analysis for peptides equivalent to that of the current generation of gas phase protein sequenators (5,6). In addition, tandem mass spectrometers can be used to quickly analyze the products of proteolytically digested proteins, since mixtures of peptides can be sequenced directly without

Techniques in Protein Chemistry II

extensive preseparation (7,8). Although MS/MS data can be generated rapidly, interpretation of the resulting mass spectra can be slow. An obvious solution to increase the efficiency of MS/MS peptide sequencing is computer aided interpretation of the data. Real time or near real time interpretation of mass spectra would make MS/MS sequencing of peptides superior to chemical microsequencing. We have written a computer program to aid in the interpretation of the low energy (10-100 eV) MS/MS mass spectra from a triple quadrupole. The program is based, in part, on previous approaches by Ishikawa and Niwa (9) and Johnson and Biemann (10) with modifications for the fragmentation patterns of peptides observed under low energy, multiple collision conditions of triple quadrupole mass spectrometers. We have applied this program to the sequence analysis of peptides of known amino acid sequence, to peptides of unknown amino acid sequence derived from snake venom toxins and to the Protein Society mass spectromery workshop peptide.

II. EXPERIMENTAL

A. Methods

Proteins obtained from Sigma, β-lactoglobulin and cytochrome c (0.5 nmol), were digested with trypsin (1 ug) in 2 M urea, 0.1 M ammonium bicarbonate, pH 7.0 at 37° C for 4 hrs. Snake toxins were obtained from crude venom of the eastern green mamba snake (*Dendroaspis angusticeps).* The venom was partially fractionated by gel filtration chromatography followed by subfractionation using HPLC. One protein toxin of particular interest, denoted as Da.4137, was treated with trypsin (2 ug) in 0.1 M ammonium bicarbonate, pH 8.5 at 37° C for 12 hrs. Dithiotheritol (4 ul of 0.1 M DTT) was then added to the peptide mixture and the solution was allowed to stand for 1 hr at 37° C.

Peptide mixtures were fractionated by HPLC on an RP-300 aquapore reverse phase HPLC column (C8, 300 A, 2.1mm x 100 mm, Applied Biosystems, Foster City, CA) using a Brownlee Microgradient HPLC equipped with a Rheodyne 7125 injector containing a 50 uL sample loop.

Mass spectra were recorded on a Finnigan MAT TSQ 70 triple quadrupole mass spectrometer equipped with a 40 keV cesium ion gun and 20 keV conversion dynode as previously described (6). Mass analysis was performed by adding 1 uL of a 0.1% trifluoroacetic acid solution containing the peptide or mixture of peptides (10-50 pmol) to 1 uL of monothioglycerol on a gold plated, copper probe tip (1 mm diameter). Peptides were sputtered out of the liquid matrix and into the gas phase by bombardment with 8-15 keV Cs+ ions generated with the cesium ion gun. Typically a mass range of m/z 450-4000 was scanned 10 times at 390 daltons/s and the resulting mass spectra were summed together.

Sequence analysis of peptides was performed using a second 1 uL aliquot of the peptide or peptide mixture (10-50 pmol). In this experiment Q1 served as an ion selection device, Q2 as a collision chamber and Q3 as a mass separation device. Ions were selected through a 3-6 dalton mass window in Q1 and allowed to pass into Q2, which was filled with argon to a pressure of 3-4 x 10^{-5} torr. Collision energies varied between -20 and -60 V. Mass spectra were obtained at a scan rate of 390 daltons/s.

Peptides were converted to their corresponding methyl esters and acetylated as previously described (8,11). The derivatized peptides were analyzed as outlined above.

B. Computer Hardware

The computer algorithm was written in C under the Ultrix 2.1 (Unix) operating system and the DecWindows (X-windows) graphical user interface for the DecStation 2100. The computer is equipped with 16 Mb of internal memory, 330 Mb of disk space and runs at 10 MIPS. Mass spectra were transferred from the TSQ 70's PDP 11/73 over a DECnet to the DecStation 2100. Display and manipulation of mass spectra was accomplished with the DecStation implementation of Finnigan's ICIS data system. The interpretation algorithm has been implemented within PepMatch, an ICIS based protein and peptide sequence analysis program. PepMatch also contains graphical tools for analyzing predicted peptide amino acid sequences. Interpretation of a mass spectrum requires about 10-30 seconds of computer time.

III. RESULTS

A. Computer Aided Interpretation

Under typical operating conditions a 3-6 dalton window in Q1 is used to select the parent ion. This window will also pass the (M+H)+ ^{13}C containing ions giving a cluster appearance to the parent ion region and the fragment ions. Fragment ions produced under low energy multiple collision conditions of the triple quadrupole are dominated by cleavages of the amide bonds of the peptide backbone (8). When the charge is retained on the N-terminal cleavage fragment the resulting ion is labelled as a B- type ion. If the charge is retained on the C-terminal fragment it is labelled a Y"- type ion. Neutral losses of water , ammonia and carbon monoxide from B- type ions are frequently observed. Clusters of ions 17/18 and 28 daltons below another ion are frequently diagnostic for B- type ions. Y"-type ions can also be accompanied by neutral losses of water and ammonia. These sets of fragment ions are the primary ions used for the determination of a peptide's amino acid sequence.

The program can accept the MS/MS spectra of underivatized peptides and/or peptides which have been converted to the methyl ester. Raw mass spectral data is processed before the interpretation by automatically calculating the ion intensity threshold and subtracting the background.

B. Interpretation Strategy:

1) Partial Sequence Extension.

The strategy used in the interpretation program is to sequentially subtract the weights of each of the amino acids from the (M+H)+ ion and compare the predicted fragment ions with the mass spectrum of the underivatized peptide, or the peptide methyl ester, or both mass spectra. A score is generated on the basis of the intensity of the sequence ions A, B, Y", and their neutral losses, eg. -17 (ammonia) and -18 (water). The scores are then sorted to generate a list of partial sequences for the next subtraction cycle. If proline or histidine are found in the sequence, the program checks for BnY"n ion series (8). The cycles continue on a particular amino acid sequence until subtraction of the mass of an amino acid leaves a difference between the predicted and observed (M+H)+ of less than 0.5 but more than -0.5. This sequence is saved for final scoring.

2) Final Sequence Scoring:

A final score is calculated for each completed sequence obtained from step 1. This is calculated from three factors: the number of peaks in the mass spectrum (free acid and methyl ester, if used) accounted for by sequence ions, the average intensities of predicted fragment ions, and the continuity of sequence ion occurences within a given fragmentation scheme. Any known sequence information can be provided to the program such as, the presence or absence of amino acids, the extension directions (N-terminal to C-terminal or C-terminal to N-terminal) and any known terminal residue(s).

The program can be used in an automated, semi-automated or manual mode. In the automated mode the user does not interact with the program while the interpretation is in progress. In the semi-automated mode, the top 324 answers are displayed after each extension and the user is given the choice of continuing on to the next cycle or stopping. A partial answer can then be used as C-terminal or N-terminal input for the automated mode. The manual mode simply provides an interactive tool to allow extension of one amino acid at a time. The result of each extension is viewed in a separate DecWindow with the predicted fragment ions in the mass spectrum highlighted in a different color. The program provides an interface to allow comparison of the final predicted peptide sequence with the actual mass spectrum. Predicted fragment ions which match peaks in the mass spectrum are highlighted in a different color in order to be easily identifed. In this fashion a quick visual check of the mass spectrum can be made to assess the interpretation.

MS/MS mass spectra of a native tryptic peptide from β-lactoglobulin and its corresponding methyl ester are shown in Figure 1. The mass spectra were input to the computer program without additional information. The output of the automated interpretation of this peptide is shown in Figure 2. Clearly the predicted amino acid sequence of XPAVFK has the highest score and this peptide would be unambiguously assigned the amino acid sequence of XPAVFK (X indicates Leu or Ile).

Results from the application of the interpretation program to peptides of known and previously unknown amino acid sequence are listed in table 1. In cases where the correct sequence has the highest score only that sequence is shown. Otherwise, several of the best answers are displayed and the correct sequence is underlined. Note that we have made no attempt to differentiate

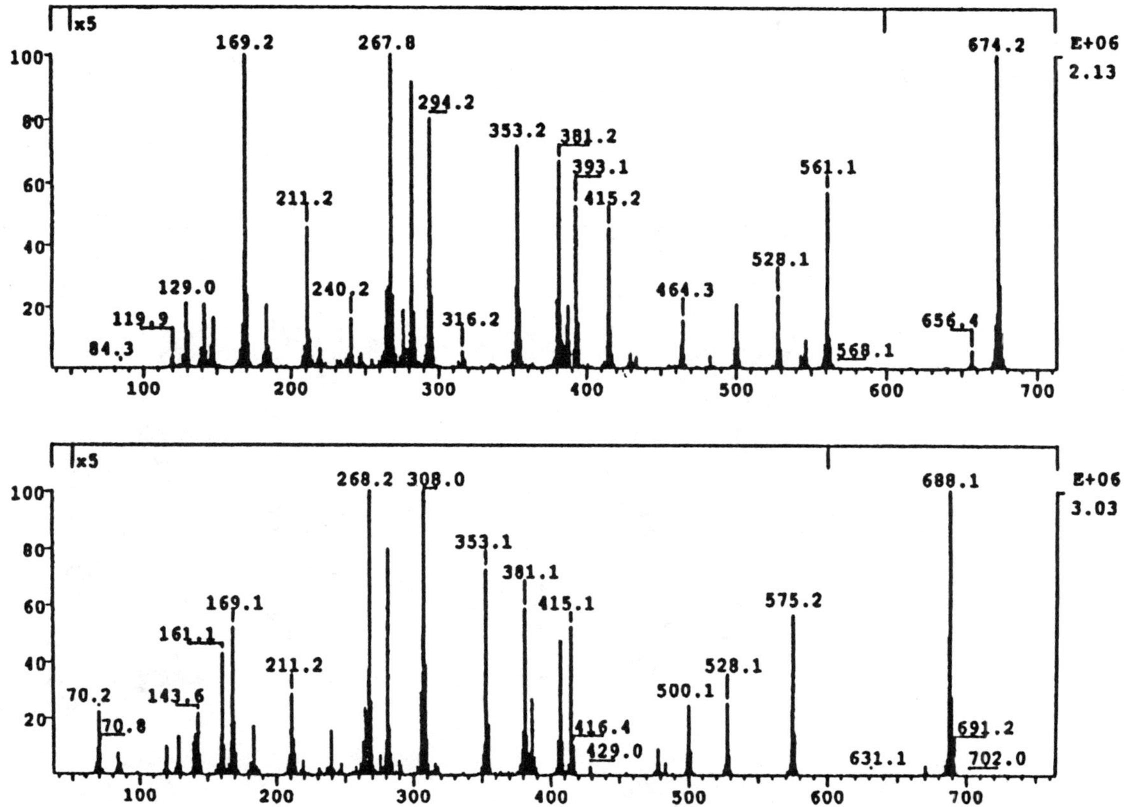

FIGURE 1: LSIMS/MS/MS spectra of the native (top) and methyl esterified (bottom) hexapeptide, IPAVFK.

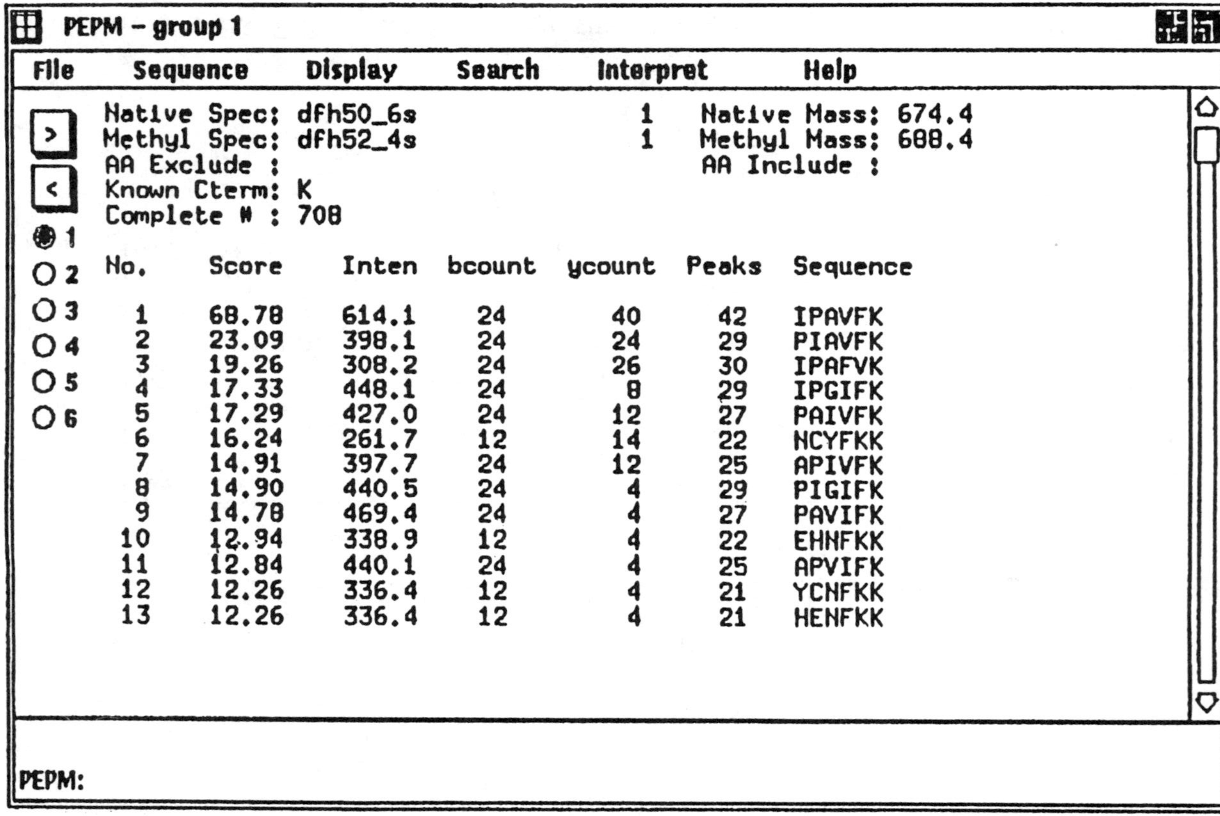

FIGURE 2: Computer generated list of the most probable sequences from LSIMS/MS/MS of the hexapeptide, IPAVFK.

TABLE I. Application of the computer algorithm to peptides of known and previously unknown sequence.

Peptides[a]	Size	Results	Comment[b]
GLDIK	672.4	GXDXK	FA
IPAVFK	673.4	XPAVFK	FA
IPAVFK	687.4	XPAVFK	OME
ALPMHIR	850.5	SPPMHXR AXPMHXR	OME
GFFYTPK	858.4	GFFYTPK	FA
LDAINENK	958.5	XDAXNENK	OME
VLVLDTDYK	1107.6	VXVXDTDYK	OME
VLVLDTDYKK	1192.7	VXVXDTDYQK	FA and OME
VGINYWLAHK	1213.7	VHFNYWXAHK	OME
TPEVDDEALEKFDK	1634.8	TPEVDNEANEQFDK	FA and OME
LSFNPTQLEEQCHI	1715.8	XHPNPTQIEEQCHI	FA and OME
TIGMVVIHK[c]	1010.6	TXGMVVXHK	OME
WYYXVPR[d]	995.5	WYYXVPR	FA
SXFGXTTEXCPDPSNX**CFK**[d]	2085.0	SXQTSTTEXCPTDYFCFK	FA and OME
ALFHGRVSWAMFPNGK[e]	1818.6	ALFHGRWWAMFPWI ALFHGRWWAMFPIW ALFHGRVSWAMFPIW ALFHGRVSWAMFPNGQ	FA and OME

[a]Correct amino acid sequence of the peptides used in the interpretation. Amino acids in bold lettering were input to the program.

[b]FA indicates the underivatized peptide was used. OME indicates the peptide methyl ester mass spectrum was used.

[c]N-terminal peptide from α-glycoasparaginase, courtesy of Dr. Ilka Mononen, Los Angeles Children's Hospital, verifed by Edman degradation.

[d]Peptides derived from *Dendroaspis angusticeps*.

[e]Protein Society mass spectrometry workshop peptide, the top four answers are shown.

Leucine and Isoleucine or Lysine and Glutamine. The assignment of Lysine and Glutamine in the sequence was based on specificity of trypsin. In some cases the underivatized peptides are used with the peptide methyl esters and in others only one of the mass spectra was used.

Also shown in table 1 are the results from the application of the computer program to the structure elucidation of the Protein Society mass spectrometry workshop peptide. In this example the interpretation was conducted from the N-terminus to the C-terminus and the first three N-terminal amino acids were given as input to the program. Two of these amino acids were derived from cycles of subtractive manual Edman degradation and the third from inspection of the MS/MS mass spectrum with knowledge of the first two amino acids. Differentiation of lysine from glutamine at the C-terminus was accomplished by acetylation of the peptide. The top two answers contain a Trp at residue 7 where it should be Val-Ser. However, the combined mass of the latter two amino acids has the same mass increment as Trp, consequently the interpretation continued on correctly for the next 5 residues. All of the top three answers incorrectly identifed the C-terminal region of the peptide with variations of Trp-Ile. The correct answer, as scored and sorted, is in the fourth position. Since the scoring is a function of the average intensity of the fragment ions and the correct sequence is one amino acid longer, it is reasonable for the average intensity of this sequence to be lower than the other sequences. If the total intensity of the fragment ions is considered as the sole criteria for scoring (this value is displayed in the computer output but is not the basis on which answers are sorted) the correct answer would be at the top of the list. Clearly, the computer program provided reasonable answers on which further reflection and use of the graphical tools helped determine the correct amino acid sequence of the workshop peptide.

IV. DISCUSSION

These preliminary results indicate that computer aided interpretation of low energy collision-activated dissociation mass spectra of peptides shows considerable promise for the rapid interpretation of MS/MS data. Application of the computer algorithm to a number of examples has shown that the program always gives the correct answer for small or medium size peptides (<1200 daltons) and correct partial sequences for larger peptides (>1500 daltons). For larger peptides the results are sensitive to the quality of the mass spectrum. Poor quality mass spectra or spectra that are devoid of some fragment ions can result in the misidentification of amino acids in the sequence. While the peptide methyl ester mass spectrum is not needed for interpretation, the confidence level in the predicted sequence greatly increases when it is used in combination with the free acid spectrum. Further refinements to the interpretation algorithm may make it possible to determine the amino acid sequence of a 20-25 residue peptide in just 10 minutes time.

V. ACKNOWLEDGEMENT

This work was supported by Finnigan MAT and, in part, by the National Science Foundation, Science and Technology Center Cooperative agreement no. 8809710.

VI. REFERENCES

1. Hewick, R. M., Hunkapillar, M. W., Hood, L. E.,and Dreyer, W. J. (1981) *J. Biol. Chem.* **256**, 7990-7997.
2. Aebersold, R. H., Teplow, D. B., Hood, L. E.,and Kent, S. B. H (1986) *J. Biol. Chem.* **261**, 4229-4238.
3. Aebersold, R. H., Leavitt, J., Saaverdra, R. A., Hood, L. E., and Kent, S. B. H. (1987) *Proc. Natl. Acad. Sci. USA* **84**, 6970-6974.
4. Jahnen, W., Ward, L. D., Reid, G. E., Moritz, R. L., and Simpson, R. J. (1990) *Biochem. Biophys. Res. Comm.* **166**, 139-145.
5. Burlingame, A. L.(1988), in "Proceedings of the 36th Annual Conference on Mass Spectrometry and Allied Topics", pp 727-728.
6. Griffin, P. R., Martino, P. A., McCormick, A. L., Shabanowitz, J., and Hunt, D. F. (1990) in "Current Research in Protein Chemistry: Techniques, Structure, and Function", J. Villafranca, Ed., Academic Press, N. Y., 117-126.
7. Hunt, D. F., Bone, W. M., Shabanowitz, J., Rhodes, J., and Ballard, J. M., (1981) *Anal. Chem.* **53**, 1704-1708.
8. Hunt, D. F, Yates, III, J. R.,Shabanowitz, J, Winston, S.and Hauer, C.R.(1986) *Proc. Natl. Acad. Sci USA* **83**, 6233-6237.
9. Ishikawa, K and. Niwa, Y. (1986) *Biomed. Env. Mass Spec.* **13**, 373-380.
10. Johnson, R. S. and Biemann, K. (1989) *Biomed. Env. Mass Spec.* **18**, 945-957.
11. Hunt, D. F, Griffin, P. R, Yates, III, J. R.,Shabanowitz, J., Fox, J. W., and Beggerly, L. K. (1989) in "Techniques in Protein Chemistry", T. Hugli, Ed., Academic Press, N. Y., 580-588.

47

Electrospray Ionization Mass Spectrometry and Tandem Mass Spectrometry for the Characterization of Proteins and Proteolytic Digests

Charles G. Edmonds, Joseph A. Loo, Rachel R. Ogorzalek Loo, and Richard D. Smith
Pacific Northwest Laboratory, Richland, Washington

I. Introduction

Recent advances in desorption/ionization methodologies have rendered nonvolatile and thermally labile biomolecules of large (> 10 kilodaltons) molecular weight amenable to analysis by mass spectrometry (MS). The most successful ionization techniques can be characterized as "soft" ionization methods, those which minimize thermal degradation of the molecular species. Electrospray ionization (ESI) mass spectrometry (Meng *et al.*, 1988; Fenn *et al.*, 1989; Loo *et al.*, 1989a; Smith *et al.*, 1990a) has attracted the attention of protein chemists with its ability to efficiently ionize polypeptides and proteins formerly beyond the range of other ionization techniques. The ESI method shows extremely high ionization efficiencies (i.e., very high [molecular ions produced]/[molecules consumed]) and the capability for very precise measurement of molecular weight or relative molecular mass (M_r). ESI produces a distribution of multiply charged (generally protonated) molecular ions from highly charged droplets formed at atmospheric pressure. Sufficient charge is typically retained by the analyte molecule to allow conventional quadrupole mass spectrometers of limited mass-to-charge (*m/z*) range to analyze proteins. To date, samples with M_r of up to 100,000 have been successfully analyzed (Loo *et al.*, 1989a).

A major characteristic of ESI mass spectra is the predominance of molecular ion species and the absence of significant fragmentation. While the presence of one or several molecular ion species allows measurement of accurate molecular weight values, little additional data regarding molecular structure is obtained. Coupling electrospray ionization with tandem mass spectrometry (MS/MS) (Smith *et al.*, 1990a, McLafferty, 1983; Busch *et al.*, 1988; Smith *et al.*, 1990a; Loo *et al.*, 1990a) is an excellent method for obtaining additional molecular structure information.

Techniques in Protein Chemistry II

Electrospray ionization occurs during the electrostatic nebulization of liquids from a capillary in a large electrostatic field, yielding highly charged droplets. These droplets are generally formed in a dry bath gas at atmospheric pressure and shrink by evaporation of neutral solvent until charge repulsion overcomes the cohesive droplet forces, leading to a "Coulombic explosion." ESI was originally described (Dole *et al*, 1968; Mack *et al.*, 1970) in studies based upon gaseous ion mobility measurements of intact ions from synthetic and natural polymers with molecular weights in excess of 100,000 daltons. After a ten year hiatus these experiments were extended by Fenn and coworkers (Meng *et al.*, 1988; Fenn *et al.*, 1989; Yamashita *et al.*, 1984a,b; Whitehouse *et al.*, 1985) employing a quadrupole mass spectrometer, and essentially simultaneously by researchers in the Soviet Union (Alexsandrov *et al.*, 1984) using a magnetic sector instrument. These workers outlined the fundamental aspects of ESI, demonstrating its utility for the analysis of biomolecules of modest molecular weight. The multiple charging phenomenon has been demonstrated to apply to molecules of over 130,000 daltons in molecular weight (Loo *et al.*, 1989a; Loo *et al.*, 1990b) and has permitted the measurement of relative mass in the range of 5-40 kDa with precision of better than 0.01% (Loo *et al.*, 1989a; Covey *et al.*, 1988; Mann *et al.*, 1989).

II. Experimental Methods

The electrospray ionization source used in our laboratory has been described in the literature (Loo *et al.*, 1989a; Smith *et al.*, 1990a; Smith *et al.*, 1988). An organic liquid sheath (typically pure methanol or acetonitrile) flows in the annular space between a 100 μm i.d. fused silica capillary, which delivers analyte solution to the ion source, and a cylindrical stainless steel electrode. An electrospray voltage of +4-6 kV is applied (for positive ion ESI) to this surrounding electrode. For experiments in which sample solution is directly infused to the ESI source, syringe pumps deliver controlled flows of analyte and sheath liquids at rates of 0.1-1 μl/min and 2-4 ml/min, respectively. Sample solutions of proteins and peptides are typically prepared in 5% v/v acetic acid in deionized water.

Ions are sampled from atmospheric pressure through nozzle-skimmer elements (+100 to +400 V potential difference for positive ion studies) in front of a radiofrequency (RF)-only quadrupole. The nozzle-skimmer region is maintained at approximately 2 Torr, while the quadrupole chamber typically reaches 10^{-6} Torr. The ability to "collisionally heat" ions by applying electric fields in the nozzle-skimmer region is useful for both desolvation and, at higher voltages, collisional activation and collisionally activated dissociation of analyte ions.

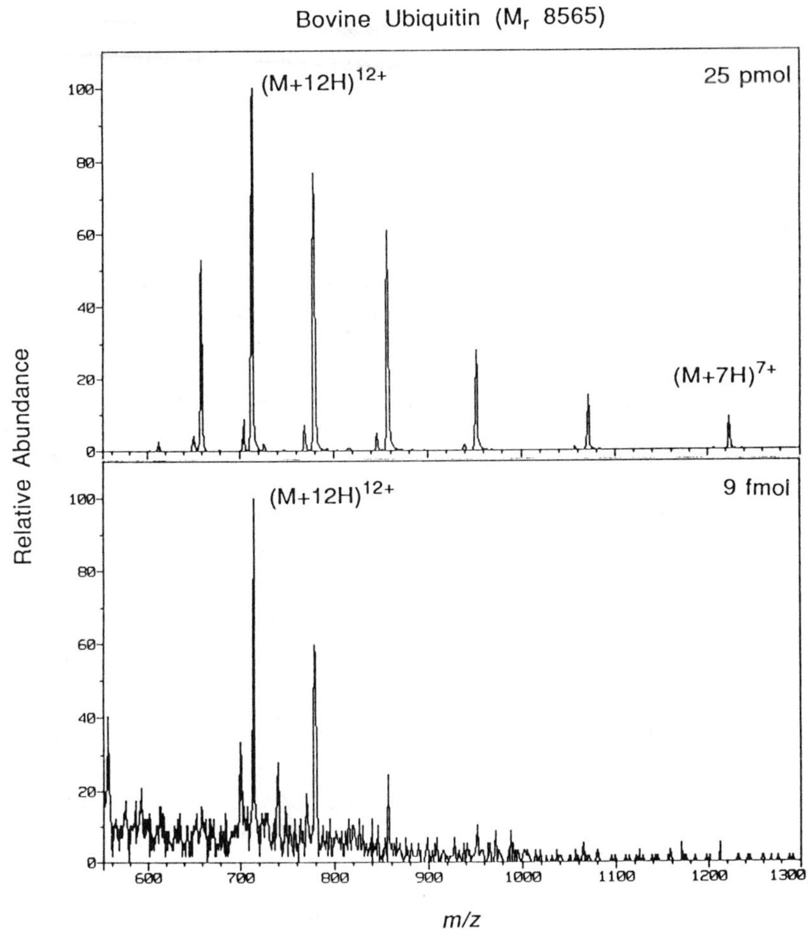

Figure 1. Electrospray ionization mass spectra of bovine erythrocyte ubiquitin from (top) 25 pmol and (bottom) 9 fmol of material consumed during the analysis.

III. Results and Discussion

A. ESI-MS of Peptides and Proteins

Large oligopeptides and small proteins so far examined by ESI-MS typically show a distribution of multiply charged molecular ions arising (in positive ion mode) by either proton or alkali ion attachment and no evidence of fragmentation unless dissociation is induced during transport into the mass spectrometer vacuum by higher energy collisions. ESI mass spectra of bovine ubiquitin (M_r 8565) at the 25 picomole and 9 femtomole (9×10^{-15} moles) levels are shown in Figure 1.

The largest protein so far examined successfully (i.e., where a molecular weight measurement is obtained) is the native covalent dimer of bovine serum albumin with M_r of more than 130 kilodaltons (Loo *et al.*, 1989a, Loo *et al.*, 1990b). It is uncertain what processes give rise to the distinctive bell-shaped distribution of charge states, but the spacing between peaks clearly reflects the quantum nature of the charge; i.e., adjacent peaks differ by one charge. For the example of ubiquitin illustrated, the highest *m/z* ion observed (with a quadrupole-MS *m/z* limit of 1400) is at *m/z* 1224, corresponding to a net charge of 7+. The most abundant multiply protonated ion is observed at *m/z* 715 for the 12+ ion and the highest charge state observed is 14+ (a small peak at *m/z* 613 in Figure 1). A striking feature of ESI mass spectra of most peptides and proteins is that the average charge state increases in an approximately linear fashion with molecular weight, extending the mass range of the MS experiment by the addition of compensating charge.

The number of possible protonation sites is the principal factor affecting the multiple charging observed in ESI mass spectra. For most compounds so far examined (prepared in aqueous solution, pH<4), an approximate linear correlation is observed between the maximum number of charges and the number of basic amino acid residues (e.g., arginine, lysine, histidine, and the NH_2-terminus) (Smith *et al.* 1990a; Loo *et al.*, 1990b). However, many proteins contain disulfide bridges in their native structure that significantly affect higher order structure. The native structure of bovine α-lactalbumin (M_r 14,175) is composed of 17 basic residues and 4 cysteine-cysteine disulfide bridges, yet a maximum of only 13+ charges is observed in the ESI mass spectrum. Reduction of disulfide bonds with 1,4-dithiothreitol (Cleland's reagent) results in higher charged molecular ions, up to 16+ (Loo *et al.*, 1990b). The number of disulfide bonds can be determined by the increase in M_r upon reduction. For α-lactalbumin, an 8 Da increase is consistent with 4 Cys-Cys bonds. Since protein conformation in solution depends upon intramolecular forces dictated by the amino acid sequence, one expects that cleavage of disulfide bonds should greatly affect the higher order structure of the protein, allowing the molecule to "relax" into more extended conformations.

ESI mass spectra of proteins composed of noncovalently bound subunits show multiply charged ions characteristic of the subunit M_r. Lactate dehydrogenase from rabbit muscle (M_r 140 kDa) is an enzyme composed of four identical subunits of M_r 36 kDa. Over 50 positive charges indicative of the subunit molecular weight are detected by ESI-MS. Human hemoglobin is a tetrameric protein consisting of 2 α and 2 β chains; its ESI mass spectrum shows multiply charged ions for both the α (M_r 15,126) and the β (M_r 15,865) chain units.

For positive ion ESI mass spectra of proteins, the determination of M_r is straightforward given two assumptions: (1) adjacent peaks of a series differ by only one charge and (2) charging is due to cation attachment (a proton or rarely, for lower M_r oligopeptides, mixed H^+ and Na^+) to the molecular ion. With these assumptions it follows that any two peaks are sufficient to uniquely determine M_r and charge. Several workers have presented more or less elaborate methods for extracting M_r from ESI-MS data with high precision and accuracy (Loo *et al.*, 1989a; Covey *et al.*, 1988; Mann *et al.*, 1989). Such measurements are generally two to three orders of magnitude more accurate than those obtained using electrophoretic methods, and perhaps an order of magnitude better than time-of-flight based plasma or laser desorption methods.

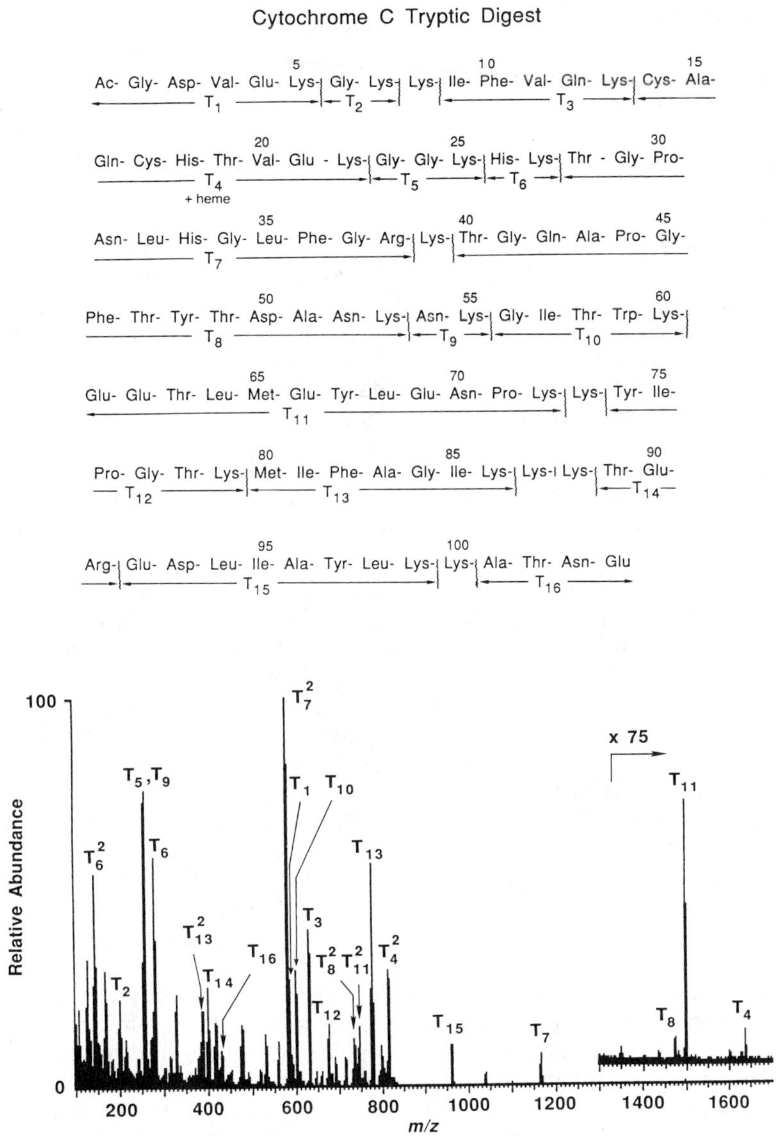

Figure 2. (Top) Proteolytic digestion of equine cytochrome *c* with trypsin and (bottom) the corresponding ESI mass spectrum from 10 pmol of the resulting mixture. A superscript denotes the charge state of the ion. (Lack of a superscript indicates a singly charged ion.)

B. ESI-MS of Proteolytic Digests

Classical investigations of protein primary structure generally require the chemical or enzymatic decomposition of the native material. In this context exhaustive digestion with highly specific proteases permits flexible reduction of the protein to a mixture of relatively small oligopeptide fragments. The mixture of products from such a digestion may be examined directly by liquid chromatography and/or mass spectrometry (e.g., fast atom bombardment or ^{252}Cf plasma desorption ionization), producing a characteristic distribution of products whose chromatographic retention and/or mass spectra "map" the original structure. This map can rapidly confirm the protein sequence, reveal disulfide bridging, highlight the presence of unrecognized modifications or degradations, etc. Should the sequence of the protein be unknown, its determination may proceed with a combined separation/mass spectrometry or mass spectrometry/mass spectrometry experiment, and/or by preparative isolation of the constituents of the digest with subsequent examination of these by manual or automated Edman degradations.

We have evaluated electrospray ionization mass spectrometry and tandem MS for the direct analysis of unfractionated mixtures of peptides obtained by exhaustive proteolysis. We have observed, for example, that the ESI-MS analysis of the mixture of tryptic fragments from equine cytochrome *c* (104 residues, M_r 12,360) reveals **all** its constituents as both singly and doubly charged molecular ions (except for the C-terminal fragment, T101-104, which contains only one basic residue at its N-terminus) (Figure 2). These peptides vary widely in hydrophobicity (e.g., using the indices of Bull and Breese (Breese, 1974) T9-13, +635 and T92-99, -499). Synthetic equimolar mixtures of peptides with varying Bull and Breese hydrophobicity indices show no ionization bias by ESI-MS. Thus, surface activity does not appear to play a major role in the electrospray ionization process. This result is in contrast to those obtained by fast atom bombardment ionization where these differences in surface activity produce differential response.(Naylor *et al.*, 1988). Figure 3 illustrates the product ion MS/MS spectra obtained on two of the tryptic peptides from the mixture. Complete primary structure analysis from such methodologies should be straightforward.

We are testing the generality of the lack of hydrophobicity based differential response in electrospray ionization in experiments on mixtures of small peptides differing by the substitution of a hydrophilic for a very hydrophobic residue. In preliminary experiments over a range of concentrations no significant difference in response between constituents of an equimolar binary mixture and the isolated single compounds was observed. These results suggest the generality of ESI-MS mapping for revealing complete primary structure (or substantial portions) of small proteins from a single experiment on an unfractionated proteolytic digest. We are continuing to evaluate the utility of such procedures for the analysis of proteins of known and unknown structure.

Additionally, partial proteolysis may be used as a method for investigating the higher order construction and the substructural organization of a protein. In this case the digestion is interrupted at an early stage to afford a relatively less complicated mixture of larger fragments. These may be isolated with more or less difficulty and their primary structure characterized by classical or mass

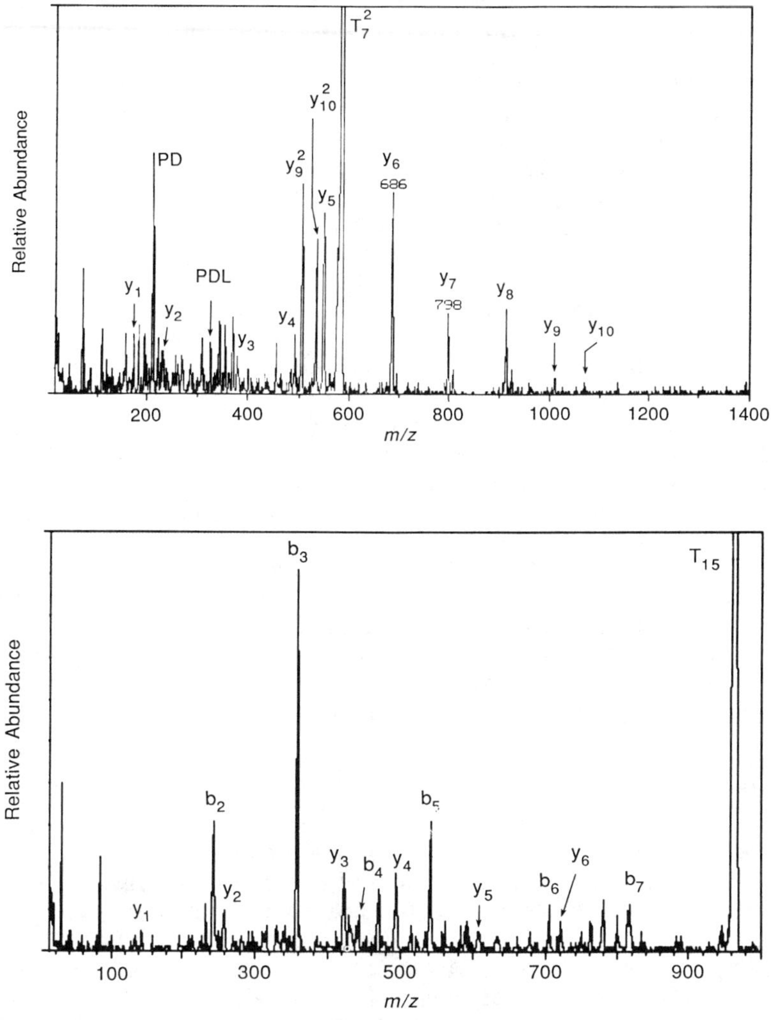

Figure 3. Product ion MS/MS mass spectra for the tryptic fragments (top) T_7 (2+ charge, M_r 1168) and (bottom) T_{15} (1+ charge, M_r 964) from a tryptic digest of quine cytochrome *c* during the analysis of the unfractionated mixture of tryptic peptides by ESI-MS/MS. (See Figures 2 for sequence.)

spectrometric means. Where the primary sequence is already known, proteolytic fragments may be identified permitting the location of the most sensitive bonds and providing insight into the spatial disposition of these residues in the higher order configuration of the protein. In collaboration with biochemical colleagues we have examined mixtures of large polypeptides (77 to 126 residues, M_r 8525-13,900) derived by partial trypsinolysis of calmodulins isolated from wild type and mutant strains of *Saccharomyces cerevisiae*. Sensitivity to proteolysis of these altered proteins has been employed as a probe to the molecular details of a temperaturre sensitive phenotype in these mutant yeast strains. The direct examination of these unfractionated mixtures of tryptic peptides by electrospray ionization mass spectrometry and mass spectrometry/mass spectrometry has permitted the rapid recognition of the principal constituents of these mixtures without their laborious preparative isolation.

III. Future Outlook

With the emergence and rapid expansion of biochemical and biotechnology related fields, new and improved analytical methodologies become vital to assess important parameters such as purity and quality control. Electrospray ionization-mass spectrometry is already proven as a sensitive detection method for a variety of biomolecular classes and molecular weight ranges and may potentially offer rapid and efficient sequencing methods when combined with capillary electrophoresis (CE) and tandem mass spectrometry. Capillary electrophoresis in its various manifestations (free solution, isotachophoresis, isoelectric focusing, polyacrylamide gel, micellar electrokinetic "chromatography") is attracting attention as a method for rapid high resolution separations of very small sample volumes of complex mixtures. Combination with the inherent sensitivity and selectivity of mass spectrometry makes CE-MS (Udseth *et al.*, 1989; Loo *et al.*, 1989b) a potentially powerful bioanalytical technique.

Acknowledgments

We are very pleased to acknowledge the collaboration of Dr. T.N. Davis and Ms. S. Brokerhoff of the Department of Biochemistry, University of Washington, Seattle, Washington, for their samples of mutant calmodulins and calmodulin digests. We are grateful for the support of the U.S. Department of Energy, Exploratory Research of the Molecular Science Research Center (DE-AC06-76RLO 1830), the National Science Foundation, Instrumentation and Instrument Development Program (DIR 8908096) and the National Institutes of Health, National Center for Human Genome Research (HG 00327). Pacific Northwest Laboratory is operated by Battelle Memorial Institute.

References

Alexsandrov, M. L., Gall, L. M., Krasnov, N. V., Nikolaev, V. K., Pavlenko, V. A., and Shkurov, V. A. (1984). *Dokl. Akad. Nauk SSSR*. **277**, 379.

Bull, H. B., and Breese, K. (1974). *Arch. Biochem. Biophys.* **161**, 665.

Busch, K. L., Glish, G. L., and McLuckey, S. A. (1988). "Mass Spectrometry/Mass Spectrometry: Techniques and Applications of Tandem Mass Spectrometry." VCH, New York.

Covey, T. R., Bonner, R. F., Shushan, B. I., and Henion, J. (1988). *Rapid Commun. Mass Spectrom.* **2**, 249.

Fenn, J. B., Mann, M., Meng, C. K., Wong, S. F., and Whitehouse, C. M. (1989). *Science* **246**, 64.

Loo, J. A., Udseth, H. R., and Smith, R. D. (1989a). *Anal. Biochem.* **179**, 404.

Loo, J. A., Jones, H. K., Udseth, H. R., and Smith, R. D. (1989b). *J. Microcolumn Sep.* **1**, 223.

Loo, J. A., Edmonds, C. G., and Smith, R. D. (1990a). *Science* **248**, 201.

Loo, J. A., Edmonds, C. G., Udseth, H. R., and Smith, R. D. (1990b). *Anal. Chem.* **62**, 693.

Mack, L. L., Kralik, P., Rheude, A., and Dole, M. (1970). *J. Chem. Phys.* **52**, 4977.

Mann, M., Meng, C. K., and Fenn, J. B. (1989). *Anal. Chem.* **61**, 1702.

McLafferty, F. W., ed. (1983) "Tandem Mass Spectrometry." John Wiley & Sons, New York.

Meng, C. K., Mann, M., and Fenn, J. B. (1988). *Z. Phys. D-Atoms, Molecules and Clusters* **10**, 361.

Naylor, S., Findeis, A. F., Gibson, B. W., and Williams, D. H. (1986). *J. Am. Chem. Soc.* **108**, 6359.

Naylor, S., Moneti, G., and Guyan, S. (1988). *Biomed. Environ. Mass Spectrom.* **17**, 393.

Smith, R. D., Barinaga, C. J., and Udseth, H. R. (1988). Anal. Chem. **60**, 1948.

Smith, R. D., Loo, J. A., Edmonds, C. G., Barinaga, C. J., and Udseth, H. R. (1990a). *Anal. Chem.* **62**, 882.

Smith, R. D., Loo, J. A., Barinaga, C. J., Edmonds, C. G., and Udseth, H. R. (1990b). *J. Am. Soc. Mass Spectrom.* **1**, 53.

Udseth, H. R., Loo, J. A., and Smith, R. D. (1989). *Anal. Chem.* **61**, 228.

Whitehouse, C. M., Dreyer, R. N., Yamashita, M., and Fenn, J. B. (1985). *Anal. Chem.* **57**, 675.

Yamashita, M., and Fenn, J. B. (1984a) *J. Phys. Chem.* **88**, 4451.

Yamashita, M., and Fenn, J. B. (1984b) *J. Phys. Chem.* **88**, 4671.

48

SEQUENCE-ORDERED PEPTIDE MAPPING BY TIME-COURSE ANALYSIS OF PROTEASE DIGESTS USING LASER DESORPTION MASS SPECTROMETRY

Richard M. Caprioli and Brenda Whaley

The Analytical Chemistry Center and
the Department of Biochemistry and Molecular Biology

The University of Texas Medical School, Houston, Texas 77030

and

Kuldip K. Mock and John S. Cottrell

Finnigan MAT Ltd., Hemel Hempstead HP2 4TG, UK

I. INTRODUCTION

Mass spectrometry has been used for many years to identify the peptides produced from a protease digest. Direct analysis of these digests using fast atom bombardment (FAB) mass spectrometry has been particularly effective and has been termed FAB mapping (1). The major use of the technique has been to help verify the primary structures of proteins whose sequences have been proposed from DNA sequences, Edman-based methods, or by comparison to homologous proteins, and to identify peptides that contain post-ribosomal modifications of various types (2,3). Usually, a limit digest is analyzed and the masses of the molecular species found in the digest are mapped along the expected protein sequence.

Several problems can be encountered in such a FAB mapping study, including i) identification of peptide fragments which do not map as expected, even when considering the masses of common modifying groups such as acetyl, carbohydrate, phosphate, or other groups, ii) the absence

Techniques in Protein Chemistry II

of one or more expected fragments, iii) the presence of a peptide with a molecular weight beyond the effective mass range of the FAB mapping technique, and iv) the possibility of redundant masses especially if modifying groups are considered. Often, one can utilize alternative techniques and strategies to obviate these difficulties. For example, digestion of a protein with a second protease to obtain sequence overlaps has been used to obtain additional information and help place a fragment in the original sequence. However, these procedures are time-consuming, require additional sample, have major limitations in themselves, and provide only specific bits of information.

We have developed a procedure to provide a 'sequence-ordered' map from a protease digest by using the overlap information that is inherent in the time-course protease digest of a protein. Fragment peptides appear at different times during the hydrolysis because all potential sites of cleavage are not equally accessible to the enzyme and the specific rates of cleavage of these sites can differ widely. The sum of the measured masses of two product peptides from a hydrolytic cleavage is related to the mass of the parent fragment by the additional mass of a water molecule plus a proton. The latter is necessary because the mass spectrometer measures the $(MH)^+$ value of the various peptides. Preliminary work involving such sequence-ordered maps for the hydrolysis of relatively low molecular weight polypeptides, less than 4000 Da, has been reported elsewhere (4,5). For the application of this procedure to proteins, we have employed laser desorption mass spectrometry (LDMS) for the identification of peptide intermediates because many of these are relatively high in molecular weight (6). We illustrate the technique of generating sequence-ordered maps using the protease clostripain to hydrolyze ribonuclease A, a protein of molecular weight of about 14,000. In addition, we have assessed the analytical utility of this approach for glycoproteins using ribonuclease B as a test sample.

II. EXPERIMENTAL

Clostripain Digests

Clostripain digests were performed in 10 mM Tris/HCl, 1 mM $CaCl_2$, 2.5 mM dithiothreitol (freshly prepared), pH 7.5 with clostripeptidase B purified from clostridium hystolyticum (Promega) at an enzyme:substrate ratio of 1:100 (w/w). Approximately 1.36 mg (95 nmol) of reduced, pyridylethyl ribonuclease A (7) was dissolved in 680 μL of the above buffer and 13.6 μL of activated enzyme solution, containing 1 μg/μL, were mixed at time zero. Aliquots of 50 μL were removed at 0, 1, 2, 4, 8, 16, 32, 51,

107 and 151 minutes, and 23 h. The aliquots were immediately mixed with 5 μL of 6 N HCl to stop the reaction.

Deglycosylation

Native ribonuclease B type III (1.26 mg) was dissolved in 0.1 M Tris pH 9.3. Peptide:N-Glycosidase F (10 units) was added and the mixture incubated at 37°C for 55 hours. The reaction was stopped using 1 M HCl and the mixture separated by HPLC with gradient elution at 200 μL/min on a reversed phase C-18 column, 2.1 mm i.d. x 22 cm. The gradient was linear from 0 to 100% B over 60 min, where solvent A was 0.1% TFA in water and B was 0.085% TFA in 70% acetonitrile. The column was maintained at 30°C, peaks were detected at 214 nm, and fractions were collected every 60 s.

Cyanogen Bromide Digestion

Reduced, pyridylethyl ribonuclease B type III (650 μg) was dissolved in 300 μL of 60% trifluoroacetic acid. Cyanogen bromide was added at 50 fold molar excess over methionine residues. The reaction mixture was flushed with nitrogen, incubated in the dark at room temperature for 24 h, and then diluted with water and lyophilized.

Mass Spectrometry

The laser desorption time-of-flight mass spectrometer was built using commercially available components (Figure 1). Light from a nitrogen laser (337 nm) was focused onto the sample target using a single fused silica lens of 50 mm focal length. Neutral density filters were used to adjust the power density to around 10^6 W/cm^2. Desorbed ions were accelerated to 20 keV energy and focused onto the detector by a two stage cylindrical extraction lens. The linear drift length was 0.5 m. The detector was a chevron microchannel plate of 40 mm active diameter. The detector signal was taken directly to a LeCroy 9450 oscilloscope and digitized at a sampling rate of 400 MHz. Spectra from multiple laser shots could be summed in the oscilloscope memory. The averaged data were transferred to a PC for storage and report generation.

Instrument operating conditions and procedures from previously published work were used (8,9). Approximately 5 μL of the reaction aliquots was diluted with 0.1% TFA to give a concentration of 10 pmol/μL, relative to the starting material. To 1 μL of this solution was added 1 μL of 5 x 10^{-2} molar sinapinic acid solution (10). The mixture was placed on a stainless steel target and dried. Spectra from multiple laser shots were summed until an acceptable signal to noise ratio was achieved, typically 30 to 60 shots at a repetition rate of 2 Hz. Sinapinic acid fragment adduct

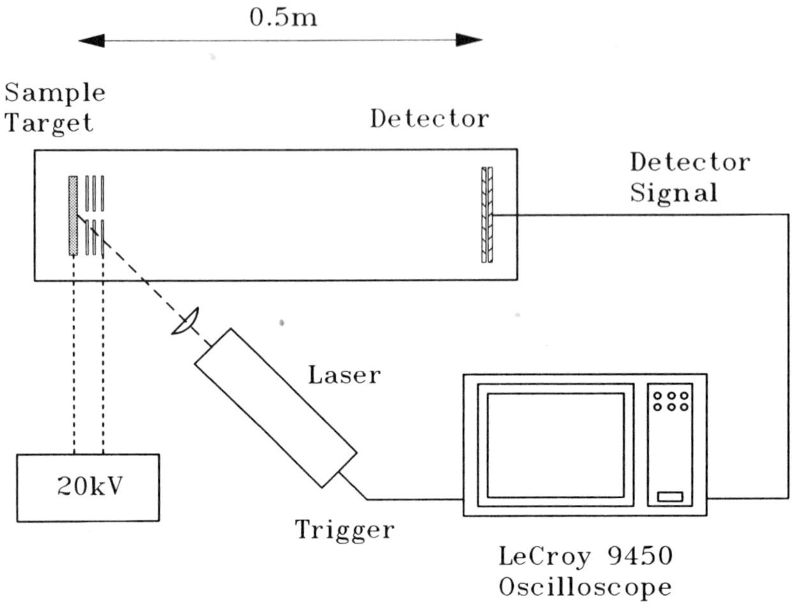

Fig. 1. LDMS instrumental setup.

peaks appear at (M + 206) mass units. The instrument was calibrated using the molecular ion species of the ribonuclease A substrate as a reference. The average mass measurement accuracy was better than 0.02%, i.e., less than 2 daltons in 10,000.

For the analysis of the low molecular weight fragment peptides (<1500 Da), 30 μL of the aliquots were pooled and fractionated by HPLC using gradient elution on a 7 μm reversed phase C18 column, 2.1 mm id x 10 cm. Solvent A: 0.1% TFA, solvent B: 70% acetonitrile containing 0.085% TFA. The gradient was linear from 0 to 100% B over 45 minutes. After concentration to a small volume, individual fractions containing peptides were analyzed by LDMS in the manner described above.

III. RESULTS AND DISCUSSION

Clostripain hydrolyzes proteins primarily on the carboxyl side of arginyl residues (11). This enzyme was used to limit the number of fragments produced in the hydrolysis reaction and to maximize the intensity of the transient intermediate fragments formed. Figure 2 shows the primary structure of ribonuclease A and the sites of cleavage of both clostripain and trypsin. Table I shows the mass spectrometric identifications of most of the expected clostripain fragments, giving the calculated average mass, the

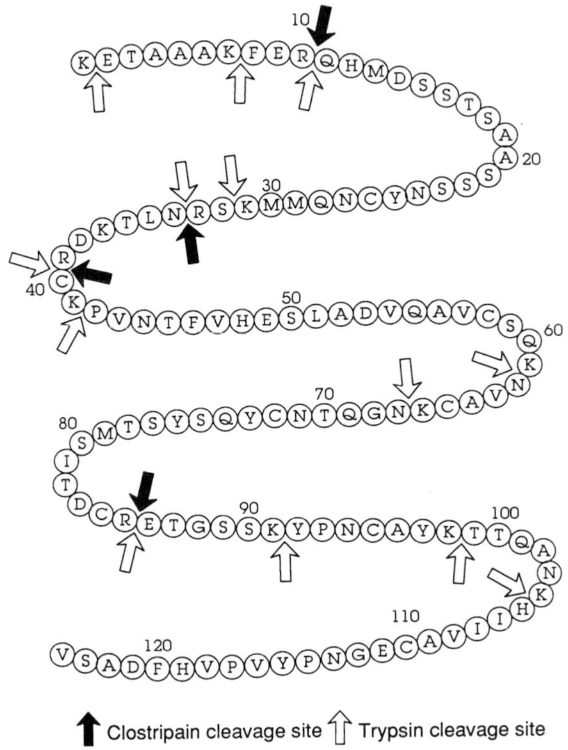

Clostripain cleavage site Trypsin cleavage site

Fig. 2. Sequence of bovine ribonuclease A showing sites of cleavage by clostripain and trypsin.

Table I. Clostripain Fragments

Peak Label	Predicted Fragment	Calc $(M+H)^+$	Measured $(M+H)^+$	% Error
A	(1-10)	1151.3	1151.1	-0.02
B	(11-33)	2657.9	2657.9	0.00
C	(34-39)	746.8	746.8	0.00
D	(40-85)	5600.4	5600.1	-0.01
E	(86-124)	4452.0	4453.7	0.04
F	(1-33)	3790.2	3790.2	0.00
G	(11-39)	3385.7	3386.3	0.02
---	(34-85)	6328.2	6325.2	-0.04
H	(40-124)	10033.4	10035.3	0.02
I	(1-39)	4518.0	4518.8	0.02
---	(11-85)	8967.1	---	---
J	(34-124)	10761.2	10763.3	0.02
---	(1-85)	10099.4	---	---
K	(11-124)	13400.1	13402.9	0.02
L	(1-124)	14532.4	14532.4	0.00

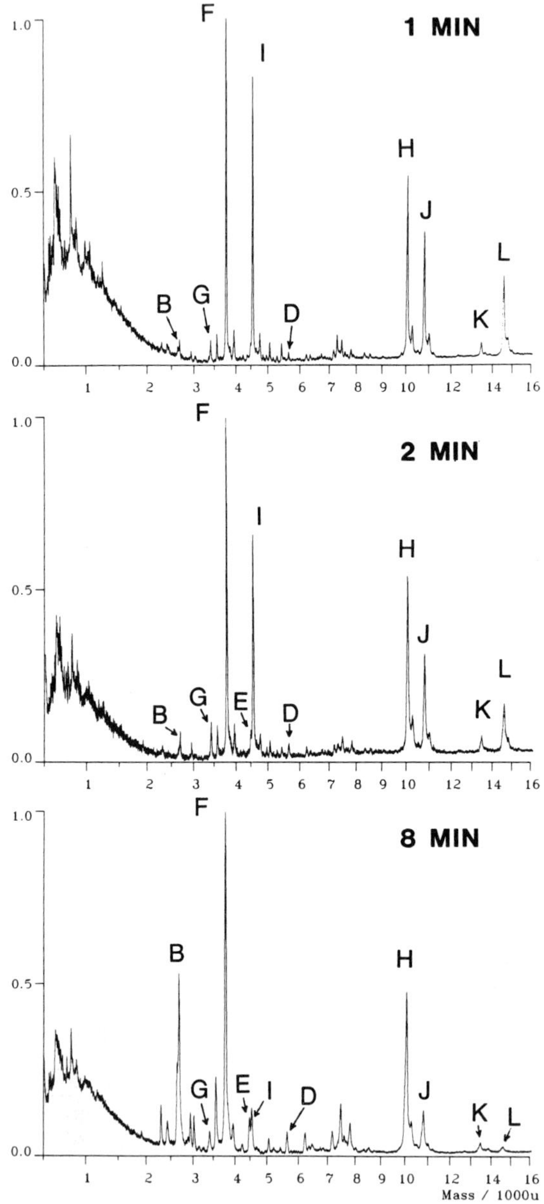

Fig. 3. Laser desorption mass spectra of time-course aliquots taken from the hydrolysis of pyridylethyl ribonuclease A with clostripain. The peak labels refer to ions listed in Table I.

measured average mass, and the percentage error of the measurements. Aliquots of the time-course reaction were analyzed directly, without removal of buffer and salt constituents. Fifteen fragments could theoretically be produced during the digest, with 5 fragments (labeled A-E in Table I) remaining at the limit digest (22 hours). The molecular species of the pyridylethyl ribonuclease A molecule is clearly seen in the early time aliquots. One can also use the mass shift of the pyridylethyl derivative, relative to the native molecule, to determine the number of cysteinyl residues in the molecule.

Peptides labeled F through L in Table I are intermediates which should be transient and not remain in the limit digest. The LDMS mass spectra of three of the reaction aliquots (1, 2 and 8 min) are given in Figure 3. The peaks in the spectra are labeled with letters listed in Table I. Seven intermediate peptide products were identified in the time-course aliquots with the fragments ranging in molecular weight from 3,384 to 13,403. The three intermediate peptides not seen in the reaction aliquots all involved potential hydrolysis of arginyl residue 85, apparently a bond which is only slowly hydrolyzed. It is probable that the pyridylethyl cysteine in the N-terminal position adjacent to residue 85 is sufficiently bulky to inhibit hydrolysis of this bond. Peptides B, D, and E, three of the five limit digest peptides, were detected by LDMS directly in the reaction aliquots after 2 min of reaction, but the two smaller fragments, having molecular weights of 746 and 1,151, were not observed. A pooled sample of the time-course aliquots that was fractionated by HPLC and analyzed by LDMS was found to contain peptides A and C, as shown in Figure 4. Although it is uncertain precisely why these two small peptides were not observed by direct analysis of the aliquots, a probable explanation would be a non-uniform distribution of these fragments in the sample drying process. Although ion suppression effects by other peptides are possible, we feel it is unlikely in this case because no other significant ion suppression phenomena were observed in these analyses where over forty peptide ions were recorded.

The clostripain fragments identified in the time-course analysis are shown in the top portion of Figure 5a listed according to size. A PC computer program was used to find appropriate matches in these data, i.e., in the hydrolysis of a given bond, the sum of the masses of the MH^+ ions of the two peptide products, minus 19 mass units for a water molecule plus a proton, equals that of the precursor peptide. This program takes into account uncertainties in mass measurement accuracy so that matches can be performed for a small mass window and not just for the discrete value measured. Usually, the mass measurement accuracy is set at 0.02% for the purposes of these calculations. Figure 5b lists eleven such correlations made in this manner, i.e., the direct precursor/ product correlations which could be produced from the various molecular ions identified. Although it is possible that mathematically, given the mass range allowed for each peptide resulting from the mass measurement inaccuracy, a fit might be

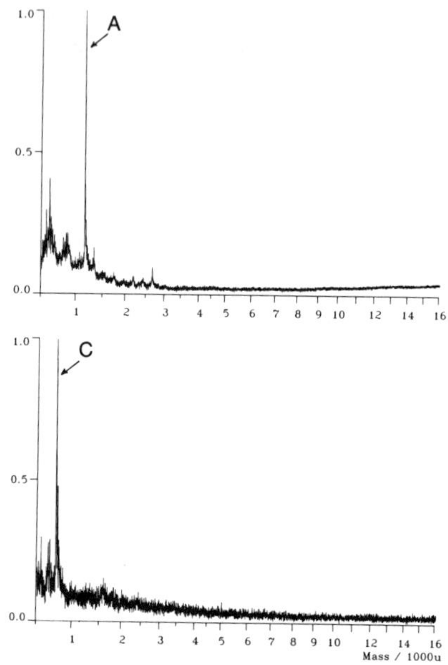

Fig. 4. Laser desorption mass spectra of low mass peptides from the clostripain hydrolysis of pyridylethyl ribonuclease A. Sample aliquots were purified by HPLC prior to analysis. The peak labels refer to the peptides listed in Table I.

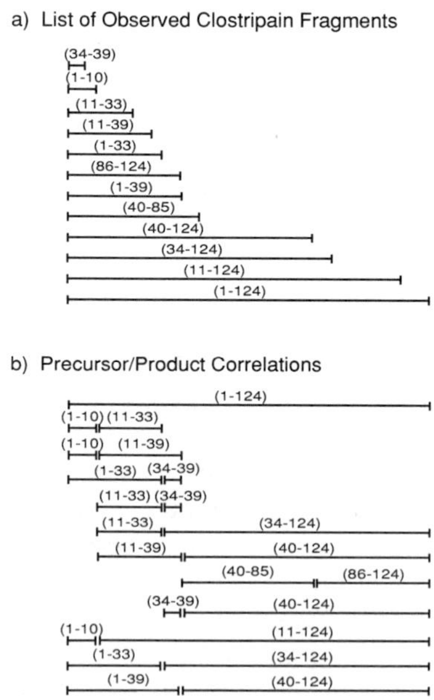

Fig. 5. a) Peptide fragments identified throughout the time-course reaction by LDMS. b) Precursor/product correlations constructed from data in (a).

obtained from two fragments not adjacent in the sequence, there are many redundancies which should make such an error obvious. In the analysis performed with ribonuclease A, however, only correct matches were made. In addition, given a list of peptides identified during the hydrolysis, one could look for fits of three, four, five, or more groups of peptides, e.g., do the sum of any three, four, or five fragments minus 38, 57, or 76 mass units, respectively, equal the mass of another peptide. In this example, one finds that a five fragment fit adds up to the molecular weight of the original protein and no combination of six or more fragments provides a fit. The sequence-order is specified by the correlations shown in Figure 5b. It should be emphasized that in making these correlations, their reliability is directly dependent on the mass measurement accuracy.

Rate curves can be produced from the LDMS data when the relative intensities of the various fragments are plotted as a function of time. This is a valuable aspect of this technique that has not been previously demonstrated. For example, Figure 6 shows the rate curves obtained from the ribonuclease A digest, with ion intensities expressed as a ratio of the total fragment ion current in each spectrum. This normalization method was used because no internal standard was included in the original digest since quantitative analysis was not required for the mapping procedure. The examples of rate curves shown in Figure 6 have the expected profiles; the

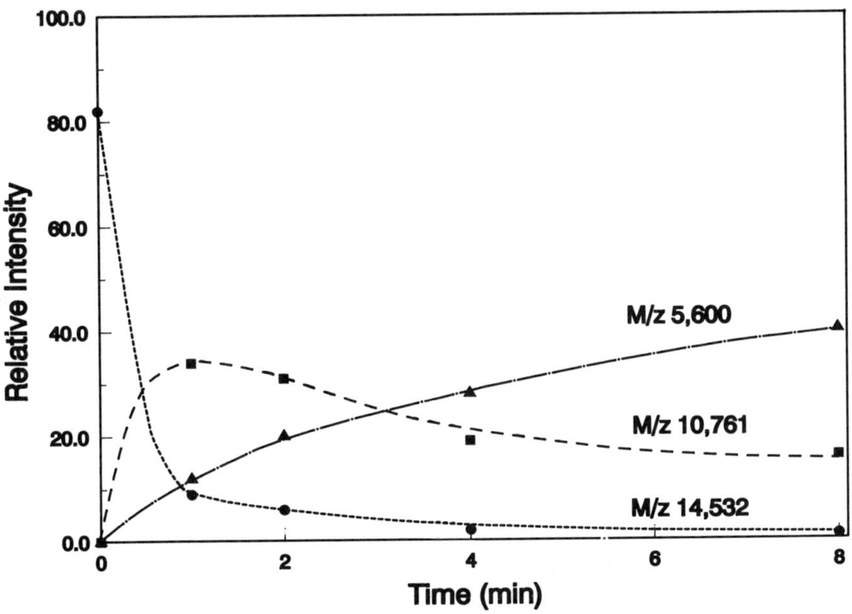

Fig. 6. Rate curves from the time-course LDMS analysis of the hydrolysis of pyridylethyl ribonuclease A with clostripain. The ion at m/z 14,532 is the MH^+ of the original protein; m/z 10,761, an intermediate (residues 34-124); m/z 5600, a final product peptide (residues 40-85).

exponential elimination of the molecular species at m/z 14,532, the rise and subsequent fall of peptide fragment intermediates (m/z 10,761), and the gradual increase in the limit digest fragments (m/z 4,452). Of course, these curves are complex in that they are the net result of a combination of different rate constants.

The construction of a sequence-ordered protease fragment map for ribonuclease A is an example of the verification of a known sequence. In the case of an unknown, the same procedure would be used to identify protein fragments generated by a protease. However, when the fragments are arranged in sequence order, one would not know the reading direction of the fragment map from these data alone, i.e., one would generate two possible sequence-ordered maps that are the reverse order of each other. Identification of the N-terminal or C-terminal fragments, or other single bit of sequence information that identifies the reading direction is required. For example, one approach used over the years would be to perform the protease digestion in oxygen-18 enriched water; only the C-terminal peptide will be unlabeled (12). Assuming a mass measurement accuracy of 0.02% and the average limit digest peptide fragment of 2,000 to 4,000 daltons, the incorporation of oxygen-18 label could be identified in the C-terminal fragment.

The clostripain preparation used also has significant trypsin activity, giving rise to many tryptic fragments during the analyses. In the digest of the ribonuclease A sample given above, thirty tryptic fragments in addition to the eleven clostripain fragments were identified in the analyses of the digest aliquot taken at 151 min. A number of these are apparent in the mass spectrum of the aliquot taken from the 8 min digest shown in Figure 3, increasing in intensity as the digest proceeded. Sequence correlations can also be made from these data, although these were not needed in the present example. Of course, one of the difficulties with using tryptic fragments of a protein is the large number of fragments generated and their relatively low molecular weight. Correlations of appearance of pairs of fragments with time would become much more important, necessitating drawing and analyzing more sample aliquots. In addition, within the variables of mass measurement accuracy, usually about 0.02%, many incorrect combinations and fragment arrangements could fit mathematically. For these reasons, enzymes such as clostripain, endo-Lys C, or other specific enzymes which produce larger fragments are preferred for this application.

An interesting and important aspect of the LDMS analysis is that ion suppression effects do not appear to constitute a major problem. Suppression effects, major disadvantages found with the use of fast atom bombardment ionization, Cf-252 plasma desorption, and other desorption ionization techniques, were not apparent in the spectra obtained here. This lack of significant ion suppression is also clearly seen in the rate curves shown in Figure 6. Figure 3 shows relatively complex mixtures of both tryptic and clostripain digest fragments in which most if not all of the expected peptides are observed. This observation is consistent with other reports of the analysis of mixtures of proteins by LDMS (13). However, it is too early to state in any quantitative sense and with any certainty that ion suppression effects do not occur in such analyses. In the data presented here, only qualitative identifications were necessary and it is not known whether the ion intensities recorded were a true reflection of the concentrations of the peptides in the digest sample solution.

We have also performed preliminary studies with glycoproteins to determine whether these compounds can be analyzed by LDMS in amounts similar to that of unglycosylated proteins and whether the heterogeneity that is usually present in the carbohydrate portion of the compound will present special analytical difficulties. In the LDMS analysis of native ribonuclease B, the mass spectrum (Figure 7a) shows an intense peak at m/z 13,683 corresponding to the non-glycosylated molecule and a cluster of peaks starting at m/z 14,900 corresponding to the glycosylated molecule. The mass difference of 1217 Da is consistent with the known glycosylation of Asn-34 with the carbohydrate moiety having the composition of five

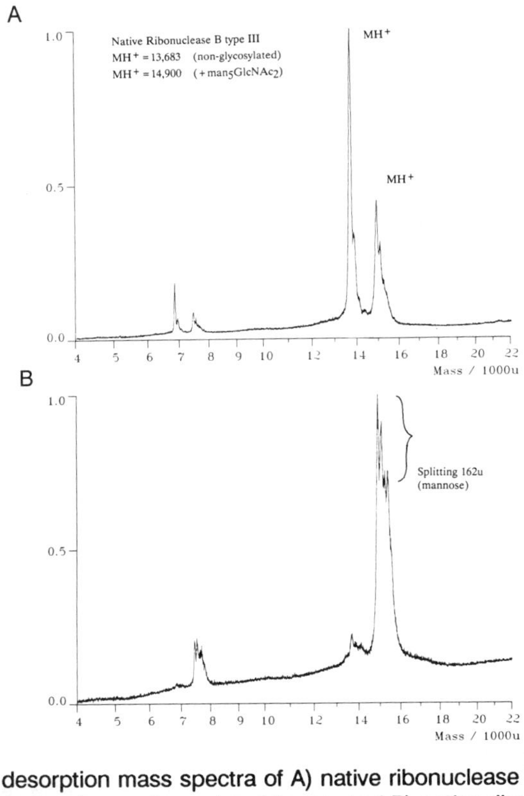

Fig. 7. Laser desorption mass spectra of A) native ribonuclease B type III, showing both the glycosylated and non-glycosylated forms, and B) native ribonuclease B type III after purification by HPLC.

mannosyl (man) and two N-acetylglucosaminyl (GlcNAc) residues. Doubly-charged ions of these molecules can be seen in the lower mass range. Figure 7b shows the mass spectrum of native ribonuclease B that had been purified by HPLC. Here, the splitting of the molecular species can be seen with each peak being 162 mass units apart, the mass of a hexose residue.

Two additional strategies were used to further characterize and corroborate the structure of the carbohydrate moiety. First, the protein was treated with glycosidase F, an enzyme which cleaves between the Asn side chain and the first GlcNAc residue. The LDMS spectrum is shown in Figure 8a. The major peaks in the spectrum correspond to the MH^+ ions of the unglycosylated molecule and that for a molecule having one and two GlcNAc residues. From this pattern, it is apparent that the enzyme preparation contained some endo H or other similar enzymic activity. The second approach in characterizing the glycosylated material would be to perform a proteolytic or cyanogen bromide (CNBr) reaction in order to examine smaller fragments. The latter was performed on the reduced, pyridylethyl ribonuclease B molecule and the mass spectrum of the reaction products, purified by HPLC, is shown in Figure 8b. The peptide of interest

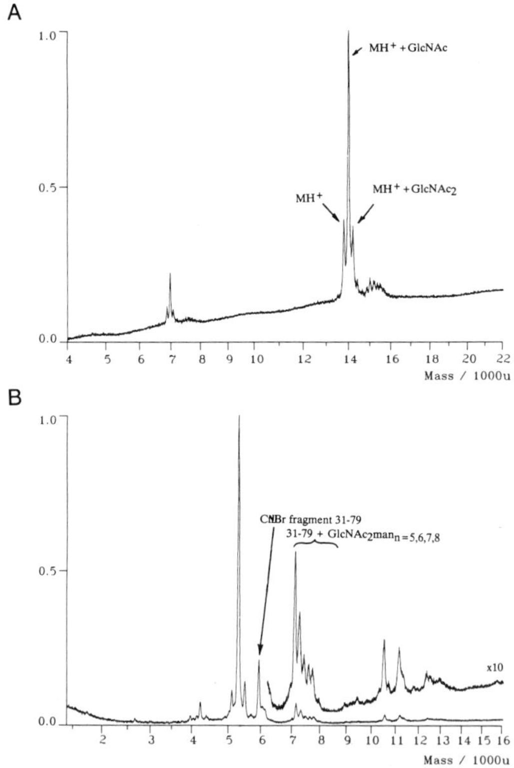

Fig. 8. Laser desorption mass spectra of pyridylethyl ribonuclease B type III A) treated with glycosidase F, and B) peptide fragments generated with cyanogen bromide treatment.

is fragment 31-79 because it contains glycosylated Asn-34. The mass spectrum clearly shows the unglycosylated fragment 31-79 and the glycosylated analog, with splitting indicating the carbohydrate moiety is $(GlcNAc)_2(Man)_n$ where n equals 5-8 residues.

IV. CONCLUSIONS

The ability to construct sequence-ordered protease digest maps is the result of inherent sequence overlap information present in the transient peptide intermediates generated by proteases over the time-course of the reaction and the capability of mass spectrometry to accurately identify the masses of these fragments. This allows one to generate a great number of product/precursor correlations and thereby deduce the sequence order of the fragments in a nearest neighbor type of analysis. Such a map provides unique information in sequence-checking proteins using a technique that requires about a picomole of sample, is easy to use in that it requires almost no sample preparation, and performs the analyses in several minutes. LDMS was also of great value in investigating the structure of glycoproteins. In combination with enzymic cleavage of the carbohydrate side chains and proteolytic or chemical cleavage of the peptide backbone, the composition of these side chains can be deduced.

V. ACKNOWLEDGEMENTS

The authors thank Paul Caprioli for writing the PC program for making precursor/product matches.

VI. REFERENCES

1. Morris, H.R., Parcico, M., Taylor, G.W., *Biochem. Biophys. Res. Commun.*, (1983), **117**, 299.
2. Carr, S.A., Biemann, K., *Methods in Enzymol.*,(1984), **106**, 29.
3. Biemann, K., Martin, F.A., *Mass Spectrom. Rev.*, (1987), **6**, 1.
4. Whaley, B.S., Caprioli, R.M., Proc. 37th ASMS Conference on Mass Spectrom. and Allied Topics, May 21-26 (1989), Miami Beach, FL, p. 722.
5. Whaley, B.S., Caprioli, R.M., *Biological Mass Spectrometry*, (1991), in press.
6. Whaley, B.S., Caprioli, R.M., Mock, K.K., Cottrell, J.S., presented at the 4th Symposium of the Protein Society, Aug. 11-15, (1990), San Diego, CA.
7. Raftery, M.A., Cole, R.D., *Methods in Enxymol.*, **11**, (1967), 199.

8. Karas, M., Bachmann, D., Bahr, U., Hillenkamp, F., *Intl. J. Mass Spectrom. Ion Phys.*, **78**, (1987), 53.
9. Karas, M., Bachmann, D., Bahr, U., Hillenkamp, F., *Intl. J. Mass Spectrom. Ion Phys.*, **92**, (1989), 53.
10. Beavis, R.C., Chait, B.T., *Rapid Commun. Mass Spectrom.*, **3**, (1989), 432.
11. Mitchell, W.M., Harrington, W.F., *Methods in Enzymol.*, **19**, (1970), 635.
12. Rose, K., Simona, M.G., Offord, R.E., Prior, C.P., Otto, B., Thatcher, D.R., *Biochem. J.* **215**, (1983), 273.
13. Chait, B.T., Beavis, R.C., presented at the Fourth Symposium of the Protein Society, August 11-15 (1990), San Diego, CA.

49

MONITORING PEPTIDE SYNTHESIS STEPWISE BY MASS SPECTROMETRY

William T. Moore and Richard M. Caprioli

The Analytical Chemistry Center and
the Department of Biochemistry and Molecular Biology

The University of Texas Medical School, Houston, Texas 77030

I. INTRODUCTION

Automated peptide syntheses may run into problems for numerous reasons: instrument malfunction, reagent quality, sequence specific difficulties (1), innate limitations of chosen chemistries (2) and even operator error. A successful synthesis is often compromised by the resultant of many of these factors. Thus, there is a real need to evaluate the progression of a synthesis "stepwise" for appropriate fine tuning of a synthetic strategy so that the amount of the desired product is maximized. A preferable method would permit the accurate detection of synthetic products obtained after each amino acid addition cycle. Such a method would be applicable while the synthesis is in progress or at least permit a "stepwise" retrospective analysis. It would also be able to detect the presence of by-products that could arise from deletion, insertion, termination, branching, or any other chemical modification reaction that may occur. Finally, an appropriate evaluation method should detect incomplete deblocking reactions and chemical modification reactions that may occur upon acidolytic cleavage of the peptide from resin, the final stage chemistry of peptide synthesis.

Existing monitoring methodologies involve several different schemes. Most rapid methods are indirect, exploiting either a reagent uptake (3) or product formation that requires potentiometric (4), spectrophotometric (5), or radiometric assay (6). The quantity of unreacted amino component is usually the only property assessed and may give an incomplete picture of

Techniques in Protein Chemistry II

a particular stage of synthesis. A slower but more thorough method, preview sequencing (7,8), has been developed to provide a more direct analysis of the progression of a synthesis. This method requires the access to an automated protein sequencer and the chromatographic resolution and identification of the various PTH unblocked and blocked amino acid derivatives. This method is definitive but constitutes a lengthy analysis that does not lend itself as a "real synthetic time" assay system.

Mass spectrometry (MS), with fast atom bombardment (FAB) (9,10), ^{252}Cf-plasma desorption (PD) (11,12), laser desorption (LD) (13,14) and electrospray (ES) (15,16) ionization methods, is an attractive analytical technology for the assessment of peptide and protein structure. The main feature that makes a mass spectrometric approach attractive is the ability to determine the molecular weight of a peptide at an accuracy approaching 1 amu. Therefore amino acid deletions, insertions, chemical modifications can be readily deduced. In addition, the short time frame for the analysis is such that an ongoing synthesis could be halted or altered in progress.

In the present study, we use continuous-flow fast atom bombardment (CF-FAB) mass spectrometry (17) as the method to evaluate the progress of a peptide synthesis using the Boc-benzyl strategy (18). This type of analysis is direct and permits a stepwise assessment. To accomplish this it was necessary to develop a micro-TFMSA cleavage and deblocking method which could release sufficient peptide for analysis from 1 mg of peptide-resin. Approximately 10 mg of this resin is automatically removed from the reaction vessel at the completion of each amino acid cycle. Two examples of single run, single peptide syntheses are presented to illustrate the consequences of using the micro-TFMSA/CF-FABMS technique.

The CF-FAB MS capability permitted the assessment of more complex synthetic problems using the automated synthetic instrumentation. We attempted "simultaneous multipeptide synthesis" in an automated instrument designed to conduct the synthesis of one peptide at time. CF-FAB MS when used together with other techniques such as the micro-TFMSA chemistry, reverse phase high performance liquid chromatography (LC), and capillary electrophoresis (CE), permitted the evaluation of the production and characterization of two different runs constituting synthesis of four and six component mixtures of singly substituted angiotensin II analogs.

II. EXPERIMENTAL

SINGLE PEPTIDE SYNTHESIS:

All peptides were synthesized using an Applied Biosystems Inc. (ABI) Model 430A peptide synthesizer. The starting material was a BOC-C-terminal amino acid-OCH_2-Pam-copoly(styrene-1%-divinylbenzene) resin obtained from ABI. The standard ABI programmed BOC protecting group strategies and DCC derived preformed symmetrical anhydride couplings and HOBT-ester couplings that are similar to those developed by Merrifield et al. (19) were used. Approximately 10 mg peptide-resin aliquots were automatically collected after each coupling. All BOC amino acids were obtained from Peninsula Laboratories Inc. The side chain protecting groups were: Bzl (Ser & Thr), Br-Z (Tyr), Cl-Z (Lys), CBZ (His), 4-CH_3BzL (Cys), CHO (Trp), OBzl (Asp & Glu), Tos (Arg). All of the peptide synthesis chemicals were from ABI except trifluoroacetic acid (Halocarbon Products Corp., Hakensack, NJ), DCM and DMF (Burdick and Jackson).

MULTIPLE PEPTIDE SYNTHESIS:

The analog mixtures of Angiotensin II variable at a single residue position were prepared in a single synthetic run using small-scale (0.1 mmole) rapid cycles (system software version 1.40). For the cycle responsible for the multiple residue incorporation, equimolar amounts of vessel cycle compatible t-BOC amino acids were loaded into a single BOC-amino acid vial. The particular mixture of BOC-amino acid's chosen was dictated by each compound having identical or similar activator, concentrator, and reaction vessel programs written for small scale rapid cycle syntheses. BOC amino acids and run file opt 21 r vessel cycles were matched as follows: A, C, D, E, F, G, H, I, M, S, T, & V - rboc21r cboc21 aboc21; K, L, P, W, & Y - rboc21r cboc22 aboc22; and N, R, & Q - rboc22r cboc23 aboc24.

MICRO-TFMSA CLEAVAGE:

The micro-TFMSA cleavage method is an adaptation of a method described by ABI (20). One mg of peptide-resin is transferred as a DCM/MeOH (60:40) slurry to a 0.5 mL capacity polypropylene microcentrifuge tube (Fisher brand 05 406-7) and dried in a Speed-vac centrifuge. Ten μL of a scavenger mixture consisting of thioanisole:ethanedithiol (2:1) is added followed by 40 μL of trifluoroacetic acid and lastly 15 μL of trifluromethanesulfonic acid. The μL volumes were delivered using a Pipetman pipettor equipped with 83 mm long polypropylene pipet tips. Immediately after the addition of TFMSA, the reaction mixture becomes burgundy in color and fades to colorless or pale yellow within 5 minutes. The reaction is allowed to proceed for 15 min at room temperature with intermittent vortexing. The reaction mixture is placed in crushed ice for 5 minutes and then 200 μL of cold diethyl ether is added to precipitate cleaved and deprotected peptide. The precipitate is concentrated by a 2-3 second spin in a microfuge and the pellet is extracted one time by vortex resuspension in an additional 300 μL of cold diethyl ether. The precipitate is again pelleted and the precipitated peptide is separated from residual resin beads by dissolving the peptide in a suitable solvent (usually 100 μL of 0.1% TFA). An additional centrifugation may be necessary to clarify the solution. In those cases where peptide is not immediately precipitable by the addition of 200 μL of diethyl ether, e.g., in the case of either a short peptide less than 10 residues in length or a very hydrophobic peptide, 10 μL of the ether-reaction mixture is added to 100 μL of 0.1% TFA and vortexed. Peptide extracted in this way in the presence of diluted organic reactants is also suitable for direct analysis.

Trifluoromethanesulfonic acid (15,853-4), thioanisole (T2,800-2), and ethanedithiol (E360-0) were obtained from Aldrich Chemical Co Inc. (Milwaukee, WI). Trifluoroacetic acid was obtained from Halocarbon Products Corp. (Hackensack, NJ) and ethyl ether was Fisher certified ACS grade.

CONTINUOUS-FLOW - FAST ATOM BOMBARDMENT MASS SPECTROMETRY:

CF-FAB MS was performed with a Finnigan MAT 90 high resolution instrument equipped with an ion gun using Xe as the bombarding atom. Samples were introduced by the flow-injection technique. The carrier solution, consisting of 5% glycerol and 3% acetonitrile in water, was pumped into the ion source of the mass spectrometer at a flow rate of 5 μL/min. When necessary for signal enhancement 9 parts of sample were mixed with 1 part of a 1:1 mixture of glycerol and thioglycerol before injection into the continuous flow probe stream. The mass spectrometer was generally operated at a resolving power of 1500 and all mass ranges were scanned at approximately 5-10 sec/scan. CsI was used to calibrate the instrument.

FULL-BORE HPLC:

Mixtures of synthetic peptides were separated and analyzed by conventional (full bore) HPLC using a reverse phase, 4.6 X 30 mm C-8 Aquapore RP300 (7 μm) column (ABI). The column was eluted with a linear acetonitrile gradient of 0% B to 100% B developed over 30 min. Solvent A was 0.1% TFA in water and solvent B, 70% acetonitrile, 30% water, and 0.085% TFA. An ABI model 151A HPLC system was used at a flow rate of 1 mL/min with UV detection at 230 nm.

MICROBORE HPLC:

A microbore Model 130A Separation System (ABI) was employed for CF-FAB LC/MS analyses using a 1 X 50 mm Aquapore RP-300 (7 μm) C-8 bonded reverse phase column as described previously (21). A linear gradient from 0% B to 100% B was developed at a flow rate of 0.5 mL/min.

This flow rate was split 1:19 to provide a 25 μL/min flow rate through the column. The post column 1:4 split permitted a 5 μL/min flow into the mass spectrometer and 20 μL/min into the UV detector (230 nm). Mobile phases consisted of acetonitrile / glycerol / water / TFA: Solvent A - 0/5/95/0.1 (v/v); Solvent B -60/5/35/0.1 (v/v).

CAPILLARY ELECTROPHORESIS:

Capillary electrophoresis (CE) coupled to CF-FABMS was performed as described previously (22). CE was conducted in a 90 cm x 50 μm (ID) capillary with UV detection at 215 nm located 60 cm away from the anode compartment. The cathodic end of the capillary was placed in a liquid junction interface that we developed for this kind of analysis. The interface was a 1 in. X 1 in. plexiglass block consisting of four ports and two 90 degree intersecting pathways (1/16 in. ID). Two of the opposing ports are inlets for the CE capillary (50 um ID) and the CF-FABMS capillary (75 um ID, 85cm long). The two capillary ends meet juxtaposed in a Teflon tubing sleeve (0.15 mm ID). Cathodic compartment solvent, which is also the CF-FABMS matrix, is introduced into the interface through the other two opposing ports. One port has a solvent filled syringe attached via a short segment of stainless steel tubing that serves as the electrode (cathode) for attachment of the ground lead from the power supply. The opposing port is attached to a length of tubing that is introduced into a beaker containing cathodic solvent. This arrangement permits the syringe to serve the dual purpose of a flushing device to remove bubbles that may form in the interface during electrophoresis and a sample loading device. A 2 mL draw on the syringe permits a 30 nL injection volume to be taken up into the anode end of the CE capillary when a sample vial is in place. After introducing sample, the anodic end of the capillary is transferred to electrophoresis buffer and high voltage (20 KV) is applied. The difference in flow rate within the CF-FABMS capillary (5 μL/min) derived by atmospheric pressure and the electrophoretically derived flow rate (about 0.1 μL/min) in the CE capillary permits efficient transfer of CE effluent to the mass spectrometer. The CE buffer (anode compartment solvent) was 40 mM ammonium citrate (pH 3.6) and the CF-FABMS solvent, which is also cathode compartment buffer was 5% glycerol, 3% acetonitrile, 2.4 mM acetic acid (pH 3.7).

III. RESULTS

MICRO-TFMSA/CF-FABMS ANALYSIS:

Two examples are given to demonstrate the capabilities of the micro-TFMSA/CF-FABMS method to assess stepwise the progress of an

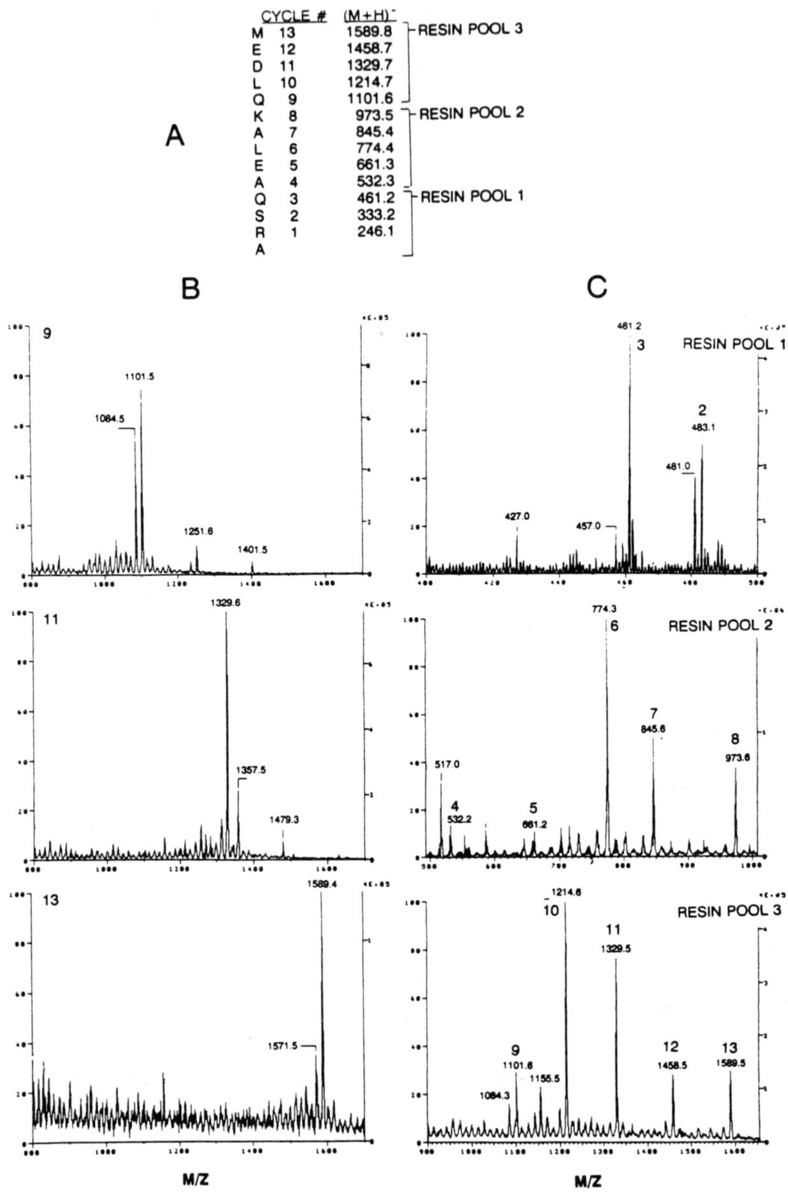

Fig. 1. Stepwise assessment of the automated solid phase synthesis of a 14 residue peptide (example I) by micro-TFMSA/CF-FAB MS analysis. Panel A - the peptide sequence, synthetic cycle numbers, and the $(M+H)^+$ values for the expected products. Panel B - the mass spectra of the products derived from cycles 9, 11, and 13. Panel B - the respective mass spectra obtained from the analysis of the three peptide-resin pools indicated in Panel A. The larger bold-faced numbers above the ion signals represent cycle numbers. The smaller numbers represent observed mass assignments.

automated peptide synthesis. The first is the synthesis of a 14 residue peptide MEDLQKALEAQSRA (peptide I) and the second the synthesis of a 46 residue peptide, KIQFTVPLLEPHLDPEAAEQIRRRR-PTPATLVLTSDQS-SPEVDEDR (peptide II).

Evaluation of peptide I began at cycle 8 after the C-terminal half of the peptide had been synthesized, i.e., on the N-terminal half of the peptide. Figure 1A shows the sequence and the expected $(M+H)^+$ values for the peptide product obtained after each amino acid addition cycle. Figure 1B indicates selected mass spectra for micro-TFMSA derived products obtained after cycles 9, 11, and 13. The results indicate that, generally, the synthesis proceeded very well, all expected masses were observed at each step. The cycle 8 spectrum alone (data not shown) indicated that the early and often the most critical part of the synthesis (23) was successful. The spectrum from cycle 9 was the only one showing a major unexpected signal and this was 17 amu less than the desired product. This was interpreted to be the result of the deamidation and cyclization of the N-terminal glutaminyl residue to form pyroglutamic acid. The unique existence of this signal in the cycle 9 spectrum indicates that the pyroglutamic acid formation occurred as a consequence of micro-TFMSA cleavage method and did not occur during the automated synthesis.

A retrospective synthetic history of the synthesis of peptide I can be defined using a slightly different procedure. Three resin pools were prepared containing mixtures of 4 or 5 peptide-resin aliquots each (Figure 1A). The three peptide-resin pools representing the early, middle, and final synthetic steps were processed and analyzed using the micro-TFMSA/CF-FABMS method. As shown in Fig. 1C, the mass spectra from the three samples contain all the expected ion signals from every step of synthesis except cycle 1. (The $(M+H)^+$ value for the product of cycle 1 fell outside the mass range analyzed). Cycle 2 product was detected as a TFMSA adduct (483.1 = 333.2 + 150, TFMSA m/z) because it contained Arg. This occurs when excess TFMSA is present in the sample, a condition encountered when the TFMSA hydrolyzed product is not precipitable or extractable, as in the case of resin pool 1 product.

Since the analysis of the peptide-resin pools is stepwise and "time based", the position of a deletion could have been unambiguously and quickly located. The premise is that any deletion would be detected as a mass shift in descendants derived only from the cycle where the event occurred. In contrast, although the analysis of the final peptide produce may indicate that a deletion has occurred, the specific location of the deletion can not be defined if that residue appears more than once in the original intended product.

A second example gives the results obtained from a more challenging single peptide synthesis. Figure 2 (upper panel) shows the desired sequence and expected $(M+H)^+$ values for some of the products derived from the 46 residue peptide (peptide II) synthesis. The synthesis of this

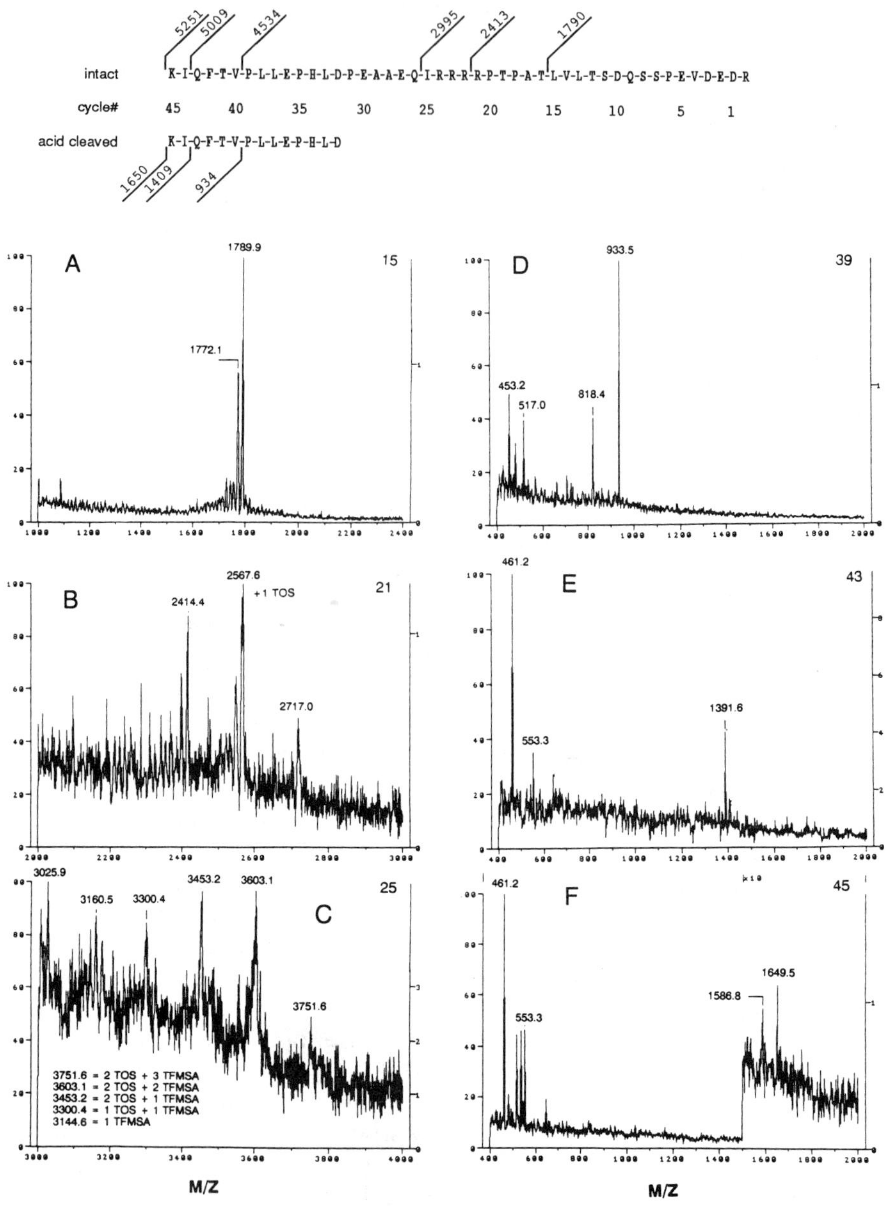

Fig. 2. Stepwise MS analysis of the synthesis of a 46 residue peptide (example II). Upper Panel - the amino acid sequence for the intact peptide and the mild acid hydrolysis generated fragment and the expected masses of the products obtained for the indicated cycle numbers. Lower Panels A, B, and C - micro-TFMSA/CF-FAB MS derived spectra for synthetic cycles 15, 21, and 25 respectively. Lower Panels D, E, and F - micro-TFMSA/CF-FABMS derived mass spectra for cycle products subjected to mild acid hydrolysis (cycles 39, 43, and 45, respectively).

peptide was interrupted at cycle 31 for removal of 30% of the peptide-resin from the reaction vessel. The peptide-resin was treated with HF to provide an N-terminal truncated derivative of the peptide that would be representative of the first major part of the synthesis and have a mass within the sensitive range of the mass spectrometer. Cycle 31 was also chosen on account of the mild acid labile Asp-Pro bond located at that position, a condition that was later exploited to help in further analyses. The CF-FABMS spectrum for the HF-derived cycle 31 product (data not shown) showed a strong signal for the expected product. The synthesis was continued using the remaining 70% portion of the peptide-resin for the final 14 cycles. Although peptide-resins were obtained at the end of each synthetic cycle, stepwise analysis was performed after cycles 7, 11, 15, 19, 21, and 25 to minimize processing effort. The mass spectra obtained from samples of cycles 15, 21 and 25 are presented in Figure 2. The observed molecular ion signals represent either the free peptide (e.g., cycle 15, Figure 2, panel A), or that of an adduct or incompletely de-blocked form of the peptide (e.g., Figure 2, cycles 21, panel B, and 25, panel C). The latter mass spectra indicate that TFMSA adducts and incomplete deblocking becomes problematic when Arg is encountered in the synthesis. Since Arg is the C-terminus and the cycle 7 product (data not shown) was not ether precipitable, excess TFMSA in the sample led to adduct formation (m/z 946 + 150 = m/z 1096) here also. The next three products examined from cycles 11, 15, and 19 were ether precipitable and extractable and produced the expected ions without adduct formation (only cycle 15 is presented in Figure 2).

As the peptide chain approached cycle 21 where the first of four adjacent Arg residues is encountered, the mass spectrometric results become complicated. The mass spectrum from cycle 21 (Figure 2, panel B) indicates the expected product as well as a tosylated derivative (BocArg(TOS) was used in the synthesis). Although the Tosyl group was efficiently removed by HF treatment, cleavage is known to be incomplete with TFMSA treatment (24). The mass spectrum for the product derived after the completion of cycle 25 is shown in Fig. 2, (panel C) and shows a complex molecular pattern. The desired molecular ion was detected as a weak signal at m/z 2994.2 in another mass range (data not shown), as well as multiple peaks due to tosyl derivatives and TFMSA adducts. The $(M+H)^+$ values for the various tosyl derivative and TFMSA adduct combinations are listed below the spectrum in panel C. However, these data indicate that peptide assembly was correct up to that point. The presence of these peptide derivatives, even though the peptide was ether precipitable and extractable, suggests that longer peptides require a more vigorous TFMSA hydrolysis and clean-up procedure. Because of this and the fact that the sensitivity of FAB mass spectrometry decreases substantially beyond m/z 3000-4000, continuation of the evaluation of the peptide II synthesis by this approach alone was halted. However, validation of correct

A

DRXYIHPF	$(M+H)^+$
-------Q------------	1075.5
-------N------------	1061.5
-------Q---------- (-Ile)	962.4
-------N---------- (-Ile)	948.4

B

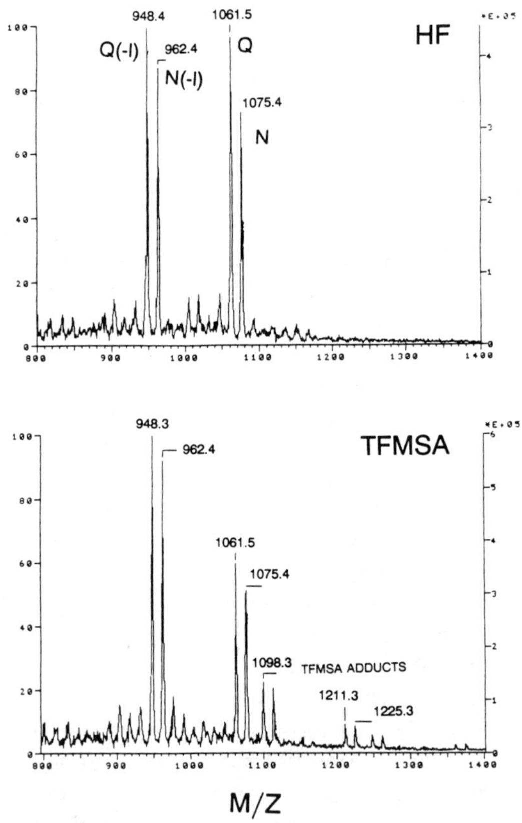

Fig. 3. CF-FABMS analysis the Angiotensin II analog mixture containing four products all derived from a single automated run. Panel A - sequences and expected masses of the product mixture. Panel B (upper) - CF-FABMS derived mass spectrum derived from HF derived final product. Panel B (lower) - the spectrum derived from the micro-TFMSA method.

synthesis up to cycle 31 was established by CF-FABMS analysis of the HF-derived peptide.

Evaluation of the synthesis of the peptide beyond step 31 using the micro-TFMSA/CF-FABMS technique was accomplished by taking advantage of an Asp-Pro bond present at the cycle 31 region (see Figure 2). The Asp-Pro bond is susceptible to mild acid hydrolysis (25), permitting the release of N-terminal peptides devoid of the problematic Arg region. The expected masses of the N-terminal products are shown in Figure 2. Aliquots of micro-TFMSA derived peptides were adjusted to 1.75M TFA and heated for 1 hr at 80°C. The mass spectra of products derived from cycles 39, 43, and 45 are shown in Figure 2 (panels D, E, and F respectively). The major observed ions present in the spectra for these cycles indicate that synthesis proceeded well up through cycle 43. The ion observed 115 amu less than the major ions in cycle 39, which was also seen in spectra for 37 and 41 (data not shown), is curious because they suggest that an Asp deletion product may have formed. Since the presence of an Asp deletion peptide in this way is unlikely, the origin of these peaks is unknown. An ion signal in the spectrum for cycle 43 is 17 amu less than the expected ion and most likely the result of formation pyro-glutamic acid from the N-terminal Gln residue. It is believed that the pyro-Glu formation occurred during either TFMSA cleavage or FABMS analysis, since this product was unique to step 43 and was not observed in the last cycle #45. The spectra from the last few cycles (i.e., 43-45) are weak, and further characterization is needed for clarification of the yield and correctness of assembly for these residues.

SIMULTANEOUS MULTIPEPTIDE SYNTHESES:

CF-FABMS alone, micro-TFMSA/CF-FABMS, and CF-FABMS coupled to fractionation methods permitted the evaluation of several quite complex mixtures of synthetic peptides. We used these techniques to assess the simultaneous synthesis of several peptides in a single automated peptide synthesis run. In one example, the simultaneous synthesis of angiotensin II analogs was performed to generate multisubstrates in protease substrate-structure function studies (26).

Four derivatives of angiotensin II (Figure 3A) were prepared by loading an equimolar mixture Boc-GLN and Boc-ASN in the BOC vial for cycle 5. Although not intended, the coupling of Ile to His was compromised, resulting in two additional products with the Ile deleted. The CF-FABMS spectra for both HF-derived and TFMSA-derived products are shown in Figure 3B. The mass spectrum for the HF-derived material (upper panel, Figure 3B) shows the existence of four products. The Ile-deleted Gln and Asn substituted analogs are observed at lower mass and the full length Gln and Asn substituted versions at higher mass. The spectrum from the micro-TFMSA/CF-FABMS analysis (lower panel, Figure 3B) shows essentially the identical pattern with the inclusion of a minor amount of the respective TFMSA adduct ions.

A rough estimate of the amount of the four derivatives produced is reflected in the intensities of the respective ion signals in the mass spectra, indicating that the Ile-deleted Gln and Asn analogs were synthesized in greater abundance than the full-length Q an N analogs. This quantitative relationship was also supported by the results of an amino acid analysis that indicated that the Ile-related products represented 15% each of the total and the full length derivatives 35% each of the total. However, it is noted that quantitative estimates from FAB spectra alone are tenuous because of differing ion yields from different peptides.

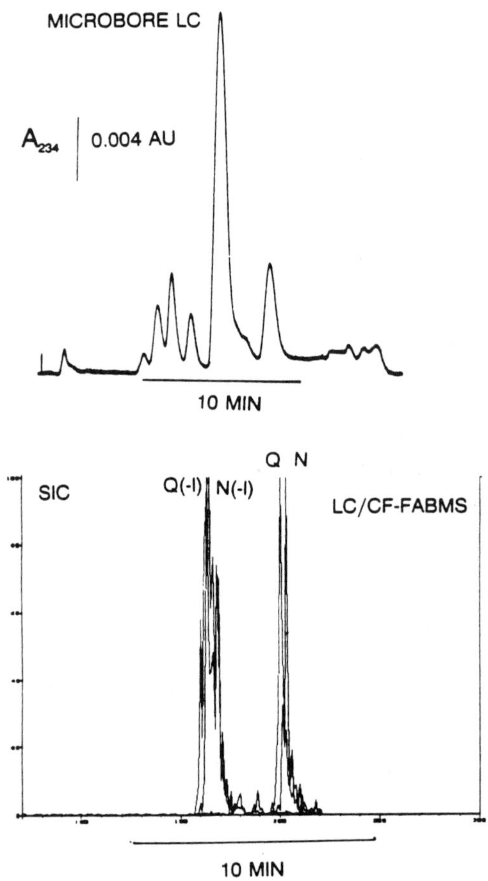

Fig. 4. Analysis of the four product mixture by microbore LC/CF-FABMS. Upper panel - UV detector trace showing the behavior of the microbore LC column. Lower panel - the mass spectral generated specific ion chromatograms for each peptide in the four product mixture microbore LC separation.

Figure 4 shows the comparison of the UV trace of a microbore LC run (upper) with specific ion chromatogram profiles derived from an on-line LC/CF-FABMS analysis of the multi-substrate mixture derived from a micro-TFMSA cleavage (lower panel). The profiles are similar except for some additional early eluting peaks present in the UV trace. The specific ion chromatogram profiles shown in Figure 4 (lower panel) indicate, as expected, that the shorter Ile-deletion Gln and Asn analogs elute first and are not well resolved. The full length Gln and Asn derivatives are fully separated from the Ile-deletion peptides but are not well resolved from one another. The selected ion chromatograms indicate that the Gln derivatives elute slightly ahead of the Asn derivatives. The mass spectra (data not shown) derived from scans across the selected ion chromatogram peaks contain major ion signals for the peptides analogs only. The TFMSA adducts seen in the CF-FABMS spectrum for the starting mixture (Figure 3B, lower panel) are absent, indicating that the RP-HPLC run effectively removed interfering TFMSA.

A second example of the synthesis and analysis of a multi-substrate mixture involves an octa-peptide mixture containing 6 products. The synthetic goal in this case was the preparation of six different analogs of angiotensin II substituted variably at position 3 (see Figure 5, upper Panel). The synthetic step responsible for the incorporation of the residue at position 3 is cycle #5. The six BOC amino acids chosen for cycle #5 were selected on the basis that they had similar or identically written instrument vessel programs (see Experimental section). The six tBOC amino acids were aliquoted in equimolar amounts to an empty cycle 5 vial. The sequences and expected masses of the protonated peptides $[(M+H)^+]$ are listed in Figure 5, upper Panel. The CF-FABMS spectrum for the HF-derived final product mixture (data not shown) indicated that all the desired peptides were synthesized. The signal intensities for the respective peptides suggest considerable variability in individual yields. However signal intensity in a FABMS spectrum may not be a true quantitative reflection of the individual synthetic yields due to ion suppression effects (9,27). A more reliable picture of the actual individual analog yields can be quickly deduced from coupling a separation method such as HPLC with CF-FABMS. Figure 5 compares an UV absorbance profiles for a microbore HPLC run (middle) and the corresponding selected ion chromatograms of an LC/MS run (lower) on the 6 component analog mixture. Although three major peaks were noted in the microbore run, the first two eluting peaks possessed shoulders suggesting a partially resolved components. Note that the respective selected ion chromatograms (Figure 5, lower) define the actual resolution of the separation and indicate the RP-HPLC (Figure 5, middle) elution order for the 6 component mixture. As expected, the hydrophilic analogs elute before the more hydrophobic analogs.

Capillary electrophoresis is an alternative fractionation technique that has been used to assess peptide mixtures (28,29). An aliquot of the same

D R X Y I H P F	*(M + H) +*
----F----------	*1094.5*
----H----------	*1084.5*
----M----------	*1078.5*
----E----------	*1076.5*
----D----------	*1062.5*
----I----------	*1060.5*

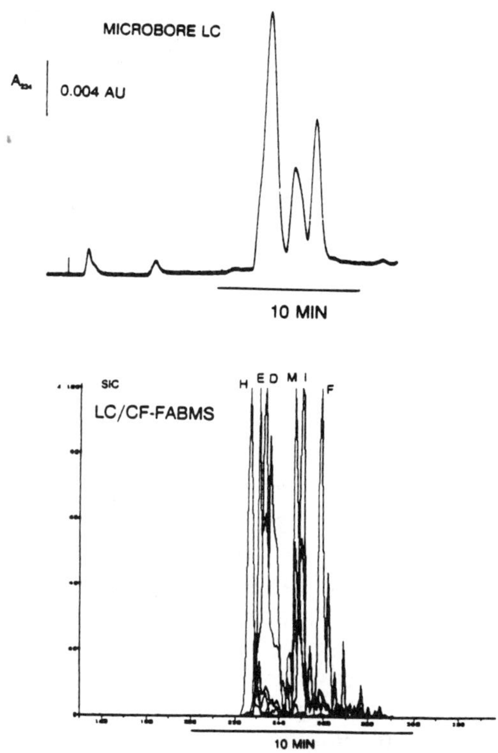

Fig. 5. Microbore LC/CF-FABMS analysis of the six components Angiotensin II analog mixture. Upper panel - sequences and expected masses of the product mixture. Middle panel - the UV detector trace for a microbore LC/CF-FABMS run. Lower panel - the overlapping mass spectral specific ion chromatograms for the individual peptides in the mixture. The letters represent the amino acid substitutions present in the six Angiotensin II analogs.

sample of the octapeptide mixture was analyzed by CE/CF-FABMS. The results of this analysis are shown in Figure 6. The upper panel of Figure 6 shows the UV trace of the electrophoretic separation of the peptides at point 2/3 the length of the CE capillary. The mass spectral specific ion chromatograms shown in the lower panel represent the electrophoretic pattern of the peptides after they reach the ion source of the mass spectrometer following additional electrophoretic separation in the last third of the CE capillary and a transfer through the liquid junction interface (22). As expected, the selected ion chromatogram data indicates that the most basic residue substituted analog emerged first, followed by an unresolved neutral residue substituted three component mixture, and lastly, the

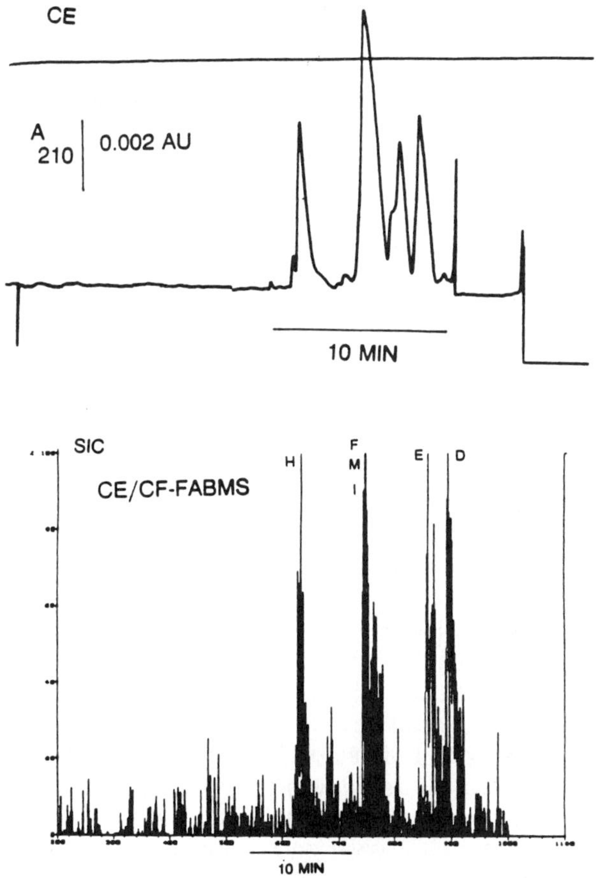

Fig. 6. CE/CF-FABMS analysis of the Angiotensin II analog mixture containing six products all derived from a single automated run. Upper panel - UV detector trace of a CE run performed on the peptide mixture. Lower panel - the overlapping mass spectral specific ion chromatograms for each peptide in the mixture. The single letters represent the amino acid substitutions present in the six Angiotensin II analogs.

resolved acidic residue substituted analogs. The four selected ion chromatogram peaks shown in the bottom panel of Figure 6 are representations of the four peaks seen in the UV trace shown in the upper panel.

IV. DISCUSSION

The use of CF-FABMS coupled techniques (micro-TFMSA/MS, LC/MS, and CE/MS) permit a thorough examination of automated peptide syntheses. Using the micro-TFMSA/CF-FABMS technique alone, the progress of an automated run can be examined "stepwise" during a synthesis and retrospectively at completion. The technique is a rapid and simple analytical procedure that can also be used to evaluate a final-product peptide-resin before it is subjected to a more involved preparative-scale cleavage treatment. In progress analysis, which can be performed within the time span of a current synthetic cycle, allows one to immediately abort a run if some specific problems are detected, i.e., residue deletion, termination, or modification.

A broader stepwise assessment strategy is to pause a synthesis periodically after every 5 cycles or after positions where difficult couplings are suspected and perform a micro-TFMSA/CF-FABMS analysis on either individual peptide-resin aliquots or pooled aliquots. Depending on the outcome, the operator may alter the run strategy and proceed with confidence knowing that preceding steps were successful and that recent instrument performance is adequate. A retrospective stepwise analysis performed on either individual peptide-resins (Figure 1B) or peptide-resin pools (Figure 1C) after a run is also informative. Sites of sequence dependent problematic couplings can be identified so that future strategic steps can be chosen if re-synthesis is an option.

The micro-TFMSA cleavage method described here was found to be sufficiently potent to remove most side chain blocking groups. The only problematic side chain blocking group appeared to be the tosyl group of arginyl residues. It is well documented that this group is difficult to remove with TFMSA treatment. Substitution of the tosyl Boc-Arg derivative with the Mts Boc-Arg derivative would obviate this problem (20,24). Another minor problem encountered with the micro-TFMSA/CF-FABMS was the detection of TFMSA adduct ions associated with some fully deblocked-Arg peptides. For the most part, the adducts are detected when the peptides are neither ether precipitable (Figure 3B, lower panel) nor adequately ether extractable (Figure 2B and C), leaving relatively high concentrations of TFMSA in the sample. However, subjecting such a sample to RP-HPLC prior to CF-FABMS analysis resulted in the disappearance of TFMSA adduct peptide ions.

Because of possible ion suppression events that may occur during the FAB process (9,27), the information should only at best be considered semi-quantitative. Ion signal abundances usually only reflect the quantitative picture when the peptide mixtures are simple and when the physical and chemical properties of the peptides are similar, e.g., charge, hydrophilicity, size, polarity, etc.). The quantitative picture of a complex synthetic peptide mixture is more realistically assessed when a two-dimensional analysis is performed such as LC/MS or CE/MS. The UV trace usually obtained in such analyses can also provide information about the separation process and gives a more accurate quantitative picture. The chromatographic or electrophoretic peak height information coupled with mass spectral identification is of enormous utility in assessing the relative amounts of each peptide present in the mixture.

The single run "multipeptide synthesis runs" are of interest from at least two different perspectives. Firstly from the perspective of the peptide chemist who is interested structure function studies, the results presented here that indicate that several closely related analog peptides can be prepared simultaneously in an instrument designed to synthesize one peptide at a time. CF-FABMS coupled techniques quickly permit the evaluation of such an uncommon approach. The chromatographic mass identification information reveals a preparative strategy to obtain each peptide in pure form if so desired. Secondly from the perspective of the peptide chemist interested in the mechanisms of synthetic chemistry, the method offers an experimental approach to investigate pre-formed anhydride or HOBT ester reactivities towards the amino functions of different amino acid residues. The methodology described here may permit an experimental means to quickly catalog and define predictive sequence specific difficult couplings.

To conclude, the CF-FABMS coupled methods described here provide a powerful combination of methods that the peptide chemist can use to design, improve, and validate automated peptide synthesis strategies.

V. ACKNOWLEDGEMENT

We would like to thank Dr. Marc Suter for offering his mass spectrometry expertise and his help with some of the experimentation. We would also like to thank Drs. Sherish Shenolikar and Jerry Wolinsky for permitting us to present data concerning the assembly of synthetic peptides prepared by us for their research programs. This work was supported by a grant from the National Institutes of Health, Grant # 1 RO1 GM 43783-01.

VI. REFERENCES

1. Meister, S.M., Kent, S.B.H., in Peptides: Structure and Function, Hruby, V.J., and Rich, D.H., Eds., Pierce Chem. Co., Rockford, IL, p. 103, 1983.
2. Merrifield, R.B., Singer, J., and Chait, B.T., Anal. Biochem. **174**, 139, 1988.
3. Gisin, B.F., Anal. Chim. Acta., **58**, 248, 1972.
4. Hancock, W.S., Prescott, D.J., Vagelos, P.R., and Marshall, G.R., J. Org. Chem., **38**, 774, 1973.
5. Sarin, V.K., Kent, S.B.H., Tam, J.P., and Merrifield, R.B., Anal. Biochem. **117**, 147, 1981.
6. Beyerman, H.C., van der Kamp, P.R.M., de Leer, E.W.B., Massen van den Brink, W., Parmentier, J.H., and Westerling, J., in Peptides 1971, Nesvadba, H., Ed., North Holland, Amsterdam, p. 162, 1973.
7. Niall, H.D., Tregear, G.W., Jacobs, J., in Chemistry and Biology of Peptides, Meienhofer, J., Ed., Ann Arbor Science Pub., Ann Arbor, MI, 1972, 695.
8. Applied Biosystems, Inc. Protein Sequencer User Bulletin No. 13, Sequence analysis of synthetic, side-chain protected, resin-bound peptides, 1985.
9. Naylor, S., Findeis, F., Gibson, B.W., and Williams, D.H., J. Am. Chem. Soc., **108**, 6359, 1986.
10. Barber, M., Bordoli, R.S., Elliot, G., Sedgwick, R.D., and Tyler, A.N., Anal. Chem., **54**, 645A, 1982.
11. MacFarlane, R.D., and Torgenson, D.F., Science, **191**, 920, 1976.
12. Chait, B.T., and Field, F.H. BBRC **134**, 420, 1986.
13. Karas, M., and Hillenkamp, F., Anal. Chem., **60**, 2299, 1988.
14. Beavis, R.C., and Chait, B.T., Rapid Comm. in Mass Spec., **3**, 233, 1989.
15. Meng, C.K., Mann, M., and Fenn, J.B., Z. Phys. D. - Atoms, Molecules and Clusters, **10**, 361, 1988.
16. Covey, T.R., Bonner, R.F., Shushan, B.I., and Henion, J.D., Anal. Chem., Rapid Commun. Mass Spectrom., **2**, 249, 1988.
17. Caprioli, R.M., Anal. Chem., **62**, 477A, 1990.
18. Merrifield, R.B., Science, **150**, 178, 1965.
19. Merrifield, R.B., Vizioli, L.D., and Boman, H.G., Biochemistry, **21**, 5020, 1982.
20. Bergot, B.J., Noble, R.L., and Geiser, T., User Bulletin No. 16, Peptide Snythesizer, Applied Biosystems, Inc., Foster City, CA, 1986.
21. Caprioli, R.M., DaGue, B., Fan, T., Moore, W.T., Biochem. Biophys. Res. Commun, **146**, 291, 1987.
22. Caprioli, R.M., Moore, W.T., Wilson, K.B., and Moring, S., J. Chromatogr., **480**, 247, 1989.
23. Meister, S.M,, and Kent, S.B.H. in Peptides: Structure and Function, Hruby, V.J., and Rich, D.H., Eds. Pierce Chem. Co., Rockford, IL, p. 103, 1983.
24. Tam, J.P., in Macromolecular Sequencing and Synthesis Selected Methods and Applications, Schlesinger, D.H., Ed., Alan R. Liss, Inc., NY, p. 153, 1988.
25. Landon, M., Methods Enzymol., **47**, 145, 1977.
26. Seifert, W.E., Moore, W.T., Ballatore, A., and Caprioli, R.M., Abstr. #ThP 9, Proc. 38th ASMS Conf. on Mass Spectrom. and Allied Topics, Tucson, AZ, June 3-8, 1990.
27. Caprioli, R.M., Moore, W.T., Petrie, G., and Wilson, K., Int. J. Mass Spectrom. Ion Proc., **86**, 187, 1988.
28. Jorgenson, J.W., and Lukacs, K.D., Science, **222**, 266, 1983.
29. Grossman, P.D., Wilson, K.J., Petrie, G., and Lauer, H.H., Anal. Biochem., **173**, 265, 1988.

50

MASS SPECTROMETRY ANALYSIS FOR DETECTING SIDE PRODUCTS IN PEPTIDE SYNTHESIS

Conway C. Chang, Primo A. Babayan, Ling Chen, and Anita L. Hong

Applied Biosystems, Inc., Foster City, California 94404

I. INTRODUCTION

In the past few years, solid-phase peptide synthesizers have facilitated the synthesis of peptides that are used for biological or physiochemical investigations. It is extremely important to verify the integrity of the synthesized peptides as well as to detect and identify undesired side products. The most commonly used methodologies for characterizing synthetic peptides are HPLC, amino acids analysis, sequence analysis and spectrometric analysis. However, these techniques may not easily detect side products that result from incomplete deprotections, racemizations, truncations or reactions with scavengers. Mass analysis has become an indispensable tool for confirming the integrity of a peptide product and for identifying side products.

In this study, we have used plasma desorption mass spectrometry (PDMS) for identifying the sequence and structure of synthetic peptides. Peptides were synthesized using Fmoc and Boc strategies.

Techniques in Protein Chemistry II

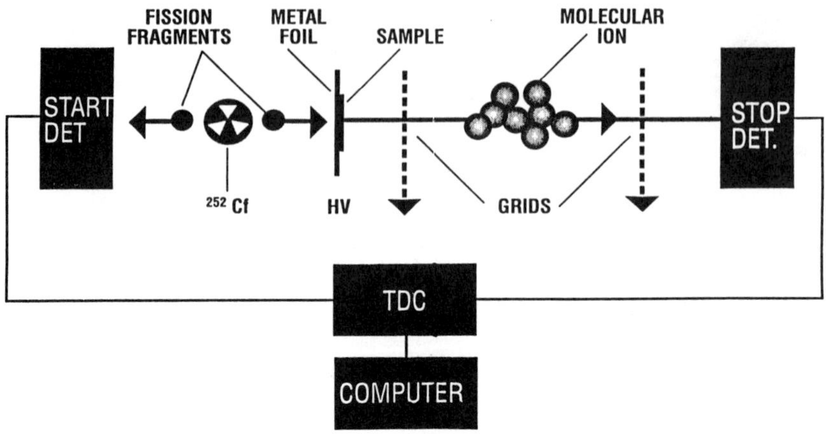

Figure 1. The general features of the BioIon^{TM}20 Biopolymer Mass Analyzer (Applied Biosystems, Inc.)

II. INSTRUMENTATION

The general features of BioIon^{TM}20 Biopolymer Mass Analyzer (Applied Biosystems, Inc.) is shown in Figure 1. It is based on plasma desorption time-of-flight mass spectrometry (PDMS). The protein/peptide sample is adsorbed on the target, which is held at a high potential of 10 to 20 KeV. The ionization source, Californium^{252}, emits two colinear fission fragments, each with an energy of 100 MeV range. One of the fission fragments strikes the "start" detector. The other fragment (primary ion) bombards the sample target, desorbing and ionizing the sample (secondary ions). The ionized secondary ions are immediately accelerated by the electric field toward the grid, which is held at ground potential. At this point, all charged ions are accelerated to the same kinetic energy and begin drifting in a field-free region toward a "stop" detector. Heavier ions move relatively slower and reach the stop detector at a later time. The "start" signal triggers the timing for each fission event and the times-of-flight for all ions are recorded as they reach the "stop" detector. The time-of-flight spectral data represents accumulated ion counts for each ion at its respective flight time. The time of flight spectrum can be easily converted into a mass spectrum using the time centroids of any two known mass ions as internal calibration peaks (e.g. H^+ and NO^+).

III. SAMPLE PREPARATION AND SYNTHESIS

On the BioIon^{TM}20, the peptide/protein sample solution is deposited onto a solid support to form a uniform sample layer. The most commonly used sample target is a thin layer of nitrocellulose-coated, aluminized, polyester support with a surface area of 1 sq cm. Typically, 10 pmol to 1 nmol of sample is dissolved in 1 to 10 μL of solvent and deposited on the nitrocellulose surface. The target is dried before it is inserted into the mass analyzer.

A peptide HCP-1 corresponding to a fragment of human carboxypeptidase was synthesized. The sequence of HCP-1 is

H-Pro-Gly-Arg-His-Ser-Tyr-Ala-Lys-
Tyr-Asn-Asn-Trp-Glu-OH

We have used both the Fmoc and Boc chemistries on the Applied Biosystems Peptide Synthesizer Models 430A and 431A to synthesize HCP-1. Stepwise solid-phase syntheses were performed using N-methylpyrrolidinone (NMP) as the coupling medium. Fmoc synthesis was performed starting with 0.25 mmol of Fmoc HMP resin. TFA cleavage was carried out at 25°C for 90 minutes with a mixture of EDT and thioanisole as the scavengers. Boc synthesis was performed starting with 0.5 mmol of t-Boc-Glu-Pam resin. HF cleavage was carried out at -5° to 0°C for 90 minutes using anisole as the scavenger.

The crude synthetic products were desalted by passing through a short C-18 cartridge and concentrated in a Speed Vac. The residues were redissolved in 0.1% TFA/H_2O, deposited onto a solid support, dried and inserted into BioIon^{TM}20.

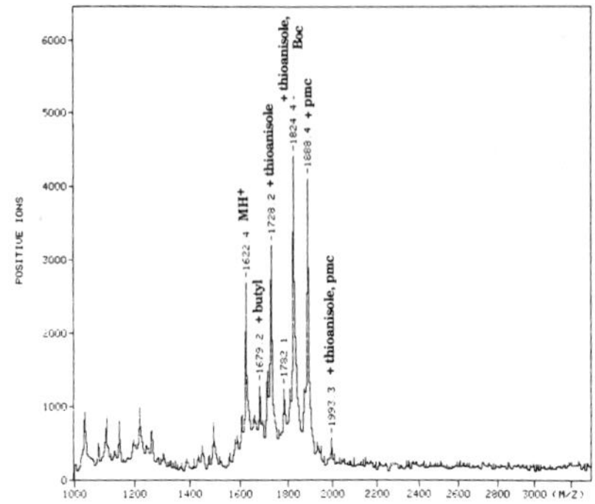

Figure 2a.

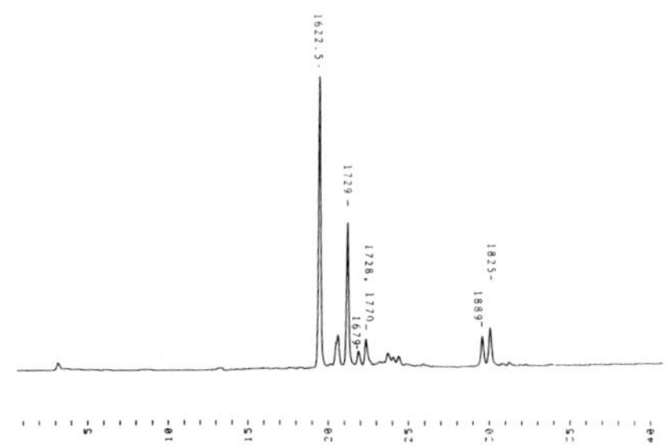

Figure 2b

Figure 2. (a) BioIon^{TM}20 spectra of desalted crude HCP-1 synthesized by Fmoc chemistry.
(b) HPLC chromatogram of HCP-1 synthesized by Fmoc chemistry. Instrument: Applied Biosystems Model 130A Separation System. Column: Brownlee Lab Aquapore^{TM} RP-300, 7 um, 22 cm X 2.1 mm ID. Mobile phase: A: 0.1% TFA in water; B: 0.08% TFA in 100% CH_3CN Gradient: 0 to 60% B in 45 min; Flow rate: 0.23 mL/min. Detection: 220 nm

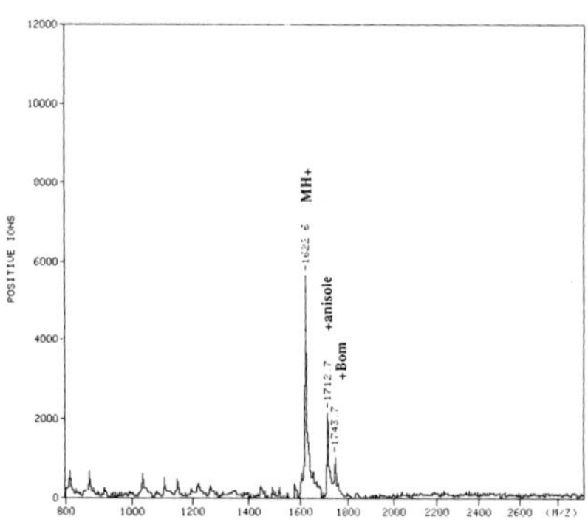

Figure 3a

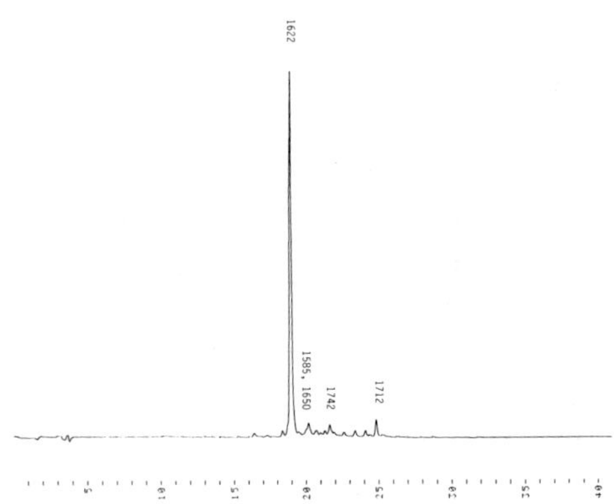

Figure 3b

Figure 3. (a) BioIon^{TM}20 spectra of desalted crude HCP-1 synthesized by Boc chemistry.
(b) HPLC chromatogram of HCP-1 synthesized by Boc chemistry. See Figure 2 for HPLC conditions.

IV. RESULTS AND DISCUSSIONS

The MH^+ ion peaks at m/z 1622, confirmed that the desired peptide is the major product in both syntheses, as shown in Figure 2a, 3a. Results of BioIon^{TM}20 analysis in conjunction with HPLC (Figure 2b, 3b) show that the Boc synthesis chemistry gave a better product than the Fmoc chemistry.

From the BioIon^{TM}20 analysis of the crude mixtures, we detected the presence of several side products.
To determine the structures of the side products, the crude peptide mixtures were purified using HPLC. Fractions were collected and analyzed on the BioIon^{TM}20. Analysis of peaks in the mass spectrum at their specific mass provided information about the origin and nature of these side products.

The results of BioIon^{TM}20 analyses of HCP-1 synthesized by Fmoc chemistry are summarized in Table II. Based on the fragmentation pattern of thioanisylated HCP-1 (m/z 1728), it was determined that the C-terminal Glu was thioanisylated. The assignment was further substantiated by comparing the fragmentation pattern of thioanisylated HCP-1 with that of HCP-1 (Figure 4a, 4b).

Table II. BioIon^{TM}20 analysis of HCP-1 synthesized by Fmoc chemistry

Compound	m/z
$(HCP\text{-}1 + H)^+$	1622 (M+H)
$(HCP\text{-}1{+}H{+}butyl)^+$	1678 (M+H+56)
$(HCP\text{-}1{+}H{+}Thioanisole)^+$	1728 (M+H+106)
$(HCP\text{-}1{+}H{+}Boc{+}Thioanisole)^+$	1824 (M+H+100+106)
$(HCP\text{-}1{+}H{+}Pmc)^+$	1888 (M+H+266)
$(HCP\text{-}1{+}H{+}Pmc{+}Thioanisole)^+$	1993 (M+H+266+106)

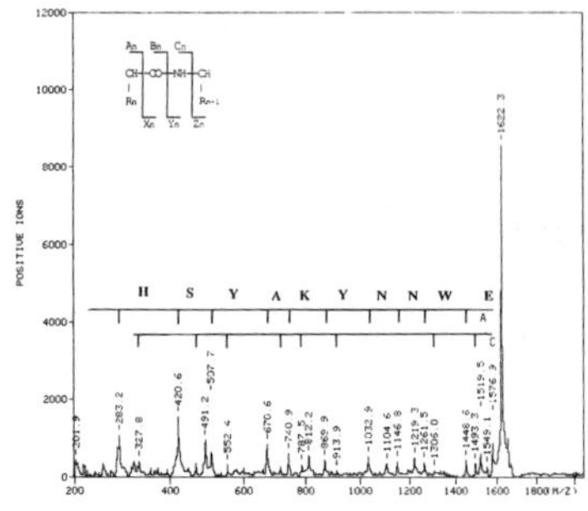
Figure 4a

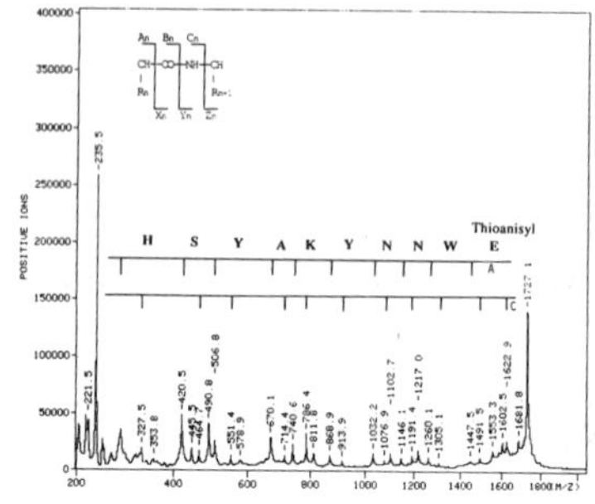
Figure 4b

Figure 4. (a) BioIon^{TM}20 spectra of purified HCP-1 synthesized by Fmoc chemistry.
(b) BioIon^{TM}20 spectra of purified thioanisylated HCP-1 synthesized by Fmoc chemistry.

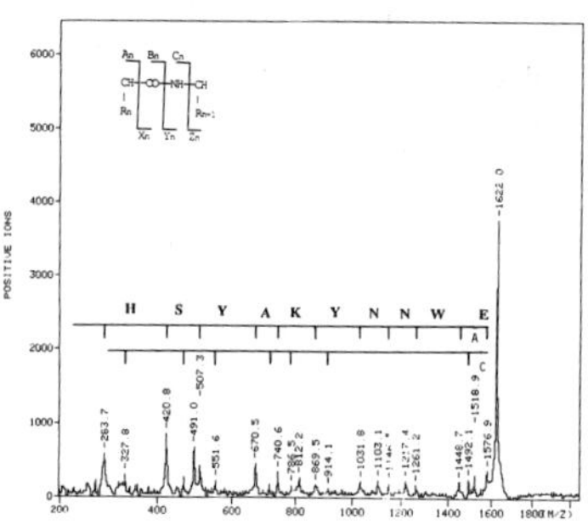
Figure 5a

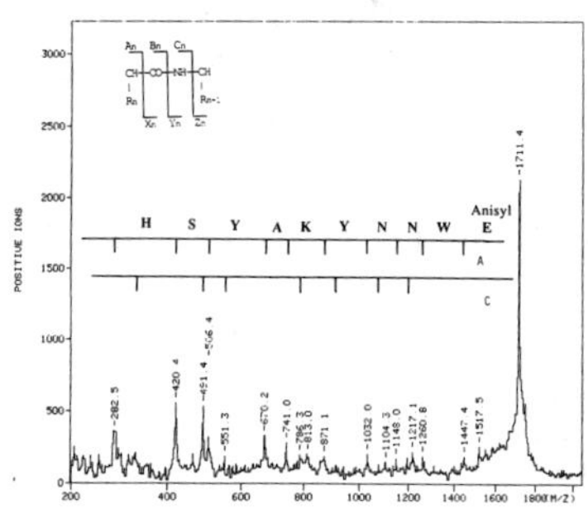
Figure 5b

Figure 5. (a) BioIon^{TM}20 spectra of purified HCP-1 synthesized by Boc chemistry.
(b) BioIon^{TM}20 spectra of purified anisylated HCP-1 synthesized by Boc chemistry.

The results of BioIon^{TM}20 analyses of HCP-1 synthesized by Boc chemistry are summarized in Table III. By comparing the fragmentation pattern of anisylated HCP-1 (m/z 1712) with that of HCP-1 (m/z 1622), we deduced that the C-terminal Glu was anisylated (Figure 5a, 5b).

Table III. BioIon^{TM}20 analysis of HCP-1 synthesized by Boc chemistry

Compound	m/z
$(HCP\text{-}1 + H)^+$	1622 (M+H)
$(HCP\text{-}1+H+anisole)^+$	1712 (M+H+90)
$(HCP\text{-}1+H+Bom)^+$	1743 (M+H+121)

The common amino acid protecting groups and the "Add On" mass unit due to other side reactions are summarized in Tables IV & V. The tables can be used to calculate the molecular weights of the side products.

Table IV.
Common protecting groups and the corresponding "Add On" mass unit.

Name	Abbreviation	"Add On" Mass unit
Acetyl	Ac	42
Benzyl	Bzl	90
Benzyloxymethyl	Bom	121
2-Bromobenzyloxycarbonyl	Br-Z	213
t-Butyl	t-Bu	56
Butyloxycarbonyl	Boc	100
2-Chlorobenzyloxycarbonyl	Cl-Z	168
Formyl	CHO	28
4-Methylbenzyl	Meb	104
2,2,5,7,8-pentamethylchroman-6-sulphonyl	Pmc	266
Tosyl	Tos	154
Trityl	Trt	242

Table V
Mass unit modification due to side reaction in Peptide Synthesis

Side Reaction	Formula	"Add On" Mass unit
Anisylation	$+C_7H_7O$	90
EDT adduct	$+C_2H_5S_2$	93
Thiocresylation	$+C_7H_7S$	106
Thioanisylation	$+C_7H_7S$	106
Dehydration	$-H_2O$	-18

V. CONCLUSION

Using mass analysis:

1. The integrity of a synthetic product can be confirmed by the presence of the desired molecular ion peak.
2. The side products resulting from incomplete deprotections, racemizations, truncations or reactions with scavenger can be deduced by adding the "Add On" mass unit to the molecular ion.
3. The modified amino acid product of the side reaction can be located by comparing its fragmentation pattern with that of the authentic product.

51

Identification by Mass Spectrometry of the Site of Modification of Creatine Kinase by the Affinity Label Epoxycreatine

Douglas D. Buechter, Katalin F. Medzihradszky, Alma L. Burlingame and George L. Kenyon
Department of Pharmaceutical Chemistry, School of Pharmacy
University of California, San Francisco, CA 94143

I. Introduction

The developments of the past decade in protein chemistry and molecular biology require rapid, sensitive and reliable methods for protein analysis. Mass spectrometry is one of the fastest developing fields in this area and is now commonly used in the analysis of molecules of biological significance [1]. In particular, soft ionization techniques, such as fast atom bombardment (FAB) and liquid secondary ion mass spectrometry (LSIMS), are valuable methods for determining the molecular ions of peptides obtained from the proteolytic digests of proteins of known sequence. Peptide mapping by mass spectrometry is a rapid and reliable method for amino acid sequence confirmation [1] and a useful tool for determining post-translational [2] and other types of covalent modifications of proteins and peptides [3,4]. Tandem mass spectrometry complements soft ionization techniques by allowing the fragmentation pattern of selected ions to be obtained. This method enables the complete sequence of a peptide to be deduced and makes possible the sequencing of peptides and proteins of previously unknown structure [1,5]. Tandem mass spectrometry is the method of choice for determining the exact site and nature of covalent modification in proteins and peptides. We have used both methods to identify the amino acid residue in the enzyme creatine kinase which is modified by the affinity label, epoxycreatine.

Creatine kinase (CK, EC 2.7.3.2) plays a critical role in the cellular utilization of ATP, particularly in those cells that have a high energy demand [6]. CK is a dimeric enzyme of subunit molecular weight approximately 43,000 Da. There are two known major cytosolic CK subunits: brain (B) and muscle (M). These subunits associate to form the muscle (MM), brain (BB) and heterodimer (MB) isozymes. Numerous chemical and spectroscopic studies have implicated several amino acids involved in binding and/or catalysis. These include lysine [7], histidine [8], and cysteine [9] residues. Only one of these, Cys282, is known with any certainty to be at or near the active site [10].

Techniques in Protein Chemistry II

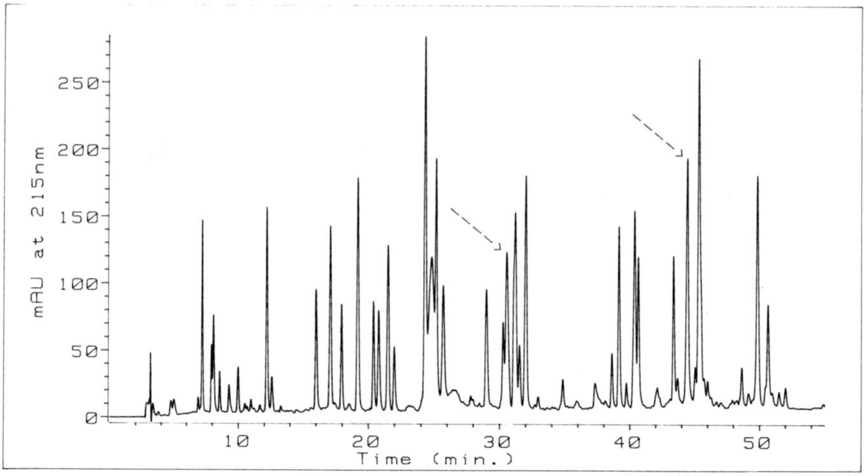

Figure 1. Reverse phase HPLC chromatogram of the tryptic digest of creatine kinase labeled with epoxycreatine. The column was a Vydac C_{18} Protein/Peptide eluted with 0.1% trifluoroacetic acid in water with a 0.7%/min increasing gradient of 0.08% trifluoroacetic acid in CH_3CN. The flow rate was 1 ml/min. The arrows point to the peaks which contained radioactivity.

Epoxycreatine [N-(2,3-epoxypropyl)-N-amidinoglycine] is an affinity label of creatine kinase [11] which covalently modifies the enzyme and results in the complete loss of activity. The inactivation follows saturation kinetics and gives a 1:1 stoichiometry.

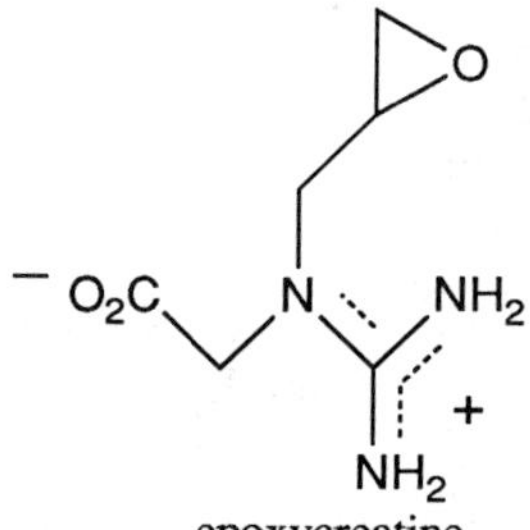

epoxycreatine

Affinity labels such as epoxycreatine are important as probes of an enzyme's active site and may lead to the identification of residues important for catalysis. To determine the residue of creatine kinase which is modified by epoxycreatine we performed a tryptic digestion of rabbit muscle CK covalently modified with [^{14}C]-labeled epoxycreatine. The molecular weights of the labeled peptides were determined by LSIMS, and comparison of these to the molecular weights of the expected tryptic peptides suggested sequences for these peptides. These sequences, as well as the identity of the modified amino acid, were then established by tandem mass spectrometry.

II. Materials and Methods

A. *Inactivation of Creatine Kinase with Epoxycreatine and Proteolytic Digestion.*

Rabbit muscle creatine kinase was purified by ethanol fractionation, blue sepharose CL-6B chromatography and anion exchange Fast Protein Liquid Chromatography (FPLC) as described previously [12-14]. [^{14}C]-Epoxycreatine was synthesized by the method of Marletta and Kenyon [11], except that the label was introduced using [^{14}C]-cyanamide. Creatine kinase was inactivated with [^{14}C]-epoxycreatine following the protocol of Marletta and Kenyon [11]. Carboxymethylation, tryptic digestion of epoxycreatine-labeled CK, and separation by high performance liquid chromatography (HPLC) of the resultant peptides were performed as described elsewhere [15]. For mass spectrometry, a second sample of creatine kinase was inactivated with cold epoxycreatine as described for [^{14}C]-epoxycreatine. The resultant HPLC chromatogram was identical to that obtained when radiolabeled epoxycreatine was used. Epoxycreatine labeled peptides were collected and taken to dryness on a Speed Vac Concentrator (Savant). Fractions were re-dissolved in 5 μl of 0.1% trifluoroacetic acid (TFA) in water and analyzed by either LSIMS or tandem mass spectrometry.

B. *Liquid Secondary Ion Mass Spectrometry and Tandem Mass Spectrometry.*

LSIMS analyses were carried out on a Kratos MS50S double focusing mass spectrometer equipped with a high field magnet and a cesium ion source [16]. The sample solution (1 μl) was added to 1 μl of glycerol:thioglycerol (1:1)/1% TFA as a matrix. Collision induced dissociation (CID) analyses were performed on a Kratos Concept IIHH four sector mass spectrometer equipped with a cesium ion source and an electro-optical multichannel array detector able to record 4% of the mass range simultaneously [17]. The pressure of the helium collision gas was so adjusted that the intensity of the selected precursor ion was decreased to 30% of its initial value [17]. The collision cell was floated at 4 keV.

III. Results and Discussion

Peptides from a five hour tryptic digestion of carboxymethylated creatine kinase which had previously been inactivated with [^{14}C]-epoxycreatine were separated by C_{18}-HPLC (Fig. 1) to yield two fractions which contained radioactivity (identified by the arrows). One of these, with a retention time of 44.5 min, was a new peak not seen in the chromatogram of unmodified enzyme (chromatogram not shown). It contained 75% of the bound radioactivity. The other, with a retention time of 31.5 min, appeared unchanged from the digest of unlabeled CK, but contained the remaining counts.

LSIMS of the peptide with retention time of 44.5 min gave a protonated molecular ion at m/z 3043.8. This molecular weight was consistent with epoxycreatine labeling of the non-carboxymethylated Cys282-containing peptide:

^{266}AGHPFMWNEHLGYVLT^{282}CPSNLGTGLR
Peptide A (expected MH^+ 3043.5)

These results suggested that epoxycreatine was bound at or near Cys282 and prevented the subsequent carboxymethylation of this residue. To obtain a smaller peptide more suitable for CID analysis, this peptide was subjected to subdigestion with endoproteinase Glu-C. Cleavage of Peptide A by this protease was expected to give two peptides of sequences:

^{266}AGHPFMWNE and ^{275}HLGYVLT^{282}CPSNLGTGLR
Peptide B Peptide C

HPLC analysis of the products of this subdigestion gave two peaks. LSIMS of the earlier eluting peak gave the expected protonated molecular ion for Peptide B (MH^+ at m/z 1088.5). LSIMS of the other peak gave an ion at m/z 1974.1, consistent with Peptide C being labeled with epoxycreatine, but not carboxymethylated.

To determine the exact site of labeling we performed CID analysis of the labeled Peptide C. In the high energy CID spectrum (Fig. 2) abundant C-terminal sequence ions (nomenclature according to Roepstorff and Fohlman [18], as modified by Biemann [19]), were observed, as is usually the case for peptides containing arginine at the C-terminus [20]. A 173 mass unit shift in the sequence ions was observed beginning at Cys282. For example, the expected $\mathbf{z}_{10}$ ion for this sequence, if Cys282 were not modified, would be at m/z 1000; however, the observed $\mathbf{z}_{10}$ ion, with epoxycreatine bound at Cys282, is at m/z 1173 (Fig 2). Ions arising from side chain fragmentation are observed in the N-terminal sequence series, $\mathbf{d}_{10}$, $\mathbf{d}_{11}$, $\mathbf{d}_{12}$, $\mathbf{d}_{14}$, and $\mathbf{d}_{16}$. This is consistent with the introduction of the basic guanidinium group of epoxycreatine at Cys282. The occurrence of **d**-ions is often indicative of the presence of a basic amino acid on the N-terminal side of the site of cleavage [20].

Having determined the major site of labeling, the radioactivity-containing fraction with retention time 31.5 min was analyzed (Fig 1). LSIMS showed this peak to contain at least three peptides. The only peptide not in common with the digest of unlabeled CK was a low abundance protonated molecular ion at m/z 1503.8. Although a variety of HPLC conditions were tried, complete purification of this peptide was not achieved. Tandem mass spectrometry made it possible to identify and obtain a sequence of it in the mixture. The CID spectrum of this peptide (data not shown) gave the sequence:

^{279}VLT^{282}CPSNLGTGLR

This peptide is evidently a chymotryptic fragment of the labeled tryptic Peptide A shown above. CID fragmentation again showed that it was labeled with epoxycreatine at Cys282.

In summary, we have been able to use successfully a combination of LSIMS and tandem mass spectrometry to determine the specfic amino acid residue of creatine kinase which is covalently modified by the affinity label, epoxycreatine. From the fragmentation patterns seen in the CID spectra it was possible to

v						1347				785	671	558		400			
w		1778				1346		968	871	784	670	557				229	
x				1693	1530	1431		1217	940	843	756	642	529	472	371		
y			1724					1191	914	817	730	616	503		345		175
z			1708	1650	1487	1389	1275	1173		800	713		486	429		272	159
	His –	Leu –	Gly –	Tyr –	Val –	Leu –	Thr –	Cys* –	Pro –	Ser –	Asn –	Leu –	Gly –	Thr –	Gly –	Leu –	Arg
a	110	223		443	542	655		1032	1130	1217	1331	1444				1772	
b	138	251					784	1060	1158		1359	1472			1687		
c								1077		1262	1376	1489	1546				
d										1201	1288	1402		1585		1730	

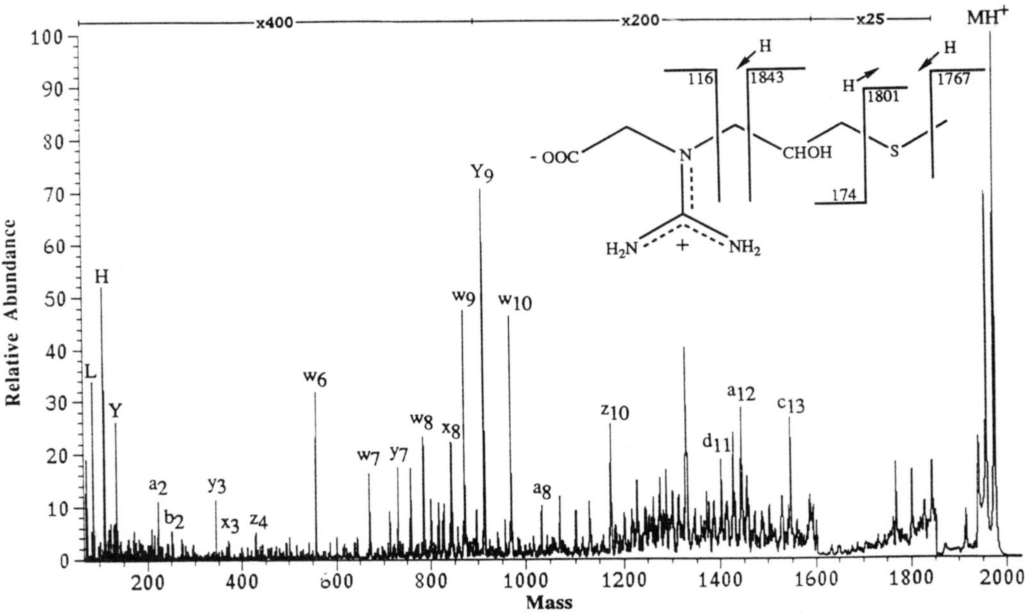

Figure 2. CID spectrum of Peptide C (^{275}HLGYVLT^{282}CPSNLGTGLR). Observed fragment ions are listed above the spectrum. Ions belonging to the same series are listed in rows. The insert shows the proposed structure for the label bound at Cys282, as well as the ions expected from the fragmentation of the label.

determine the sequences of the modified peptides and precisely which residue was labeled. Tandem mass spectrometry was indispensable in sequencing the minor labeled peptide; sequencing mixtures of peptides by conventional Edman techniques usually leads to ambiguous results, especially if the components are present in comparable amounts.

Acknowledgments

This work was supported by USPHS Grant AR17323 (to GLK) and NSF Grant DIR 8700766 and USPHS Grant RR01614 (to ALB). The authors would like to thank F.C. Walls for obtaining the CID spectra.

References

1) Burlingame, A.L., Millington, D.S., Norwood, D.L., and Russell, D.H. (1990) *Anal. Chem.* **62**, 268R.
2) Settineri, C.A., Medzihradszky, K.F., Chu, C., George-Nascimento, C., Masiarz, F.R., and Burlingame, A.L. (1990) *Biomed. Environ. Mass Spectrom.* in press.
3) Kaur, S., Hollander, D., Haas, R., and Burlingame, A.L. (1989) *J. Biol. Chem.* **264**, 16981.
4) DeWolf, Jr., W.E., Carr, S.A., Varrichio, A.S., Goodhart, P.J., Mentzer, M.A., Roberts, G.D., Sonthan, C., Dollem R.E., and Kruse, L.I. (1988) *Biochemistry* **27**, 9093.
5) Biemann, K., and Scoble, H.A. (1987) *Science* **237**, 992.
6) Seraydarian, M.W., and Abbott, B.C. (1976) *J. Mol. Cell. Cardiol.* **8**, 741.
7) James, T.L., and Cohn, M. (1974) *J. Biol. Chem.* **249**, 2599.
8) Rosevear, P.R., Desmeules, P., Kenyon, G.L., and Mildvan, A.S. (1981) *Biochemistry* **21**, 6155.
9) Thomson, A.R., Eveleigh, J.W., and Miles, B.J. (1964) *Nature (Lond.)* **4942**, 267.
10) Cohn, M., Diefenbach, H., and Taylor, J.S. (1971) *J. Biol. Chem.* **246**, 6037.
11) Marletta, M.A., and Kenyon, G.L. (1979) *J. Biol. Chem.* **254**, 1879.
12) Kuby, S.A., Noda, L., and Lardy, H.A. (1954) *J. Biol. Chem.* **209**, 191.
13) Miller, J., Johnson, M., and Wei, R. (1982) *Clinica Chimica Acta* **118**, 67.
14) Quest, A.F.G., Eppenberger, H.M., and Walliman, T. (1989) *Enzyme* **41**, 33.
15) Buechter, D.D., Medzihradszky, K.F., Burlingame, A.L., and Kenyon, G.L., *manuscript in preparation.*
16) Falick, A.M., Wang, G.H., and Walls, F.C. (1986) *Anal. Chem.* **58**, 1308.
17) Walls, F.C., Baldwin, M.A., Falick, A.M., Gibson, B.W., Kaur, S., Maltby, D.A., Gillece-Castro, B.L., Medzihradszky, K.F., Evans, S., and Burlingame, A.L. (1990) in *Biological Mass Spectrometry* (A.L. Burlingame and J.A. McCloskey, eds) Elsevier, Amsterdam, p. 197.
18) Roepstorff, P., and Fohlman, J. (1984) *Biomed. Mass Spectrom.* **11**, 601.
19) Biemann, K. (1988) *Biomed. Environ. Mass Spectrom.* **16**, 99.
20) Johnson, R.S., Martin, S.A., and Biemann, K. (1988) *Intl. J. Mass Spectrom. Ion Processes* **86**, 137.

52

Analysis of Oxidized Human Relaxin by Reverse Phase HPLC, Mass Spectrometry and Bioassays

David C. Cipolla and Steven J. Shire

Dept. of Pharmaceutical Research and Development
Genentech, Inc., South San Francisco, CA 94080

I. Introduction

Relaxin is a protein hormone which plays a major role in the reproductive biology of various species (1,2). The relaxin molecule which is structurally homologous to insulin consists of two polypeptide chains which are linked by inter- and intrachain disulfide bonds (Fig. 1). The positions of the disulfide links are similar in both proteins and there is a conservation of the gly residues immediately adjacent to the cysteine residues in the B chain (1). Since these residues are crucial to maintaining the proper structural folding for insulin, it is likely that relaxin has a similar tertiary structure to that of the insulins (3). Although there are striking similarities between insulin and relaxin, the proteins exhibit very different physiological and biochemical properties. Whereas insulin regulates the concentration of blood glucose, relaxin appears to modulate the restructuring of connective tissues in order to generate the required changes in organ structure during pregnancy and parturition. In particular, relaxin regulates a number of biological responses of reproductive tissues in pregnant animals including softening of cervical tissue and inhibition of uterine myometrial contractions (4). Some of the important roles for relaxin during pregnancy include inhibition of premature labor and onset of cervical ripening at parturition. The latter activity has been investigated in a clinical setting using relaxin isolated from porcine corpus lutea and ovaries (5-7). The preparation of porcine relaxin has been well documented (8-10), and since the sow ovary serves as a relatively plentiful source of relaxin, most biochemical and clinical studies on this hormone have been conducted with porcine relaxin.

The development of human relaxin as a protein pharmaceutical agent requires an investigation of protein stability in a chosen formulation. The stability of chemically synthesized human relaxin and the formation of

Techniques in Protein Chemistry II

degradation products have been monitored primarily by a reverse phase HPLC assay. In particular, the relaxin molecule undergoes extensive degradation when exposed to intense light. In this report we discuss the identification of these degradation products by tryptic maps and mass spectrometric analysis as well as an analysis of the biological activity of these degradation products.

II. Materials and Methods

Relaxin Preparation

Human relaxin with a pyroglutamic acid residue at the N-terminus of the A chain was prepared from purified A and B chains, which had been synthesized by solid phase peptide synthesis techniques. The solid phase synthesis, assembly of the purified chains and purification of synthetic relaxin have been previously described (11).

Reverse Phase HPLC

Analysis of each sample was performed by injecting 7 μl onto a Vydac C4 reverse phase column equilibrated at 35°C. The initial equilibration buffer was 18% acetonitrile, 0.1% trifluoroacetic acid (TFA). After a one minute hold at 18% acetonitrile, the elution was carried out by running a linear acetonitrile gradient to a final concentration of 45% acetonitrile after 30 min. The flow rate was held constant at 1.0 ml/min and the absorbance

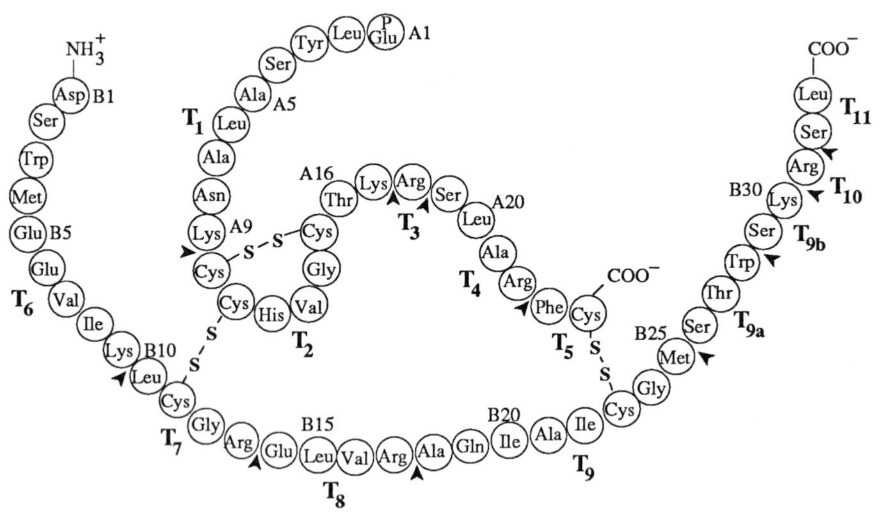

Figure 1. Representation of human relaxin including the sites of trypsin cleavage (indicated by arrows), and the resulting tryptic fragments generated (indicated by T_1, T_2, . . . T_{11}).

was monitored at 214 nm. The main relaxin peak and new peaks, which were generated after relaxin was exposed to light, were collected from the column effluent and stored at 4°C until subsequent analysis.

Tryptic Digestion

The sample to be digested was first exchanged into 10 mM Tris, 10 mM sodium acetate, 0.016 mM $CaCl_2$, pH 7.3, by ultrafiltration at 4,000 rpm in an Amicon Centricon 10 microconcentrator (10,000 MW cutoff). The final volume was typically 100 µl of relaxin at approximately 1 mg/ml which was subsequently digested for four hr at 30°C. Two aliquots of TPCK-treated trypsin (Worthington Biochemical) at 1/100 (µg trypsin/µg relaxin) were added to the relaxin sample: one at time zero and the other after two hours of digestion. The reaction was terminated by adjusting the pH to below 4 with 0.1 N HCl.

RP-HPLC of the Tryptic Digested Relaxin

The tryptic fragments were separated by RP-HPLC on a Nucleosil C18 (4.6 x 15 mm) column equilibrated at room temperature. The initial equilibration buffer was 0.08% TFA in 3% acetonitrile. The linear elution gradient was generated from 3% to 30% acetonitrile over 54 min. The flow rate was held constant at 1.0 ml/min, and the absorbance was monitored at 214 nm. Tryptic peptides were collected from the HPLC column effluent and stored at 4°C until further analysis.

Amino Acid Analysis

The collected tryptic peptides were lyophilized under vacuum in a Savant Speed-Vac for approximately 1 hr. The sample tubes were sealed under vacuum after the addition of 110 µl of 6 N HCl and 1 µl of phenol, and then heated for 24 hr at 110°C. Following hydrolysis, samples were dried under vacuum and analyzed for amino acid composition on a Beckman 6300 Amino Acid Analyzer.

Mass Spectrometric Analysis

The collected tryptic peptides were lyophilized under vacuum in the Savant Speed-Vac for approximately 1 hr, and then reconstituted in 10% acetic acid. One microliter of each sample was mixed with 0.6 µl of thioglycerol on the probe and analyzed by fast atom bombardment mass spectrometry to determine the mass to charge ratio. Tandem mass spectrometry was utilized to determine complete sequence information of selected fragments. Spectra were obtained with a JEOL JMS-HX110HF/HX110HF Tandem Double-Focusing Mass Spectrometer (E1B1-E2B2 configuration), equipped with a JEOL DA-5000 data system.

Mouse Pubic Symphysis Assay (MPS)

Female albino CFW mice (approximately 20 g) were primed with subcutaneous injections of 5 μg of estradiol 17-β-cyclopentylpropionate in 0.2 ml of peanut oil. Seven days later the mice were dosed with 200 μl of samples ranging from 0 to 12 μg/ml of relaxin. All samples contained 1% benzopurpurine-4B. After sacrifice the interpubic ligament was exposed and measured with a calibrated caliper (12).

Cyclic AMP Assay

Relaxin samples were first diluted into assay buffer to a concentration of 3.2 and 1.6 ng/ml. Samples (250 μl) were incubated with normal human uterine endometrium cells in 96 well tissue culture plates at 37°C for 30 min. After incubation the medium was discarded and 0.1 N HCl was added to each well. Intracellular cAMP released into solution was measured by radioimmunoassay and expressed as pmol/ml. A relaxin standard curve was also generated and used to correlate cAMP concentration to relaxin concentration (13).

Exposure of Relaxin to Light

Two experiments were performed; the first to generate degraded relaxin for characterization, and the second to obtain degraded relaxin for analysis of activity. In the first experiment, a 3 ml vial containing 1 ml of relaxin at 1 mg/ml was placed in a Forma 3890 Light Stability Cabinet and exposed to 1800-2200 foot candles of light. Samples were removed and assayed by RP-HPLC to determine the rate of degradation. In the second experiment, two 3 ml vials containing 1 ml of relaxin at 2 mg/ml were placed in the light box and exposed to ≈3600 foot candles of light. As a control, one vial was enclosed in aluminum foil and shielded from the light. Samples were removed (≈100 μl) and assayed by RP-HPLC after 3, 5, 7, 10, 12, 14, and 17 days of storage. Bioactivity was determined by the cyclic AMP assay after 5, 10, 12, 14, and 17 days of exposure to light and by the MPS bioassay after 7 and 17 days of exposure to light.

III. Results and Discussion

Characterization of Relaxin Exposed to Light

Samples of relaxin exposed to light were analyzed by reverse phase HPLC and observed to form degradation products which are more hydrophilic than the native relaxin molecule. A characteristic pattern of degradation has been reproducibly observed (Fig. 2), although the rate of formation of these products is dependent on the intensity of the light. The amount of undegraded relaxin, peak 4 eluting at 19.1 min, decreases with time of exposure to light. Concomitant with the decrease in the undegraded relaxin

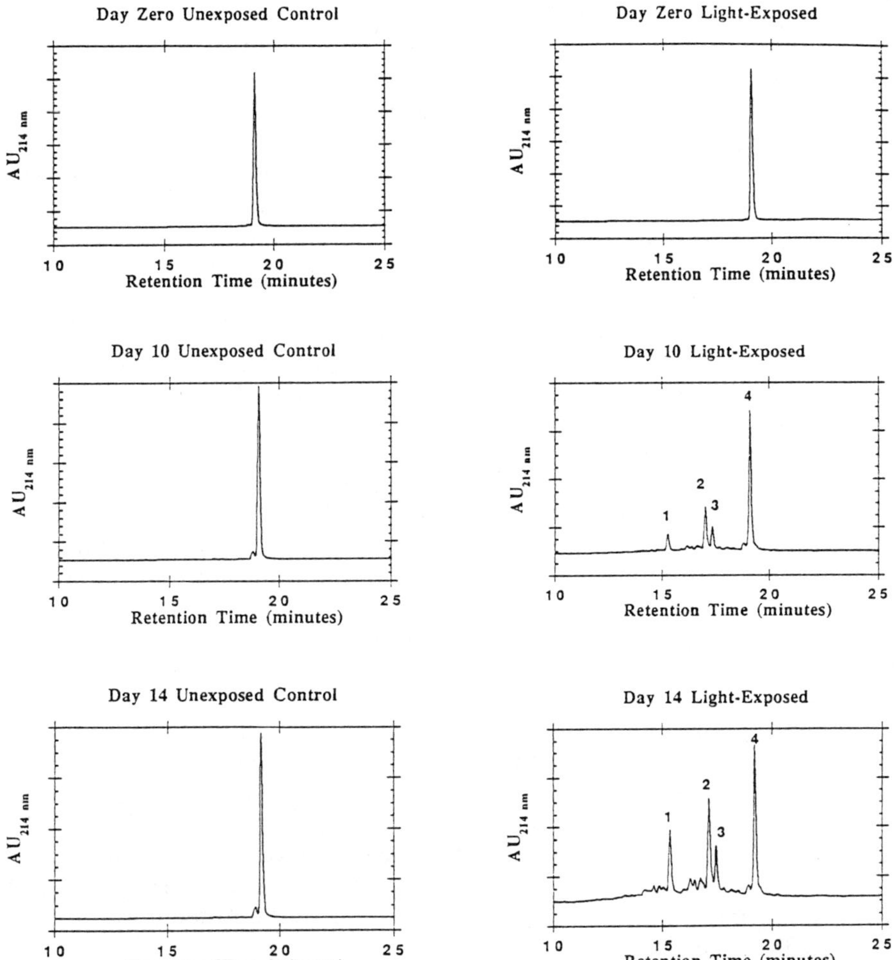

Figure 2. Reverse Phase HPLC analysis of synthetic human relaxin exposed to 3600 foot-candles of light for 0, 10, and 14 d. The ordinate is an arbitrary scale, and each chromatogram is adjusted for full scale expansion in order to clearly show all peaks. The four labelled fractions were collected for further analysis. The percent of total area for each peak is as follows: peak 1 is 6% at d 10 and 12.8% at d 14; peak 2 is 17.4% at day 10 and 15.3% at d 14; peak 3 is 9.3% at d 10 and 7.9% at d 14; peak 4 is 50.3% at d 10 and 22% at d 14. The unexposed control sample was wrapped in foil to shield it from light.

peak is an increase in two earlier-eluting peaks (labelled 2 and 3, Fig. 2). These two peaks increase in intensity reaching a maximum after 10 to 12 d of exposure to light. The subsequent decrease in the amount of these two peaks is followed by an increase in peak 1 and the appearance of a variety of smaller peaks. It would appear that undegraded relaxin (peak 4) is first converted into two intermediates (peaks 2 and 3) which are then further degraded (peak 1).

The control sample which was shielded from light but not from temperature fluctuations was for the most part unaffected. However, the undegraded relaxin peak decreases from 99% to 91% of the total area during the 17-day experiment, and a small earlier-eluting peak is observed. This degradation peak, which forms at the same rate in samples exposed to light, is mainly caused by exposure to the 37°C temperature inside the light box (unpublished observations). No further characterization of this route of degradation will be discussed.

The four labelled peaks were collected from the column effluent and the protein digested with trypsin. The resulting tryptic digests were analyzed by reverse phase HPLC (Fig. 3). Previous characterization of the tryptic peptides by amino acid analysis, N-terminal sequencing and FAB-mass spectrometry has led to an assignment of the tryptic peaks (11). The nomenclature used to label the expected tryptic peptides is shown in Fig. 1. This method of analysis allows for the rapid identification of those tryptic fragments of the relaxin molecule affected by exposure to light. As expected, the reverse phase chromatogram of the tryptic digest of protein in peak 4 is identical to that of undegraded relaxin (data not shown). However, the tryptic map of protein in peak 3 differs from that of peak 4. In peak 3 the T_6 peptide has almost entirely disappeared, and a new peptide, labelled "altered T_6", is seen to elute much earlier. Chromatography of the tryptic digest of the protein in peak 2 also reveals the loss of the T_5-T_9 peak and the appearance of an earlier eluting peak, labelled "altered T_5-T_9". Note that the T_6 and the T_5-T_9 peptides do not completely disappear in the chromatograms of peak 3 and 2, respectively. This is most likely due to cross contamination during the collection process. Finally, the chromatogram of the tryptic digest of protein in peak 1 is entirely void of both the T_6 and the T_5-T_9 peaks but contains both the "altered T_6" and the "altered T_5-T_9" peaks.

Apparently the exposure to light produces two modifications, one which occurs in the T_6 peptide and one which occurs in the T_5-T_9 peptide. These modifications occur independently of each other as evidenced by the generation of both species (peaks 2 and 3). However, each species can also undergo further modification resulting in the formation of the final degradation product observed as peak 1 in Fig. 2. Interestingly, the T_5-T_9 peptide appears to be more susceptible to modification than the T_6 peptide as illustrated by the faster formation of peak 2 compared to peak 3 in the light degradation experiment.

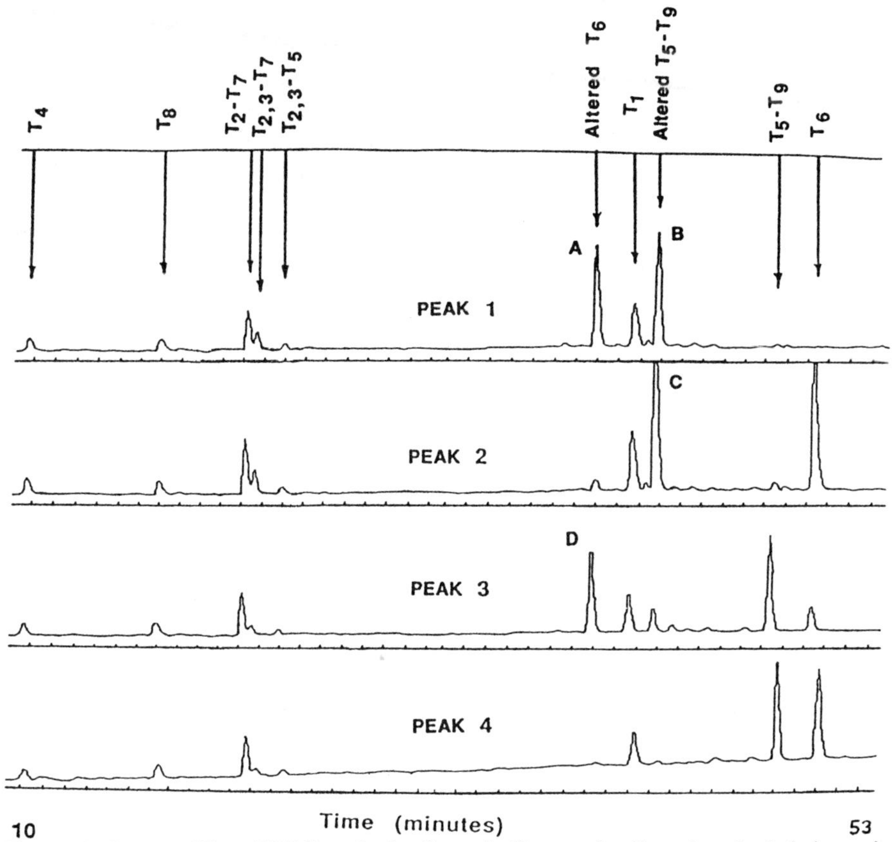

Figure 3. Reverse Phase HPLC analysis of tryptic digests of collected peaks 1-4 shown in Fig. 2. The four labelled components, A, B, C, and D, were collected for amino acid and mass spectrometric analysis.

The four unusual tryptic peptides, labelled with the letters A, B, C, and D were collected from the column effluent for amino acid and mass spectrometric analysis. Amino acid analysis revealed that the compositions of these samples are nearly identical to the theoretically expected values for the T_5-T_9 and T_6 peptides (data not shown). However, the tryptophan content of these peptides was not determined, since tryptophan is degraded during acid hydrolysis. It is therefore possible that the single tryptophan residue in the altered tryptic peptides could have undergone oxidation resulting in the formation of kynurenine or N-formylkynurenine (14). Similarly, the oxidation of methionine in each peptide to the sulfoxide would also be undetectable by amino acid analysis since this oxidation product is easily reduced under conditions used for acid hydrolysis (15). The presence of the sulfone, which would be detected by amino acid analysis was not observed, and is, therefore not a potential degradation product.

Mass spectrometry was utilized to determine the masses of the four tryptic peptides, labelled A, B, C, and D. The molecular ions for samples A and B are shown in Figure 4. A comparison between the observed molecular ions for all four collected samples A, B, C, and D, and those predicted for the postulated degradation components is made in Table I.

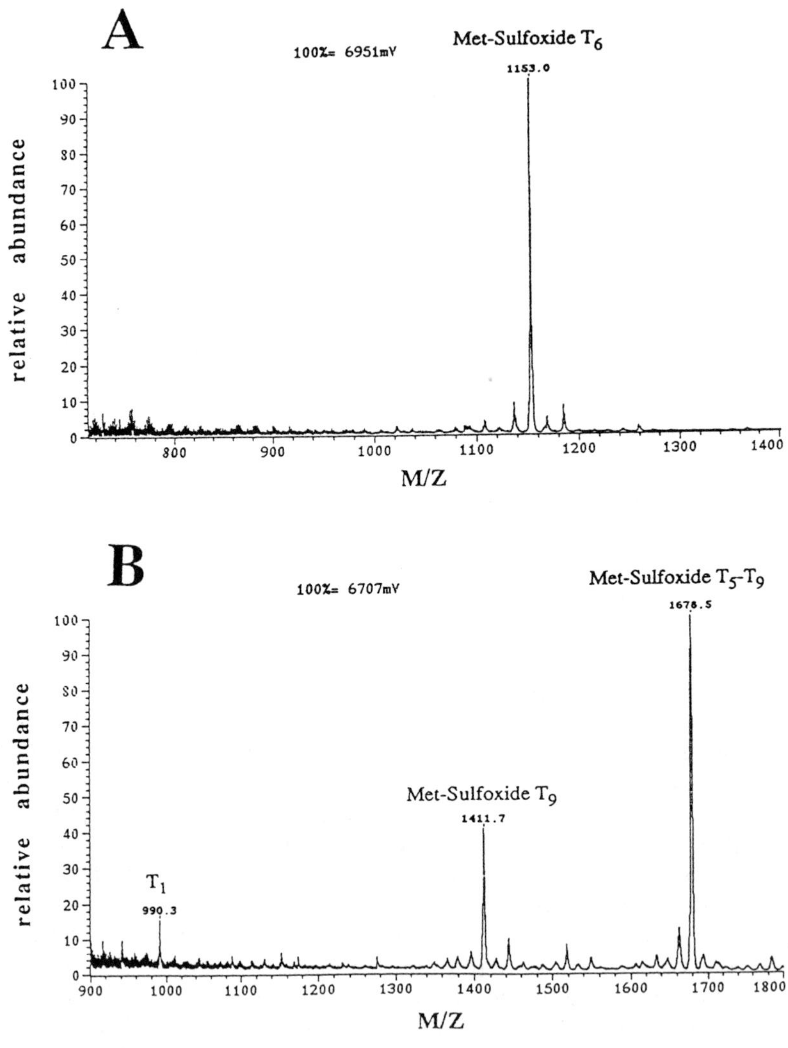

Figure 4. FAB-MS spectra of the tryptic fragments A, labelled "altered T_6", and B, labelled "altered T_5-T_9".

Table I. Major MH^+ ions of tryptic digest fragments of light-exposed human relaxin

Tryptic fragment	Theoretical Mass	Sample A Alt. T_6	Sample B Alt. T_5-T_9	Sample C Alt. T_5-T_9	Sample D Alt. T_6
T_6	1137.3				
T_6 Met-sulfoxide	1153.3	1153.0			1152.4
T_6 kynurenine	1141.3				
T_6 N-formylkynurenine	1169.3				
T_1 (pyroglu)	990.2		990.3	991.0	
T_5-T_9	1662.9				
T_5-T_9 Met-sulfoxide	1678.9		1678.5	1678.5	
T_5-T_9 kynurenine	1666.9				
T_5-T_9 N-formylkynurenine	1694.9				
T_9	1395.6				
T_9 Met-sulfoxide	1411.6		1411.7	1411.8	

The two "altered T_5-T_9" peptides originally isolated from peaks 1 and 2 have identical masses. The two "altered T_6" peptides originally isolated from peaks 1 and 3 also have identical molecular ions. The main component in the "altered T_6" sample has a molecular ion which is consistent with the oxidation of methionine to the sulfoxide. The main component in the "altered T_5-T_9" sample also corresponds to the presence of methionine sulfoxide in the T_9 peptide. No photo-oxidation products of tryptophan residues are apparent in these samples. This information is entirely consistent with the hypothesis that peaks 2 and 3 are intermediates in the formation of peak 1, implying that first one and then the second methionine is oxidized. Note that the presence of the T1 molecular ion in the "altered T_5-T_9" sample is most likely due to sample carryover during the collection process.

The collisionally-induced-dissociation (CID) daughter spectrum of the T_5-T_9 peptide fragment (Figure 5) shows a b-ion series (b_2-b_8) which clearly identifies the presence of the methionine sulfoxide (Table II). The x-ion series (x_7-x_{11}) and the y-ion series (y_2-y_5, y_8, y_9, and y_{11}) also confirm that every fragment containing the methionine residue has a 16 mass unit increase consistent with the oxidation of methionine to the sulfoxide, while every fragment lacking the methionine residue agrees with the expected mass. Similar data (data not shown) also confirms the presence of methionine sulfoxide in the T_6 tryptic peptide. No other alterations in any other residues, including the tryptophan, have been observed.

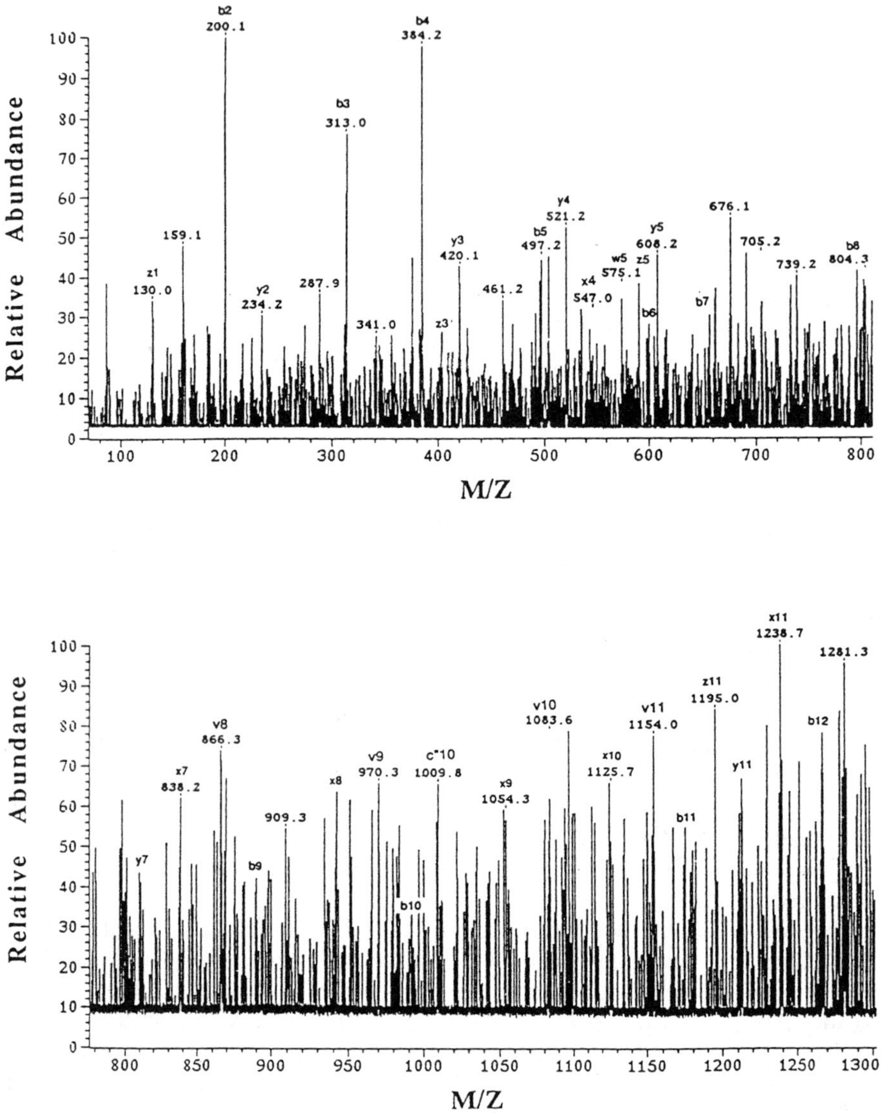

Figure 5. Collisionally-induced-dissociation (CID) daughter ion spectrum of the "altered T_5-T_9" tryptic peptide.

Table II. Fragment molecular ions from tandem mass spectrometry of tryptic peptide T9 isolated from light-exposed human relaxin.[a]

Residue	Theoretical x	x+16	Observed x	Theoretical y	y+16	Observed y	Theoretical b	b+16	Observed b
K	173.1		-	147.1		-			
S	260.1		-	234.2		234.2		1265.6	-
W	446.2		-	420.2		420.1		1178.5	-
T	547.3		547.0	521.3		521.2		992.5	-
S	634.3		-	608.3		608.2		891.4	-
M		781.3	-		755.3	-		804.4	804.3
G		838.4	838.2		812.4	-	657.3		657.0
C		941.4	941.5		915.4	915.4	600.3		600.2
I		1054.4	1054.3		1028.5	1028.0	497.3		497.2
A		1125.5	1125.7		1099.5	-	384.2		384.2
I		1238.6	1238.7		1212.6	1213.0	313.2		313.0
Q		1366.6	-		1340.6	-	200.1		200.1
A							72		-

[a] Nomenclature of sequence ions as described by J. T. Stults (16).

Biological Activity of Relaxin Exposed to Light

The experiments described have lead to a clear identification of the oxidation degradation products which are formed when relaxin is exposed to light. The biological activity of degraded relaxin was assessed by two assays. The mouse pubic symphysis assay was used to determine the bioactivity of relaxin after 17 d of exposure to light. Although only 9% of the light-exposed relaxin was undegraded, this sample appears to have equal potency when compared with an unexposed control (Fig. 6). There also appears to be no effect on the cyclic AMP bioactivity of light-degraded relaxin (Fig. 7) during 17 d exposure to light.

IV. Summary and Conclusions

This paper has identified the major route of degradation of human relaxin which appears in the incipient stages of the protein's exposure to intense light. Evidence of this degradation has been observed by reverse phase HPLC analysis. After 17 d exposure to 3600 foot-candles of light, reverse phase HPLC analysis reveals that >91% of the relaxin has been altered (Figure 2). However, full activity has been retained in both the mouse pubic symphysis bioassay (Fig. 6) and the cyclic AMP bioassay (Fig. 7). Subsequent characterization of the major degradation products reveals that the methionine residues at positions B4 and B25 undergo oxidation to the sulfoxide form. This has been confirmed by tandem mass spectrometric

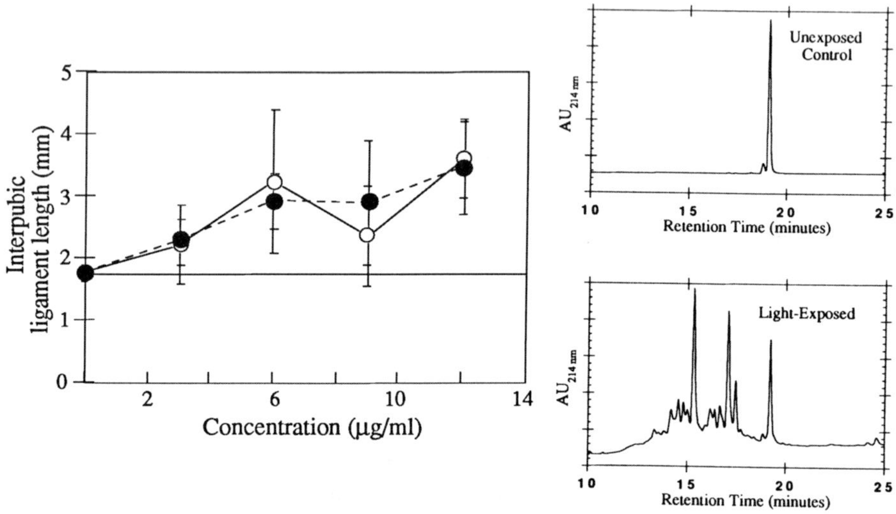

Figure 6. Dose response in the mouse pubic symphysis bioassay of 17-day light exposed (○) and unexposed control (●) relaxin samples. Each value is the mean for seven mice, and the error bars are the standard deviations. Reverse phase HPLC chromatograms of each sample illustrate the degree of relaxin degradation.

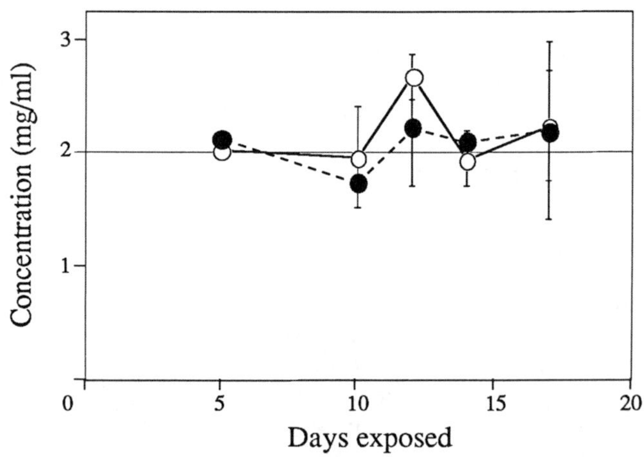

Figure 7. The concentration of the 2.0 mg/ml light-exposed (○) and unexposed control (●) relaxin samples as determined by release of cyclic AMP after binding to normal human uterine cells. Each value is the mean of replicate determinations, and the error bars are the standard deviations.

analysis of tryptic digests of the degraded protein (Fig. 5, Table 2). This oxidation occurs in two discrete steps, as evidenced by the formation of both intermediate species (Fig. 3). However, it appears that the B25 residue is more easily oxidized than the B4 residue (Fig. 2 and 3; Table 1). Although there was no evidence of tryptophan oxidation in the collected fractions, there were many small peaks which were not collected which may be due to tryptophan oxidation products or other as yet undetermined degradation products (Fig. 2).

Acknowledgements

We wish to acknowledge Reed Harris for both his assistance in performing the amino acid analysis and his insight in interpreting the results. Also appreciated were the efforts of the Mass Spectrometry Group: John Chakel, Jim Bourell, and especially John Stults for his special attention to deciphering the tandem mass spectrometry analysis. In addition, thanks to Ernst Rinderknecht for preparing human relaxin, David Fei and Richard Vandlen for their assistance in the cyclic AMP bioassay, and finally Mary Chestnut for her assistance with the MPS bioassay.

References

1. Sherwood, O. D., (1988). *In* "The Physiology of Reproduction" (E. Knobil, J. Neill, eds.) pp. 585-673. Raven Press, New York.
2. Bryant-Greenwood, G. D. (1982). *Endocrine Rev.* **3**, 62-90.
3. Isaacs, N., James, R., Niall, H., Bryant-Greenwood, G., Dodson, G., Evans, A., and North, A. C. T. (1978). *Nature* **271**, 278-281.
4. Steinetz, B. G., Beach, V. L., and Kroc, R. L. (1959). *In* "Recent Progress in the Endocrinology of Reproduction" (C. W. Lloyd, ed.) pp. 389-427. Academic Press, New York.
5. Evans, M. I., Dougan, M. B., Moawod, A. H., Evans, W. J., Bryant-Greenwood, G. D., and Greenwood, F. C. (1983). *Am. J. Obstet. Gynecol.* **147**, 410-414.
6. MacLennan, A. H., Green, R. C., Grant, P., and Nicolson, R. (1986). *Obstet. Gynecol.* **68**, 598-601.
7. MacLennan, A., Green, R. C., Bryant-Greenwood, G. D., Greenwood, F. C. and Seamark, R. F. (1981). *Obstet. Gynecol.* **58**, 601-604.
8. Sherwood, C. D., and O'Byrne, E. M. (1974). *Arch. Biochem. Biophys.* **160**, 185-196.
9. Bullesbach, E. E., and Schwabe, C. (1985). *Biochemistry* **24**, 7717-7722.
10. Bullesbach, E. E., and Schwabe, C. (1986). *Biochemistry* **25**, 5998-6004.
11. Canova-Davis, E., Baldonado, I. P., and Teshima, G. M. (1990). *J. Chromat.* **508**, 81-96.
12. Steinetz, B. G., Beach, V. L., and Kroc, R. L. (1969). *In* "Methods in Hormone Research", (R. I. Dorfman, ed.) pp. 481-513. Academic Press, New York.
13. Fei, D. T. W., Gross, M. C., Lofgren, J. L., Mora-Worms, M., and Chen, A. B. (1990). *Biochem. Biophys. Res. Commun.* **170**, 214-222.
14. Means, G. E., and Feeney, R. E. (1971). *In* "Chemical Modification of Proteins" Holden-Day, Inc., San Francisco.
15. Brot, N., Weissbach, H. (1983). *Arch. Biochem. Biophys.* **223**, 271-281.
16. Stults, J. T. (1990). *In* "Biomedical Applications of Mass Spectrometry" (C. H. Suelter, J. T. Watson, eds.) pp. 145-201. John Wiley & Sons, Inc., New York.

53

Identification of Variant Hemoglobins by Tandem Mass Spectrometry[1]

A. M. Falick*, H. E. Witkowska†, B. H. Lubin†, R. L. Nagel§, and C. H. L. Shackleton†

*Department of Pharmaceutical Chemistry, University of California, San Francisco, CA 94143, †Children's Hospital Oakland Research Institute, Oakland, CA 94609 and §Department of Medicine, Albert Einstein College of Medicine, Bronx, NY 10461

I. Introduction

Human hemoglobin is a non-covalently bound tetramer consisting of two α chains (15,126 Da, 141 amino acid residues) and two non-α chains: β, δ or γ (~16,000 Da, 146 residues), each associated with a heme group. The types of non-α globin chains present are age-specific and developmentally regulated; for example, β occurs in normal adult hemoglobin ($\alpha_2\beta_2$, Hb A), $^G\gamma$ and $^A\gamma$ in fetal hemoglobin ($\alpha_2\gamma_2$, Hb F), and δ occurs in a minor component of adult hemoglobin ($\alpha_2\delta_2$, Hb A_2) which constitutes about 2% of the total hemoglobin. Over 400 known genetic variants of human hemoglobin have been described so far (1) and undoubtedly there are many more that have not yet been discovered. The vast majority of hemoglobin variants have simple amino acid substitutions in the protein chain, which arise from (usually single) base substitutions in the globin coding gene. In Hb S (sickle cell) for instance, a single base mutation, A→T, in the triplet codon for glutamic acid results in the substitution β6Glu→Val (2). The majority of known variants are not associated with any apparent clinical manifestations (2). However, the severity of the conditions produced by the clinically important variants, especially when present in a homozygous state (for example Hb SS) or in a heterozygous mixture with non-compatible counterparts (for instance, Hb SC, Hb SD-Los Angeles, or Hb SO-Arab), justify the growing interest in variant hemoglobin detection and identification. The marked decrease in mortality of infants with sickle cell anemia that has resulted from early detection and treatment has prompted many states and countries to establish mandatory screening programs for newborns for the disease, which, in turn, has accelerated the rate of discovery of new variants.

---

[1]This work was supported by the National Institutes of Health (Grant RR01614 to A.L. Burlingame and Grants HL 20985 to BHL and RR03300 to CHLS) and the National Science Foundation Biological Instrumentation Program (Grant DIR8700766)

Techniques in Protein Chemistry II

Characterization of globin chains is carried out using the same general approaches as are applied to other soluble proteins. Individual α and β chains are separated by HPLC when necessary. The resulting protein is enzymatically cleaved (e.g. with trypsin) and the peptides are separated by HPLC and identified by mass spectrometric molecular weight measurements. In practice, since the normal peptide sequences and HPLC retention patterns are so well known, usually only one or two peptides are identified for further study by their unusual HPLC retention times. In some instances, molecular weight measurement of these peptides is sufficient to identify the mutation; however when this is not the case, the detailed sequence of the peptide must be determined. Up until the last few years, abnormal peptides were sequenced by Edman degradation, but mass spectrometry has now become the method of choice for analyzing the tryptic peptides (3-7). Recently, two new mass spectrometric techniques, 4-sector tandem mass spectrometry (MS/MS)(8-10) and electrospray mass spectrometry (ESMS) (11-16) have emerged as powerful new tools for peptide and protein characterization. ESMS is an effective technique for the accurate measurement of molecular weights of intact proteins up to about 100 kDa. Applications of this methodology to the analysis of variant hemoglobins has been described elsewhere (15, 16).

Tandem mass spectrometry allows rapid analysis of peptide primary sequences, and is particularly fast when only minor variations exist, as in the hemoglobins studied here. The method also possesses the advantages that it can provide the masses of all residues present in a peptide, modified or unmodified (including N-terminally blocked residues) and that it can be applied directly to simple mixtures of peptides (8-10). A tandem mass spectrometer consists of two mass spectrometers (MS1 and MS2) attached end to end, with a collision cell between them. The latter acts essentially as the ion source for MS2. The precursor ions to be studied (typically peptide MH^+ ions) are mass-selected in MS1, then allowed to collide with He gas in a gas cell located between MS1 and MS2. The fragment ions formed by collision-induced dissociation (CID) are analyzed in MS2. The use of two double-focusing instruments (resulting in a "4-sector" machine) allows both the precursor and the CID fragment ions to be selected and measured with excellent resolution, a property that can be essential when distinguishing between, for example Asp and Asn. In this chapter we describe the application of this technique to the identification and characterization of several uncommon and/or previously undiscovered hemoglobin variants.

II. Materials and Methods

Peptide molecular weights were measured by means of liquid secondary ion mass spectrometry (LSIMS). Mass measurements were carried out on a Kratos (Manchester, UK) MS-50S double-focusing mass spectrometer equipped with a high field magnet (mass range 3000 Da at 8 kV), a 10 keV Cs^+ LSIMS source (17, 18) and a cooled sample introduction probe (19) was used. In some instances, peptide masses were obtained on a VG (Altrincham, UK) 30-250 quadrupole mass spectrometer controlled by a Digital PDP-11/73 data system, using an 8 kV Xe beam. Collected HPLC fractions were evaporated by vacuum centrifugation and the peptides dissolved in 0.1 M heptafluorobutyric acid or

0.1% trifluoroacetic acid. An aliquot of this solution was added to the glycerol:thioglycerol (1:1)/ 0.1 M HCl matrix on the probe tip.

MS/MS experiments were carried out on a Kratos Concept IIIH 4-sector instrument of EBEB geometry, (20) equipped with an LSIMS source and a multichannel array detector (21) capable of acquiring data simultaneously over a 4% mass window. Precursor ions were generated using an LSIMS source employing an 18 keV Cs^+ primary beam. The collision energy for collision induced dissociation (CID) was 4 keV. The collision gas (He) was used at a pressure sufficient to suppress the precursor ion beam to about 30% of its initial value. The instrument was controlled and data were acquired with a DS-90 data system. Calibration and data display were carried out with the aid of a Mach 3 data processing system.

S-aminoethylation of globins was performed according to a published method (22). S-derivatized and non-derivatized globins were digested with TPCK-trypsin (EC 3.4.21.4) (6 μg per 150 μg protein) in 50 mM ammonium bicarbonate (pH 7.8). Hemolysates were prepared from blood by washing erythrocytes three times with normal saline, then lysing packed red cells with 1-2 volumes of water. Cell membranes were removed by centrifugation and globins were precipitated with acetone/0.7% HCl. Tryptic digests (pH adjusted to 3.0 with glacial acetic acid) were stored at -20°C before further analyses.

Hemoglobins were separated by DE52 column chromatography, eluted with 0.05 M Tris HCl, pH 8.2 (23). Separation of individual globin chains (24) was done on a reverse-phase semi-preparative C-4 column (Vydac, 1.0 x 25 cm) with detection at 220 nm. Tryptic peptides (7) were separated with a reverse phase analytical C-18 column (Vydac, 0.46 x 25 cm) using an acetonitrile:water/TFA gradient with detection at 214 nm.

III. Results and Discussion

A. *G-Accra (Korle-Bu) variant*

Screening by isoelectric focusing showed an unusual mobility for this sample, and suggested the presence of a β-chain variant. Heme was removed by acetone precipitation of the globin chains. The resulting mixture of α and β chains was analyzed by ESMS; the resulting measured masses of the intact α and β chains were 15,125.4 ± 1.5 Da (normal = 15,126.4) and 15,865.9 ± 1.5 Da (normal = 15,867.2), respectively. These were both close to normal values, but modification of either chain by 1-2 Da could not be ruled out. A tryptic digest of the separated β chain was carried out and the resulting peptides were separated by HPLC. Only the βT9 peptide was found to have an unusual retention time, and when the fraction containing this peptide was analyzed by LSIMS, a peak corresponding to MH^+ = 1668.9 Da was found, one dalton less than the normal βT9 mass. The sequence of this peptide in normal hemoglobin is VLGAFS^{73}DGLAHL^{79}D^{80}NLK. The one dalton shift in mass could be accounted for by any of three possible mutations, namely, β73Asp→Asn, β79Asp→Asn, or β80Asn→Ile. Each of these could have been caused by a single base change in the DNA coding for this protein. Replacement of Asn by Leu was also numerically possible, but unlikely because two bases would have had to have changed simultaneously for it to occur. A tandem CID fragment ion spectrum of the variant peptide was obtained (Figure 1) and compared with the calculated fragment ions expected from each of the three possible variants (Table I). As the data show, the variant was quickly identified as β73Asp→Asn, a variant which had been identified previously as G-Accra or Korle-Bu (1).

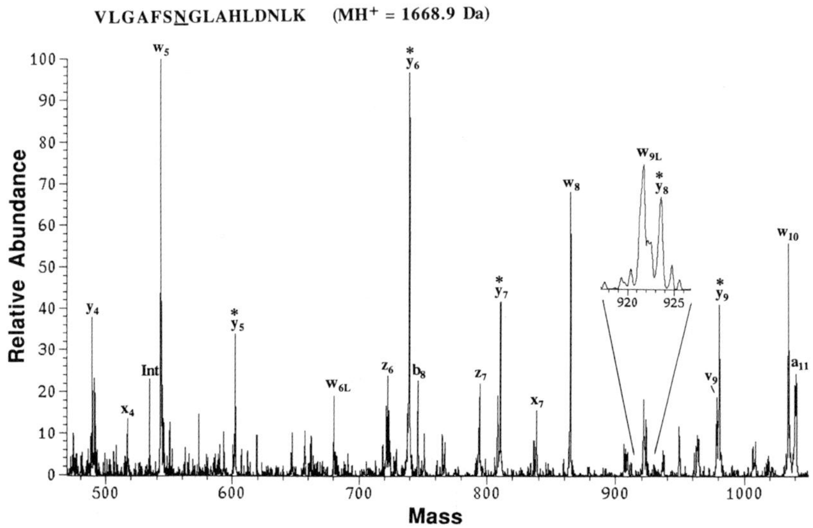

Figure 1. Partial tandem CID fragment ion mass spectrum of the βT9 peptide isolated from a hemoglobin variant subsequently identified as "Korle-Bu." The fragment ion nomenclature follows Johnson, *et al.* (25). The peak labeled "Int" corresponds to an internal fragment, and the peaks labeled $\mathbf{w_{6L}}$ and $\mathbf{w_{9L}}$ correspond to β-γ cleavage of the Leu side chain (26). The $\mathbf{y_n}$ ions marked with asterisks were critical in the sequence determination (see Table I).

Table I. Calculated and measured CID fragment ion masses (Da) for a peptide isolated from a variant hemoglobin. The measurements clearly show that the variant is β73Asp→Asn, and not β79Asp→Asn. The tandem CID mass spectrum of the variant peptide is shown in Figure 1.

	Calculated for:		
Fragment type	β73Asp→Asn	β79Asp→Asn	Found
y_4	489.3	488.3	489.2
y_5	602.4	601.4	602.2
y_6	739.4	738.4	739.4
y_7	810.4	809.4	810.5
y_8	923.5	922.5	923.6
y_9	980.6	979.6	980.5

The data in Fig. 1 and Table I demonstrate the good resolution and mass measurement accuracy for both the precursor ion (in MS1) and the fragment ions (in MS2) that the 4-sector machine provides. This ability is necessary in order to clearly distinguish between residues differing by one Da in mass, such as Ile/Leu (113), Asn (114) and Asp (115).

B. Hb β87 Thr→Ile

A three year old child symptomatic for sickle cell disease was initially diagnosed only with sickle cell trait (i.e., heterozygous) on the basis of routine electrophoretic techniques. As a result of clinical indications of more serious hemoglobinopathy, the mixture of precipitated globins was analyzed by ESMS, which showed the presence of *two* variant β chains: one with molecular mass 30 Da lower than normal β, consistent with the expected presence of β^S, and one with molecular mass 12 Da higher than the normal β chain. The normal β chain was completely absent. A tryptic digest of the aminoethylated whole globin mixture was prepared and separated by HPLC. LSIMS examination of the resulting peptides revealed the presence of an aberrant peptide (MH^+ = 1476.7 Da) whose mass corresponded to 12 Da more than that of the normal aminoethylated βT10 fragment ($G^{84}TFA^{87}TL^{89}SELHCDK$). The HPLC fraction containing this peptide also contained the coeluting normal αT5 fragment (MH^+ = 1071.6 Da). An increase of 12 Da could be the result of substitution of Ile or Leu for Thr, or substitution of Val for Ser. However, the substitutions Thr→Leu and Ser→Val both require at least two base changes in the codon. On the other hand, the normal βT10 fragment contains two Thr residues at positions 84 and 87, whose codons (ACC and AUC, respectively) can be transformed to isoleucine codons (AUC and AUA) by a one-point mutation, making one of these mutations the most likely. To identify which of the two Thr residues was modified, tandem mass spectrometry was used. Further separation to remove the αT5 peptide present in the βT10 fraction was not necessary prior to the tandem MS analysis.

Part of the tandem CID fragment ion mass spectrum of the variant protonated peptide (MH^+ = 1476.7 Da) is shown in Figure 2. To facilitate rapid identification of the modified residue, predicted fragment ion masses were calculated for each of the two most likely modifications (Table II) (27). Comparison of the experimental and calculated masses quickly identified the substitution as β87Thr→Ile. The peaks marked with an asterisk are specifically characteristic of the peptide resulting from this substitution and would be expected to be absent in the case of the β84Thr→Ile peptide. The peaks labeled $\mathbf{w_{9a}}$ and $\mathbf{w_{9b}}$ in Figure 2 also confirm that the residue at position 87 is Ile and not Leu (26). The full spectrum also contains a nearly complete series of ions corresponding to **v-17**, which we attribute to ammonia loss from the aminoethylcysteine residue.

Table II. Calculated masses (Da) of possible CID fragment ions from two closely related peptides, GTFAILSELHXDK (upper) and GIFATLSELHXDK (lower) (X is aminoethylcysteine). Both peptides have MH^+ = 1476.8 Da. The underlined masses are those that allow the two peptides to be distinguished.

	d	a	b	c	v	x	y	z	w
G	–	30.0	58.0	75.1	–	–	–	1460.8	–
T	115.1	131.1	159.1	176.1	1373.7	1445.8	1419.8	1403.8	1386.8
F	202.1	278.2	306.1	323.2	1226.7	1344.7	1318.7	1302.7	1225.7
A	–	349.2	377.2	394.2	–	1197.6	1171.7	1155.6	1154.6
I	434.2	462.3	490.3	507.3	1042.5	1126.6	1100.6	1084.6	1055.6
L	533.3	575.4	603.4	620.4	929.5	1013.5	987.5	971.5	928.5
S	646.4	662.4	690.4	707.4	842.4	900.4	874.5	858.4	841.4
E	733.4	791.4	819.4	836.5	713.4	813.4	787.4	771.4	712.4
L	862.5	904.5	932.5	949.5	600.3	684.4	658.4	642.4	599.3
H	975.6	1041.6	1069.6	1086.6	463.2	571.3	545.3	529.3	462.3
X	1112.6	1187.7	1215.7	1232.7	317.1	434.2	408.2	392.2	316.2
D	1258.7	1302.7	1330.7	1347.7	202.1	288.1	262.1	246.1	201.1
K	1373.7	1430.8	1458.8	–	–	173.1	147.1	131.1	–

	d	a	b	c	v	x	y	z	w
G	–	30.0	58.0	75.1	–	–	–	1460.8	–
I	115.1	143.1	171.1	188.1	1361.7	1445.8	1419.8	1403.8	1374.7
F	214.2	290.2	318.2	335.2	1214.6	1332.7	1306.7	1290.7	1213.6
A	361.2	361.2	389.2	406.2	–	1185.6	1159.6	1143.6	1142.6
T	446.3	462.3	490.3	507.3	1042.5	1114.6	1088.6	1072.6	1055.6
L	533.3	575.4	603.4	620.4	929.5	1013.5	987.5	971.5	928.5
S	646.4	662.4	690.4	707.4	842.4	900.4	874.5	858.4	841.4
E	733.4	791.4	819.4	836.5	713.4	813.4	787.4	771.4	712.4
L	862.5	904.5	932.5	949.5	600.3	684.4	658.4	642.4	599.3
H	975.6	1041.6	1069.6	1086.6	463.2	571.3	545.3	529.3	462.3
X	1112.6	1187.7	1215.7	1232.7	317.1	434.2	408.2	392.2	316.2
D	1258.7	1302.7	1330.7	1347.7	202.1	288.1	262.1	246.1	201.1
K	1373.7	1430.8	1458.8	–	–	173.1	147.1	131.1	–

Tandem mass spectrometry was also used to confirm that the second variant β chain present was indeed β^S. The tandem CID mass spectrum of this peptide (MH^+ 922.5), shown in Figure 3, clearly demonstrates that the modification is β6Glu→Val. This confirmation was necessary because the βT1 peptide, which contains the usual β^S modification, has a second Glu residue at position 7, which could have been modified instead of the one at position 6. A full report describing this previously unknown β chain mutation β87Thr→Lys as well as its interaction with the concurrent β^S mutation in the other β chain will appear elsewhere (28).

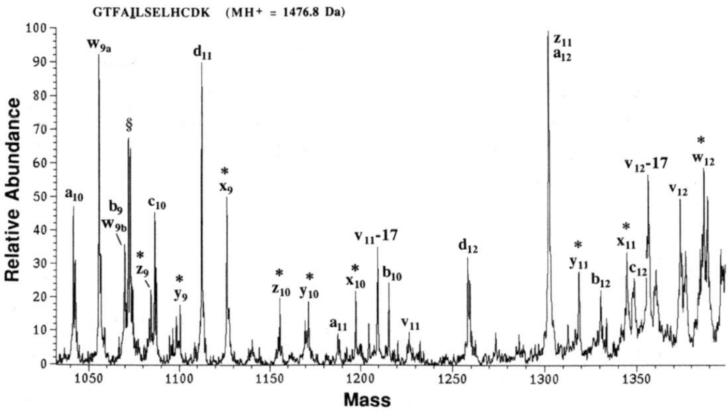

Figure 2. Partial tandem CID fragment ion spectrum of variant βT10 peptide. The major sequence ion peaks are indicated with the conventional nomenclature (25). Those marked with an asterisk show that the peptide contains the mutation β87Thr→Ile and not β84Thr→Ile (see Table II). The peaks marked "§" are artifacts due to the presence of a large amount of the normal αT5 peptide (MH+ = 1071.5) in the sample (29).

	Val	His	Leu	Thr	Pro	Val	Glu	Lys
a	72	209	322	423	520	619	748	
b		237	350	451	548	647	776	904
c		254	367			664	793	
d			280	407		605	690	819
w	891	740	627	540/2	429	344		
x		849	712	599	498	401		
y		823	686	573	472		276	
z		807	670	557		359	260	

Hb S βT1

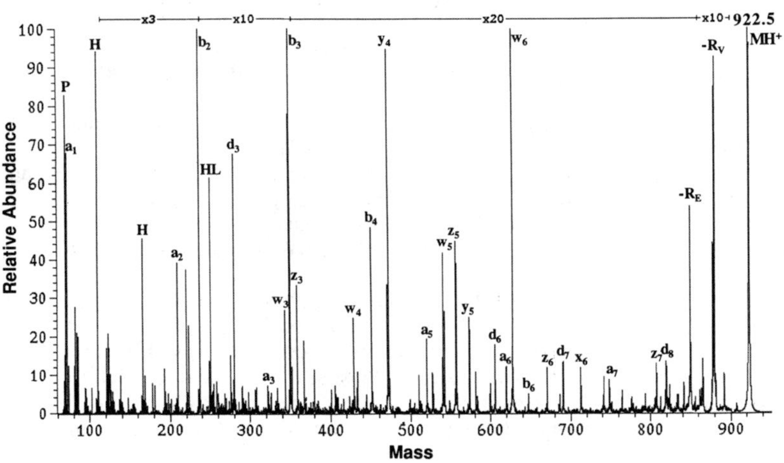

Figure 3. Tandem CID mass spectrum of βT1 peptide characteristic of Hb S. The tandem spectrum confirms that the mutation is β6Glu→Val, and not β7Glu→Val. The peak marked "HL" is due to an internal dipeptide ion, and those marked "-R$_V$" and "-R$_E$" correspond to side chain losses from Val and Glu, respectively. The table at the top of the figure shows the nominal masses and assignments of all of the sequence ions observed in the CID spectrum.

Acknowledgements

The authors wish to thank A. L. Burlingame for his support and encouragement of this study, and Klara Kleman, Esther Roitman and Linda Pinkoski for excellent technical assistance. We are indebted to Fred Walls for help in obtaining the tandem mass spectra and to Brian Green (VG Instruments) for obtaining the electrospray spectra.

References

1. International Hemoglobin Information Center (1988) *Hemoglobin* **12**, 209-282
2. Bunn, H. F., and Forget, B. G. (1986) *Hemoglobin: Molecular, Genetic and Clinical Aspects,* W.B. Saunders, Philadelphia
3. Wada, Y., Hayashi, A., Fujita, T., Matsuo, T., Katakuse, I., and Matsuda, H. (1981) *Biochem. Biophys. Acta* **667**, 233
4. Wada, Y., Matsuo, T., and Sakurai, T. (1989) *Mass Spectrom. Rev.* **8**, 379-434
5. Prome, D., Prome, J., Blouquit, Y., Lacombe, C., Rosa, J., and Robinson, J. D. (1987) *Spectros. Int. J.* **5**, 157-170
6. Katakuse, I., Ichihara, T., Nakabushi, H., Matsuo, T., Matsuda, H., Wada, Y., and Hayashi, A. (1984) *Biomed. Mass Spectrom.* **11**, 386
7. Rahbar, S., Lee, T. D., Baker, J. A., Rabinowitz, L. T., Asmeron, Y., Legesse, K., and Ranney, H. M. (1986) *Hemoglobin* **10**, 379-400
8. Kaur, S., Hollander, D., Haas, R., and Burlingame, A. L. (1989) *J. Biol. Chem.* **264**, 16981-16984
9. Biemann, K., and Scoble, H. (1987) *Science* **237**, 992-998
10. Martin, S. A., Johnson, R. S., Costello, C. E., and Biemann, K. (1989) in *The Analysis of Peptides and Proteins by Mass Spectrometry.* (McNeal, C. J., ed) pp. 135-150, J. Wiley & Sons, New York
11. Fenn, J. B., Mann, M., Meng, C. K., Wong, S. F., and Whitehouse, C. M. (1990) *Mass Spectrom. Rev.* **9**, 37-70
12. Loo, J. A., Udseth, H. R., and Smith, R. D. (1989) *Anal. Biochem.* **179**, 404-412
13. Meng, C. K., Mann, M., and Fenn, J. B. (1988) *Z. Phys. C. Atoms Molecules Clusters* **10**, 361-368
14. Prome, D., Prome, J.-E., Pratbernou, F., Blouquit, Y., Galacteros, F., Lacombe, C., Rosa, J., and Robinson, J. D. (1988) *Biomed. Environ. Mass Spectrom.* **16**, 41-44
15. Green, B. N., Oliver, R. W. A., Falick, A. M., Shackleton, C. H. L., Roitman, E., and Witkowska, H. E. (1990) in *Biological Mass Spectrometry.* (Burlingame, A. L. and McCloskey, J. A., ed) pp. 129-146, Elsevier, Amsterdam
16. Shackleton, C. H. L., Falick, A. M., Green, B. N., and Witkowska, H. E. (1990) *J. Chromatog.* in press
17. Aberth, W. A., Straub, K. M., and Burlingame, A. L. (1982) *Anal. Chem.* **54**, 2029-2034
18. Falick, A. M., Wang, G. H., and Walls, F. C. (1986) *Anal. Chem.* **58**, 1308-1311
19. Falick, A. M., Walls, F. C., and Laine, R. A. (1986) *Anal. Biochem.* **159**, 132-137
20. Walls, F. C., Baldwin, M. A., Falick, A. M., Gibson, B. W., Kaur, S., Maltby, D. A., Gillece-Castro, B. L., Medzihradszky, K. F., Evans, S., and Burlingame, A. L. (1990) in *Biological Mass Spectrometry.* (Burlingame, A. L. and McCloskey, J. A., ed) pp. 129-146, Elsevier, Amsterdam
21. Cottrell, J. S., and Evans, S. (1987) *Anal. Chem.* **59**, 1990-1995
22. Schwartz, W. E., Smith, P. K., and Royer, G. P. (1980) *Anal. Biochem.* **106**, 43-48
23. Abraham, E. C., Reese, A., Stallings, M., and Huisman, T. H. J. (1976) *Hemoglobin* **1**, 24-27
24. Shelton, J. B., Shelton, J. R., and Schroeder, W. A. (1984) *J. Liquid Chromat.* **7**, 1969-1977
25. Johnson, R. S., Martin, S. A., and Biemann, K. (1988) *Int. J. Mass Spectrom. Ion Processes* **86**, 137-154

26. Johnson, R. S., Martin, S. A., Biemann, K., Stults, J. T., and Watson, J. T. (1987) *Anal. Chem.* **59**, 2621-2625
27. Hines, W. (1990) unpublished software
28. Witkowska, H. E., Esseltine, D. W., Beuzard, Y., Lubin, B., Vichinsky, E., Kleman, K., Pinkoski, L., Roitman, E., Falick, A. M., Green, B. N., and Shackleton, C. H. L. (1990) manuscript in preparation
29. Falick, A. M., Medzihradszky, K., and Walls, F. C. (1990) *Rapid Commun. Mass Spectrom.* in press

54

CHARACTERISATION AND QUALITY ASSURANCE OF RECOMBINANT HUMAN SERUM ALBUMIN BY ELECTROSPRAY MASS SPECTROMETRY

Michael J Geisow, Roy Harris and Neil Dodsworth

Delta Biotechnology Ltd. Nottingham U.K.

Brian Green and Therese Hutton

VG BIOTECH Ltd. Altrincham WA14 5RZ U.K.

I. INTRODUCTION

In the last two years the value of Electrospray ionization mass spectrometry (ESMS) for accurate mass determination of labile biopolymers of medium to high molecular weight has been proven(1). This article emphasises the value of the method to biotechnology companies. Recombinant proteins of predicted amino acid sequence can easily be checked, not only for post-translational modifications, but also for purity and quality. The last is very important for the lengthy process of product registration as well as for continuing product quality assurance. For this alone, use of ESMS is certain to increase as larger and more complex recombinant proteins are produced by industry.

Human serum albumin (HSA) is being expressed and secreted by yeast in a program to develop a plasma expander to replace the material usually obtained by plasma fractionation. There is a world-wide effort to produce safe, consistent human plasma proteins by genetic engineering and HSA represents the most extensively used product. Consumption runs into 100's of tonnes per annum. Since the per person dose is relatively high (12 to 17 gm infusion) product quality is critically important.

Techniques in Protein Chemistry II

The size, conformational flexiblity and numerous binding sites of HSA make it a difficult molecule to characterise at the confidence level required for product registration. This article describes our experiences with ESMS used to determine the purity and quality of recombinant product and to support our claim to have made a "nature-identical" human albumin.

II. MATERIALS AND METHODS

A. Proteins Full-length rHA(1-585), double rHA(1-419) and single rHA(1-194) domain human albumins were expressed as alpha mating factor 'pre'; HSA 'pro' fusion proteins in S. cerevisiae ssp. A minor, naturally-occuring rHA fragment (47KDa approx.) was recovered from cultures expressing full-length rHA. Proteins were purified either in the pilot plant or in the laboratory by ion-exchange and gel-permeation chromatography. HSA was obtained from eight individual plasma fractionation companies. All the proteins, plasma and recombinant, gave identical amino terminal sequence and PTH recoveries using a gas phase protein sequencer. Proteins were desalted by exhaustive dialysis against de-ionized water and, in most cases, by reversed-phase HPLC on butyl-silica using 0.1% v/v TFA and an acetonitrile gradient. Eluted proteins were vacuum - dried in a centrifuge. All proteins were 98% or better homogeneous by non-reducing SDS-PAGE.

B. Electrospray Mass Spectrometry Proteins were dissolved in 2% v/v acetic acid and mixed with an equal volume of methanol to give a final concentration of 1 to 25 pmol/µl. Sample was introduced into the electrospray source of a VG BIO-Q quadrupole mass analyser at 5 µl/min. Mass was calibrated with horse heart myoglobin (16951.5 Da) and bovine trypsinogen (23977.9 Da). HSA molecular weight was calculated using the atomic weights C=12.011, H=1,00794, N=14.0007, O=15.9994 and S=32.06. Sample molecular weights were determined from a selection of the multiply-charged molecular ions as previously reported and used to compute a mean and standard deviation (2).

III. RESULTS AND DISCUSSION

Desalted samples of rHA and eight different human serum albumin preparations were analysed. The most abundant molecular ions carried 40 - 48 positive charges. This relatively low degree of charging of HSA

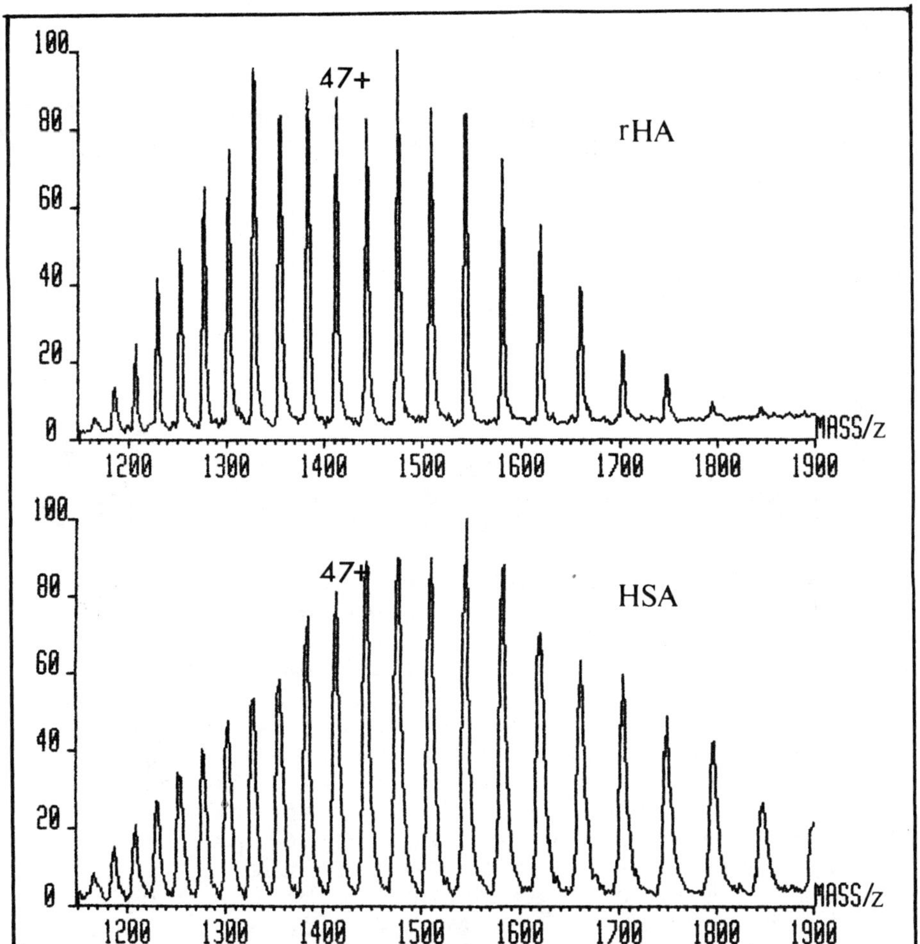

Figure 1 ESMS of rHA and HSA.

is due to the extensive intrachain disulphide bridging. The mass determined for full length rHA(1-585) was 66 440±2 Da, within 3 Da of the expected molecular weight (66 437 Da); an error of 0.0045% (Figure 1). This was much closer than the average mass values determined using the human plasma material. One evident reason for this was the presence of multiple, overlapping molecular ions in all the HSA samples (Figure 2).

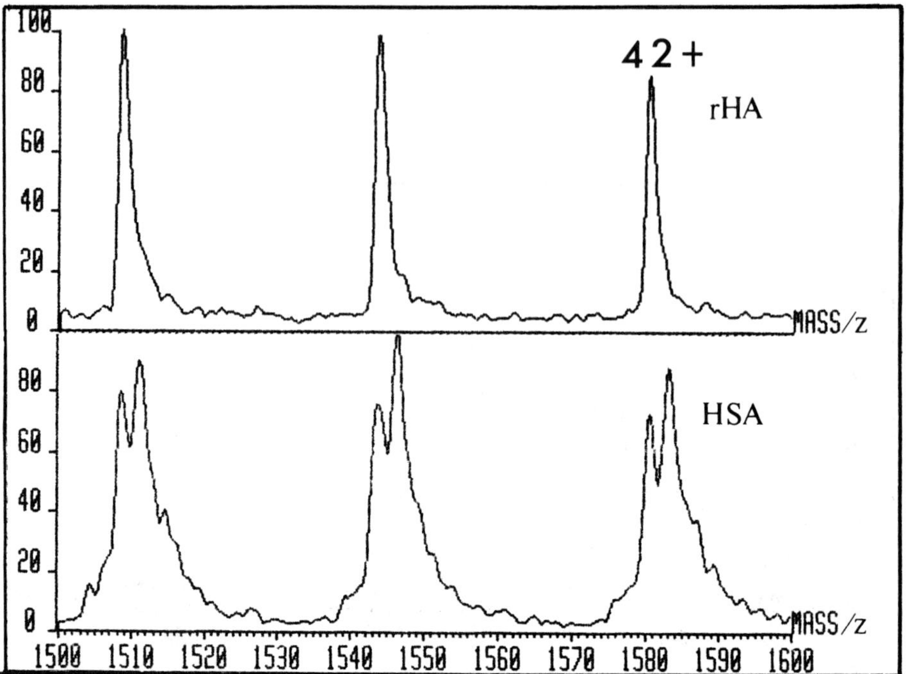

Figure 2. ESMS of 3 molecular ions of rHA and HSA

The heterogeneity of one of the HSA samples is shown by comparison with the more Gaussian distribution of mass in the rHA molecular ion peaks. The effect was to increase the apparent mass of the human protein.

The deviation from the ideal can be quantitated. From the natural isotopic distribution in proteins, the theoretical peak width at half height should be about 18 Da. The peak widths for the rHA and HSA samples shown are 99 and 300 Da respectively and this deviation becomes more extreme at 10% peak height. The partly-resolved species present in the HSA sample are probably due to partial occupancy of the single unpaired cysteine of HSA at position 34. This is known to form up to 50% mixed disulphide esters in plasma. Another possible modification is glycation of protein lysine. HSA solutions are straw coloured. We have been unsuccessful in removing the chromophores responsible by denaturing the protein and our view is that this represents covalent modification of HSA. If so, then the material could also contribute to the broad molecular ions observed for HSA samples.

An amino-terminal fragment of rHA was present in culture supernatants (2) and it was important to determine the exact nature of its carboxy-terminus to assist in identifying the protease(s) responsible.

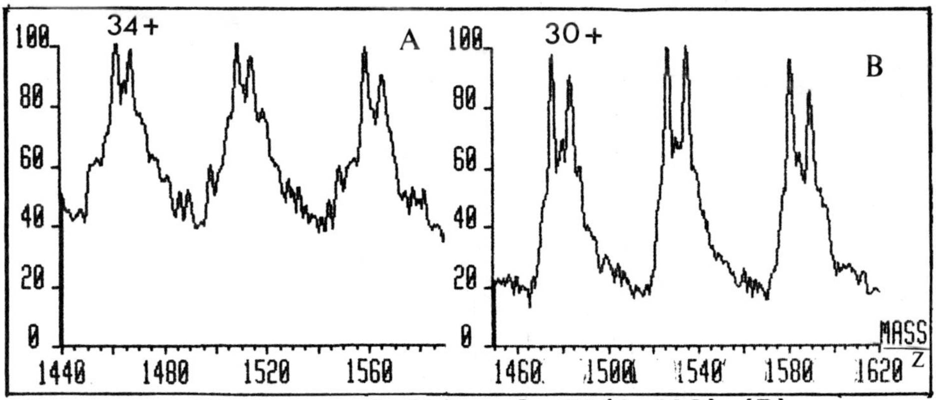

Figure 3. rHA fragment (A) and rHA(1-419) (B)

Carboxypeptidase Y digests indicated valine and leucine as probable terminal amino acids. ESMS showed that the material was heterogeneous (Figure 3A), with major species as shown in Table 1. Although there was a large standard deviation, the data allowed assignment of rHA(1-407 Leu) and rHA(1-409 Val), in agreement with the carboxypeptidase result. Since the homologous site in BSA has been reported to be susceptible to pepsin at pH 3 to 4, it is likely that a yeast-specified acid protease with pepsin-like specificity is responsible. Mild reduction decreased the masses of the principal species by about 120 Da and it has therefore been assumed in Table 1 that cysteine was linked to a proportion of Cys_{34}.

Double and single rHA domains were expressed in yeast during our R. and D. programme. rHA(1-419) after isolation proved to have ragged carboxy-terminal ends (Figure 3B and Table 1) with rHA(1-387 Leu) and rHA(1-389 Lys) principal components. Cys_{34} was esterified as with fragment and the Leu_{387}-Ile_{388} bond is known to be labile to pepsin.

Expression of the single domain rHA(1-194) produced a low abundance molecular ion series of the expected mass (R) and a major component (O) 119Da heavier (Figure 4A). The former ions disappeared after incubation of with cystine at pH 8 and the latter after mild reduction with dilute thiol (Table 1). It was concluded that free cysteine from the culture medium had formed a disulphide bond with Cys_{34}.

Most desorption mass spectrometries do not permit ready quantitative comparison of ions on the basis of peak height and for this a known, isotopically labelled internal standard is usually necessary.

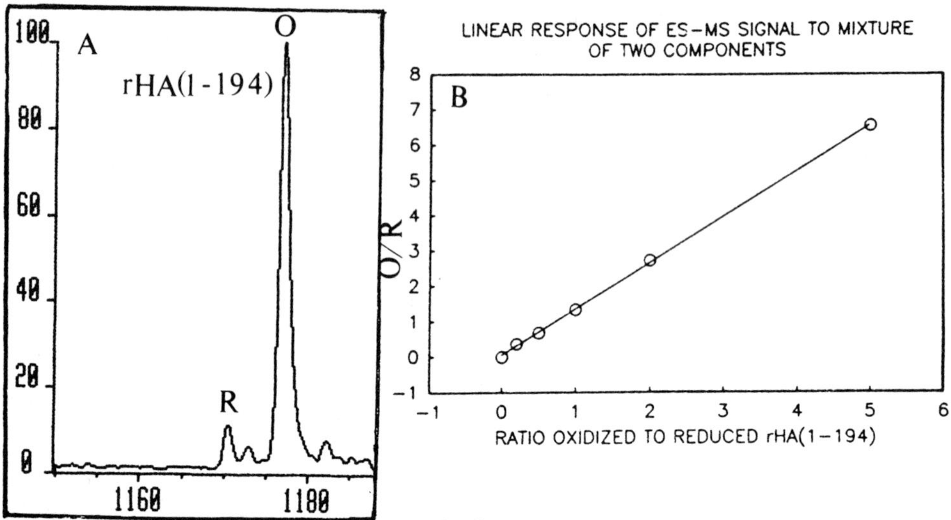

Figure 4. rHA(1-194) O=oxidized, R=reduced forms.

TABLE I. Summary of ESMS rHA mass determinations

Product (or assignment)	Observed (Da)	Expected (Da)	Error (Da)
rHA(1-585)	66440	66437	3
rHA(1-387)cys	44307	44291	16
rHA(1-389)cys	44539	44533	6
rHA(1-407)cys	46807	46792	15
rHA(1-409)cys	46979	47019	30
rHA(1-194)cys	22352	22353	1
rHA(1-194)	22233	22234	1

The oxidized and reduced forms of rHA(1-194) are similar, so the peak height ratios of mixtures of known composition were compared. Figure 4B indicates that the relationship is perfectly linear up to 25 pmol/µl of the oxidized form. This gives a measure of confidence that peak heights can be used as an estimate of the amount of a modified protein present in a sample. In this case 1% or less of either form can be determined.

REFERENCES

1 Fenn, J.B., Mann, M., Meng, C.K., Wong, S.F. and Whitehouse, C.M. (1989) Science, 246, 64 - 71.
2 Dorsselaer, A.V., Bitsch, F., Green, B., Jarvis, S. Lepage, P., Bischoff, R., Kolbe, U.V. and Roitsch, C. (1990) Biomed. Environ. Mass Spec. (in press).

Index

C

D

E

N

O

P

R

S

T

U

W

X

Y